Proceedings of the Conference in Honor of C. N. Yang's 85th Birthday

Statistical Physics, High Energy, Condensed Matter and Mathematical Physics

Proceedings of the Conference in Honor of C. N. Yang's 85TH Birthday

Singapore, 31 October – 3 November 2007

Statistical Physics, High Energy, Condensed Matter and Mathematical Physics

Editors

M.-L. Ge *(Nankai University, China)*

C. H. Oh *(National University of Singapore, Singapore)*

K. K. Phua *(Nanyang Technological University, Singapore)*

World Scientific

NEW JERSEY • LONDON • SINGAPORE • BEIJING • SHANGHAI • HONG KONG • TAIPEI • CHENNAI

Published by

World Scientific Publishing Co. Pte. Ltd.

5 Toh Tuck Link, Singapore 596224

USA office: 27 Warren Street, Suite 401-402, Hackensack, NJ 07601

UK office: 57 Shelton Street, Covent Garden, London WC2H 9HE

British Library Cataloguing-in-Publication Data
A catalogue record for this book is available from the British Library.

STATISTICAL PHYSICS, HIGH ENERGY, CONDENSED MATTER AND MATHEMATICAL PHYSICS
Proceedings of the Conference in Honor of C N Yang's 85th Birthday

ISBN-13 978-981-279-417-8
ISBN-10 981-279-417-4

Printed by Fuisland Offset Printing (S) Pte Ltd. Singapore

PREFACE

Professor Chen Ning Yang is a living legend in physics — undoubtedly one of the greatest physicists of our times. The depth and breadth of his contributions in physics are simply immense and striking. Many fundamental developments and new directions in physics, such as parity violation (1957 Nobel prize), Yang–Mills field theory (basis of the standard model in particle physics), Yang–Baxter equation in the theory of integrable systems, and the applications of differential geometry as well as topology in physics, are attributable to him. At the age of 86, he is still actively engaged in research, publishing original research papers in physics.

Professor Yang is closely associated with universities in Singapore, in particular the Nanyang Technological University ("Nantah"). He first visited Singapore in 1967, subsequently frequenting the country numerous times. He has made significant contributions to the advancement of science in Singapore in various capacities ranging from the scientific advisor to the government to an external examiner for Nantah as well as the National University of Singapore ("NUS"). We were truly proud and honoured to be able to mark the joyous occasion of Prof. Yang's 85th birthday in Singapore, with many physicists and other well-wishers from all over the world travelling here to join in the celebrations.

This Proceedings record most of all the invited papers and abstracts of papers presented in the parallel sessions.

The generous supports of NTU, NUS, Lee Foundation and A*STAR are gratefully acknowledged. We would also like to thank Professor Ngee Pong Chang for his assistance in the preparation of this Proceedings.

Editors

CONTENTS

Statistical Physics, Condensed Matter and Biophysics

*Abstract.

†Abstract and slides.

Photographs

Chen Ning Yang.

From left: Chen Ning Yang, Choo Hiap Oh and Tharman Shanmugaratnam (Minister for Education of Singapore).

From left: Chen Ning Yang, Anthony S. C. Teo, Kok Khoo Phua and Claude Cohen-Tannoudji.

Martin L. Perl.

Walter Kohn.

Ludwig D. Faddeev.

Chen Ning Yang and Akira Tonomura.

Chen Ning Yang and David Sherrington.

Chen Ning Yang and Paul C. W. Chu.

Anthony Zee.

Andrew C.-C. Yao.

David J. Thouless.

Fa Yueh Wu.

Bailin Hao.

Maw-Kuen Wu.

Michael E. Fisher.

Kerson Huang.

Chen Ning Yang and Michiji Konuma.

From left: Bang-Fen Zhu, Chen Ning Yang and Tai Tsun Wu.

From left: Vladimir E. Korepin, Mo-Lin Ge and Bruce H. J. McKellar.

From left: Tai Tsun Wu, Mo-Lin Ge, Chen Ning Yang and Kok Khoo Phua.

From left: Claude Cohen-Tannoudji, Chen Ning Yang, Martin L. Perl and Walter Kohn.

Lawrence J. Lau and Su Guaning.

Birthday cake of C. N. Yang.

High Energies and Field Theories

DEVELOPING CREATIVITY AND INNOVATION IN ENGINEERING AND SCIENCE

MARTIN L. PERL

Stanford Linear Accelerator Center, Stanford University, 2575 Sand Hill Road Menlo Park, CA 94025, USA

In this talk I discuss a range of topics on developing creativity and innovation in engineering and science: the constraints on creativity and innovation such as the necessity of a fitting into the realities of the physical world; necessary personal qualities; getting a good idea in engineering and science; the art of obsession; the technology you use; and the technology of the future.

1. Creativity and Innovation in Engineering and Science

1.1. *Creativity*

Creativity is sought everywhere: in the arts, entertainment, business, mathematics, engineering, medicine, the social sciences, and the physical sciences. Common elements of creativity are originality and imagination. Creativity is intertwined with the freedom to design, to invent and to dream. In engineering and science, however, creativity is useful only if it fits into the realities of the physical world.

1.2. *Examples of Constraints on Creativity and Innovation*

A creative idea in science or engineering must conform to the law of conservation of energy (including the mass energy mc^2). An inventor that thinks that she or he knows how to violate the conservation of energy will have to disprove a vast amount of laboratory measurements and accepted theory.

Figure 1 shows a traditional design for a perpetual motion machine, a rotating wheel with moving weights that seem to always give the wheel a non-zero torque. A direct dynamical analysis of the forces is tedious, therefore physicists and engineers simply use conservation of energy to negate the scheme, but this does not convince perpetual motion zealots who vary

Fig. 1. A traditional proposal for a perpetual motion machine.

the design. At present many perpetual motion seekers are thinking about obtaining energy from the quantum mechanical fluctuations of the vacuum. This is a sophisticated proposal and negating it is complicated, perhaps at some deeper level the proposal has validity, but a radical change in our physics theory and new experiments are required before one should talk about building a machine to extract energy from the vacuum.

A creative idea in science or engineering must conform to our present knowledge of the nature of matter as shown in Fig. 2, unless we invent or find a new form of matter. Of course we have created new structures out of the known forms of matter such as nanotubes and layered materials.

1.3. *Observations and Rules of Thumb*

If your idea is in an area where the basic science or mathematics is not known, begin by paying attention to the known observations and rules of thumb in that area. Keep in mind, however, that observations and rules of thumb may be wrong. Remember when doctors thought that stomach ulcers were caused by stress or spicy food, now it is known that most ulcers are caused by bacterial infection.

1.4. *Practicality and Feasibility Constraints*

Creativity in science, engineering and computer science is constrained by feasibility and practicality. Consider the work in the US on a nuclear reactor powered airplane in the 1950's. Before the development of intercontinental missiles there was a desire to build a bomber that could fly around the world and perhaps even keep circling [1]. There were three severe problems faced

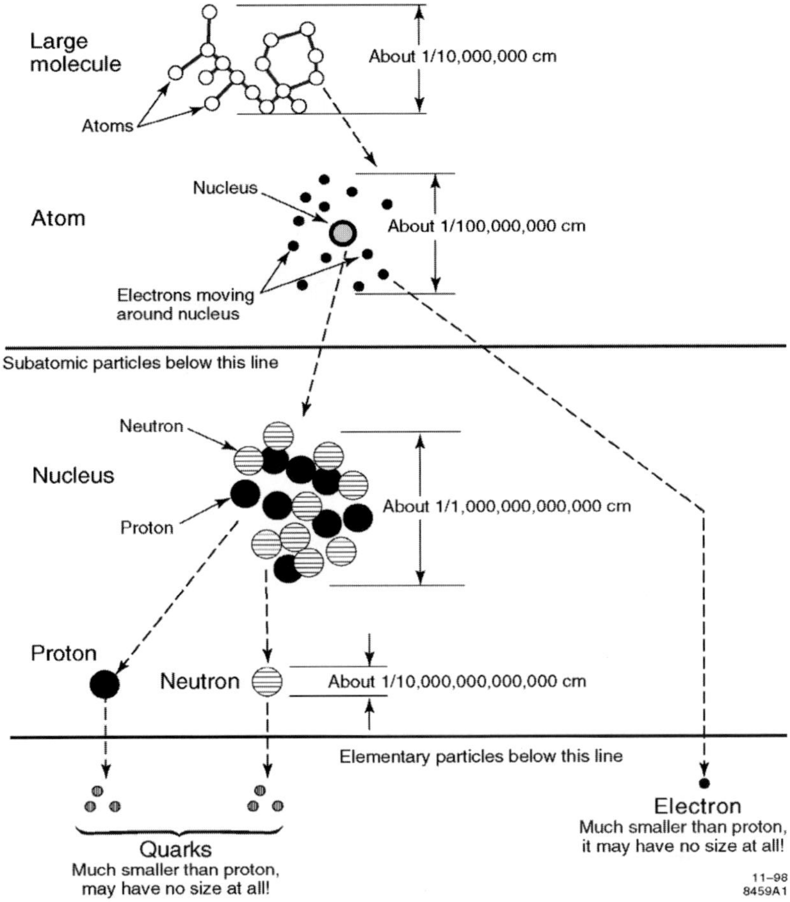

Fig. 2. All objects are made of the ordinary matter outlined here. There are other known types of matter such as unstable quarks and leptons, force carrying particles such as the Z^0, and dark matter. But we do not know how to make objects of these particles.

by the designers: the weight of the reactor and the shielding, the shielding of the crew from the reactor radiation, and the contamination of an area if the plane crashes. Tests went as far as connecting a nuclear reactor to an engine. But the plane was never built. This idea violated the constraint of feasibility.

Since the maturation of automobile technology and powered aircraft technology, inventors have dreamed of a flying car, a vehicle used by the public that could be driven on the road or flown. The vehicle would have easy convertibility between the two modes, Fig. 4. There have been a few

Fig. 3. An artist's conception of a nuclear reactor powered airplane. The crew's cabin would be in the rear, far from the reactor in the front.

temporary successes but the concept does not meet the constraint of practicality. How is the airspace to be regulated? Where are the wings when the vehicle is used as an automobile. What is the cost of purchase and maintenance?

2. Necessary Personal Qualities for Creativity in Engineering and Science

2.1. *Be Competent in Mathematics*

You don't have to be a mathematical genius. While there are positions in scientific and technical fields that don't require much mathematics, you should be competent in mathematics so that you can understand new developments.

2.2. *Visualization*

In engineering and scientific work it is crucial to be able to visualize how the work can be accomplished. The intended work might be the invention of a mechanical or electronic device, the synthesis of a complicated molecule, the design of an experiment to evaluate the efficacy of a new drug, or the full modeling of how proteins fold and unfold.

Different kinds of work require different kinds of visualization. Spread sheets or flow charts may work best in some cases. Drawings might be more suitable in others. Whatever the project, the value of visualization is

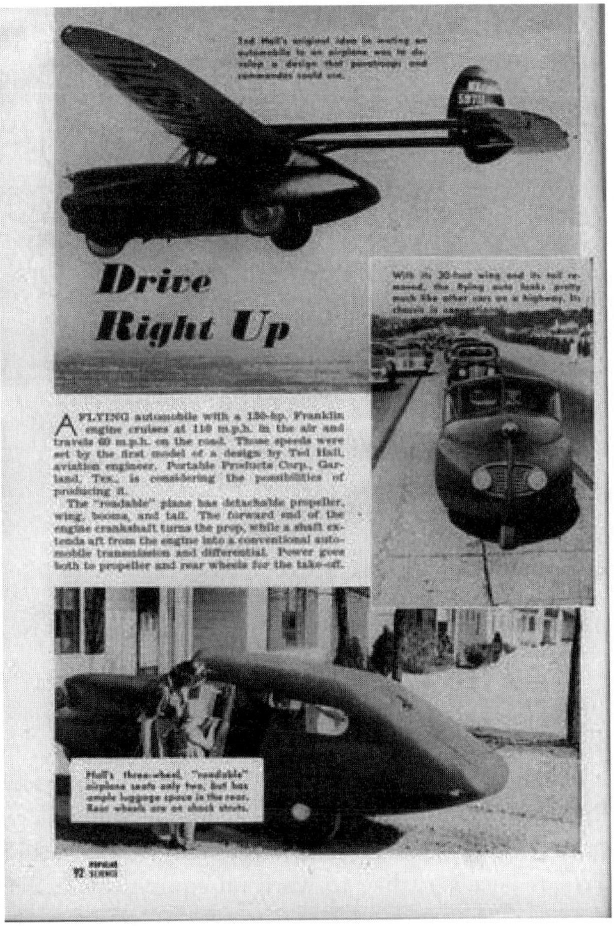

Ted Hall's original idea in mating an automobile to an airplane was to develop a design that paratroops and commandos could use.

Drive Right Up

With its 30-foot wing and its tail removed, the flying auto looks pretty much like other cars on a highway. Its chassis is

A FLYING automobile with a 150-hp. Franklin engine cruises at 110 m.p.h. in the air and travels 60 m.p.h. on the road. Those speeds were set by the first model of a design by Ted Hall, aviation engineer. Portable Products Corp., Garland, Tex., is considering the possibilities of producing it.

The "roadable" plane has detachable propeller, wing, booms, and tail. The forward end of the engine crankshaft turns the prop, while a shaft extends aft from the engine into a conventional automobile transmission and differential. Power goes both to propeller and rear wheels for the take-off.

Hall's three-wheel "roadable" airplane seats only two, but has ample luggage space in the rear. Rear wheels are on shock struts.

92 POPULAR SCIENCE

Fig. 4. A typical popular magazine forecast of a flying car.

in finding the best way to proceed while avoiding mistakes and perhaps even finding alternative solutions or good related ideas. Do not go into engineering or science if you do not have a basic ability to visualize. Visualization is crucial for creativity in engineering and science!

2.3. *Imagination*

Imagination is another crucial ability required to be creative in engineering and science — imagination with respect for the constraints I have talked about: known physical laws, correct observation and experimentation, feasibility, practicality. Begin with the far reaches of your imagination at the science fiction level, then gradually apply these constraints. Figure 5 shows the change from Jules Verne's science fiction space vehicle to the space shuttle.

Fig. 5. The change from a science fiction space vehicle to the Enterprise space shuttle.

2.4. *Evaluate Your Laboratory Skills*

Evaluate the extent of your hands-on skills and laboratory skills. Are you good at working with tools, at building equipment, at running equipment — electronics, microscopes, telescopes ...? This is my strength. I am an experimenter in physics because I like to work on equipment, am mechanically handy and get great pleasure when an experiment works. But hands-on skills do not have to be your strength. Isadore Rabi, my doctoral research supervisor at Columbia University in the 1950's, had no laboratory skills. Yet Rabi won a Nobel Prize for advancing experimental atomic physics. When choosing what to work on in engineering and science, honestly evaluate the extent of your hands-on and laboratory skills.

3. Getting a Good Idea in Engineering and Science

3.1. *Personality and Temperament*

You must take into account your personality and temperament when choosing a technical field, or particular field of science. Be yourself. Creative scientists and engineers have many different types of personalities.

3.2. *It is Much Easier to Get Bad Ideas than Good Ideas*

In science and engineering for every good idea expect five or ten or twenty wrong ideas, or useless ideas, or obsolete ideas. Consider some of the following obsolete, bad ideas:

- The phlogiston model of combustion.
- Lamarckian evolution.
- A physical electromagnetic ether.
- Steam powered automobiles that can be competitive with internal combustion automobiles.

There are other ideas that appear to be wrong but are still pursued:

- Cold fusion.
- Using zero point energy from fluctuations of the vacuum.
- Telepathy.

3.3. *Great Engineers and Scientists Have Bad Ideas as well as Good Ideas*

Nikola Tesla was a pioneer and inventor in electrical technology. He was one of the first to understand alternating current phenomena and its use. He was one of the first to demonstrate the feasibility of long distance wireless, indeed in this field he is the equal of Marconi. But he also thought he could use the same wireless transmitting tower, Fig. 6, to transmit efficiently, large amounts of low frequency power to an antenna very far away. At the radio frequencies used by Tesla this was not possible because the power spreads out rapidly. Of course, substantial amounts of power can be transmitted at high frequencies using microwave beams. I don't understand how Tesla, who understood radio theory so well and could visualize alternating current phase diagrams in his head, could be confused here.

3.4. *Reduce the Frequency of Bad Ideas*

There are several rules for reducing the frequency of your bad ides. Make sure that you understand the physical laws and the neighboring technology relevant to your new idea. Colleagues, the literature, and the Web can be of help. Sometimes you have to keep going until *you* are the expert on the idea and *you* discover the show-stopper! Try to avoid the "dam the torpedoes, full speed ahead" state of mind. Several times I have rushed into a project even though it didn't feel quite right, just hoping that it would work out in the end. It never did work out.

Fig. 6. A tower used by Tesla to transmit wireless signals. He also hoped to use such a tower to transmit large amounts of electrical power over long distances.

3.5. *Sorting Out Good and Bad Ideas*

On the other hand you may turn a bad idea into a good idea — don't kill the bad idea prematurely. A bad idea can evolve into a good idea. This evolution into a good idea can be a short process, like turning a bug into a feature, to quote my colleague Eric Lee. Or the evolution from bad to good can be long with many intermediary steps. It is rare for the complete development of a good idea to occur quickly. Be prepared for a winding road of research, development and prototyping or for a maze with many wrong turns.

3.6. *Can Creativity and Innovation Skills in Engineering and Science Be Taught in the Classroom?*

I believe the pressure of reality is important — a product *must* be improved or an experiment must work or a more efficient computer algorithm is *needed*. I don't think these skills can be taught in the classroom. For a contrary view see Ref. 2.

3.7. *Helpful Hints*

Keep your eyes and ears open by scanning the literature, usually through the internet these days. Also eat lunch with colleagues, don't eat at your desk. Avoid the "not invented here" prejudice. If you find an available technology that is superior to your own, use it!

You can learn from many people with different talents and different technical specialties. Five years ago we wanted to make a colloidal solution of powdered meteorite, my academic friends in colloid science were of little help, they knew a great deal about the theory and behavior of colloidal solutions of pure substances. But our meteoritic material consisted of a mixture of minerals such as silicates plus small metallic nodules. We learned how to make a colloidal solution of powered meteorite from an engineer who was a specialist in the lubrication of automobile engines. One of the functions of engine oil is to suspend small particles that come from engine wear and incomplete combustion.

3.8. *Limit Your Working Hours*

These days there is pressure in engineering and science to work very long hours, a "24/7" work-week. But creativity and innovation require relaxation time and non-technical activities.

3.9. *Luck*

The importance of Good Luck is overrated for discovery and innovation in engineering. For a contrary view see Ref. 3.

But it is important to avoid Bad Luck. The basic avoidance principle is the same as being careful when crossing a freeway. In engineering and science most bad luck is caused by mistakes in calculations, design, measurements, or experiments.

4. Colleagues

In the modern world the highly productive solitary engineer or scientist is rare. Find colleagues who are smarter than you and know more. I always look for such colleagues. The obvious advantage is that she or he may be able to solve the problem that has produced a dead end in your work. But more importantly, smart and knowledgeable colleagues can save you a lot of time!

You don't have to be a fast thinker or a fast talker. In fact, it is best to avoid having such people as colleagues.

5. Obsession

5.1. *Obsession is Important When You Have a Good Computing, Engineering, or Scientific Idea*

When you are imagining and visualizing an idea that you expect to be fruitful it is important to be obsessed with the idea. Think about the idea as much as possible — perhaps even to the extent of neglecting boyfriends, girlfriends, children or spouses. Obsession, immersing yourself in the problem, will enable you to focus and thoroughly explore all the aspects of the idea: what has been done on related ideas, compatibility with physical laws and mathematics and logic, feasibility, practicality, extensions, and variations.

Obsession involving an entire field often leads to great new technology. Serious efforts to build a powered, heavier-than-air airplane occupied decades before the Wright brothers flew in 1903, Fig. 7. They were the first to make a controlled flight using design principles that are the foundation of present airplane design.

Fig. 7. An early airplane built by the Wright brothers and the brothers themselves: scientists, inventors, and entrepreneurs.

5.2. *Ending Obsession*

But, if in the course of the work you find that you run out of money, or someone else has a better idea, or your idea has a serious flaw — give up the obsession and move on!

An entire field can also be involved in a hopeless obsession. A good example is the concept of using a rigid, lighter-than-air dirigible to complete with airplanes. This was exciting technology in the early decades of the twentieth century. But almost all dirigibles, whether commercial or military, crashed or were dismantled within a decade of their construction. The building of commercial or large military dirigibles ended in 1938 when the Hindenburg zeppelin exploded and crashed in New Jersey, Fig. 8. The obsession was over.

Fig. 8. In New Jersey in1938, during the process of mooring the Hindenburg zeppelin to the mast, the Hindenburg exploded and burned. The Hindenburg used hydrogen for lift. There are many explanations of the cause of the explosion.

5.3. *Ambiguous Obsession — Power from Controlled Fusion*

Since the 1950's a substantial scientific and engineering community has been working on using controlled nuclear fusion to produce power. The physics and engineering is understood in outline. There is no violation of any known laws of nature. There are several reasons for the long gestation period. The plasma physics is very complicated and details may not yet be understood, even prototype apparatus are enormous and expensive. There are severe engineering materials problems to be solved.

Yet the controlled fusion scientific and engineering community believes that it is feasible to build such a power plant. The question is practicality. How much will it cost to build and operate a fusion power plant? I think the fusion powder community is obsessed. It may be a good obsession that will lead to final success, or it may be an obsession that should be ended.

6. The Technology You Use

You must be interested in — perhaps even enchanted by — some of the technology, software, or mathematics you use. Then the bad days when the project or the research is stalled or moves backwards are not so bad, at least you have enjoyed the technology.

My 1955 Ph.D. thesis [4] made use of the atomic beam resonance apparatus of Fig. 9 for measuring the nuclear quadrupole moment of sodium.

The apparatus was beautiful — a shining brass vacuum vessel with a glass McLeod vacuum gauge filled with mercury. The current for the beam deflecting magnets came from surplus submarine batteries that were recharged every night from an ac-dc motor generator set. The sodium beam was produced by a pinhole oven that could produce a beam for about eight hours. I loved the technology and had myself built the smaller parts. But if the oven clogged I had to stop the experiment, clean and refill the oven, recharge the submarine batteries, *that was a bad day.*

Another advantage of being enchanted by the technology, programming, or mathematics that you work with is that you will be more likely to think of improvements and variations.

You should be fond of the technology, mathematics, or programs that you use, but not so much in love that you are blind to the possibility that there may be a better way.

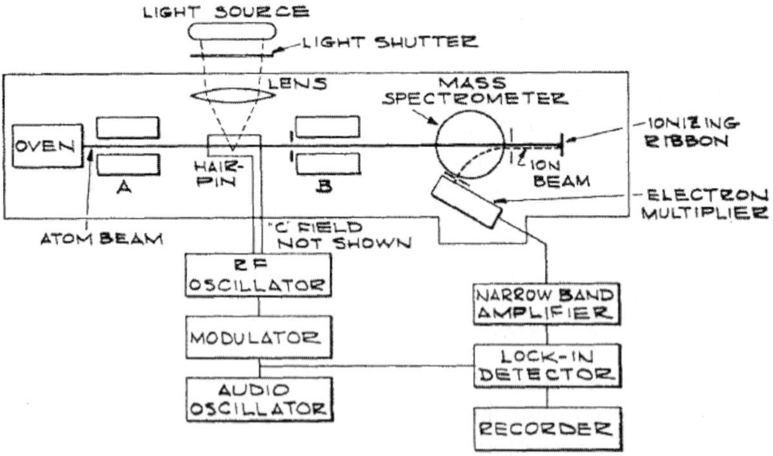

Fig. 9. A schematic of the apparatus I used in my 1955 Ph.D. thesis. I drew this, in 1955 there were no computer drawing programs.

7. The Technology of the Future — Replacement of Technologies

It is often impossible to predict the future of a technology. Some technologies are replaced again and again by new technologies serving the same function. An example is sound reproduction.

- Invented in 1890's: Gramophone and phonograph mechanical inscription of a physical trace of sound on a disc or a cylinder with mechanical reproduction.
- Introduced in 1920's: Electrical amplification used for mechanical inscription and sound reproduction. Gradually replaced purely mechanical system. Cylinders no longer in general use.
- Introduced in 1950's: Long playing records.
- Introduced in 1960's: Radical change in technology by recording on magnetic tape, cassette format dominant.
- Introduced in 1980's: Another radical change in technology — development of the Compact Disc with digital recording. Until then all widely used systems for sound reproduction were analog.
- Present: Prevalent use of digital sound recording on magnetic hard drives and flash memories.

8. The Technology of the Future — Incremental Improvements

Some technologies persist through incremental improvements. A good example is the reciprocating internal gasoline engine, developed in practical form largely in Germany in the last few decades of the nineteenth century. Many efforts have been made to replace the use of internal gasoline engines in automobiles and small trucks; for example the Wankle rotary engine has been tried commercially. But the reciprocating internal gasoline engine is continually improved with the use of new auxiliary technologies such as computer control, Fig. 10.

9. The Technology of the Future — Some Promising Technologies Go Nowhere

Early in my technical career I learned that some promising technologies go nowhere. In 1950 I was a chemical engineer working in a radio tube factory of the U.S. General Electric Company. My boss had a special interest in developing very small radio tubes for use in portable radios and hearing aids. The smaller filaments used to heat the cathode would take less

Fig. 10. The reciprocating internal combustion engine has a hundred year successful history.

power allowing a longer lasting battery. other tube companies had the same interest. My boss also pointed out that very small vacuum tubes would be an advantage in home radio sets because the time between turning on the set and hearing the sound would be shorter since the filament would heat up faster.

While we and other companies worked developing smaller vacuum tubes, the transistor was invented at Bell Laboratories in 1949 by John Bardeen, Walter Brattain, and William Shockley. The transistor age had arrived and the small vacuum tube was relegated to a few special uses, Fig. 11.

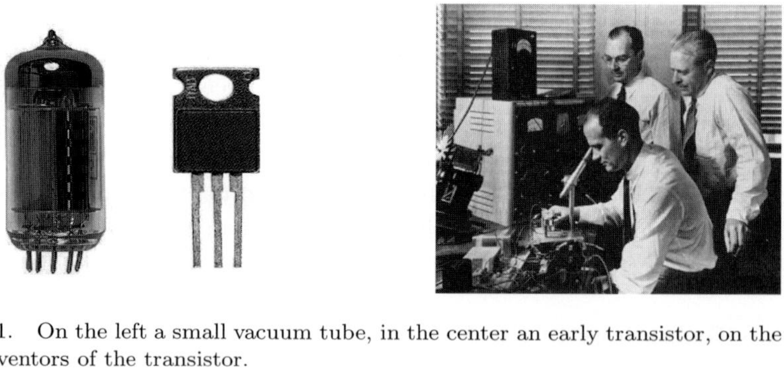

Fig. 11. On the left a small vacuum tube, in the center an early transistor, on the right the inventors of the transistor.

Acknowledgments

I am grateful to Professor K. K. Phua for encouragement in the writing of this paper.

References

1. See: *Review of Manned Aircraft Nuclear Propulsion Program* http://en.wikipedia.org/wiki/Flying_car.
2. J. L. Adams, *Conceptual Blockbusting* (Perseus Publishing, Cambridge, 2001).
3. J. H. Austin, *Chase, Chance and Creativity* (Columbia University Press, New York, 1978).
4. M. L. Perl, I. I. Rabi and B. Senitzky, *Phys. Rev.* **97**, 835 (1955).

KNOTS AS POSSIBLE EXCITATIONS OF THE QUANTUM YANG-MILLS FIELDS

L. D. FADDEEV

St. Petersburg Department of Steklov Mathematical Institute

It is a great honour and pleasure for me to participate in the conference, dedicated to 85 years celebration for Professor C. N. Yang.

The influence of C. N. Yang on my own research is very strong. Two of my directions — quantization of the Yang-Mills field and theory of solitons stem from his works. I am proud to remind, that the term "Yang-Baxter equation" was introduced by L. Takhtajan and me and now covers extensive research in integrable models and quantum groups.

In my talk I shall describe the subject, which to some extend connects solitons and Yang-Mills quantum field theory. As is reflected in the title, it is still not well established. However, I believe, that work on it will be continued in the future.

1. Introduction

Quantum Yang-Mills theory [1] is most probably the only viable relativistic field theory in 4-dimensional space-time. The special property, leading to this conviction, is dimensional transmutation [2] and related property of asymtotic freedom [3]. However the problem of description of corresponding particle-like excitations is still not solved. The question, posed by W. Pauli in 1954 during talk of C. N. Yang at Oppenheimer seminar at IAS [4], waits for an answer for more than 50 years. In this talk I shall present a hypothetical scenario for this picture: particles of Yang-Mills field are knot-like solitons.

The idea is based on another popular hypothese, according to which the confinement in QCD is effectuated by gluonic strings, connecting quarks. Thus a natural question is what happenes to these strings in the absence of quarks, i. e. in the pure Yang-Mills theory. The strings should not disappear, they rather become closed, producing rings, links or knots. This idea was leading in my recent activity in collaboration with Antti Niemi.

Our approach is based on a soliton model, which I proposed in the mid-70ties in the wake of interest to the soliton mechanism for particle-like excitations. My proposal was mentioned in several talks, partly refered to in [5].

The model is a kind of nonlinear σ-model with nonlinear field $n(x)$ taking values in the two-dimensional sphere $\mathbb{S}^2$. It does not allow complete separation of variables, so practical research was to wait until mid-90ties when computers strong enough became available. It was Antti Niemi, who was first to sacrifice himself for complicated numerical work with the great help of supercomputer center at Helsinki. The first result, published in [5], attracted attention of two groups [6], [7]. Their work revealed rich structure of knot-like solitons, confirming my expectations. Thus a candidate for dynamical model with knot-like excitations was found. Next step was to find a place for this field among the dynamical variables of the Yang-Mills field theory.

We developed consequentively two approaches for this. The first one was based on the proposal of Y. M. Cho [8] to construct kind of the magnetic monopole connection, described by means of the n-field [9]. This approach is still discussed by several groups [10]–[11]. In fact Cho connection was found before in [12]. However, now we do not consider this approach as promising anymore and in the beginnning of new century developed another one. The short announcement [13] was developed in a detailed paper [14]. In this talk I shall briefly describe our way to this proposal and give its exposition. I shall begin with the description of the σ-model, then propose its application in the condenced matter theory and finally explain our approach to the Yang-Mills theory.

2. Nonlinear σ-model

The field variable is n-field — a unit vector

$$\vec{n} = (n_1, n_2, n_3), \quad \sum n_i^2 = 1.$$

In other words the target is a sphere $\mathbb{S}^2$. For static configurations the space variables ran through $\mathbb{R}^3$. Boundary condition

$$n|_\infty = (0, 0, 1)$$

compactifies $\mathbb{R}^3$ to $\mathbb{S}^3$, so n-field realizes the map

$$n : \mathbb{S}^3 \to \mathbb{S}^2.$$

Such maps are classified by means of the topological charge, called Hopf invariant, which is more exotic in comparison with more usual degree of map, used when space and target have the same dimension.

To describe this topological charge consider the preimage of the volume form on $\mathbb{S}^2$ — 2 form on $\mathbb{R}^3$ (or $\mathbb{S}^3$)

$$H = H_{ik} dx^i \wedge dx^k$$

with

$$H_{ik} = (\partial_i \vec{n} \times \partial_i \vec{n}, \vec{n}) = \epsilon_{abc} \partial_i n^a \partial_k n^b n^c,$$

which is exact

$$H = dC.$$

Then Chern-Simons integral

$$Q = \frac{1}{4\pi} \int_{\mathbb{R}^3} H \wedge C$$

acquires only integer values and is called Hopf invariant.

The formulas above have natural interpretation in terms of magnetic field. Indeed, the Poincare dual of H_{ik}

$$B_i = \frac{1}{2} \epsilon_{ikj} H_{kj}$$

is divergenceless

$$\partial_i B_i = 0$$

and can be taken as a description of magnetic field. The preimage of a point on $\mathbb{S}^2$ is a closed contour, describing a line of force of this field. The Hopf invariant is an intersection number of any two such lines.

It is instructive to mention, that n-filed gives a way to describe the magnetic field alternative to one based on the the vector potential. In particular the configuration

$$\vec{n} = \frac{\vec{x}}{|x|}$$

describes the magnetic monopole without annoying Dirac string.

There are two natural functional, which can be used to introduce the energy. The first is the traditional σ-model hamiltonian

$$E_1 = \int_{\mathbb{R}^3} (\partial n)^2 d^3x.$$

The second is the Maxwell energy of magnetic field

$$E_2 = \int_{\mathbb{R}^3} (H_{ik})^2 d^3x.$$

Functional E_1 is quadratic in the derivatives of n-field and E_2 is quartic in them. Thus they have opposite reaction to scaling $x \to \lambda x$

$$E_1 \to \lambda E_1, \quad E_2 = \frac{1}{\lambda} E_2,$$

which is reflected in their different dimensions

$$[E_1] = [L], \quad [E_2] = [L]^{-1}.$$

We take for the energy their linear combination

$$E = aE_1 + bE_2,$$

where

$$[a] = [L]^{-2}$$

and b is dimensionless. Derric theorem – the well known obstruction for the existence of localized finite energy solutions (solitons) – does not apply here.

The estimate

$$E \geq c|Q|^{3/4},$$

found in [15], supports the belief that such solutions do exist. Unfortunately the relevant mathematical theorem is not proved until now, so we are to refer to numerical evidence [6], [7]. The picture of solutions looks as follows. The lowest energy $Q = 1$ soliton is axial symmetric; it is concentrated along the circle $n_3 = -1$; the magnetic surfaces (preimages of lines $n_3 = \text{const}$) are toroidal, wrapped once by by magnetic lines of force. For $Q = 4$ minimal solution is a link and for $Q = 7$ it is trefoil. Beautiful computer movies, illustrating the calculations based on the descent method, can be found in [16].

There is a superficial analogy of the σ-model with the Skyrme model [17] for the principal chiral field $g(x)$ with values in the manifold of compact Lie group G. Skyrme lagrangian is expressed via the Maurer-Cartan current

$$L_\mu = \partial_\mu g g^{-1}$$

as follows

$$\mathcal{L} = a \,\mathrm{tr}\, L_\mu^2 + b \,\mathrm{tr}[L_\mu, L_\nu]^2,$$

which also contains terms quadratic and quartic in derivatives of g. Corresponding topological charge

$$Q = \int \mathrm{tr}[L_i, L_k] L_j \epsilon_{ikj} d^3x$$

coinsides with the degree of map for $G = SU(2)$. There is an estimate for static Hamiltonian

$$E \geq c|Q|.$$

The minimal excitation for $Q = 1$ is spherically symmetric and concentrated around a point.

So there are two important differences between two models. First, the excitations of Skyrme model are point like, whereas those for nonlinear σ-model are string-like. Second the term E_2 has natural interpretation as Maxwell energy whereas the quartic term in the Skyrme model is rather artificial.

This concludes the description of the nonlinear σ-model and I must turn to its applications. Before the main one to Yang-Mills field, I shall consider more simple example, developed together with Niemi and Babaev [18].

3. Two Component Landau-Ginsburg-Gross-Pitaevsky Equation

The equation from the title appears in the theory of superconductivity (LG) and Bose gas (GP). The main degree of freedom is a complex valued function $\psi(x)$ — gap in the superconductivity or density in Bose-gas. Magnetic field is described by vector potential $A_k(x)$. There is a huge literature dedicated to the LGGP equation. Our contribution consists in using two components ψ

$$\psi = (\psi_1, \psi_2),$$

corresponding to a mixture of two materials.

The energy in the appropriate units is written as

$$E = \sum_{\alpha=1}^{2} |\nabla_i \psi_\alpha|^2 + F_{ik}^2 + v(\psi),$$

where

$$\nabla_i \psi = \partial_i \psi + i A_i \psi$$

and

$$F_{ik} = \partial_i A_k - \partial_k A_i.$$

The functional E is invariant with respect to the abelian gauge transformation

$$A_i \to A_i + \partial_i \lambda, \quad \psi_\alpha \to e^{-i\lambda}\psi_\alpha$$

with an arbitrary real function λ. In the case of one component ψ the change of variables

$$\psi = \rho e^{i\theta}, \quad A_k = B_k + \frac{1}{\rho^2} J_k,$$

where

$$J_k = \frac{1}{2i}\left(\overline{\psi}\partial_k \psi - \overline{\partial_k \psi}\psi\right),$$

transforms E to the gauge invariant form

$$E = (\partial\rho)^2 + \rho^2 B^2 + (\partial_i B - \partial_k B)^2 + v(\rho),$$

eliminating phase θ and leaving gauge invariant density ρ and supercurrent B. The potential $v(\rho)$ is supposed to produce the nonzero mean value for ρ

$$< \rho > = \Lambda,$$

vector field B becomes massive (Meissner effect with finite penetration length).

In the case of two components $\psi_\alpha, \alpha = 1, 2$ the analogous change of variables, proposed in [18], looks as follows

$$\rho^2 = |\psi_1|^2 + |\psi_2|^2,$$

$$\vec{n} = \frac{1}{\rho^2}(\bar{\psi}_1, \bar{\psi}_2)\vec{\tau}\begin{pmatrix}\psi_1 \\ \psi_2\end{pmatrix}$$

$$A_k = B_k + \frac{1}{\rho^2} J_k$$

$$J_k = \frac{1}{2i}\sum_\alpha \left(\bar{\psi}_\alpha \partial_k \psi_\alpha - \partial_k \bar{\psi}_\alpha \psi_\alpha\right).$$

Here $\vec{\tau} = (\tau_1, \tau_2, \tau_3)$ is set of Pauli matrices

$$\tau_1 = \begin{pmatrix} 0 & 1 \\ 1 & 0 \end{pmatrix}, \quad \tau_2 = \begin{pmatrix} 0 & -i \\ i & 0 \end{pmatrix}, \quad \tau_3 = \begin{pmatrix} 1 & 0 \\ 0 & -1 \end{pmatrix}.$$

Variables ρ, B and n are gauge invariant. Thus the difference with the case of one component is appearence of the n-field.

The energy in new variables looks as follows

$$E = (\partial\rho)^2 + \rho^2(\partial n)^2 + \left(\partial_i B_k - \partial_k B_i + H_{ik}\right)^2 + \rho^2 B^2 + v(\rho)$$

and contains both ingredients of the nonlinear σ-model from section 1. If due to the Meissner effect massive vector field B vanishes in the bulk, only n-field remains there and should produce knot-like excitations. This is our main prediction and we wait for the relevant experimental work.

I want to stress the difference of our excitations with Abrikosov vortices. Our closed strings have finite energy in 3-dimensional bulk, whereas Abrikosov vortices are two-dimensional. Moreover, the corresponding topological charges are distinct — Hopf invariant in our case and degree of map $\mathbb{S}^1 \to \mathbb{S}^1$ in the case of Abrikosov vortices.

Now it is time to turn to the main subject — Yang-Mills field.

4. $SU(2)$ Yang-Mills Theory

The field variables are 3 vector fields $A_\mu^a, a = 1,2,3$, describing connection in the fiber bundle $M_4 \times SU(2)$, where M_4 is a space-time, which for definiteness we shall take as euclidean $\mathbb{R}^4$. Let τ^a be Pauli matrices and

$$A_\mu = A_\mu^a \tau^a .$$

The gauge tranformation is given by

$$A_\mu \to g A_\mu g^{-1} + \partial_\mu g g^{-1}$$

with arbitrary 2×2 unitary matrix g. The curvature (field strength) $F_{\mu\nu}$

$$F_{\mu\nu} = \partial_\mu A_\nu - \partial_\nu A_\mu + [A_\mu, A_\nu]$$

transforms homogeneously

$$F_{\mu\nu} \to g F_{\mu\nu} g^{-1}$$

and Largangian

$$\mathcal{L}_{\mathrm{YM}} = \frac{1}{4} \operatorname{tr}(F_{\mu\nu})^2$$

is gauge invariant.

The maximal abelian partial gauge fixing (MAG), which we shall use, put restriction on the offdiagonal components A_μ^1 and A_μ^2. We shall use the complex combination

$$B_\mu = A_\mu^1 + i A_\mu^2$$

and MAG condition looks as follows

$$\nabla_\mu B_\mu = 0,$$

where

$$\nabla_\mu = \partial_\mu + iA_\mu, \quad A_\mu = A^3_\mu.$$

The fact, that we use a distinguished (diagonal) direction in the charge space is not essential, see [14] for details.

The remaining gauge freedom is the abelian one

$$B_\mu \to e^{-i\lambda} B_\mu, \quad A_\mu \to A_\mu + \partial_\mu \lambda.$$

MAG condition can be realized by adding the quadratic form $\frac{1}{2}|\nabla_\mu B_\mu|^2$ to $\mathcal{L}_{\mathrm{YM}}$, leading to

$$\begin{aligned}\mathcal{L}_{\mathrm{MAG}} &= \mathcal{L}_{\mathrm{YM}} + \frac{1}{2}|\nabla_\mu B_\mu|^2 \\ &= \frac{1}{2}|\nabla_\mu B_\nu|^2 + \frac{1}{4}(F_{\mu\nu} + H_{\mu\nu})^2 + \frac{1}{2}F_{\mu\nu}H_{\mu\nu},\end{aligned}$$

where

$$F_{\mu\nu} = \partial_\mu A_\nu - \partial_\nu A_\mu, \quad H_{\mu\nu} = \frac{1}{2i}(\bar{B}_\mu B_\nu - \bar{B}_\nu B_\mu).$$

The last term appears after the integration by parts, used to eliminate the unwanted term $\bar{\nabla}_\mu \bar{B}_\nu \nabla_\nu B_\mu$.

Now I come to the main trick. Observe, that two vector fields A^1_μ and A^2_μ define 2-plane in M_4. Let us parametrize this 2-plane by the orthogonal zweibein e_μ

$$\begin{gathered} e_\mu = e^1_\mu + ie^2_\mu \\ \bar{e}_\mu e_\mu = 1, \quad e^2_\mu = \bar{e}^2_\mu = 0 \end{gathered}$$

and express B_μ as

$$B_\mu = \psi_1 e_\mu + \psi_2 \bar{e}_\mu,$$

introducing two complex coefficients ψ_1 and ψ_2. Altogether the set e_μ, ψ_1, ψ_2 contains 9 real functions and B_μ has only 8 real components. The discrepancy is resolved by comment, that expression for B_μ is invariant with respect to the abelian gauge transformation

$$e_\mu \to e^{i\omega} e_\mu, \quad \psi_1 \to e^{-i\omega}\psi_1, \quad \psi_2 \to e^{i\omega}\psi_2.$$

Corresponding $U(1)$ connection is given by

$$\Gamma = \frac{1}{i}(\bar{e}_\nu \partial_\mu e_\nu), \quad \Gamma_\mu \to \Gamma_\mu + \partial_\mu \omega.$$

Now having ψ_1, ψ_2 we can repeat trick from section 2, introducing n-field. However in our case we can do more. Indeed, the combination $H_{\mu\nu}$, entering $\mathcal{L}_{\mathrm{MAG}}$, can be written as

$$H_{\mu\nu} = \rho^2 n_3^2 g_{\mu\nu}$$

with

$$\rho^2 n_3^2 = |\psi_1|^2 - |\psi_2|^2, \quad g_{\mu\nu} = \frac{1}{2i}(\bar{e}_\mu e_\nu - \bar{e}_\nu e_\mu).$$

Putting

$$p_i = g_{0i}, \quad q_i = \frac{1}{2}\epsilon_{ijk} g_{jk}$$

we get two vectors p_i, q_i satisfying conditions

$$p^2 + q^2 = 1, \quad (p,q) = 0,$$

thus defining two spheres $\mathbb{S}^2$. Indeed, what we get here is a particular parametrization of the Grassmanian $G(4,2)$. In static case p_i disappears and we are left with one unit 3-vector q, which evidently could be used to introduce the magnetic monopoles.

Now we can put the new variables into $\mathcal{L}_{\mathrm{MAG}}$. All details are to be found in [14]. Here I shall write explicitely the static energy

$$E = (\partial_i \rho)^2 + \rho^2\big((\nabla_k n)^2 + (\partial_k q)^2\big) + \rho^2 C_k^2 + \\ + \frac{1}{4}\big((\partial_i n \times \partial_k n, n) + (\partial_i q \times \partial_k q, q) + 2H_{ik} + \partial_i C_k - \partial_k C_i\big)^2 - \frac{3}{4}\rho^4 n_3^2,$$

where C is supercurrent

$$C_k = A_k + \frac{1}{2\rho^2}\big(\bar{\psi}_1(\partial_k + iA_k + i\Gamma_k)\psi_1 + \bar{\psi}_2(\partial_k + iA_k - i\Gamma_k)\psi_2 - \text{c.c.}\big)$$

and

$$\nabla_k n = \partial_k n + i\Gamma_k n.$$

We see, that the structure of nonlinear σ-model appears twice — via fields n and q. We can interprete it as a new manifestation of electromagnetic duality in the nonabelian Yang-Mills theory.

The expression for E can be taken as a point of departure for speculations on the knot-like excitations for the $SU(2)$ Yang-Mills field. The corresponding transformation for $SU(3)$ case, done in [19], is more complicated due to difference of rank and number of roots.

I want to stress, that by no means I propose to use the new variables to make a change of variables in the functional integral. Rather they should be put into the renormalized effective action, which should be found in the background field formalism. The variant of this method, where the background field is not classical, but is a solution of the quantum modified equation of motion is given in [20]. In the course of renormalization this effective action should experience the dimensional transmutation. We still do not know, how it happens, so we can only use speculations. The main hope, that in this way the mean value of ρ^2 will appear. At the same time the classical lagrangian should be main part of the effective action, as it represents the only possible local gauge invariant functional of dimension -4. The condensate of $< \rho^2 >$ of dimension -2 is the subject of many papers in the last years (see e. g. [21], [22]). Thus all this makes the picture of string-like excitations for the Yang-Mills field more feasible. However the real work only begins here. I hope, that this subject will take fancy of some more young researchers.

References

[1] C. N. Yang and R. Mills, "Conservation of Isotopic Spin and Isotopic Gauge Invariance," *Phys. Rev.* **96**, 191–195, (1954).

[2] S. Coleman, "Secret Symmetries: An Introduction to Spontaneous Symmetry Breakdown and Gauge Fields," Lecture given at 1973 Intern. Summer School in Phys. Ettore Majorana. Erice (Sicily), 1973, *Erice Subnucl. Phys.*, 1973.

[3] D. J. Gross and F. Wilczek, "Ultraviolet Behavior of non-abelian Gauge Theories," Phys. Rev. Lett. **30**, 1343 (1973); H. D. Politzer, "Reliable Perturbative Results for Strong Interactions?," Phys. Rev. Lett. **30**, 1346 (1973).

[4] See commentary by C. N. Yang in C. N. Yang, *Selected papers 1945-1980*, Freeman and Company, (1983) 19-21.

[5] L. D. Faddeev and A. J. Niemi, "Knots and particles," Nature **387**, 58 (1997) [arXiv:hep-th/9610193]

[6] J. Hietarinta, P. Salo, "Faddeev-Hopf Knots: Dynamics of Linked Unknots," Phys. Lett., 1999, **B451**, 60.

[7] R. Battye, P. M. Sutcliffe, "Knots as Stable Soliton Solutions in a Three-Dimensional Classical Field Theory," Phys. Rev. Lett., 1998, **81**, 4798.

[8] Y. M. Cho, "A Restricted Gauge Theory," Phys. Rev. D **21**, 1080 (1980).

[9] L. D. Faddeev and A. J. Niemi, "Partially dual variables in SU(2) Yang-Mills theory," Phys. Rev. Lett. **82**, 1624 (1999) [arXiv:hep-th/9807069].

[10] K. I. Kondo, T. Murakami and T. Shinohara, "Yang-Mills theory constructed from Cho-Faddeev-Niemi decomposition," Prog. Theor. Phys. **115**, 201 (2006) [arXiv:hep-th/0504107].

[11] Y. M. Cho, "Knot topology of QCD vacuum," Phys. Lett. B **644**, 208 (2007) [arXiv:hep-th/0409246].

[12] Y.S. Duan and M.L. Ge, Sinica Sci. **11** (1979) 1072.

[13] L. D. Faddeev and A. J. Niemi, "Decomposing the Yang-Mills field," Phys. Lett. B **464**, 90 (1999) [arXiv:hep-th/9907180].

[14] L. D. Faddeev and A. J. Niemi, "Spin-charge separation, conformal covariance and the SU(2) Yang-Mills theory," Nucl. Phys. B **776**, 38 (2007) [arXiv:hep-th/0608111].
[15] A. F. Vakulenko, L. V. Kapitansky, *Dokl. Akad. Nauk USSR*, **248**, (1979), 840–842.
[16] http://users.utu.fi/hietarin/knots/index.html
[17] T. H. R. Skyrme, "A Nonlinear field theory," Proc. Roy. Soc. Lond. A **260**, 127 (1961).
[18] E. Babaev, L. D. Faddeev and A. J. Niemi, "Hidden symmetry and duality in a charged two-condensate Bose system," Phys. Rev. B **65**, 100512 (2002) [arXiv:cond-mat/0106152]
[19] T. A. Bolokhov and L. D. Faddeev, "Infrared variables for the SU(3) Yang-Mills field," Theor. Math. Phys. **139**, 679 (2004) [Teor. Mat. Fiz. **139**, 276 (2004)].
[20] L. D. Faddeev, "Notes on divergences and dimensional transmutation in Yang-Mills theory," Theor. Math. Phys. **148**, 986 (2006) [Teor. Mat. Fiz. **148**, 133 (2006)].
[21] F.V. Gubarev, L. Stodolsky and V.I. Zakharov, Phys. Rev. Lett. **86**, 2220 (2001); L Stodolsky, P. van Baal and V.I. Zakharov, Phys. Lett. **B552**, 214 (2003).
[22] H. Verschelde, K. Knecht, K. Van Acoleyen and M. Vanderkelen, "The non-perturbative groundstate of QCD and the local composite operator A(mu)**2," Phys. Lett. B **516**, 307 (2001) [arXiv:hep-th/0105018].
[23] K. I. Kondo, "Vacuum condensate of mass dimension 2 as the origin of mass gap and quark confinement," Phys. Lett., 2001, **B514**, 335.

A TORSIONAL TOPOLOGICAL INVARIANT

H. T. NIEH

Center for Advanced Study, Tsinghua University,
Beijing 100084, China
and
C. N. Yang Institute for Theoretical Physics,
State University of New York at Stony Brook,
Stony Brook, New York 11794, USA
nieh@tsinghua.edu.cn

Curvature and torsion are the two tensors characterizing a general Riemannian space–time. In Einstein's general theory of gravitation, with torsion postulated to vanish and the affine connection identified to the Christoffel symbol, only the curvature tensor plays the central role. For such a purely metric geometry, two well-known topological invariants, namely the Euler class and the Pontryagin class, are useful in characterizing the topological properties of the space–time. From a gauge theory point of view, and especially in the presence of spin, torsion naturally comes into play, and the underlying space–time is no longer purely metric. We describe a torsional topological invariant, discovered in 1982, that has now found increasing usefulness in recent developments.

Keywords: Gravity; Nieh–Yan class; torsional topological invariant.

1. Introduction

Professor C. N. Yang played a leading role in laying the foundation for the development of gauge theories in particle physics. His hallmark contribution is the Yang–Mills non-Abelian gauge theory.[1] My first learning of the Yang–Mills gauge theory came in 1963 when Professor Julian Schwinger, my thesis advisor, assigned me to investigate the question of mass of the Yang–Mills gauge particle; apparently he himself was thinking about the mass problem of gauge bosons during that period.[2] After struggling for about a year, of course, nothing came out of it. It was also around that time Schwinger taught in his quantum field theory class Herman Weyl's 1929 formulation of Dirac electrons in the gravitational field,[3–7] in which Einstein's gravitational theory is cast as a gauge theory of local Lorentz gauge symmetry. In the mid-1970's, after the standard model of Glashow–Weinberg–Salam had more or less settled down and after the appearance of Professor Yang's 1974 paper on gravitation,[8] I became interested in learning again about the theory of gravitation. In the spirit of gauge theory, in which the connection, or the gauge field, plays central role, Weyl's and Fock's pioneering formulation[3–7] together with its revival by Kibble[9] and Sciama[10] in the early 1960's clearly indicate that torsion (for a review, see Ref. 11), in addition to curvature, could play an important role in the development of gravitational theory. It was during this learning process that an identity relating the totally antisymmetric part of the curvature tensor with torsion was found, which M. L. Yan and I later published in 1982.[12] This leads to a topological invariant that characterizes the torsional property of the space–time, and has found increasing usefulness in recent developments. It is this work I will report on.

2. Riemann–Cartan Space–Time

The Yang–Mills gauge theory is based on the premise of local freedom in defining isospin. To make "compatible" the definitions of isopin at neighboring space–time points, a connection field, or the Yang–Mills gauge field, is introduced. Einstein's theory of general relativity is based on Riemannian geometry, in which local definitions of vectors at neighboring space–time points are correlated by the familiar Christoffel connection. The Christoffel connection, though expressed in terms of the metric tensor, thus plays the role of a gauge field. Other than electromagnetic theory, the theory of general relativity is indeed the earliest gauge theory, and with a very rich structure. The Dirac spinor, on the other hand, is a two-valued representation of the Lorentz group SO(3, 1), which has very different transformation properties from vector representations. To accommodate local freedom of defining spinor property with respect to local Lorentz frames at neighboring space–time points, Weyl and Fock[3–7] introduced a new connection, the Lorentz connection or spin-connection, to relate local definitions of a Dirac spinor. Again, this is very much a play of the gauge concept. Under Lorentz transformations of the local Cartesian frame, which is represented by the four orthonormal vectors $e^a{}_\mu(x)$, the "vierbeins," the Dirac spinor $\Psi(x)$ transforms according to[a]

$$\Psi(x) \to e^{-i\varepsilon^{ab}(x)\sigma_{ab}/4}\Psi(x)\,, \tag{1}$$

[a]We follow the notation in Ref. 13.

where

$$\sigma_{ab} = \frac{i}{2}[\gamma_a, \gamma_b], \quad \{\gamma_a, \gamma_b\} = 2\eta_{ab}, \quad \eta_{ab} = (1, -1, -1, -1).$$

The Roman letters a, b, etc. are the frame indices, while the Greek letters μ, ν, etc. are the coordinate indices. The corresponding connection field ${\omega^{ab}}_\mu(x)$, commonly called the spin connection or Lorenz connection, is introduced such that the covariant derivative

$$D_\mu \Psi \equiv \left(\partial_\mu - \frac{i}{4}\, {\omega^{ab}}_\mu\, \sigma_{ab} \right) \Psi \tag{2}$$

transforms in a covariant way:

$$D_\mu \Psi(x) \to e^{-i\varepsilon^{ab}(x)\sigma_{ab}/4} \Psi(x). \tag{3}$$

This requires that the spin connection field ${\omega^{ab}}_\mu(x)$ transforms according to

$$\omega_\mu(x) \to {\omega'}_\mu(x) = e^{-i\varepsilon(x)} \omega_\mu(x) e^{i\varepsilon(x)} - \left[i\partial_\mu e^{-i\varepsilon(x)} \right] e^{i\varepsilon(x)}, \tag{4}$$

where

$$\omega_\mu(x) \equiv \frac{1}{4} {\omega^{ab}}_\mu(x) \sigma_{ab}, \qquad \varepsilon(x) \equiv \frac{1}{4} \varepsilon^{ab}(x) \sigma_{ab}.$$

The space–time metric is defined by

$$g_{\mu\nu} = \eta_{ab} {e^a}_\mu\, {e^b}_\nu \tag{5}$$

while the proper definition for the GL(4) connection, which plays the role of the gauge field for general coordinate transformations, is given by[9]

$${\Gamma^\lambda}_{\mu\nu} \equiv {e_a}^\lambda ({e^a}_{\mu,\nu} + {\omega^a}_{b\nu}\, {e^b}_\mu). \tag{6}$$

The "minimum" combination

$$\Psi i\gamma^a {e_a}^\mu \left(\partial_\mu - \frac{i}{4}\, {\omega^{bc}}_\mu\, \sigma_{bc} \right) \Psi + \text{h.c.}, \tag{7}$$

where ${e_a}^\mu$ is the inverse of ${e^a}_\mu$, being invariant under local Lorentz transformations as well as under the GL(4) general coordinate transformations when the transformation properties of the vierbein fields are correspondingly defined, is the natural choice as the Dirac Lagrangian in curved space–time.

We note that, in this formulation, there are two sets of field variables: the spin-connection fields ${\omega^{ab}}_\mu(x)$ and the "vierbein" fields ${e^a}_\mu$. In defining covariant derivatives, ${\omega^{ab}}_\mu(x)$ and ${\Gamma^\lambda}_{\mu\nu}$ are the gauge fields for the local Lorentz transformations and the general coordinate transformations, respectively. The corresponding "curvature" tensors or field strengths are given, respectively, by

$$R^{ab}{}_{\mu\nu} \equiv \omega^{ab}{}_{\mu,\nu} - \omega^{ab}{}_{\nu,\mu} - \omega^{ac}{}_{\mu}\,\omega_c{}^b{}_\nu + \omega^{ac}{}_\nu\,\omega_c{}^b{}_\mu\,, \tag{8}$$

$$R^{\lambda\rho}{}_{\mu\nu} \equiv g^{\rho\sigma}(\Gamma^\lambda{}_{\sigma\mu,\nu} - \Gamma^\lambda{}_{\sigma\nu,\mu} - \Gamma^\lambda{}_{\alpha\mu}\Gamma^\alpha{}_{\sigma\nu} + \Gamma^\lambda{}_{\alpha\nu}\Gamma^\alpha{}_{\sigma\mu})\,. \tag{9}$$

Having the property

$$R^{\lambda\rho}{}_{\mu\nu} = e_a{}^\lambda\, e_b{}^\rho\, R^{ab}{}_{\mu\nu}\,, \tag{10}$$

these two curvature tensors are closely related.

There are the following two possibilities, yielding different theories:[14] (i) both $\omega^{ab}{}_\mu(x)$ and $e^a{}_\mu$ are taken to be independent field variables in the theory and are to be determined by the theory. (ii) $\omega^{ab}{}_\mu(x)$ is given by the Ricci coefficients of rotation in terms of $e^a{}_\mu$ and has the property of satisfying the required transformation property:[9]

$$\omega_{ab\mu} = \frac{1}{2}\, e^c{}_\mu(\gamma_{cab} - \gamma_{abc} - \gamma_{bca})\,, \tag{11}$$

where

$$\gamma^c{}_{ab} = (e_a{}^\mu\, e_b{}^\nu - e_b{}^\mu\, e_a{}^\nu)e^c{}_{\mu,\nu}\,.$$

When one opts for the possibility (ii), only $e^a{}_\mu$ are basic field variables of the theory and the GL(4) connection field $\Gamma^\lambda{}_{\mu\nu}$ defined by (7) is the Christoffel connection:

$$\Gamma^\lambda{}_{\mu\nu} = \frac{1}{2} g^{\lambda\rho}(g_{\rho\mu,\nu} + g_{\nu\rho,\mu} - g_{\mu\nu,\rho})\,, \tag{12}$$

which is symmetric and yields vanishing torsion tensor $C^\lambda{}_{\mu\nu}$, where the torsion tensor is defined by

$$C^\lambda{}_{\mu\nu} \equiv \Gamma^\lambda{}_{\mu\nu} - \Gamma^\lambda{}_{\nu\mu} = e_a{}^\lambda(e^a{}_{\mu,\nu} - e^a{}_{\nu,\mu} + \omega^a{}_{b\nu}e^b{}_\mu - \omega^a{}_{b\mu}e^b{}_\nu)\,. \tag{13}$$

If, on the other hand, we accept spinors as physically fundamental and regard Lorentz group as the fundamental gauge group, then the spin connection $\omega^{ab}{}_\mu(x)$ should be regarded as basic field variables. Namely, one would opt for the possibility (i) mentioned above. The GL(4) connection field $\Gamma^\lambda{}_{\mu\nu}$ as defined by (7) is then, in general, not symmetric: $\Gamma^\lambda{}_{\mu\nu} \neq \Gamma^\lambda{}_{\nu\mu}$, giving rise to nonvanishing torsion tensor. It is clear from the Dirac Lagrangian (5) that *when the connection field or the gauge field* $\omega^{ab}{}_\mu(x)$ *is taken to be an independent variable in the theory*, it receives contribution from the Dirac field in the form of

$$\frac{1}{4}\, e^{a\mu}\, \bar{\Psi}(\gamma_a\sigma_{bc} + \sigma_{bc}\gamma_a)\Psi\,,$$

or,

$$\frac{1}{2}\,\eta_{abcd}\,e^{a\mu}\,\Psi\gamma_5\gamma^d\Psi\,. \tag{14}$$

This is the origin of a nonvanishing torsion tensor. It is thus seen that torsion could play an important role in gauge theories of gravitation when the Dirac field is brought into the system of consideration.

3. Gauss–Bonnet Identities

In purely metric Riemannian space–time, the Euler and Pontryagin 4-forms

$$\sqrt{-g}\varepsilon_{\alpha\beta\gamma\delta}\varepsilon^{\mu\nu\lambda\rho}R^{\alpha\beta}{}_{\mu\nu}R^{\lambda\delta}{}_{\lambda\rho} \quad \text{(Euler)}\,,$$

$$\sqrt{-g}\varepsilon^{\mu\nu\lambda\rho}R^{\alpha\beta}{}_{\mu\nu}R_{\alpha\beta\lambda\rho} \quad \text{(Pontryagin)}$$

are well known to satisfy the Gauss–Bonnet identities. In the case of nonvanishing torsion, the identification (10) allows verification of these identities[15] by making use of the Clifford algebra satisfied by the Dirac matrices.

Define

$$\varpi_\mu \equiv \frac{1}{4}\,\omega^{ab}{}_\mu\sigma_{ab}\,, \tag{15}$$

$$\bar{R}_{\mu\nu} \equiv \varpi_{\mu,\nu} - \varpi_{\nu,\mu} - i[\varpi_\mu, \varpi_\nu] = \frac{1}{4}\,R^{ab}{}_{\mu\nu}\sigma_{ab}\,, \tag{16}$$

where in the last step use has been made of the Lorentz algebra

$$\frac{i}{2}\,[\sigma_{ab}, \sigma_{cd}] = \eta_{ac}\sigma_{bd} - \eta_{ad}\sigma_{bc} + \eta_{bd}\sigma_{ac} - \eta_{bc}\sigma_{ad}\,. \tag{17}$$

Denoting by η_{abcd} the totally antisymmetric Minkowski tensor, with $\eta_{0123} = -1$, and noticing

$$\eta_{abcd}e^a{}_\mu e^b{}_\nu e^c{}_\lambda e^d{}_\rho = \varepsilon_{\mu\nu\lambda\rho}\,, \quad (\det e^a{}_\mu)^2 = -\det g_{\mu\nu} = -g\,, \tag{18}$$

the Euler 4-form can be expressed in the form

$$\begin{aligned}\sqrt{-g}\varepsilon_{\alpha\beta\gamma\delta}\varepsilon^{\mu\nu\lambda\rho}R^{\alpha\beta}{}_{\mu\nu}R^{\lambda\delta}{}_{\lambda\rho} &= \sqrt{-g}\eta_{abcd}\varepsilon^{\mu\nu\lambda\rho}R^{ab}{}_{\mu\nu}R^{cd}{}_{\lambda\rho} \\ &= 4i\sqrt{-g}\varepsilon^{\mu\nu\lambda\rho}\,\mathrm{Tr}[\gamma_5\bar{R}_{\mu\nu}\bar{R}_{\lambda\rho}]\,,\end{aligned} \tag{19}$$

where, with

$$\gamma_5 = i\gamma^0\gamma^1\gamma^2\gamma^3\,,$$

use has been made of

$$\mathrm{Tr}[\gamma_5\sigma_{ab}\sigma_{cd}] = -4i\eta_{abcd}\,.$$

On account of (14) and

$$\varepsilon^{\mu\nu\lambda\rho}\,\mathrm{Tr}[\gamma_5\varpi_\mu\varpi_\nu\varpi_\lambda\varpi_\rho] = 0\,, \qquad [\gamma_5, \varpi_\mu] = 0\,,$$

(17) becomes

$$\begin{aligned}
&\sqrt{-g}\varepsilon_{\alpha\beta\gamma\delta}\varepsilon^{\mu\nu\lambda\rho}R^{\alpha\beta}{}_{\mu\nu}R^{\lambda\delta}{}_{\lambda\rho} \\
&\quad = 16i\sqrt{-g}\varepsilon^{\mu\nu\lambda\rho}\,\mathrm{Tr}\{\lambda_5[\varpi_{\mu,\nu}\varpi_{\lambda,\rho} - i\varpi_{\mu,\nu}\varpi_\lambda\varpi_\rho - i\varpi_\mu\varpi_\nu\varpi_{\lambda,\rho}]\} \\
&\quad = 16i\sqrt{-g}\varepsilon^{\mu\nu\lambda\rho}\,\mathrm{Tr}\left\{\gamma_5\left[(\partial_\mu\varpi_\nu)(\partial_\lambda\varpi_\rho) + \frac{2i}{3}\partial_\mu(\varpi_\nu\varpi_\lambda\varpi_\rho)\right]\right\} \\
&\quad = \partial_\mu\left\{16i\sqrt{-g}\varepsilon^{\mu\nu\lambda\rho}\,\mathrm{Tr}\left[\gamma_5\left(\varpi_\nu\partial_\lambda\varpi_\rho + \frac{2i}{3}\varpi_\nu\varpi_\lambda\varpi_\rho\right)\right]\right\}.
\end{aligned} \tag{20}$$

This is the Gauss–Bonnet formula for the Euler 4-form. It is due to this property of being a total derivative that the volume integral of the Euler 4-form is a topological invariant. Along the same vein, we can verify that in the case of nonvanishing torsion the Pontryagin 4-form satisfies the following Gauss–Bonnet-type identity:

$$\begin{aligned}
\sqrt{-g}\varepsilon^{\mu\nu\lambda\rho}R^{\alpha\beta}{}_{\mu\nu}R_{\alpha\beta\lambda\rho} &= 2\sqrt{-g}\varepsilon^{\mu\nu\lambda\rho}\,\mathrm{Tr}[\bar{R}_{\mu\nu}\bar{R}_{\lambda\rho}] \\
&= \partial_\mu\left[8\sqrt{-g}\varepsilon^{\mu\nu\lambda\rho}\,\mathrm{Tr}\left(\varpi_\nu\partial_\lambda\varpi_\rho + \frac{2i}{3}\varpi_\nu\varpi_\lambda\varpi_\rho\right)\right].
\end{aligned} \tag{21}$$

We can extend the above derivation to a larger set of field variables. The larger set[12,15] contains the spin connection fields $\omega^{ab}{}_\mu(x)$ and the vierbein fields $e^a{}_\mu$, which we group together to form antisymmetric $\omega^{AB}{}_\mu$ $(A, B = 0, 1, 2, 3, 5)$:

$$\omega^{AB}{}_\mu = \omega^{ab}{}_\mu \quad (\text{for } A, B = 0, 1, 2, 3)\,, \tag{22}$$

$$\omega^{a5}_\mu = \frac{1}{l}e^a{}_\mu \quad (a = 0, 1, 2, 3)\,, \tag{23}$$

where a constant l with the dimension of length is added to match the dimension of $\omega^{ab}{}_\mu(x)$.[16,17] As γ^0, γ^1, γ^2, γ^3, γ^5 form a set of five anti-commuting matrices, we can easily construct a set of antisymmetric X_{AB} $(A, B = 0, 1, 2, 3, 5)$ satisfying the de Sitter or anti-de Sitter algebra:

$$\frac{i}{2}[X_{AB}, X_{CD}] = \eta_{AC}X_{BD} - \eta_{AD}X_{BC} + \eta_{BD}X_{AC} - \eta_{BC}X_{AD}\,, \tag{24}$$

with $\eta_{AB} = (1, -1, -1, -1, \pm 1)$. Letting

$$\Omega_\mu \equiv \frac{1}{4} X_{AB} \varpi^{AB}{}_\mu \,, \tag{25}$$

we define $\bar{F}_{\mu\nu}$ and $F^{AB}{}_{\mu\nu}$ according to

$$\bar{F}_{\mu\nu} \equiv \Omega_{\mu,\nu} - \Omega_{\nu,\mu} - i[\Omega_\mu, \Omega_\nu] = \frac{1}{4} X_{AB} F^{AB}{}_{\mu\nu} \,. \tag{26}$$

The $F^{AB}{}_{\mu\nu}$ so defined is given by, according to (24),

$$F^{AB}{}_{\mu\nu} = \omega^{AB}{}_{\mu,\nu} - \omega^{AB}{}_{\nu,\mu} + \eta_{CD}(\omega^{AC}{}_\mu \omega^{BD}{}_\nu - \omega^{AC}{}_\nu \omega^{BD}{}_\mu) \,, \tag{27}$$

which has the contents $(a, b = 0, 1, 2, 3)$:

$$F^{ab}{}_{\mu\nu} = R^{ab}{}_{\mu\nu} + \frac{1}{l^2} \eta_{55} (e^a{}_\mu e^b{}_\nu - e^a{}_\nu e^b{}_\mu) \,, \tag{28}$$

$$F^{a5}{}_{\mu\nu} = \frac{1}{l} e^a{}_\lambda C^\lambda{}_{\mu\nu} \,, \tag{29}$$

where $C^\lambda{}_{\mu\nu}$ is the torsion tensor defined in (13). Following the same procedure as in deriving the identities (20) and (21), we can derive a similar identity for $F^{AB}{}_{\mu\nu}$:

$$\begin{aligned} \sqrt{-g}\varepsilon^{\mu\nu\lambda\rho} F^{AB}{}_{\mu\nu} F_{AB\lambda\rho} &= 2\sqrt{-g}\varepsilon^{\mu\nu\lambda\rho}\, \mathrm{Tr}[\bar{F}_{\mu\nu} \bar{F}_{\lambda\rho}] \\ &= \partial_\mu \left[8\sqrt{-g}\varepsilon^{\mu\nu\lambda\rho}\, \mathrm{Tr}\left(\Omega_\nu \partial_\lambda \Omega_\rho + \frac{2i}{3} \Omega_\nu \Omega_\lambda \Omega_\rho \right) \right] . \end{aligned} \tag{30}$$

On account of (28) and (29), we have the difference of the two Pontryagin 4-forms:

$$\begin{aligned} &\sqrt{-g}\varepsilon^{\mu\nu\lambda\rho} F^{AB}{}_{\mu\nu} F_{AB\lambda\rho} - \sqrt{-g}\varepsilon^{\mu\nu\lambda\rho} \bar{R}^{ab}{}_{\mu\nu} \bar{R}_{ab\mu\nu} \\ &= -4\eta_{55} \frac{1}{l^2} \sqrt{-g}\varepsilon^{\mu\nu\lambda\rho} \left(R_{\mu\nu\lambda\rho} + \frac{1}{2} C^\lambda{}_{\mu\nu} C_{\lambda\mu\nu} \right) . \end{aligned} \tag{31}$$

Substracting (30) from (21) yields

$$\sqrt{-g}\varepsilon^{\mu\nu\lambda\rho} \left(R_{\mu\nu\lambda\rho} + \frac{1}{2} C^\alpha{}_{\mu\nu} C_{\alpha\lambda\rho} \right) = \partial_\mu (-\sqrt{-g}\varepsilon^{\mu\nu\lambda\rho} C_{\nu\lambda\rho}) \,, \tag{32}$$

or, in the more compact form:

$$-R_{ab} \wedge e^a \wedge e^b + C_a \wedge C^a = d(C_a \wedge e^a) \,.$$

The 4-form

$$\sqrt{-g}\varepsilon^{\mu\nu\lambda\rho} \left(R_{\mu\nu\lambda\rho} + \frac{1}{2} C^\alpha{}_{\mu\nu} C_{\alpha\lambda\rho} \right) \tag{33}$$

is thus seen to be the exterior derivative of the Chern–Simons-type term:

$$-\sqrt{-g}\varepsilon^{\mu\nu\lambda\rho}C_{\nu\lambda\rho}\,. \tag{34}$$

It follows from the identity (32) that

$$\int d^4x\,\sqrt{-g}\varepsilon^{\mu\nu\lambda\rho}\left(R_{\mu\nu\lambda\rho}+\frac{1}{2}\,C^{\lambda}_{\mu\nu}C_{\alpha}\right)=-\int d\sigma_{\mu}\,\sqrt{-g}\varepsilon^{\mu\nu\lambda\rho}C_{\nu\lambda\rho}\,. \tag{35}$$

Like the Euler class and the Pontryagin class, the left-hand side of the above equation is a topological invariant.[16,17] It is an invariant that characterizes the torsional topology of the underlying space–time; with vanishing torsion, each individual term on both sides of (32) and (35) vanishes.

The identity (32) can also be derived directly[12] from the Bianchi identity for nonvanishing torsion.[15] But, the derivation presented here has the advantage of making the meaning of the topological invariant more transparent. The geometric properties of the invariant have been studied by Chandia and Zanelli[16,17] and others.[18,19] It is well known that the Pontryagin or Chern–Weil class for the Lorentz group,

$$\int d^4x\,\sqrt{-g}\varepsilon^{\mu\nu\lambda\rho}R^{ab}{}_{\mu\nu}R_{ab\lambda\rho}\,, \tag{36}$$

is a topological invariant with an integral spectrum of values. Likewise, the Pontryagin class for the de Sitter or anti-de Sitter group,

$$\int d^4x\,\sqrt{-g}\varepsilon^{\mu\nu\lambda\rho}F^{AB}{}_{\mu\nu}F_{AB\mu\nu}\,, \tag{37}$$

is also a topological invariant with an integral spectrum of values. It is seen from (31) that the torsional topological invariant

$$\int d^4x\,\sqrt{-g}\varepsilon^{\mu\nu\lambda\rho}\left(R_{\mu\nu\lambda\rho}+\frac{1}{2}\,C^{\alpha}{}_{\mu\nu}C_{\alpha\lambda\rho}\right), \tag{38}$$

is proportional to the difference of the two Pontryagin classes (36) and (37), and thus has a discrete spectrum of values.[16,17] The dimensional constant l appearing in (31), however, is an unknown parameter with no clear physical identification. In de Sitter-type gravitational theories, the constant l could conceivably play the role of the length characterizing the breakdown of the de Sitter symmetry.

Torsion is a geometric property not well investigated in mathematics. It is physicists' search for an extension of Einstein's general theory of relativity that torsion naturally appears. The properties of the torsional topological invariant (35) remain to be studied.

References

1. C. N. Yang and R. L. Mills, *Phys. Rev.* **96**, 191 (1954).
2. J. Schwinger, *Phys. Rev.* **128**, 2425 (1962).
3. H. Weyl, *Proc. Nat. Acad. Sci.* **15**, 323 (1929).
4. H. Weyl, *Z. Phys.* **56**, 330 (1929).
5. H. Weyl, *Phys. Rev.* **77**, 699 (1950).
6. V. Fock, *Z. Phys.* **57**, 261 (1929).
7. R. Utiyama, *Phys. Rev. D* **101**, 1957 (1956).
8. C. N. Yang, *Phys. Rev. Lett.* **33**, 445 (1974).
9. T. W. B. Kibble, *J. Math. Phys.* **2**, 212 (1961).
10. D. W. Sciama, in *Recent Developments in General Relativity* (Pergamon, Oxford, 1962).
11. F. W. Hehl, P. van der Heyde and G. D. Kerlick, *Rev. Mod. Phys.* **48**, 393 (1976).
12. H. T. Nieh and M. L. Yan, *J. Math. Phys.* **23**, 373 (1982).
13. H. T. Nieh and M. L. Yan, *Ann. Phys.* **138**, 237 (1982).
14. H. Weyl, *Phys. Rev.* **77**, 699 (1950).
15. H. T. Nieh, *J. Math. Phys.* **21**, 1439 (1980).
16. O. Chandia and J. Zanelli, *Phys. Rev. D* **55**, 7580 (1997).
17. O. Chandia and J. Zanelli, *Phys. Rev. D* **58**, 45014 (1998).
18. S. Li, *J. Phys. A* **32**, 7153 (1999).
19. H. Y. Guo, K. Wu and W. Zhang, *Commun. Theor. Phys.* **32**, 381 (1999).

KNOT TOPOLOGY OF CLASSICAL VACUUM SPACE-TIME

Y. M. CHO

Center for Theoretical Physics and School of Physics, College of Natural Sciences, Seoul National University, Seoul 151-742, Korea
ymcho@yongmin.snu.ac.kr

We present a topological classification of vacuum space-time. Viewing Einstein-Cartan theory as a gauge theory of Lorentz group and identifying the gravitational connection as the gauge potential of Lorentz group, we construct all possible vacuum gravitational connections which give a vanishing curvature tensor. With this we show that, when the space-time admits a global chart, the vacuum connection has the same topology which describes the multiple vacua of $SU(2)$ gauge theory. This tells that the vacuum space-time can be classified by the knot topology $\pi_3(S^3) = \pi_3(S^2)$. We discuss the physical implications, in particular the space-time tunneling, of our result in quantum gravity.

Keywords: Topology of vacuum space-time; classification of vacuum space-time.

1. Introduction

An important goal in theoretical physics is to construct a decent quantum gravity. For this purpose, one must understand the structure of the classical vacuum space-time first. To find the classical vacuum space-time, it is not enough to solve the vacuum Einstein's equation,

$$R_{\mu\nu} - \frac{1}{2} R\, g_{\mu\nu} = 0. \tag{1}$$

Clearly we need to solve the vacuum equation

$$R_{\mu\nu\rho\sigma} = 0, \tag{2}$$

because the full curvature tensor (not just the Ricci tensor) must vanish in vacuum space-time. Obviously (2) is more difficult to solve, because it has more equations than (1). On the other hand, the space of the solutions of (2) is much smaller compared to (1), precisely because it has more equations. So if we are clever enough, we could construct all possible solutions of (2) without ever solving it. *The purpose of this Letter is to show*

how to construct the most general solutions of the vacuum space-time in terms of the gravitational connection, and to demonstrate that the vacuum space-time can be classified by the knot topology $\pi_3(S^3) = \pi_3(S^2)$. Viewing Einstein-Cartan theory as a gauge theory of Lorentz group and imposing the vacuum isometry to the gauge potential, we construct all possible vacuum connections in Einstein-Cartan theory, and show that in R^4 space-time the vacuum has identical topological structure as the vacuum potential in $SU(2)$ gauge theory, the knot topology.

In doing so we will also construct the restricted gravity which is much simpler than Einstein's gravity but which has the full general invariance, and thus inherits the full topological properties of Einstein's theory. Because of this the restricted gravity describes the core dynamics of Einstein's theory which can play the important role in quantum gravity. We show that there are two types of restricted gravity, the space-like (or time-like) A_2 gravity and the light-like (or null-like) B_2 gravity. We obtain the restricted gravity imposing the A_2-isometry or the B_2-isometry to Einstein's theory.

Gauge theories and general relativity have many things in common. This is not accidental; they are based on the same principle, the general invariance or equivalently the gauge invariance. The gauge theory is well-known to be a part of a higher-dimensional gravity which originates from the extrinsic crvature of a non-trivial embedding of the 4-dimensional space-time to the $(4+n)$-dimensional unified space [1, 2]. Conversely, Einstein's theory itself can be viewed as a gauge theory, because the general invariance of Einstein's theory can be viewed as a gauge invariance [3–6]. On the other hand, it has been well-known that the non-Abelian gauge theory has a non-trivial topology. It admits magnetic monopoles which can be classified by the monopole topology $\pi_2(S^2)$ [7, 8]. Moreover, it has mutiple vacua which can be classified by the knot topology $\pi_3(S^3) = \pi_3(S^2)$ [9–12]. If so, one might suspect that Einstein's theory should have simliar non-trivial topology.

To demonstrate this it is important to keep in mind that Einstein's theory can be viewed as a gauge theory [3, 4]. There are two ways to view Einstein's theory as a gauge theory [5, 6]. It can be viewed as a gauge theory of the 4-dimensional translation group [5]. In this view the (non-trivial part of) the tetrad becomes the gauge potential of translation group. Or it can be viewed as a gauge theory of Lorentz group [6]. In this view the gravitational connection (more precisely the spin connection) becomes the gauge potential of Lorentz group and the curvature tensor becomes the field strength. Here we adopt the second view, and construct all possible vacuum connection which yields a vanishing curvature tensor. With this we show that the vacuum connection is described by the vacuum potential of the $SU(2)$ subgroup of Lorentz group, which confirms that the vacuum space-time has the knot topology.

2. Restricted QCD and QCD Vacuum

To understand the topology of vacuum space-time it is crucial to understand the topology of non-Abelian gauge theory first. A best way to study the non-Abelian topology is through the magnetic symmetry [7, 13]. Consider the $SU(2)$ gauge theory for simplicity, and let $\hat{n}$ be an arbitrary unit isotriplet which selects the Abelian (color) direction. Imposing the following magnetic symmetry on the gauge potential which dictates $\hat{n}$ to be invariant under the parallel transport [7, 13],

$$D_\mu \hat{n} = (\partial_\mu + g\vec{A}_\mu \times)\, \hat{n} = 0, \qquad (\hat{n}^2 = 1) \tag{3}$$

we can select the restricted potential $\hat{A}_\mu$ which describes the "Abelian" part of the $SU(2)$ gauge potential,

$$\hat{A}_\mu = A_\mu \hat{n} - \frac{1}{g}\hat{n} \times \partial_\mu \hat{n}. \quad (A_\mu = \hat{n} \cdot \vec{A}_\mu). \tag{4}$$

With this we can retrieve the full $SU(2)$ gauge potential simply by adding the gauge covariant valence potential $\vec{X}_\mu$ of the coset space G/H to $\hat{A}_\mu$,

$$\vec{A}_\mu = \hat{A}_\mu + \vec{X}_\mu. \tag{5}$$

This way of projecting out the restricted potential of the maximal Abelian subgroup H of the gauge group G and decomposing the non-Abelian gauge potential into the Abelian (binding) part and the valence part are called the Abelian projection and the Abelian decomposition [7, 13].

Obviously (4) is restricted by (3). Nevertheless it has the full $SU(2)$ gauge degrees of freedom. Under the infinitesimal gauge transformation

$$\delta\vec{n} = -\vec{\alpha} \times \vec{n}, \qquad \delta\vec{A}_\mu = \frac{1}{g}D_\mu\vec{\alpha}, \tag{6}$$

one has

$$\delta A_\mu = \frac{1}{g}\vec{n} \cdot \partial_\mu\vec{\alpha}, \quad \delta\hat{A}_\mu = \frac{1}{g}\hat{D}_\mu\vec{\alpha}, \quad \delta\vec{X}_\mu = -\vec{\alpha} \times \vec{X}_\mu. \tag{7}$$

This shows that $\hat{A}_\mu$ by itself describes an $SU(2)$ connection which enjoys the full $SU(2)$ gauge degrees of freedom. Furthermore $\vec{X}_\mu$ transforms covariantly under the gauge transformation. This confirms that our decomposition provides a gauge-independent decomposition of the non-Abelian potential into the restricted part $\hat{A}_\mu$ and gauge covariant part $\vec{X}_\mu$.

The Abelian decomposition has played a crucial role to clarify the non-Abelian dynamics. To see this notice that from the decomposition (5) we have

$$\vec{F}_{\mu\nu} = \hat{F}_{\mu\nu} + \hat{D}_\mu \vec{X}_\nu - \hat{D}_\nu \vec{X}_\mu + g\vec{X}_\mu \times \vec{X}_\nu, \tag{8}$$

so that we can express the Yang-Mills Lagrangian as

$$\begin{aligned} \mathcal{L} = -\frac{1}{4}\vec{F}_{\mu\nu}^2 &= -\frac{1}{4}\hat{F}_{\mu\nu}^2 - \frac{g}{2}\hat{F}_{\mu\nu} \cdot (\vec{X}_\mu \times \vec{X}_\nu) \\ &\quad - \frac{1}{4}(\hat{D}_\mu \vec{X}_\nu - \hat{D}_\nu \vec{X}_\mu)^2 - \frac{g^2}{4}(\vec{X}_\mu \times \vec{X}_\nu)^2. \end{aligned} \tag{9}$$

This shows that the Yang-Mills theory can be viewed as the restricted gauge theory made of the Abelian projection, which has an additional gauge covariant charged vector field (the valence gluons) as its source. As importantly this shows that, with $\vec{X}_\mu = 0$, one can construct the restricted gauge theory which is much simpler than the original gauge theory which has the full gauge symmetry and thus has the full topological properites of the original gauge theory. The restricted gauge theory describes the core dynamics of the non-Abelian gauge theory, and has played an important role to demonstrate the Abelian dominance and magnetic confinement in QCD [7, 13–17].

Since the restricted potential $\hat{A}_\mu$ retains all topological properties of the original gauge theory, it can easily describe the multiple vacua. To see this, let $\hat{n}_i$ $(i = 1, 2, 3)$ with $\hat{n}_3 = \hat{n}$ be orthonormal isotriplets which form a right-handed basis ($\hat{n}_1 \times \hat{n}_2 = \hat{n}_3$), and impose three isometries,

$$\forall_i \quad D_\mu \hat{n}_i = 0. \tag{10}$$

Obviously this puts more restriction on the restricted potential (4). Indeed (10) requires the potential to have a vanishing field strength,

$$\forall_i \quad [D_\mu,\ D_\nu]\ \hat{n}_i = g\vec{F}_{\mu\nu} \times \hat{n}_i = 0 \Longleftrightarrow \vec{F}_{\mu\nu} = 0. \tag{11}$$

This assures that a vacuum potential must be the one which parallelizes the local orthonormal frame.

Solving (10) we obtain a most general $SU(2)$ vacuum potential $\hat{\Omega}_\mu$ [12]

$$\begin{aligned} \hat{\Omega}_\mu &= -C_\mu \hat{n} - \frac{1}{g}\hat{n} \times \partial_\mu \hat{n} = -C_\mu^k\ \hat{n}_k, \\ C_\mu^k &= -\frac{1}{2g}\epsilon_{ij}{}^k(\hat{n}_i \cdot \partial_\mu \hat{n}_j), \qquad (\hat{n} = \hat{n}_3;\ C_\mu = C_\mu^3) \\ \hat{\Omega}_{\mu\nu} &= \partial_\mu \hat{\Omega}_\nu - \partial_\nu \hat{\Omega}_\mu + g\ \hat{\Omega}_\mu \times \hat{\Omega}_\nu \\ &= -(\partial_\mu C_\nu^k - \partial_\nu C_\mu^k + g\ \epsilon_{ij}{}^k C_\mu^i C_\nu^j)\ \hat{n}_k = 0. \end{aligned} \tag{12}$$

This confirms that $\hat{\Omega}_\mu$, or equivalently $(C_\mu^1, C_\mu^2, C_\mu^3)$, describe a most general classical $SU(2)$ vacuum.

Notice that, although the vacuum is fixed by three isometries, it is essentially fixed by $\hat{n}$. This is because $\hat{n}_1$ and $\hat{n}_2$ are uniquely determined by $\hat{n}$, up to a $U(1)$ gauge transformation which leaves $\hat{n}$ invariant. With

$$\hat{n} = \begin{pmatrix} \sin\alpha\cos\beta \\ \sin\alpha\sin\beta \\ \cos\alpha \end{pmatrix}, \tag{13}$$

we have [12]

$$\begin{aligned} C_\mu^1 &= \frac{1}{g}(\sin\gamma\partial_\mu\alpha - \sin\alpha\cos\gamma\partial_\mu\beta), \\ C_\mu^2 &= \frac{1}{g}(\cos\gamma\partial_\mu\alpha + \sin\alpha\sin\gamma\partial_\mu\beta), \\ C_\mu^3 &= \frac{1}{g}(\cos\alpha\partial_\mu\beta + \partial_\mu\gamma), \end{aligned} \tag{14}$$

where the angle γ represents the $U(1)$ angle which leaves $\hat{n}$ invariant.

A nice feature of (12) is that the topological character of the vacuum is naturally inscribed in it. The topological vacuum quantum number is given by the non-Abelian Chern-Simon index of the potential $\hat{\Omega}_\mu$ [10, 12]

$$\begin{aligned} n &= -\frac{3g^2}{8\pi^2}\int \epsilon_{\alpha\beta\gamma}(C_\alpha^i\partial_\beta C_\gamma^i + \frac{g}{3}\epsilon_{ijk}C_\alpha^i C_\beta^j C_\gamma^k)d^3x \\ &= -\frac{g^3}{96\pi^2}\int \epsilon_{\alpha\beta\gamma}\epsilon_{ijk}C_\alpha^i C_\beta^j C_\gamma^k d^3x \qquad (\alpha,\beta,\gamma = 1,2,3). \end{aligned} \tag{15}$$

The index represents the non-trivial vacuum topology $\pi_3(S^3)$ of the mapping from the compactified 3-dimensional space S^3 to the $SU(2)$ space S^3. But this vacuum topology can also be described by $\hat{n}$, because $\hat{\Omega}_\mu$ is essentially fixed by $\hat{n}$ up to the $U(1)$ degree of freedom which determines $\hat{n}_1$ and $\hat{n}_2$. So we can conclude that the vacuum topology is imprinted in $\hat{n}$. With this understanding, we can conclude that the vacuum topology is fixed by the knot topology $\pi_3(S^2)$, because $\hat{n}$ (with $\hat{n}(\infty) = (0,0,1)$) defines the mapping $\pi_3(S^2)$ from the compactified space S^3 to the coset space $SU(2)/U(1)$ fixed by $\hat{n}$. Clearly this the knot topology $\pi_3(S^2)$ can be transformed to $\pi_3(S^3)$ through the Hopf fibering [12].

This tells that the $SU(2)$ vacuum can be classified by $|n\rangle$, where n is the integer. But the topologically distinct vacua are not stable under the quantum fluctuation, because the instanton allows the vacuum tunneling by connecting topologically different vacua [9, 10]. With the vacuum tunneling one can define the θ-vacuum [10]

$$|\theta\rangle = \sum_n \exp(in\theta)\ |n\rangle, \tag{16}$$

in non-Abelian gauge theory. Although the θ-vacuum does not guarantee the confinement and does not describe the physical vacuum, it has played an important role for us to understand the dynamics of non-Abelian gauge theory.

3. Einstein-Cartan Gravity: Gauge Theory of Lorentz Group

Now, regarding Einstein's theory as a gauge theory of Lorentz group, we can find the most general vacuum connection imposing the vacuum isomstry. It has been well known that Einstein's theory (more precisely Einstein-Cartan theory) can be viewed as a gauge theory of Lorentz group [6]. To see this we introduce a coordinate basis

$$[\partial_\mu,\ \partial_\nu] = 0, \quad (\mu, \nu = t, x, y, z)$$

and an orthonormal basis

$$\begin{gathered}[\xi_a,\ \xi_b] = f_{ab}{}^c \xi_c, \\ \xi_a = e_a{}^\mu \partial_\mu, \qquad \partial_\mu = e_\mu{}^a \xi_a \ \ (a, b = 0, 1, 2, 3),\end{gathered} \tag{17}$$

where $e_\mu{}^a$ and $e_a{}^\mu$ are the tetrad and inverse tetrad. Let $J^{ab} = -J^{ba}$ be the six generators of Lorentz group,

$$[J_{ab},\ J_{cd}] = \eta_{ac} J_{bd} - \eta_{bc} J_{ad} + \eta_{bd} J_{ac} - \eta_{ad} J_{bc}, \tag{18}$$

where $\eta_{ab} = diag\ (-1, 1, 1, 1)$ is the Minkowski metric. Clearly they can be expressed by the rotation and boost generators $L_i (= J^{0i})$ and $K_i (= \epsilon_{ijk} J^{jk}/2)$,

$$\begin{aligned}[L_i,\ L_j] &= \epsilon_{ijk} L_k, \quad [L_i,\ K_j] = \epsilon_{ijk} K_k, \\ [K_i,\ K_j] &= -\epsilon_{ijk} L_k \quad (i, j, k = 1, 2, 3).\end{aligned} \tag{19}$$

Let p^{ab} be an isosextet which forms an adjoint representation of Lorentz group. We can denote it by $\mathbf{p}$,

$$\begin{gathered}\mathbf{p} = \frac{1}{2} p_{ab} \mathbf{I}^{ab} = \begin{pmatrix} \vec{m} \\ \vec{e} \end{pmatrix}, \quad p^{ab} = \mathbf{p} \cdot \mathbf{I}^{ab} = \frac{1}{2} p^{mn} I_{mn}{}^{ab}, \\ \mathbf{I}^{ab} = \begin{pmatrix} \hat{a}^{ab} \\ \hat{b}^{ab} \end{pmatrix}, \quad \hat{a}_i{}^{ab} = \epsilon_{0i}{}^{ab}, \quad \hat{b}_i{}^{ab} = (\delta_0{}^a \delta_i{}^b - \delta_0{}^b \delta_i{}^a), \\ I_{cd}{}^{ab} = (\delta_c{}^a \delta_d{}^b - \delta_c{}^b \delta_d{}^a),\end{gathered} \tag{20}$$

where $\vec{m}$ and $\vec{e}$ are two isotriplets which represent the magnetic and electric components of $\mathbf{p}$ which correspond to the 3-dimensional rotation and boost.

To express Einstein-Cartan theory as the gauge theory of Lorentz group, we let $\mathbf{\Gamma}_\mu$ (or $\Gamma_\mu^{\ ab}$) be the gauge potential of Lorentz group, and $\mathbf{R}_{\mu\nu}$ (or $R_{\mu\nu}^{\ \ ab}$) be the curvature tensor

$$\mathbf{R}_{\mu\nu} = \partial_\mu \mathbf{\Gamma}_\nu - \partial_\nu \mathbf{\Gamma}_\mu + \mathbf{\Gamma}_\mu \times \mathbf{\Gamma}_\nu, \tag{21}$$

where we have normalized the coupling constant to be the unit which one can always do without loss of generality. To proceed further we now introduce the gauge covariant metric $\mathbf{g}_{\mu\nu}$ (or $g_{\mu\nu}^{\ \ ab}$), the metric which forms an adjoint representation of Lorentz group, by

$$\mathbf{g}_{\mu\nu} = e\ e_\mu^a\ e_\nu^b\ \mathbf{I}_{ab}, \quad g_{\mu\nu}^{\ \ ab} = e(e_\mu^{\ a} e_\nu^{\ b} - e_\nu^{\ b} e_\mu^{\ a}), \quad e = \mathrm{Det}\ (e_{a\mu}). \tag{22}$$

Notice that $\mathbf{g}_{\mu\nu}$ expressed in the orthonormal Lorentz frame becomes $\mathbf{I}_{ab}$ up to the scale factor e, so that $\mathbf{g}_{\mu\nu}$ is nothing but $\mathbf{I}_{ab}$ written in the coordinate frame.

Now, in the absence of the matter field, the Einstein-Hilbert action in the first order formalism can be written as

$$S[e_a^\mu,\ \mathbf{\Gamma}_\mu] = \frac{1}{16\pi G_N} \int \Big(\mathbf{g}_{\mu\nu} \cdot \mathbf{R}^{\mu\nu}\Big)\ d^4x, \tag{23}$$

where G_N is the Newton's constant. From this we have the following equation of motion

$$\begin{aligned} \delta e_{\mu a}; \quad & \mathbf{g}_{\mu\nu} \cdot \mathbf{R}^{\nu\rho} e_{\rho a} = R_{\mu a} = 0 \\ \delta \mathbf{\Gamma}_\mu; \quad & D_\mu \mathbf{g}^{\mu\nu} = (\partial_\mu + \mathbf{\Gamma}_\mu \times) \mathbf{g}^{\mu\nu} = 0, \end{aligned} \tag{24}$$

where $R_{\mu a} = e^{\nu b} R_{\mu\nu ab}$ is the Ricci tensor. The first equation assures that, in the absence of matter fields, the Ricci tensor must vanish. The second equation tells that the gauge potential of Lorentz group $\Gamma_\mu^{\ ab}$ must be the matric compatible spin connection $\omega_\mu^{\ ab}$,

$$\begin{aligned} \Gamma_\mu^{\ ab} &= \mathbf{\Gamma}_\mu \cdot \mathbf{I}^{ab} \\ &= \frac{1}{2}(e^{a\nu} e_{c\mu} \partial^b e^c_{\ \nu} + e^{a\nu} \partial_\mu e^b_{\ \nu} + \partial^b e^a_{\ \mu} - e^{b\nu} e_{c\mu} \partial^a e^c_{\ \nu} - e^{b\nu} \partial_\mu e^a_{\ \nu} - \partial^a e^b_{\ \mu}) \\ &= \omega_\mu^{\ ab}. \end{aligned} \tag{25}$$

So, the equation $D_\mu \mathbf{g}^{\mu\nu} = 0$ becomes identical to the metric-compatibility condition of the connection

$$\nabla_\alpha g_{\mu\nu} = \partial_\alpha g_{\mu\nu} - \Gamma_{\alpha\mu}^{\ \ \ \rho} g_{\rho\nu} - \Gamma_{\alpha\nu}^{\ \ \ \rho} g_{\mu\rho} = 0, \qquad g_{\mu\nu} = \eta_{ab} e_\mu^{\ a}\ e_\nu^{\ b}, \tag{26}$$

which requires the space-time metric $g_{\mu\nu}$ to be invariant under the parallel transport defined by the connection. Geometrically the condition $D_\mu \mathbf{g}^{\mu\nu} =$

0 tells that the Lorentz covariant metric $\mathbf{g}_{\mu\nu}$ must be covariant constant, but in the gauge formalizm this condition plays the role of the metric-compatibility condition of the connection $\nabla_\alpha g_{\mu\nu} = 0$.

This confirms that the above equation (24) describes general relativity. In fact it is this second equation which allows us to interpret $\mathbf{R}_{\mu\nu}$ and $R_{\mu a}$ as the curvature tensor and the Ricci tensor. Remember that in ordinary gauge theory the gauge potential is completely arbitrary, so that we treat $\mathbf{\Gamma}_\mu$ as independent variable in (23). But here the eqation of motion tells that the gauge potential is determined by the tetrad. With this preliminary, we show how to construct the restricted gravity.

4. Restricted Gravity

To obtain the restricted gravity, we have to impose the proper magnetic isometry to $\mathbf{\Gamma}_\mu$ first. To see what types of isometry is possible, it is important to remember that Lorentz group has two maximal Abelian subgroups, A_2 made of L_3 and K_3 and B_2 made of $(L_1 + K_2)/\sqrt{2}$ and $(L_2 - K_1)/\sqrt{2}$. This tells that we have two possible Abelian decompositions of the gravitational connection. And in both cases the magnetic isometry is described by two, not one, commuting sextet vector fields of Lorentz group which are dual to each other.

To see this, consider the following magnetic isometry

$$D_\mu \mathbf{p} = (\partial_\mu + \mathbf{\Gamma}_\mu \times)\ \mathbf{p} = 0. \tag{27}$$

This automatically assures

$$D_\mu \tilde{\mathbf{p}} = (\partial_\mu + \mathbf{\Gamma}_\mu \times)\ \tilde{\mathbf{p}} = 0, \tag{28}$$

where $\tilde{\mathbf{p}}$ (or $\tilde{p}^{ab} = \epsilon_{abcd}\ p^{cd}/2$) is the dual partner of $\mathbf{p}$. This is because ϵ_{abcd} is an invariant tensor of Lorentz group. This tells that the magnetic isometry in Lorentz group always contains the dual partner. To verify this we decompose the gauge potential $\mathbf{\Gamma}_\mu$ into the 3-dimensional rotation and boost parts $\vec{A}_\mu$ and $\vec{B}_\mu$, and let

$$\mathbf{\Gamma}_\mu = \begin{pmatrix} \vec{A}_\mu \\ \vec{B}_\mu \end{pmatrix}. \tag{29}$$

With this both (27) and (28) can be written as [18]

$$D_\mu \vec{m} = \vec{B}_\mu \times \vec{e}, \qquad D_\mu \vec{e} = -\vec{B}_\mu \times \vec{m},$$
$$D_\mu = \partial_\mu + \vec{A}_\mu \times . \tag{30}$$

This confirms that (27) and (28) are actually identical to each other, which tells that the magnetic isometry in Lorentz group must be even-dimensional.

Let $\hat{n}_i$ $(i = 1,2,3)$ be the right-handed orthonormal basis of $SU(2)$ subgroup of Lorentz group (with $\hat{n}_1 \times \hat{n}_2 = \hat{n}_3$), and $\mathbf{l}_i$ and $\mathbf{k}_i = -\tilde{\mathbf{l}}_i$ be the orthonormal basis of Lorentz group which describe the 3-dimensional rotation and boost,

$$\mathbf{l}_i = \begin{pmatrix} \hat{n}_i \\ 0 \end{pmatrix}, \qquad \mathbf{k}_i = -\tilde{\mathbf{l}}_i = \begin{pmatrix} 0 \\ \hat{n}_i \end{pmatrix}. \tag{31}$$

Then the A_2 isometry can be described by [18]

$$D_\mu \mathbf{l} = 0, \qquad D_\mu \mathbf{k} = -D_\mu \tilde{\mathbf{l}} = 0,$$
$$\mathbf{l} = \mathbf{l}_3 = \begin{pmatrix} \hat{n} \\ 0 \end{pmatrix}, \qquad \mathbf{k} = \mathbf{k}_3 = \begin{pmatrix} 0 \\ \hat{n} \end{pmatrix}, \qquad \hat{n} = \hat{n}_3. \tag{32}$$

The restricted connection $\hat{\mathbf{\Gamma}}_\mu$ and the restricted curvature tensor $\hat{\mathbf{R}}_{\mu\nu}$ which satisfy the isometry condition is given by

$$\hat{\mathbf{\Gamma}}_\mu = \Gamma_\mu \, \mathbf{l} - \widetilde{\Gamma}_\mu \, \tilde{\mathbf{l}} - \mathbf{l} \times \partial_\mu \mathbf{l}, \qquad \Gamma_\mu = \mathbf{\Gamma}_\mu \cdot \mathbf{l}, \quad \widetilde{\Gamma}_\mu = \mathbf{\Gamma}_\mu \cdot \tilde{\mathbf{l}},$$
$$\hat{\mathbf{R}}_{\mu\nu} = \partial_\mu \hat{\mathbf{\Gamma}}_\nu - \partial_\nu \hat{\mathbf{\Gamma}}_\mu + \hat{\mathbf{\Gamma}}_\mu \times \hat{\mathbf{\Gamma}}_\nu = (\Gamma_{\mu\nu} + H_{\mu\nu}) \, \mathbf{l} - \widetilde{\Gamma}_{\mu\nu} \, \tilde{\mathbf{l}},$$
$$\Gamma_{\mu\nu} = \partial_\mu \Gamma_\nu - \partial_\nu \Gamma_\mu, \quad \widetilde{\Gamma}_{\mu\nu} = \partial_\mu \widetilde{\Gamma}_\nu - \partial_\nu \widetilde{\Gamma}_\mu,$$
$$H_{\mu\nu} = -\mathbf{l} \cdot (\partial_\mu \mathbf{l} \times \partial_\nu \mathbf{l}), \tag{33}$$

where Γ_μ and $\widetilde{\Gamma}_\mu$ are two Abelian connections of $\mathbf{l}$ and $\tilde{\mathbf{l}}$ components which are not restricted by the isometry condition.

With this A_2 projection, we can construct the A_2 gravity. Consider the following restricted Einstein-Hilbert action,

$$S[e_a^\mu, \hat{\mathbf{\Gamma}}_\mu] = \frac{1}{16\pi G_N} \int d^4x \{ \mathbf{g}_{\mu\nu} \cdot \hat{\mathbf{R}}_{\mu\nu} + \lambda(\mathbf{l}^2 - 1) + \tilde{\lambda}(\mathbf{l} \cdot \tilde{\mathbf{l}}) + \lambda_\mu \hat{D}_\nu \mathbf{g}^{\mu\nu} \}$$
$$= \frac{1}{16\pi G_N} \int d^4x \{ G_{\mu\nu}(\Gamma^{\mu\nu} + H^{\mu\nu}) - \widetilde{G}_{\mu\nu} \widetilde{\Gamma}_{\mu\nu}$$
$$+ \lambda(\mathbf{l}^2 - 1) + \tilde{\lambda}(\mathbf{l} \cdot \tilde{\mathbf{l}}) + \lambda_\mu \hat{D}_\nu \mathbf{g}^{\mu\nu} \},$$
$$G_{\mu\nu} = e \; e_\mu^a e_\nu^b \; l_{ab}, \quad \widetilde{G}_{\mu\nu} = e \; e_\mu^a e_\nu^b \; \tilde{l}_{ab}. \tag{34}$$

The restricted action gives the following equations which describes the A_2 gravity,

$$G_{\mu\nu} \, (\Gamma^{\nu\rho} + H^{\nu\rho}) - \widetilde{G}_{\mu\nu} \widetilde{\Gamma}^{\nu\rho} = 0,$$
$$\partial_\mu G^{\mu\nu} = 0, \quad \partial_\mu \widetilde{G}^{\mu\nu} = 0,$$
$$\hat{D}_\mu \mathbf{g}^{\mu\nu} = 0. \tag{35}$$

Notice that $G_{\mu\nu}$ and $\widetilde{G}_{\mu\nu}$ are dual to each other, so that we can introduce a potential G_μ for $G_{\mu\nu}$,

$$G_{\mu\nu} = \partial_\mu G_\nu - \partial_\nu G_\mu. \tag{36}$$

This tells that the second equation is of Maxwell type, which implies that the A_2 gravity is essentially Abelian.

With the same procedure we can construct the B_2 gravity. To do this we impose the following B_2 isometry [18]

$$D_\mu \mathbf{j} = 0, \quad D_\mu \tilde{\mathbf{j}} = 0,$$
$$\mathbf{j} = \frac{e^\lambda}{\sqrt{2}}(\mathbf{l}_1 + \mathbf{k}_2) = \frac{e^\lambda}{\sqrt{2}}\begin{pmatrix} \hat{n}_1 \\ \hat{n}_2 \end{pmatrix}, \quad \tilde{\mathbf{j}} = \frac{e^\lambda}{\sqrt{2}}(\mathbf{l}_2 - \mathbf{k}_1) = \frac{e^\lambda}{\sqrt{2}}\begin{pmatrix} \hat{n}_2 \\ -\hat{n}_1 \end{pmatrix}, \tag{37}$$

where λ is an arbitrary function of space-time. The reason why the B_2 isometry allows an arbitrary function is because it is light-like, $\mathbf{j}^2 = \tilde{\mathbf{j}}^2 = 0$. To find the restricted connection $\hat{\mathbf{\Gamma}}$ which satisfies the isometry condition we introduce 4 more basis vectors $\mathbf{k}$, $\tilde{\mathbf{k}}$, $\mathbf{l}$, and $\tilde{\mathbf{l}}$ which together with $\mathbf{j}$ and $\tilde{\mathbf{j}}$ form a complete basis as before [18],

$$\begin{aligned}
\mathbf{k} &= \frac{e^{-\lambda}}{\sqrt{2}}(\mathbf{l}_1 - \mathbf{k}_2) = \frac{e^{-\lambda}}{\sqrt{2}}\begin{pmatrix} \hat{n}_1 \\ -\hat{n}_2 \end{pmatrix}, \\
\tilde{\mathbf{k}} &= -\frac{e^{-\lambda}}{\sqrt{2}}(\mathbf{l}_2 + \mathbf{k}_1) = \frac{e^{-\lambda}}{\sqrt{2}}\begin{pmatrix} -\hat{n}_2 \\ -\hat{n}_1 \end{pmatrix}, \\
\mathbf{l} &= -\mathbf{j} \times \tilde{\mathbf{k}} = -\tilde{\mathbf{j}} \times \mathbf{k} = \begin{pmatrix} \hat{n}_3 \\ 0 \end{pmatrix}, \\
\tilde{\mathbf{l}} &= \mathbf{j} \times \mathbf{k} = -\tilde{\mathbf{j}} \times \tilde{\mathbf{k}} = \begin{pmatrix} 0 \\ -\hat{n}_3 \end{pmatrix}.
\end{aligned} \tag{38}$$

With this we find the following restricted connection and the restricted curvature tensor for the B_2 isometry,

$$\begin{gathered}
\hat{\mathbf{\Gamma}}_\mu = \Gamma_\mu\, \mathbf{j} - \widetilde{\Gamma}_\mu\, \tilde{\mathbf{j}} - \mathbf{k} \times \partial_\mu \mathbf{j}, \qquad \Gamma_\mu = \mathbf{\Gamma}_\mu \cdot \mathbf{k}, \quad \widetilde{\Gamma}_\mu = \mathbf{\Gamma}_\mu \cdot \tilde{\mathbf{k}}, \\
\hat{\mathbf{R}}_{\mu\nu} = \partial_\mu \hat{\mathbf{\Gamma}}_\nu - \partial_\nu \hat{\mathbf{\Gamma}}_\mu + \hat{\mathbf{\Gamma}}_\mu \times \hat{\mathbf{\Gamma}}_\nu = (\Gamma_{\mu\nu} + H_{\mu\nu})\mathbf{j} - (\widetilde{\Gamma}_{\mu\nu} + \widetilde{H}_{\mu\nu})\tilde{\mathbf{j}}, \\
\Gamma_{\mu\nu} = \partial_\mu \Gamma_\nu - \partial_\nu \Gamma_\mu, \quad \widetilde{\Gamma}_{\mu\nu} = \partial_\mu \widetilde{\Gamma}_\nu - \partial_\nu \widetilde{\Gamma}_\mu, \\
H_{\mu\nu} = -\mathbf{k} \cdot (\partial_\mu \mathbf{j} \times \partial_\nu \mathbf{k} - \partial_\nu \mathbf{j} \times \partial_\mu \mathbf{k}), \\
\widetilde{H}_{\mu\nu} = -\tilde{\mathbf{k}} \cdot (\partial_\mu \mathbf{j} \times \partial_\nu \mathbf{k} - \partial_\nu \mathbf{j} \times \partial_\mu \mathbf{k}),
\end{gathered} \tag{39}$$

where Γ_μ and $\widetilde{\Gamma}_\mu$ are two Abelian connections of $\mathbf{j}$ and $\tilde{\mathbf{j}}$ components which are not restricted by the isometry condition.

To obtain the B_2 gravity we choose the following restricted action

$$
\begin{aligned}
S[e^\mu_a, \hat{\mathbf{\Gamma}}_\mu] &= \frac{1}{16\pi G_N} \int d^4x \{\mathbf{g}_{\mu\nu} \cdot \hat{\mathbf{R}}_{\mu\nu} + \lambda \mathbf{j}^2 + \lambda' (\mathbf{j} \cdot \tilde{\mathbf{j}}) + \lambda_\mu \hat{D}_\nu \mathbf{g}^{\mu\nu}\}, \\
&= \frac{1}{16\pi G_N} \int d^4x \{\mathcal{J}_{\mu\nu}(\Gamma^{\mu\nu} + H^{\mu\nu}) \\
&\quad - \tilde{\mathcal{J}}_{\mu\nu}(\tilde{\Gamma}^{\mu\nu} + \tilde{H}^{\mu\nu}) + \lambda(\mathbf{j}^2) + \lambda'(\mathbf{j} \cdot \tilde{\mathbf{j}}) + \lambda_\mu \hat{D}_\nu \mathbf{g}^{\mu\nu}\}, \\
\mathcal{J}_{\mu\nu} &= e\ e^a_\mu\ e^b_\nu\ j_{ab} \quad \tilde{\mathcal{J}}_{\mu\nu} = e\ e^a_\mu\ e^b_\nu\ \tilde{j}_{ab}.
\end{aligned} \tag{40}
$$

This produces the following equation of motion which describes the B_2 gravity,

$$
\begin{aligned}
\mathcal{J}_{\mu\nu}(\Gamma^{\nu\rho} + H^{\nu\rho}) - \tilde{\mathcal{J}}_{\mu\nu}(\tilde{\Gamma}^{\nu\rho} + \tilde{H}^{\nu\rho}) &= 0, \\
\partial_\mu \mathcal{J}^{\mu\nu} = 0, \quad \partial_\mu \tilde{\mathcal{J}}^{\mu\nu} &= 0, \\
\hat{D}_\mu \mathbf{g}^{\mu\nu} &= 0.
\end{aligned} \tag{41}
$$

Notice that $\mathcal{J}_{\mu\nu}$ and $\tilde{\mathcal{J}}_{\mu\nu}$ are dual to each other, so that $\mathcal{J}_{\mu\nu}$ admits a potential,

$$
\mathcal{J}_{\mu\nu} = \partial_\mu \mathcal{J}_\nu - \partial_\nu \mathcal{J}_\mu. \tag{42}
$$

So the second equation again becomes of Maxwell type. This implies that the B_2 gravity is also Abelian.

With the restricted gravity, we can recover the full Einstein-Cartan theory by adding the valence connection $\mathbf{Z}_\mu$ to the restricted connection,

$$
\mathbf{\Gamma}_\mu = \hat{\mathbf{\Gamma}}_\mu + \mathbf{Z}_\mu, \tag{43}
$$

and expressing (23) explicitly in terms of $\hat{\mathbf{\Gamma}}_\mu$ and $\mathbf{Z}_\mu$. This provides the Abelian decomposition of general relativity. This means that Einstein-Cartan theory can be viewed as the restricted gravity which has the Lorentz covariant valence connection as the gravitational source. In fact we could have obtained the restricted gravity from the Abelian decomposition of general relativity, putting $\mathbf{Z}_\mu = 0$.

It has often been said that Einstein's theory is the simplest theory of gravitation which satisfies the general invariance. But our result tells that this is not so. Just like the restricted QCD, the restricted gravities have the full Lorentz invariance and thus have the full general invariance. Nevertheless they are much simpler than Einstein's gravity, because they are restricted. This tells that there are theories of gravitation which satisfy the general invariance and yet is simpler than Einstein's theory.

Moreover the restricted gravities inherit the full topological properties of general relativity, because they enjoy the full general invariance. This tells that the restricted gravities describe the core dynamics of general relativity which can play the crucial role in quantum gravity.

5. Vacuum Space-time

Since the restricted connection enjoys the full local Lorentz degrees of freedom, it can describe the most general vacuum space-time. Let $\mathbf{l}_i$ and $\mathbf{k}_i$ (i=1,2,3) be the unit sextet vector fields which describe the rotation and the boost. Now, it must be clear that the most general vacuum connection which guarantees $\mathbf{R}_{\mu\nu} = 0$ is described by the following isometry

$$\forall_i \quad D_\mu \mathbf{l}_i = 0, \qquad D_\mu \mathbf{k}_i = -D_\mu \tilde{\mathbf{l}}_i = 0 \tag{44}$$

because this is the maximal isometry that we can impose on the gravitational connection. Indeed, with this we have

$$\forall_i \quad [D_\mu,\ D_\nu]\ \mathbf{l}_i = \mathbf{R}_{\mu\nu} \times \mathbf{l}_i = 0, \qquad [D_\mu,\ D_\nu]\ \mathbf{k}_i = \mathbf{R}_{\mu\nu} \times \mathbf{k}_i = 0, \tag{45}$$

which is possible if and only if

$$\mathbf{R}_{\mu\nu} = 0, \tag{46}$$

or $R_{\mu\nu}{}^{ab} = 0$.

But the vacuum isometry can also be written as

$$\forall_i \quad D_\mu\ \mathbf{l}_i = 0, \tag{47}$$

or equivalently

$$\forall_i \quad D_\mu\ \mathbf{k}_i = -D_\mu\ \tilde{\mathbf{l}}_i = 0. \tag{48}$$

Although we need both, either one is enough because one assures the other as we have explained.

Now, all that we have to do to find the most general vacuum is to solve (44) in terms of the connection. To do this notice that in 3-dimensional notation (44) is written as

$$\forall_i \quad D_\mu \hat{n}_i = \vec{B}_\mu \times \hat{n}_i, \quad D_\mu \hat{n}_i = -\vec{B}_\mu \times \hat{n}_i. \tag{49}$$

This has the unique solution

$$\vec{A}_\mu = \hat{\Omega}_\mu, \quad \vec{B}_\mu = 0. \tag{50}$$

where $\hat{\Omega}_\mu$ (with $g = 1$) is precisly the vacuum potential (12) of the $SU(2)$ subgroup. So the most general vacuum connection $\mathbf{\Omega}_\mu$ which yields vanishing curvature tensor is given by

$$\mathbf{\Gamma}_\mu = \mathbf{\Omega}_\mu = \begin{pmatrix} \hat{\Omega}_\mu \\ 0 \end{pmatrix}. \tag{51}$$

This proves that the gravitational connection of the vacuum space-time is fixed by the rotational part of the spin connection which describes the vacuum potential of $SU(2)$ gauge theory.

There are two ways to obtain the most general vacuum space-time. First, one could obtain most general vacuum space-time by solving the vacuum equation (2) by brute force, as we have remarked in the introduction. But this is impractical. Here we have shown that there is another way to obtain the most general vacuum space-time, by solving the vacuum isometry equation (44) in terms of the gravitational connection in stead of (2). Obviously the second way is simpler and more practical.

6. Topological Classification of Vacuum Space-time

Now, it is straightforward to make the topological classification of the vacuum space-time. Notice that in general (51) describes the vacuum locally for any open submanifold of space-time. So for the topological classification we must assume that (just as in gauge theory) the space-time has a global chart with no singularities, because once we allow the singularities it becomes impossible to classify the vacuum.

With this assumption we can conclude that the topology of the vacuum connection is identical to the topology of the vacuum in $SU(2)$ gauge theory, because the vacuum connection is identical to the vacuum potential of $SU(2)$ gauge theory. This assures that the vacuum space-time can be classified by $\pi_3(S^3)$, or equivalently by the knot topology $\pi_3(S^2)$. This tells that one can classify the vacuum space-time by $|n\rangle$, where n is an integer. As importantly this tells that general relativity is made of infinitely many topologically different sectors, which are related by topology changing singular Lorentz gauge transformations.

We have assumed the existence of a global chart for space-time to classify the vacuum space-time, and one might wonder whether this assumption is too strong to be realistic. On this we have two remarks. First, this is not an unrelistic assumption for quantum gravity, because for any quantum theory we have to assume the existence of the global chart. Indeed all existing quantum theories are based on this assumption. Secondly, even with this assumption no topological classification of vacuum space-time has been available so far. This makes our classification important.

Notice that our vacuum space-time is described in terms of the connection, not the metric. This is because our classification is based on the gauge formalism of general relativity which treats the gauge potential (i.e., the connection), not the metric, as the fundamental field. This touches a subtle but very sensitive issue. As we have pointed out, the gauge theory of Lorentz group is not identical to Einstein's theory but Einstein-Cartan theory, because in this gauge formalism the gauge potential allows torsion [6]. So precisely speaking, our classification of vacuum space-time applies to Einstein-Cartan theory, not Einstein's theory.

This raise a important question: What is responsible for the knot topology of vacuum space-time, metric or torsion? Or does the knot topology of vacuum space-time survive in Einstein's theory which has no torsion? Obviously this is an important question which should influence the future development of general relativity. Independent of the question, however, the fact that the vacuum space-time has a knot topology has a far reaching implication in general relativity. Torsion or metric, we have to take into account gravity in topologically non-trivial sectors.

Of course the vacuum topology of space-time has been discussed before in Euclidian space-time [19, 20]. In Euclidian space-time, the gauge symmetry of Einstein-Cartan theory becomes $SO(4)$, which is isomorphic to $SU(2) \times SU(2)$. So in this case the vacuum topology (again in the presence of a global chart) becomes $\pi_3(S^3) \times \pi_3(S^3)$, so that it is described by two integers. But we emphasize that in real (Minkowski) space-time the gauge symmetry becomes the Lorentz group, so that the vacuum space-time is classified by one integer, by $|n\rangle$.

7. Discussions

In this talk we have shown how to construct the most general vacuum space-time, and proved that the vacuum space-time can be classified by the knot topology. This tells that general relativity is made of infinitely many topologically non-trivial sectors. In doing so we have shown that there are two types of restricted gravities, the A_2 gravity and the B_2 gravity, which are much simpler than Einstein's gravity but which have the full general invariance.

Clearly the existence of topologically non-trivial sectors in general relativity raises more questions. A first question is the vacuum tunneling in quantum gravity. It has been well-known that the instantons in gauge theory make multiple vacua unstable against quantum fluctuation [9, 10]. If so, one may ask whether we can have the gravito-instantons in Einstein's theory which can destablize the topologically distinct vacuum space-times. The answer is not obvious, because the Einstein's eqation is not of Yang-Mills type. In spite of the fact that Einstein's theory can be viewed as a gauge theory, they have important differences. In gauge theory the Yang-Mills action is quadratic in field strength, but in Einstein's theory the Einstein-Hilbert action is linear in curvature tensor. Moreover, in gauge theory the dynamical field which propagates is the gauge potential, but in Einstein's theory it is the metric (not the connection) which propagates [6]. This, of course, does not rule out the gravito-instanton. In fact there is a real possibility that such instanton could exist. At this point we remark that candidates of the "gravitational instanton" in Einstein's theory which has finite Euclidian action have been discussed by many authors [19, 20]. On the other hand they have been proposed without any reference to the above multiple vacua and the quantum tunneling. Whether any of these can actually demonstrate the tunneling is a very interesting question which has yet to be answered [21]. It would be very intersting if one can demonstrate the quantum tunneling between vacuum space-time.

An intimately related question is the existence of the θ-vacuum in quantum gravity. Assuming the tunneling, we can certainly introduce the θ-vacuum. If so, one may wonder whether this θ-vacuum can be the physical vacuum in quantum gravity. Of course, in gauge theory we know that the θ-vacuum can not be the physical vacuum, because the physical vacuum in QCD must explain the color confinement [14, 15]. But in general relativity there is a good reason that the θ-vacuum could be the physical vacuum. First, the dynamics of general relativity is different from that of the gauge theory. In gauge theory the vacuum should explain the confinement, but general relativity does not need confinement. More importantly, the impact of the tunneling in general relativity can be far greater than in gauge theory, because a self-dual gravito-instanton (if exists) has a vanishing action and thus has the maximum tunneling probability. This makes the instantons and the θ-vacuum in general relativity much more important.

Another question is to look for the gravity in topologically non-trivial sectors. For example, one might like to find out how black holes in topologically non-trivial sectors (if there is any) should look like. Or, one might want to find out how gravitational wave propagates in topologically non-trivial sectors. To answer these it is very important to study the vacuum metric and/or torsion first.

Clearly these questions put general relativity (and the quantum gravity) in a totally new perspective. Most of the known classical and quantum gravity concern the trivial sector. But our result shows that there are infinitely many topologially different sectors in general relativity. And we have to deal with them equally. This requires a totally new approach in general relativity, in particular in quantum gravity.

Finally we like to emphasize the importance of the magnetic isometry in quantum gravity. The isometry has allowed us to construct all possible vacuum connections in Einstein-Cartan theory. As we have explained, it also allows us to make the Abelian projection of Einstein-Cartan theory and obtain the restricted gravity which is much simpler than general relativity, without compromising the general invariance [18]. The restricted gravity inherits all topological properties of Einstein-Cartan theory, and describes the core dynamics of general relativity which could play the crucial role in quantum gravity.

This strongly implies the Abelian dominance in quantum gravity. The restricted gauge theory has demonstrated the Abelian dominance in gauge theory [7, 13]. So one might expect the same in general relativity. In fact the equations of motion (35) and (41) of the restricted gravities look very similar to Maxwell's equation. They imply, among others, that the restricted gravities can be described by spin-one vector potential rather than spin-two metric. This strongly suggests the Abelian dominance in general relativity. This Abelian dominance in general relativity, if established, can play a crucial role in quantum gravity.

The detailed discussions on these and related issues will be presented in a separate paper [21].

Acknowledgments

The work is supported in part by the BSR Program (Grant KRF-2007-314-C00055) of Korea Research Foundation and in part by the ABRL Program (Grant R14-2003-012-01002-0) and the International Coorperation Program of Korea Science and Enginering Foundation.

References

[1] T. Kaluza, Sitzber. Preuss. Akad. Wiss. 966 (1921); O. Klein, Z. Physik **37**, 895 (1926).
[2] Y. M. Cho, J. Math. Phys. **16**, 2029 (1975); Y. M. Cho and P. G. O. Freund, Phys. Rev. **D12**, 1711 (1975).
[3] T. W. B. Kibble, J. Math. Phys. **2**, 212 (1961).
[4] R. Utiyama, Phys. Rev. **101**, 1597 (1956).
[5] Y. M. Cho, Phys. Rev. **D14**, 2521 (1976).
[6] Y. M. Cho, Phys. Rev. **D14**, 3335 (1976). See also, F. Hehl, P. Heyde, G. Kerlick, and J. Nester, Rev. Mod. Phys. **48**, 393 (1976).
[7] Y. M. Cho, Phys. Rev. **D21**, 1080 (1980); Phys. Rev **D62**, 074009 (2000).
[8] Y. M. Cho, Phys. Rev. Lett. **44**, 1115 (1980).

[9] A. Belavin, A. Polyakov, A. Schwartz, and Y. Tyupkin, Phys. Lett. **B59**, 85 (1975); Y. M. Cho, Phys. Lett. **B81**, 25 (1979).
[10] G. 't Hooft, Phys. Rev. Lett. **37**, 8 (1976); R. Jackiw and C. Rebbi, Phys. Rev. Lett. **37**, 172 (1976).
[11] P. van Baal and A. Wipf, Phys. Lett. **B515**, 181 (2001).
[12] Y. M. Cho, Phys. Lett. **B644**, 208 (2006).
[13] Y. M. Cho, Phys. Rev. Lett. **46**, 302 (1981); Phys. Rev. **D23**, 2415 (1981); W. S. Bae, Y. M. Cho, and S. W. Kimm, Phys. Rev. **D65**, 025005 (2002).
[14] Y. M. Cho, H. W. Lee, and D. G. Pak, Phys. Lett. **B 525**, 347 (2002); Y. M. Cho and D. G. Pak, Phys. Rev. **D65**, 074027 (2002).
[15] Y. M. Cho, D. G. Pak, and M. Walker, JHEP **05**, 073 (2004); Y. M. Cho and M. Walker, Mod. Phys. Lett. **A19**, 2707 (2004).
[16] S. Shabanov, Phys. Lett. **B458**, 322 (1999); H. Gies, Phys. Rev. **D63**, 125023 (2001); R. Zucchini, Int. J. Geom. Meth. Mod. Phys. **1**, 813 (2004).
[17] K. Kondo, Phys. Lett. **B600**, 287 (2004); S. Kato *et. al.*, Phys. Lett. **B632**, 326 (2006).
[18] Y. M. Cho, S. W. Kim, and J. H. Kim, hep-th/0702200.
[19] S. Hawking, Phys. Lett. **A60**, 81 (1977); D. Page, Phys. Lett. **B78**, 239 (1978); G. Gibbons and S. Hawking, Phys. Lett. **B78**, 430 (1978).
[20] See also, T. Eguchi, P. Gilkey, and A. Hanson, Phys. Rep. **66**, 213 (1980), and references therein.
[21] Y. M. Cho, Prog. of Theo. Phys. in press; Y. M. Cho, A. Nielsen, and D. Pak, to be published.

SOME THOUGHTS ON THE COSMOLOGICAL QCD PHASE TRANSITION

W.-Y. P. HWANG

The Leung Research Center for Cosmology and Particle Astrophysics, Center for Theoretical Sciences, Institute of Astrophysics, and Department of Physics, National Taiwan University, Taipei 106, Taiwan
E-mail: wyhwang@phys.ntu.edu.tw

The cosmological QCD phase transitions may have taken place between 10^{-5} and 10^{-4} seconds in the early Universe offers us one of the most intriguing and fascinating questions in cosmology. In bag models, the phase transition is described by the first-order phase transition and the role played by the latent "heat" or energy released in the transition is highly nontrivial and is being classified as the first-order phase transition. In this presentation, we assume, first of all, that the cosmological QCD phase transition, which happened at a time between $10^{-5}\, sec$ and $10^{-4}\, sec$ or at the temperature of about $150\, MeV$ and accounts for confinement of quarks and gluons to within hadrons, would be of first order. Of course, we may assume that the cosmological QCD phase transition may not be of the first order. To get the essence out of the first-order scenario, it is sufficient to approximate the true QCD vacuum as one of possibly degenerate vacua and when necessary we try to model it effectively via a complex scalar field with spontaneous symmetry breaking. On the other hand, we may use a real scalar field in describing the non-first-order QCD phase transition. In the first-order QCD phase transition, we could examine how and when "pasted" or "patched" domain walls are formed, how long such walls evolve in the long run, and we believe that the significant portion of dark matter could be accounted for in terms of such domain-wall structure and its remnants. Of course, the cosmological QCD phase transition happened in the way such that the false vacua associated with baryons and many other color-singlet objects did not disappear (that is, using the bag-model language, there are bags of radius 1.0 fermi for the baryons) - but the amount of the energy remained in the false vacua is negligible by comparison. The latent energy released due to the conversion of the false vacua to the true vacua, in the form of "pasted" or "patched" domain walls in the short run and their numerous evolved objects, should make the concept of the "radiation-dominated" epoch, or of the "matter-dominated" epoch to be re-examined.

1. Introduction

I'm sorry that the time limit for me forced me give a lousy talk - so I try to compensate on that. The other excuse is my experience of the cerebral haemorrhage about three and a half years ago. But this occasion of C.N.Yang's 85th Birthday Symposium, I should complain about my own lousiness and try to quote a sentence from C.N. Yang from "Einstein's Impact on Theoretical Physics in the 21st Century":[1]

"... It led to the discipline of modern cosmology which is destined to become one of the important scientific fields of the 21st century."

I should say to admire C.N. Yang in his saying his prediction - in fact, I started working on this "field" about eight years ago and start feeling maybe the same thing. I think that we, humankind, should be proud of that Cosmology is becoming the empirical science in the 21st century.

Indeed, the discovery[2] of fluctuations or anisotropies, at the level of 10^{-5}, associated with the cosmic microwave background (CMB) has helped transformed the physics of the early universe into a main-stream research area in astronomy and in particle astrophysics, both theoretically and observationally.[3] CMB anisotropies[4] and polarizations,[5] the latter even smaller and at the level of 10^{-7}, either primary (as imprinted on the last scattering surface just before the universe was $(379 \pm 8) \times 10^3$ years old) or secondary (as might be caused by the interactions of CMB photons with large-scale structures along the line of sight), are linked closely to the inhomogeneities produced in the early universe.

On the other hand, over the last three decades, the standard model of particle physics has been well established to the precision level of 10^{-5} or better in the electroweak sector, or to the level of $10^{-3} - 10^{-2}$ for the strong interactions. In the theory, the electroweak (EW) phase transition, which endows masses to the various particles, and the QCD phase transition, which gives rise to confinement of quarks and gluons within hadrons in the true QCD vacuum, are two well-established phenomena. Presumably, the EW and QCD phase transitions would have taken place in the early universe, respectively, at around $10^{-11}\,sec$ and at a time between $10^{-5}\,sec$ and $10^{-4}\,sec$, or at the temperature of about $300\,GeV$ and of about $150\,MeV$, respectively. Indeed, it has become imperative to formulate the EW and QCD phase transitions in the early universe if a quantitative theory of cosmology can ever be reached.

The purpose of this presentation is to focus our attention on cosmological QCD phase transition and to assess whether its roles in the early universe can be synthesized in a more quantitative terms. To simplify the situation, we use the bag-model language (and thus the first-order QCD phase transition) as the zeroth order approximation and, whenever necessary, try to model the possibly degenerate vacua, the lower-temperature phase, as the minima of the spontaneously-broken complex scalar fields. Because we use the bag-model language as the approximation, the phase transition in this case is of first order in nature - in other words, the latent "heat" or energy is all released at the critical temperature T_c. Whether a given amount of "latent energy" is released by the system is of importance - in bag models, the answer is definitely "yes" judging from the difference of the vacua. Owing to the high computation costs, determination of the order of QCD phase transition remains to be a great challenge in lattice calculations. To summarize the situations,[6] the order of the transition depends on the quark mass; for example, in three-flavor QCD for vanishing quark masses the transition is of first order; but for intermediate masses it is probably a crossover; however, for infinitely heavy quark masses the transition is again first order. In three-flavor unimproved staggered QCD, using a lattice spacing of about 0.28 fm, the first-order and the crossover regions are separated by a pseudoscalar mass of $m_{\pi,c} \approx 300\, MeV$. Studying the same three-flavor theory with the same lattice spacing, but with an improved $p4$ action, they[6] obtained $m_{\pi,c} \approx 70\, MeV$. So, in the first approximation, a pseudoscalar mass of $140\, MeV$ (which corresponds to the numerical value of the physical pion mass) would be first order; but in the second approximation, it would be crossover.

In lattice calculations, the minimum of the error bars is what you cannot get rid of. The conceptual errors in the "physical" quark masses might eventually enter the game since after all we cannot observe an isolated quark and cannot weigh it. It seems to me that, in some cases, it might be possible to rule out the first-order or second-order phase transition or some higher-order transition; numerically, it might be very hard to determine the crossover "transition". All in all, we are afraid that the question about the first-order QCD phase transition would remain to be controversial for some time to come. In fact, to say that the cosmological QCD phase transition is NOT first-order is equivalent to saying that bag models are wrong - this may be right but we need more in the proof. In this presentation, we use the bag-model approximation to simplify the complications of the problem in the context of Cosmology. That means that we assume the existence of the first-order QCD phase transition - perhaps with too much of latent "heat" but the amount of latent "heat", even by reduction of three or four orders of magnitude, would still make the claim of the present paper valid.

Usually we treat the ground state as being non-degenerate. This might not be the case - owing to the many degrees of freedom associated with QCD such as the sixteen gauge degrees of freedom. This is true when one tries to work out the ground state in a given model, such as trying to obtain the QCD vacuum in the instanton liquid model.[7] This aspect is different from the so-called θ vacuum associated with the strong CP problem[8] - there is no degeneracy problem there.

Owing to the complexities of our problem, we could try to tackle the problem of the phase transition by dividing it into problems in four different categories, viz.: (1) how a bubble of different vacuum grows or shrinks; (2) how two growing bubbles collide or squeeze, and merging, with each other; (3) how the Universe eventually stabilize itself later while keeping expanding for several orders of magnitude; and (4) how specific objects, such as back holes or magnetic strings, get produced during the specific phase transition. Although the system "temperature" kept decreasing, the meaning of the "temperature" may have lost meaning in the local sense during each of the sub-steps - try to consult the textbooks on non-equilibrium statistical mechanics.

One reference that is highly relevant for our paper is by Svetitsky and Yaffe,[9] where as one integrates out all degrees of freedom except this order parameter and obtains the effective theory for the order parameter, globally invariant under the center symmetry (i.e. a global symmetry group, the center of the SU(3) gauge group). Of course, when for some good reason, we keep two degrees of freedom un-integrated, we would be left out a complex scalar field instead. In what follows, we use a complex scalar field since the real scalar could be handled with relative ease by comparison.

Questions related to part (4) (i.e. that how specific objects, such as black holes or magnetic strings, get produced during the specific phase transition), which are quite complicated, will not be addressed here; see, e.g., ref.[10] Forwe could describe the intermediate solutions based on the so-called comparison, in the case of the electroweak phase transition, please consult.[11] In the framework which we consider, "pasted" or "patched" domain walls when the majority of the false vacua get first eliminated - but how it would evolve from there and how long it would evolve still uncertain. See the discussions given later.

The major result of this presentation is that the latent heat (or latent energy), which turns out to be identified as the "bag constant", is huge compared to the radiation density at the cosmological QCD phase transition (i.e. at about $3\times 10^{-5}sec$). As time evolved to the present, we assume naturally that the percentage of this quantity becomes probably the majority of dark matter (25 % of the composition of the present Universe).

2. The Basics

A prevailing view regarding our universe is that it originates from the joint making of Einstein's general relativity and the cosmological principle while the observed anisotropies associated with the cosmic microwave background (CMB), the sizes up to about one part in 100,000, might stem, e.g., from quantum fluctuations in the inflation era. In what follows, we wish to first outline very briefly a few key points in the standard scenario so that we shall have a framework which we may employ to elucidate the roles of phase transitions in the early universe.

Based upon the cosmological principle which state that our universe is homogeneous and isotropic, we use the Robertson-Walker metric to describe our universe.[12]

$$ds^2 = dt^2 - R^2(t)\{\frac{dr^2}{1-kr^2} + r^2 d\theta^2 + r^2 sin^2\theta d\phi^2\}. \tag{1}$$

Here the parameter k describes the spatial curvature with $k = +1, -1,$ and 0 referring to an open, closed, and flat universe, respectively. The scale factor $R(t)$ describes the size of the universe at time t.

To a reasonable first approximation, the universe can be described by a perfect fluid, i.e., a fluid with the energy-momentum tensor $T^{\mu}\ _{\nu} = diag\,(\rho,\ ,\ -p,\ -p,\ -p)$ where ρ is the energy density and p the pressure. Thus, the Einstein equation, $G^{\mu}\ _{\nu} = 8\pi G_N T^{\mu}\ _{\nu} + \Lambda g^{\mu}\ _{\nu}$, gives rise to only two independent equations, i.e., from $(\mu,\ \nu) = (0,\ 0)$ and $(i,\ i)$ components,

$$\frac{\dot{R}^2}{R^2} + \frac{k}{R^2} = \frac{8\pi G_N}{3}\rho + \frac{\Lambda}{3}. \tag{2}$$

$$2\frac{\ddot{R}}{R} + \frac{\dot{R}^2}{R^2} + \frac{k}{R^2} = -8\pi G_N\ p + \Lambda. \tag{3}$$

Combining with the equation of state (EOS), i.e. the relation between the pressure p and the energy density ρ, we can solve the three functions $R(t)$,

$\rho(t)$, and $p(t)$ from the three equations. Further, the above two equations yields

$$\frac{\ddot{R}}{R} = -\frac{4\pi G_N}{3}(\rho + 3p) + \frac{\Lambda}{3}, \tag{4}$$

showing either that there is a positive cosmological constant or that $\rho + 3p$ must be somehow negative, if the major conclusion of the Supernovae Cosmology Project is correct,[13] i.e. the expansion of our universe still accelerating ($\ddot{R}/R > 0$).

Assuming a simple equation of state, $p = w\rho$, we obtain, from Eqs. (2) and (3),

$$2\frac{\ddot{R}}{R} + (1 + 3w)(\frac{\dot{R}^2}{R^2} + \frac{k}{R^2}) - (1 + w)\Lambda = 0, \tag{5}$$

which is applicable when a particular component dominates over the others - such as in the inflation era (before the hot big bang era), the radiation-dominated universe (e.g. the early stage of the hot big bang era), and the matter-dominated universe (i.e., the late stage of the hot big bang era, before the dark energy sets in to dominate everything else). *In light of cosmological QCD phase transition, we would like to examine if the radiation-dominate universe and the matter-dominated universe could ever exist at all, since this has become a dogma in the thinking of our Universe.*

For the **Inflation Era**, we could write $p = -\rho$ and $k = 0$ (for simplicity), so that

$$\ddot{R} - \frac{\dot{R}^2}{R} = 0, \tag{6}$$

which has an exponentially growing, or decaying, solution $R \propto e^{\pm\alpha t}$, compatible with the so-called "inflation" or "big inflation". In fact, considering the simplest case of a real scalar field $\phi(t)$, we have

$$\rho = \frac{1}{2}\dot{\phi}^2 + V(\phi), \qquad p = \frac{1}{2}\dot{\phi}^2 - V(\phi), \tag{7}$$

so that, when the "kinetic" term $\frac{1}{2}\dot{\phi}^2$ is negligible, we have an equation of state, $p \sim -\rho$. In addition to its possible role as the "inflaton" responsible for inflation, such field has also been invoked to explain the accelerating expansion of the present universe, as dubbed as "quintessence" or "complex quintessence".[14]

Let's look at the standard textbook argument leading to the radiation-dominated universe and the matter-dominated universe:

For the **Radiation-Dominated Universe**, we have $p = \rho/3$. For simplicity, we assume that the curvature is zero ($k = 0$) and that the cosmological constant is negligible ($\Lambda = 0$). In this case, we find from Eq. (5)

$$R \propto t^{\frac{1}{2}}. \tag{8}$$

Another simple consequence of the homogeneous model is to derive the continuity equation from Eqs. (2) and (3):

$$d(\rho R^3) + pd(R^3) = 0. \tag{9}$$

Accordingly, we have $\rho \propto R^{-4}$ for a radiation-dominated universe ($p = \rho/3$) while $\rho \propto R^{-3}$ for a matter-dominated universe ($p << \rho$). The present universe is believed to have a matter content of about 5%, or of the density of about $5 \times 10^{-31} g/cm^3$, much bigger than its radiation content $5 \times 10^{-35} g/cm^3$, as estimated from the $3°$ black-body radiation. However, as $t \to 0$, we anticipate $R \to 0$, extrapolated back to a very small universe as compared to the present one. Therefore, the universe is necessarily dominated by the radiation during its early enough epochs.

For the radiation-dominated early epochs of the universe with $k = 0$ and $\Lambda = 0$, we could deduce, also from Eqs. (2) and (3),

$$\rho = \frac{3}{32\pi G_N} t^{-2}, \qquad T = \{\frac{3c^2}{32\pi G_N a}\}^{\frac{1}{4}} t^{-\frac{1}{2}} \cong 10^{10} t^{-1/2} (°K). \tag{10}$$

These equations tell us a few important times in the early universe, such as $10^{-11} sec$ when the temperature T is around 300 GeV during which the electroweak (EW) phase transition is expected to occur, or somewhere between $10^{-5} sec$ ($\cong 300\ MeV$) and $10^{-4} sec$ ($\cong 100\ MeV$) during which quarks and gluons undergo the QCD confinement phase transition.

For the **Matter-Dominated Universe**, we have $p \approx 0$, together with the assumption that $k = 0$ and $\Lambda = 0$. Eq. (5) yields

$$R \propto t^{\frac{2}{3}}. \tag{11}$$

As mentioned earlier, the matter density ρ_m scales like R^{-3}, or $\rho_m \propto t^{-2}$, the latter similar in the radiation-dominated case.

When $t = 10^9 sec$, we have $\rho_\gamma = 6.4 \times 10^{-18} gm/cm^3$ and $\rho_m = 3.2 \times 10^{-18} gm/cm^3$, which are close to each other and it is almost near the end of the radiation-dominated universe. The present age of the Universe is 13.7 billion years - for a large part of it, it is matter-dominated although now we have plenty of dark energy (65% $\sim$ 70%).

However, it is generally believed that our present universe is already dominated by the dark energy (the simplest form being of the cosmological constant; about 70%) and the dark matter (about 25%). The question is when this was so - when the dark part became dominant.

There is another basic point - trivial but important. For both the electroweak and QCD phase transitions in the early Universe, if the phase transitions are described (approximately) by the complex fields ϕ, then the density of the system is given by

$$\rho = \rho_\phi + \rho_\gamma + \rho_m + ..., \tag{12}$$

before or after or during the phase transition is being taking place. For the electroweak or QCD phase transition, we know that $\rho_m << \rho_\gamma$, but the role played by ρ_ϕ is clearly of importance in our considerations.

What would be missing in the standard textbook argument? There are two important points: First, the Einstein equation is classical - where would the quantum principle enter? Presumably, we all believe that our universe grew from a tiny bit but physicists are struggling for a hundred years that these tiny objects obey the quantum principle. Secondly, a more humble request, how do we know, for example, that our universe was once dominated by matter or was in the other epoch once dominated by radiation? The worry comes from two reasons: The dark matter nowadays occupies 25% of the universe, much larger than the (ordinary) matter (5%) and there is no reason that dark matter component would disappear suddenly. The other reason is what we will try to explain - maybe the latent heat released in the cosmological QCD or EW phase transition is so much compared to radiation at the point.

We would come back in Section 6 to the last point (on the latent heat or energy in the cosmological QCD phase transtion), after we set up the general framework and have gained enough of insights. The crucial point is whether cosmological QCD phase transition is the first-order phase transition - *if it is, there is the latent "heat" or energy released in the transition; the story would change dramatically if the amount of energy density turns out to be greater than either ρ_γ or ρ_m in the previous radiation-dominated or matter-dominated era.* We have to question what happened if ρ_γ would no longer be dominant in the radiation-dominated universe - the "new" dominant sector of the universe may not influence the "old" radiation-dominated piece but Einstein equation in principle no longer guarantee its validity (about the old, no-longer-dominant sector).

3. The Cosmological QCD Phase Transition

Let's try to focus on the QCD phase transition in the early Universe, or on the cosmological QCD phase transition.

At the temperature $T > T_c \sim 150MeV$, i.e., before the phase transition takes place, free quarks and gluons can roam anywhere. As the Universe expands and cools, eventually passing the critical temperature T_c, the bubbles nucleate here and there. These bubbles "explode", as we call it "exploding solitons". When it reaches the "supercooling" temperature, T_s, or something similar, the previous bubbles become too many and in fact most of them become touched each other - now the false vacua or "bubbles" of different kind (where quarks and gluons can move freely) start to collapse - or we call it "imploding solitons". When all these bubbles of different kind implode completely, the phase transition is now complete.

There is some specialty regarding the QCD phase transition in the early Universe. Namely, the collapse of the false vacuum does depend on the inside quark-gluon content - e.g., if we have a three-quark color-singlet combination inside, the collapse of the false vacuum would stop (or stabilize) at a certain radius (we called the bag radius, like in the MIT bag radius); of course, there are meson configurations, glueballs, hybrids, six-quark or multi-quark configurations, etc. The QCD phase transition in the early Universe does not eliminate all the false vacua; rather, the end state of the transition could have at least lots of baryon or meson states, each of them has some false vacuum to stabilize the system.

How big can a bubble grow? It is with the fastest speed which the bubble can grow is through the speed of light or close to the speed of light. The bubble could sustain from the moment it creates, say, $T \approx T_c$ to the moment of supercooling, $T_s \sim 0.95 \cdot T_c$, or during the time span $t \sim 3 \times 10^{-5} \times 0.05 sec$ (or $1.5 \times 10^{-7} sec$). So, the bubble can at most grow into $c \cdot 1.5 \times 10^{-7} sec$ or $4.5 \times 10^3\ cm$.

How big was the Universe during the cosmological QCD phase transition? Compared to the size now, an estimate is the expansion of 5.7×10^{12}, a huge factor. (See the beginning of Sect. 6.) In the simplest approximation (when the scalar fields don't couple to the other fields, such as gluons or quarks), the domain walls could not disappear - not only sometime because of the possible nontrivial topology but that there should eventually be some QCD dynamics to annihilate the walls. So, in light of the huge expansion factor, the domain wall structure cannot survive, except the strict topology (which we call "domain-wall nuggets").

As a yardstick, we note that, at $t \sim 10^{-5}\,sec$ or $T \sim 300\,MeV$, we have

$$\rho_\gamma \sim 6.4 \times 10^{10} gm/cm^3, \qquad \rho_m \sim 3.2 \times 10^3 gm/cm^3. \tag{13}$$

Or, at $t \sim 3.30 \times 10^{-5}\,sec$ or $T = T_c \sim 150 MeV$, we have

$$\rho_\gamma = 5.88 \times 10^9 gm/cm^3, \qquad \rho_m = 6.51 \times 10^2 gm/cm^3. \tag{14}$$

Slightly later when QCD phase transition has completed, at $t \sim 10^{-4}\,sec$ or $T \sim 100\,MeV$, we have

$$\rho_\gamma \sim 6.4 \times 10^8 gm/cm^3, \qquad \rho_m \sim 1.0 \times 10^2 gm/cm^3. \tag{15}$$

In what follows, we use the so-called "bag models"[15,16] to have the simplified version of quark confinement - I think it is important to use the simplified version in the complicated cosmological environment, in order to extract meaningful results.

When the low-temperature bubbles start to show up (i.e. to nucleate), it is about $T_c \approx T < T_s$. This period is to be called "exploding soliton era",[16] to be described in the next section (Sect. 4). The supercooling temperature T_s, presumably $\sim 0.95 T_c$ (to be determined more reliably in the future), refers to the situation where the bubbles begin to saturate. We call it the "colliding soliton era". This is to be described in Section 5. Presumably it would be over when $T \leq T_s - (T_c - T_s)$ or longer. So, the cosmological QCD phase transition would be over when the Universe was $10^{-4} sec$ old. The important things is that, because the phase transition is of the first order, it releases a huge amount of energy:

$$\rho_{vac} = 1.0163 \times 10^{14} gm/cm^3, \tag{16}$$

clearly much bigger than the radiation density (cf. Eq. (14)). This quantity is in fact the same as "the zero-point energy". **That is why we question the radiation-dominated universe.**

When the low-temperature bubbles fill up the space, the neighboring two bubbles would in general be labelled by different $\theta_{i,j}$ representing different but degenerate vacua - we assume that there are infinite many choices of θ; they are degenerate but complete equivalent. The domain wall is used to separate the two regions. Three different regions would meet in a line - which we call a vortex. We have to estimate the total energy associated with the domain walls and the vortices - particularly when these objects persist to live on for a "long" time - say, $\tau \gg 10^{-4} sec$. These domain walls and vortices are governed, in the QCD phase transition in the early Universe, by the QCD dynamics - this is an important point; if not, what else? It is a tough question to figure out how long the Universe would stabilize itself through QCD dynamics and others; my rough guess is from a few seconds to years, say τ.

For the moment, QCD enables us to make some estimates. Let us focus on $t \sim 10^{-4} sec$, where $\rho_m = 1.0 \times 10^2\, gm/cm^3$. Or, considering a unit volume of $1.0\, cm^3$, the amount of the matter would be $100 gm$ or $5.609 \times 10^{31} GeV/c^2$. One proton or neutron weighs about $1\, GeV/c^2$ so, in a volume $1.0\, cm^3$ at $t \sim 10^{-4} sec$, we had at least 5.609×10^{31} baryons or, in the MIT bag model language, 5.609×10^{31} bags or $R = 1.0\, fermi$ false vacua associated with the system. To begin with, all the excited baryons and mesons, including topological objects, and multi-baryons, all have equal opportunities. But, remembering $1\, cm^3 = 10^{39} fermi^3$, most space had to collapse into the true vacua with different θ_i. That is, this used to be called "the radiation-dominated era" and so the physics of matter was quite negligible.

When lots of bubbles nucleated, most of which grew rapidly or exploded, then a lot of bubble collisions and eventually turned around, and then the bubbles of false vacuum imploded, these things might happen without the meaning of the "local" temperatures - it is likely that the notion of equilibrium or near-equilibrium could lose the meaning. We believe that, at these moments, the dynamics in terms of scalar fields or QCD remains to be valid and applicable. To describe the situation in terms of the *effective* scalar fields, we simplify the genuine QCD dynamics but eventually we have to take into account the couplings between these scalar fields and the gluons/quarks. These couplings in turn determine the timings of these processes.

4. Exploding Solitons

(Note: A simplified version of this section, and the following two sections, was presented during my 30-min talk. Here I have included it for the sake of completeness. The materials are also presented in my talk at 10th Asia-Pacific Physics Conference and in Korea Journal of Physics (to be published).)

We begin our study by examining an isolated bubble - expanding, that is, inside the bubble, it is the true vacuum labelled by some θ; outside the bubble, the false vacuum; we are thinking of the Universe cooling down and expand. That is, how the bubble nucleates in the false vacuum (high-temperature). Remember that this happened in the period of time roughly when $T_c \leq T \leq T_s$.

Consider a spherical wall of radius R and thickness Δ separating the true vacuum inside from the false vacuum outside. The energy density difference of the vacua is B, the bag constant in the most simplified situation, and the energy τ per unit area associated with the surface tension on the separating wall is a quantity to be calculated but nevertheless is small compared to the latent heat. If the wall expands outward for a distance δR, then the energy budget arising from the vacuum change is

$$B \cdot 4\pi R^2 \cdot \delta R - \tau \cdot 4\pi\{(R+\delta R)^2 - R^2\} = -p\delta V, \tag{17}$$

where p is the pressure and is so defined that a negative pressure would push the wall outward. (We use the notation τ here, since σ and ρ are reserved for other purposes.)

When the surface tension energy required for making the wall bigger is much less than the latent heat required from the expansion of the bubble, the bubble of the stable vacuum inside will grow in an accelerating way, possibly resulting in explosive growth of the bubble. The scenario may be as follows: When the universe expands and cools, to a temperature slightly above the critical temperature T_c, bubbles of lower vacua will nucleate at the spots where either the temperature is lower, and lower than T_c, or the density is higher, and higher than the critical density ρ_c. As the universe continues to expand and cool further, most places in the universe have the temperature slightly below T_c; that is, the destiny arising from eternal expansion of the universe is driving the average temperature of the entire universe toward below the critical temperature. The universe must find a way to convert itself entirely into another vacuum, the true vacuum at the lower temperature.

Maybe we should add a footnote. We are afraid that the situation is much more complicated - the local temperature may lose its meaning; in other words, there are too many things happening at a spot in a short time interval so that non-equilibrium is the real situation. Thus, the above statements regarding the "temperature" are for reference only.

Therefore, we have a situation in which bubbles of true vacua pop up (nucleate) here and there, now and then, and each of them may grow explosively in the environment made of the false vacuum for now, but previously the true vacuum when the temperature was still near the critical temperature T_c. In the expanding universe which cooled down relatively rapidly, i.e. from T_c to the supercooling temperature T_s, the situation is awfully complicated. When the temperature becomes lower than T_s, the problem can be modelled, in the simplest way, by characterizing the vacuum structure by a complex scalar field interacting via the potential $V(\phi)$:

$$V(\phi) = \frac{\mu^2}{2}\phi^*\phi + \frac{\lambda}{4}(\phi^*\phi)^2, \qquad \mu^2 < 0, \quad \lambda > 0. \tag{18}$$

For $T > T_c$, we have $\mu^2(T) > 0$ and $\lambda > 0$, so it is between T_c and T_s when the situations are awfully complicated (and we try to avoid in this paper). Note also that, in the complex scalar field description, the true vacua have degeneracy described by a continuous real parameter θ. $\phi = 0$ everywhere in the spacetime describes the false vacuum for the universe at a temperature below the critical temperature T_c. Consider the solution for a bubble of true vacuum in this environment. It is required that the field ϕ must satisfy the field equation everywhere in spacetime, including crossing the wall of thickness Δ to connect smoothly the true vacuum inside and the false vacuum outside. This is why we may call the bubble solution "a soliton", in the sense of a nontopological soliton of T.D. Lee's. However, the soliton grows in an accelerating way, or the name "exploding soliton".

Of course, it is much easier to describe the situation in terms of a real scalar field in view of no degeneracy. Our solutions could easily be extended to the real scalar field case.

The situation must have changed so explosively that at a very short instant later the universe expands even further and cools to even a little more farther away from T_c and most places in the universe must be in the true vacuum, making the previously false vacuum shrink and fractured into small regions of false vacua, presumably dominantly in spherical shape, which is shrinking in an accelerating way, or "implosively". Using again the complex scalar field as our language, we then have "imploding solitons".

In what follows, we attempt to solve the problem of an exploding soliton, assuming that the values of both the potential parameters μ^2 and λ are fairly stable during the period of the soliton expansion. The scalar field must satisfy:

$$\frac{1}{r^2}\frac{\partial}{\partial r}\left(r^2\frac{\partial\phi}{\partial r}\right) - \frac{\partial^2\phi}{\partial t^2} = V'(\phi). \tag{19}$$

The radius of the soliton is $R(t)$ while the thickness of the wall is Δ:

$$\begin{aligned} \phi = \phi_0, \quad & for \quad r < R_0 + vt - \frac{\Delta}{2}, \\ = 0, \quad & for \quad r > R_0 + vt + \frac{\Delta}{2}, \end{aligned} \tag{20}$$

with $R(t) = R_0 + vt$ and v the radial expansion velocity of the soliton.

We may write

$$\phi \equiv f(r+vt); \qquad w \equiv (1-v^2)r, \tag{21}$$

so that the field equation becomes

$$\frac{d^2 f}{dw^2} + \frac{2}{w}\frac{df}{dw} = (1-v^2)^{-1}\lambda f(\mid f \mid^2 - \phi_0^2). \tag{22}$$

We will be looking for a solution of f across the wall so that it connects smoothly the true-vacuum solution inside and the false vacuum solution outside.

Introducing $g \equiv wf(w)$, we find

$$g'' = (1-v^2)^{-1}\lambda g\{\mid \frac{g}{w} \mid^2 - \phi_0^2\}, \tag{23}$$

an equation which we may solve in exactly the same manner as the colliding-wall problem to be elucidated in the next section.

We could examine the issue whether we could treat the QCD phase transition as the first-order transition. The basic equation as revealed by Equation (17) indicates that, if the pressure term is outward, the latent heat (i.e. the bag constant term) must be there - note that the sign of the surface tension is also opposite. That is, if the phase transition is not first-order, the pressure would be outward negative - the bubbles would be in general shrinking - a strange result, indeed!! In other words, it is in fact nontrivial to think of the QCD phase transition as not first-order.

5. Colliding Walls: Formation of "Pasted" Domain Walls

When bubbles of true vacua grow explosively, the nearby pair of bubbles will soon squeeze or collide with each other, resulting in merging of the two bubbles while producing cosmological objects that have specific coupling to the system. The situation is again extremely complicated. Remember that this happened when $T \sim T_s$, not too long after.

We try to disentangle the complexities by looking at between the two bubble walls that are almost ready to touch and for the initial attempt neglecting the coupling of the vacuum dynamics to the matter content. Between the two bubble walls, especially between the centers of the two bubbles, it looks like a problem of plane walls in collision - and this is where we try to solve the problem to begin with.

In fact, we have to consider one bubble first - the spherical situation as in the previous section but the bubble is "very" large we could look at the z-direction in the sufficiently good plane approximation (i.e. all bubble surfaces are just like planes). At this point, we have one wall, with thickness Δ, moving with velocity v in the z direction; on the left of the wall is the false vacuum, and on the right the true vacuum.

The wall, of thickness Δ, separates the true vacuum on one side from the false vacuum on the other side of the wall. For the sake of simplicity, the wall is assumed parallel to the $(xy)-$plane and are infinite in both the x and y directions. In addition, at some instant the wall is defined between $z = z_0 - \frac{\Delta}{2}$ and $z = z_0 + \frac{\Delta}{2}$ with the instantaneous velocity $+v$; this region connects the true vacuum with the false vacuum.

In other words, we need to consider the case that we have two parallel walls approaching each other: The left-hand one at some instant is at $-R-\frac{\Delta}{2} < z < -R+\frac{\Delta}{2}$ while the right-hand one at $R-\frac{\Delta}{2} < z < R+\frac{\Delta}{2}$; in the middle $-R+\frac{\Delta}{2} < z < +R-\frac{\Delta}{2}$, it is the false vacuum; the walls are moving toward each other, so the false vacuum gets squeeze out.

For $z > R+\frac{\Delta}{2}$ and all x and y, the complex scalar field ϕ assumes ϕ_0, a value of the true vacuum (the ground state). On the other hand, for $-R+\frac{\Delta}{2} < z < +R-\frac{\Delta}{2}$ and all x and y, the complex scalar field ϕ assumes $\phi = 0$, the false vacuum. As indicated earlier, the field ϕ must satisfy the field equation everywhere in spacetime:

$$\frac{\partial^2 \phi}{\partial z^2} - \frac{\partial^2 \phi}{\partial t^2} = V'(\phi). \tag{24}$$

We may write the wall on the right hand side but moving toward the left with the velocity v:

$$\phi = f(z - vt), \qquad for\ z - vt > 0,\ t < R/v. \tag{25}$$

so that

$$(1 - v^2) f'' = \lambda f(|\, f\,|^2 - \sigma^2), \qquad \sigma \equiv |\, \phi_0\,| > 0. \tag{26}$$

In fact, we are interested in the situation that the function f is complex:

$$f \equiv u e^{i\theta}, \tag{27}$$

so that, with $\tilde{\lambda} \equiv \lambda/(1 - v^2)$,

$$u'' - u(\theta')^2 = \tilde{\lambda} u (u^2 - \sigma^2), \tag{28}$$

$$2u'\theta' + u\theta'' = 0. \tag{29}$$

Integrating the second equation, we find

$$u^2 \theta' = K, \tag{30}$$

with K an integration constant. The equation for u is thus given by

$$u'' = \frac{K}{u^3} + \tilde{\lambda} u(u^2 - \sigma^2), \tag{31}$$

provided that the θ function is defined (in the region of the true vacuum and the wall).

Let us try to focus on the last two basic equations - for u and θ, say, as the functions of ξ (e.g. $\xi = z \pm vt$). For $\xi \geq \Delta$, we have $\phi = \sigma e^{i\theta}$ (the true vacuum) and, for $\xi < 0$, we have $\phi = 0$ (the false vacuum; with θ undetermined). We find, for $\xi \to 0^+$,

$$\theta = \frac{1}{2}\sqrt{-K}(ln\xi)(1 + F(\xi)) + C_0, \tag{32}$$

with C_0 a constant and $F(\xi)$ regular near $\xi \sim 0$. Therefore the $\theta(\xi)$ function could be "mildly singular" or blow up near $\xi \sim 0$ - this is in fact a very important point.

Of course, the equation for u can be integrated out to obtain the result. For the "wall" region (i.e. $0 < \xi < \Delta$), the solution reads as follows:

$$\xi = \frac{\sigma^2}{2} \int_0^{u^2/\sigma^2} \frac{dy}{\sqrt{-K + \alpha y - 2\beta y^2 + \beta y^3}}, \tag{33}$$

with

$$\Delta = \frac{\sigma^2}{2} \int_0^1 \frac{dy}{\sqrt{-K + \alpha y - 2\beta y^2 + \beta y^3}}. \tag{34}$$

Here $\beta \equiv \frac{\tilde{\lambda}}{2}\sigma^6$, and K and α parameters related to the integration constants. Of course, the solution in true-vacuum region can be obtained by extension.

In the wall region, we could compute the surface energy per unit area (i.e. surface tension mentioned earlier in Eq. (17)):

$$\tau = \int_0^\Delta d\xi \frac{1}{2}\{(u')^2 + u^2(\theta')^2\}, \tag{35}$$

some integral easy to calculate.

There is an important note - that is, the solution for ϕ obtained so far applies for the true vacuum and the wall, and which is continuous in the region; how about the false vacuum? This is an important question because in the false vacuum we know that $u = 0$ but θ is left undetermined. So, in first-order phase transitions we have certain function undefined in the false-vacuum region(s). This is a crucial point to keep in mind with.

As a parenthetical footnote, we note that the equation for the exploding or imploding spherical soliton, Eq. (22), may be integrated and solved in an identical manner.

Now let us focus on the merge of the two bubbles - the growing of the two true-vacuum bubbles such that the false-vacuum region gets squeezed away. This is another difficult dynamical question. In fact, we can make the false-vacuum region approaching to zero, i.e., the region with the solution $u = 0$ gets squeezed away; one true-vacuum region with θ_1 and Δ_1 (the latter for the wall) is connected with the one with θ_2 and Δ_2 - we could use (K_1, K_2) to label the new boundary; to be precise, we could call it "the pasted domain wall" or "the patched domain wall". It is in fact two walls pasted together - if we look at the boundary condition in between, we realize that the structure would persist there for a while to go. The pasted domain wall could evolve further but this may not be relevant for counting the energies involved. The evolved forms of the pasted domain walls could be determined by the topology involved - for the purpose of this paper, we can ignore this fine aspect.

Suppose that the cosmological QCD phase transition was just completed - we have to caution that, not everywhere, the false vacua be replaced by the true vacua so that in between the walls be replaced (approximately) by the pasted domain walls. There are places for color-singlet objects (i.e. hadrons) which quarks and gluons tried to hide; these places are still called by the "false vacua" with the volume energies. Thus, the volume energy, i.e. B in Eq. (17) or defined suitably via λ and μ^2 (in Eq. (18)), or at least some portion of it, may convert itself into the surface energy and others - $B = 57\, MeV/fm^3$ using the so-called "bag constant" in the MIT bag model[15] or Columbia bag model.[16]

This energy density $B = 57 MeV/fm^3 = 1.0163 \times 10^{14} gm/cm^3$ is huge as compared to the radiation density ρ_γ (which is much bigger than the matter density ρ_m) at that time, $t \sim 10^{-5} \sim 10^{-4} sec$ (see Eqs. (13)-(15)). Some exercise indicates that this quantity of energy is exactly the latent "heat" or energy released in the first-order phase transition.

The cosmological QCD phase transition should leave its QCD mark here - since the volume energy that stays with the "false vacuum" is simply reduced because the volumes with the "false vacua" are greatly reduced - but not eliminated because quarks and gluons, those objects with colors, still have some places to go (or, to hide themselves).

6. Possible Connection with the Dark Matter

Let us begin by making a simple estimate - the expansion factor since the QCD phase transition up to now. The present age of the Universe is 13.7 billion years or $13.7 \times 10^9 \times 365.25 \times 24 \times 3600$ or 4.323×10^{17} seconds.

As indicated earlier (cf. the end of Sec. 2), about the first $10^9 sec$ period of the hot big bang is previously-believed radiation-dominated. Consider the length $1.0\, fermi$ at $t \sim 10^{-5} sec$, it will be expanded by a factor of 10^7 up to $t \sim 10^9 sec$ (radiation-dominated) and expanded further by another factor of 5.7×10^5 until the present time - so, a total expansion factor of 5.7×10^{12}; changing a length of $2\, fermi$ at $t \sim 10^{-5} sec$ into a distance of $1\, cm$ now. A proton presumably of $R = 1\, fermi$ at $t \sim 10^{-4} sec$ should be more or less of the same size now; or, the bag constant or the energy associated with the false vacuum should remain the same.

What would happen to the pasted or patched domain walls as formed during the cosmological QCD phase transition? According to Eqs. (30) and (31) together with Eq. (32), we realize that the solutions in previously two different true-vacuum regions cannot be matched naturally - unless the K values match accidently. On the other hand, it is certain that the system cannot be stretched or over-stretched by such enormous factor, 10^{12} or 10^{13}.

As we said earlier, at some point after the supercooling temperature T_s, say, at $T_s - \lambda(T_c - T_s)$ (with λ an unknown factor, presumably $\lambda \gg 1$), the system (the Universe) was temporarily stabilized since most of the pasted or patched domain walls had no where to go. Remember that all these happened in a matter of a fraction of $10^{-4} sec$, as judging from the size of T_c and T_s. The next thing to happen is probably the following.

We believe that the field ϕ, being effective, cannot be lonely (i.e., non-interacting with other fields); in fact, we believe that there are higher-order interactions such as

$$c_0 \phi G_\mu^a G^{\mu,a}, \quad c_1 \phi GGG, \, ..., \qquad d_0 \phi \bar{\psi} \psi, \tag{36}$$

some maybe being absent because of the nature of ϕ. The sizes of these couplings determine the time scales needed for these interactions. In other words, we may believe that the strong interactions are primarily responsible for the phase transition in question, such that the effective field ϕ couples to the gluon and quark fields; the details of the coupling are subject to investigations.

That is, when the field ϕ responsible for the pasted or patched domain walls is *effective* - the ϕ field couples, in the higher-order (and thus weaker) sense, to the gluon and quark fields. It is very difficult to estimate what time is needed for pasted domain walls to disappear, if there are no nontrivial topology involved. If there is some sort of nontrivial topology present, there should left some kind of topological domain nugget - however, energy conservation should tell us that it cannot be expanded by too many orders (but our Universe did expand for many many orders of magnitude). I would guess that it takes about from a fraction of a second to several years (from the strong interaction nature of the problem), but certainly before the last scattering surface (i.e. 3.79×10^5 years).

To summarize, the energy associated with the cosmological QCD phase transition, mainly the vacuum energy associated with the false vacuum, disappeared in several ways, viz.: (1) the bag energies associated with the baryons and all the other color-singlet objects, (2) the energies with all kinds of topological domain nuggets or other topological objects, and (3) the decay products from pasted or patched domain walls with trivial topology.

Let us begin with the critical temperature $T = T_c \approx 150\,MeV$ or $t \approx 3.30 \times 10^{-5} sec$. At this moment, we have

$$\rho_{vac}=1.0163\times10^{14}gm/cm^3,\quad \rho_\gamma=5.88\times10^9 gm/cm^3,\quad \rho_m=6.51\times10^2 gm/cm^3. \tag{37}$$

Here the first term is what we expect the system to release - the so-called "latent heat"; I call it "latent energy" for obvious reasons. The identification of the latent "heat" with the bag constant is well-known in Coulomb bag models.[16]

This can be considered just before the cosmological QCD phase transition which took place - at the moment the energy components which we should take into consideration.

As time went on, the Universe expanded and the temperature cooled further - from the critical temperature to the supercooling temperature ($T_s \sim 0.95 \times T_c$ with the fraction 0.95 in fact unknown) and even lower, and then the cosmological QCD phase transition was complete. When the phase transition was complete, we should estimate how the energy ρ_{vac} is to be divided.

Let's assume that the QCD phase transition was completed at the point T_s (in fact maybe a little short after T_s). Let's take $T_s = 0.95\,T_c$ for simplicity. We would like to know how the energy ρ_{vac} is to be divided. First, we can estimate those remained with the baryons and other color-singlet objects - the lower limit is given by the estimate on the baryon number density (noting that one baryon weighs about $1.0GeV/c^2$):

$$\rho_m = 6.51 \times 10^2 gm/cm^3 \times 0.5609 \times 10^{24} GeV/c^2/gm = 3.65 \times 10^{26} GeV/c^2/cm^3. \tag{38}$$

So, in the volume $1.0cm^3$ or $10^{39} fermi^3$, we have at least 3.65×10^{26} baryons. One baryon has the volume energy (i.e. the bag energy or the false vacuum energy) $57MeV/fermi^3 \times \frac{4}{3}\pi(1.0fermi)^3$ (which is $238.8MeV$). So, in the volume $1.0cm^3$, we have at least $238.8MeV \times 3.65 \times 10^{26}$ or $8.72 \times 10^{25}GeV$ in baryon bag energy. Or, in different units $8.72 \times 10^{25}/(0.5609 \times 10^{24})$ gm/c^2 or $155.5gm/c^2$. *Only a tiny fraction of ρ_{vac} is to be hidden in baryons or other color-singlet objects after the QCD phase transition in the early Universe.*

So, where did the huge amount of the energy ρ_{vac} go? In the beginning of the end of the phase transition, the pasted domain walls with the huge kinetic energies seem to be the main story. A pasted domain wall is forming by colliding two domain walls while eliminating the false vacuum in between. The kinetic energies associated with the previously head-on collision become vibration, center-of-mass motion, etc. Of course, the pasted domain walls would evolve much further such as through the decaying interactions given earlier or forming the "permanent" structures. In any case, the total energy involved is known reasonably - a large fraction of ρ_{vac}, much larger than the radiation ρ_γ (with ρ_m negligible at this point).

The story is relatively simple when the cosmological QCD phase transition was just completed and most "pasted" domain walls still have no time to evolve. We return to Eqs. (2) and (3) (i.e. Einstein equations) for the master equations together with the equation of state with ρ and p determined by the energy-momentum tensor:

$$T^{\phi}_{\mu\nu} = g_{\mu\alpha}\frac{\partial L}{\partial(\partial_\alpha \phi)}\partial_\nu \phi - Lg_{\mu\nu}. \tag{39}$$

Further analysis indicates that the equation of state for the "pasted" or "patched" domain walls is nothing unusual - the reason is that we are working in the real four-dimensional space-time and all of the objects are of finite dimensions in all the directions. The "domain walls" discussed by us are for real and cannot be stretched to infinity in a certain dimension.

In fact, there is certain rule which one cannot escape. Let assume a simple equation of state, $\rho = wp$, for simplicity and come to look at Eq. (5). Let's consider the situation in which there is no curvature $k = 0$ and the cosmological constant λ is not yet important.

$$2\frac{\ddot{R}}{R} + (1+3w)\frac{\dot{R}^2}{R^2} = 0, \tag{40}$$

which yields

$$R \propto t^n, \tag{41}$$

with $n = \frac{2}{3} \cdot \frac{1}{1+w}$.

From the equation of continuity, $d(\rho R) + pd(R^3) = 0$, it is easy to obtain $\rho \propto R^{3(1+w)}$. Thus, we deduce that, under very general situations, the density behaves like

$$\rho = Ct^{-2}, \tag{42}$$

where the constant C is related to w in the simplified equation of state. It is clear that the limit to $w = -1$ (the cosmological constant) is a discontinuity.

Of course, Eq. (4) is still valid:

$$\frac{\ddot{R}}{R} = -\frac{4\pi G_N}{3}(\rho + 3p) + \frac{\Lambda}{3}. \tag{43}$$

This has an important consequence - the idea of the previous universe expansion usually based on the radiation alone from $t \sim 10^{-10}$ *sec* (after the cosmological electroweak phase transition had taken place) to $t \sim 10^9$ *sec* (when it was close that $\rho_\gamma = \rho_m$) has to be modified because the latent energy ρ_{vac} was about 2×10^5 times the radiation energy at the moment of the cosmological QCD phase transition.

Shown in Fig. 1 is our main result - even though it is a qualitative figure but it tells us a lot. At $t \sim 3.30 \times 10^{-5}$ *sec*, where did the latent energy $10^{14} gm/cm^3$ evolve into? We should know that the curve for ρ_γ, for massless relativistical particles, is the steepest in slope. The other curve for ρ_m is the other limit for matter (which $P \approx 0$). In this way, the latent energy is connected naturally with the curve for ρ_{DM} - in fact, there seems

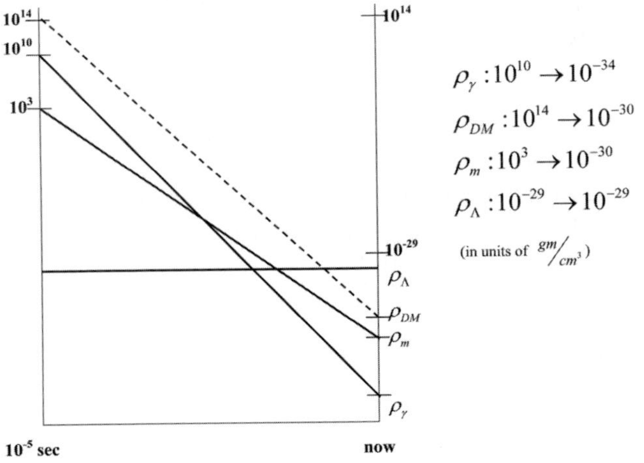

Fig. 1. The various densities of our universe versus time.

to be no other choice. Remember that $\rho \propto t^{-2}$ except the slope for different types of "matter".

Coming back to Eq. (43) or (4), we could assume for simplicity that when the cosmological QCD just took place the system follows with the relativistical pace (i.e. $P = \rho/3$) but when the system over-stretched enough and had evolved long enough it was diluted enough and became non-relativistic (i.e. $P \approx 0$). It so happens that in both cases the density to the governing equation, Eq. (43) or (4), looks like $\rho \propto t^{-2}$ although it is $R \propto t^{\frac{1}{2}}$ followed by $R \propto t^{\frac{2}{3}}$.

It is so accidental that what we call "the radiation-dominated universe" is in fact dominated by the latent energy from the cosmological QCD phase transition in the form of "pasted" or "patched" domain walls and the various evolved objects. In our case, the transition into the "matter-dominated universe", which happened at a time slightly different from $t \sim 10^9 sec$, occurred when all the evolutions of the pasted domain walls ceased or stopped. In other words, it is NOT the transition into the "matter-dominated universe", as we used to think of.

In fact, the way of thinking of the "dark matter", or the majority of it, turns out to be very natural. Otherwise, where did the 25% content of our universe come from? Of course, one could argue about the large amount of the cosmological QCD phase transition. We believe that the curves in Fig. 1 make a lot of sense.

Of course, one should ask what would happen before the cosmological QCD phase transition. It might not be the radiation-dominated. I believe that it opens up a lot of important and basic questions.

7. Outlook

To sum up, we tried to illustrate how to describe the QCD phase transition in the early Universe, or the cosmological QCD phase transition.

The scenario that some first-order phase transitions may have taken place in the early Universe offers us one of the most intriguing and fascinating questions in cosmology. In fact, the role played by the latent "heat" or energy released in the process is highly nontrivial.

In this talk, we assume that the QCD phase transition, which happened at a time $t \approx 3.30 \times 10^{-5}\, sec$ or at the temperature of about $150\, MeV$ and accounts for confinement of quarks and gluons to within hadrons in the true QCD vacuum, would be of first order. Thus, it is sufficient to approximate the true QCD vacuum as one of degenerate vacua and when necessary we try to model it effectively via a complex scalar field with spontaneous symmetry breaking. We examine how and how long "pasted" or "patched" domain walls were formed, how and how long such walls evolve further, and why the majority of dark matter might be accounted for in terms of these evolved objects. [It is much easier to examine the case of the real scalar field.]

Our central result could be summarized by Fig. 1 together with the explanations. Mainly, we are afraid that the "radiation-dominated" epoch and the "matter-dominated" epoch, in the conventional sense, could not exist once the cosmological QCD phase transition took place. That also explains why there is the 25% dark-matter content, larger than the baryon content, in our present universe.

In other words, if what we are proposing for the dark matter is largely correct, then at least the "radiation-dominated universe" terminated when the cosmological QCD phase transition took place - from there on, we have something like the "dark-matter-dominated universe". So, it is indeed important to determine whether the QCD phase transition is first-order.[6]

Acknowledgments

The Taiwan CosPA project is funded by the Ministry of Education (89-N-FA01-1-0 up to 89-N-FA01-1-5) and the National Science Council (NSC 95-2752-M-002-007-PAE and others). This research is also supported in part as another National Science Council project (NSC 96-2752-M-002-007-PAE and NSC 96-2112-M-002-023-MY3).

References

1. C.N. Yang, *AAPPS Bulletin*, p.4, Vol. **15**, No. 1, February 2005.
2. G. Smoot *et al.*, *Astrophys. J.* **396**, L1 (1992); C. Bennett *et al.*, *Astrophys. J.* **396**, L7 (1992); E. Wright *et al.*, *Astrophys. J.* **396**, L11 (1992).
3. C. L. Bennett, M. S. Turner, and M. White, *Physics Today*, November 1997, p. 32, for an early general review.
4. See, e.g., the news in *Physics Today*, April 2003, p. 21; and the references therein.
5. See, e.g., the news in *Physics Today*, May 2006, p.16; and the references therein.
6. Y. Aoki, G. Endrodi, Z. Fodor, S. D. Katz, and K.K. Szabo, *Nature* **443**, 675 (2006) [arXiv:hep-lat/0611014]; M. Cheng *et al.*, *Phys. Rev.* **D 75**, 034506 (2007) [arXiv: hep-lat/0612001].
7. Of course, the ground state in the simplest case is nondegenerate but in realistic cases the problem of the degenerate vacua is rather common. See, E. Shuryak and T. Schaefer, *Annu. Rev. Nucl. Part. Sci.* **47**, 359 (1997).
8. See, for instance, M.S. Turner, *Phys. Rept.* **197**, 67 (1990).

9. B. Svetitsky and L.G. Yaffe, *Nucl. Phys. B* **210**, 423 (1982).
10. Ariel Zhitnitsky, arXiv:astro-ph/0603064, 29 July 2006 and the references therein.
11. For example, see H. Kurki-Suonio and M. Laine, *Phys. Rev. Lett.* **77**, 3951 (1996); R.M. Haas, *Phys. Rev.* **D57**, 7422 (1998).
12. E.W. Kolb and M.S. Turner, *The Early universe, Addison-Wesley Publishing Co.* (1994).
13. S. Perlmutter *et al.* [Supernova Cosmology Project], *Astrophys. J.* **517**, 565 (1999); A. G. Riess *et al.* [Supernova Search Team], *Astron. J.* **116**, 1009 (1998).
14. R.R. Caldwell, R. Dave, and P.J. Steinhardt, *Phys. Rev. Lett.* **80**, 1582 (1998); Je-An Gu and W-Y. P. Hwang, *Phys. Lett.* **B517**, 1 (2001).
15. A. Chodos, R.L. Jaffe, K. Johnson, C.B. Thorn, and V.F. Weisskopf, *Phys. Rev.* **D9**, 3471 (1974); T.A. DeGrand, R.L. Jaffe, K. Johnson, and J. Kiskis, *Phys. Rev.* **D12**, 2060 (1975). For the bag model parameters, see, e.g., W-Y. P. Hwang, *Phys. Rev.* **D31**, 2826 (1985).
16. R. Friedberg and T.D. Lee, *Phys. Rev.* **D16**, 1096 (1977); **D18**, 2623 (1978). For generalizations, see, e.g., W-Y. P. Hwang, *Phys. Lett.* **116B**, 37 (1982) and **D29**, 1465 (1984).

ANALYTIC SCATTERING AMPLITUDES FOR QCD

DIANA VAMAN
Department of Physics, University of Virginia, Charlottesville, VA, 22904, USA
dv3h@virginia.edu

YORK-PENG YAO
Department of Physics, University of Michigan, Ann Arbor, MI, 48109, USA
yyao@umich.edu

By analytically continuing QCD scattering amplitudes through specific complexified momenta, one can study and learn about the nature and the consequences of factorization and unitarity. In some cases, when coupled with the largest time equation and gauge invariance requirement, this approach leads to recursion relations, which greatly simplify the construction of multi-gluon scattering amplitudes. The setting for this discussion is in the space-cone gauge.

Keyword: QCD.

1. Introduction

The LHC will be turned on soon. Excluding serendipitous events, it seems that signals will be seen only after complicated backgrounds have been properly subtracted out. Therefore, one must have a good account of the multi-particle processes, particularly those induced by QCD. Also, one should know the proper energy scale in a calculation to ensure stability relative to higher order effects. In other words, loops are also important, besides tree level results.

There has been much progress in perturbative evaluations,[1] especially in the past few years.[2–5] One would even venture to say that there is a new technology, which is applicable to all field theories. We shall confine our attention to QCD here.

If one is to follow the usual Feynman rules and diagrams to calculate a multi-gluon process, one will find that the algebra becomes horrendous very fast. For an n-gluon process at the tree level, if we just examine the three-point vertices, there are $n-2$ of them and each has six terms which depend on some momenta, not to mention the internal symmetry coupling. Then one has to permute these n legs over the vertices.

As we all know, a massless particle in four spacetime dimensions has at most two degrees of freedom, but a manifestly covariant formulation requires four components for a vector field. Therefore, there are tremendous amount of cancellations in the intermediate stage of a calculation to yield some simple-looking final answer. The process somehow knows that the unphysical degrees of freedom should not be there and tries its best to expel them. Lots of efforts were wasted in the old ways. It will help if one eliminates all these unwanted degrees of freedom at an early stage in some way.

The new developments in non-Abelian gauge field calculations on the whole pursue two different paths, the ideas of which are not entirely new, but the executions are much improved:

(1) Using a physical gauge, such that there are explicitly only two components for each internal symmetry index.
(2) Using an extended dispersive technique, such that an n-point amplitude will be constructed from lower point on-shell physical amplitudes.

It turns out that these two methods can be made to complement each other and give rise to recursion relations for all the tree and some of the one-loop amplitudes.[6] For the other one-loop amplitudes with more complicated helicity composition, they are very much like the dispersion method using Cutkosky rules, but of course with much better handle and insight.

You must appreciate the possibility of having recursion relations, because one can recycle whatever hard work one has already put in to build up more complicated processes, rather than to start from the scratch all over again. This is possible, much to the credit of analytic continuation into complex momenta.[5]

2. Spinors, Twistors and Complex Momenta

For a particle with zero mass, we can use two component spinors or twistors representation

$$P^{\dot{a}b} = (\tilde{\sigma} \cdot P)^{\dot{a}b} = |p]^{\dot{a}} \langle p|^{b} \tag{1}$$

and

$$P_{b\dot{a}} = (\sigma \cdot P)_{b\dot{a}} = |p\rangle_{b} [p|_{\dot{a}} \,, \tag{2}$$

where $\sigma^\mu = (-I, \boldsymbol{\sigma})$ and $\tilde{\sigma}^\mu = (-I, -\boldsymbol{\sigma})$. We use them to form scalar products of spinors

$$\langle p_i p_j \rangle = \langle p_i|^b |p_j\rangle_b = -\langle p_j p_i \rangle \tag{3}$$

and

$$[p_j p_i] = [p_j|_{\dot{a}} |p_i]^{\dot{a}} = -[p_i p_j] \,, \tag{4}$$

from which the scalar product of two vectors is

$$-2P_i \cdot P_j = \langle p_i p_j \rangle [p_j p_i] \,. \tag{5}$$

Also, we use them to build polarization vectors for gauge particles of momentum[7,8] K_i

$$\epsilon^{h=+}(Q_i, K_i)^\mu = \frac{\langle q_i|\sigma^\mu|k_i]}{\sqrt{2}\langle q_i k_i\rangle} \,, \qquad \epsilon^{h=-}(Q_i, K_i)^\mu = \frac{[q_i|\tilde{\sigma}^\mu|k_i\rangle}{\sqrt{2}[q_i k_i]} \tag{6}$$

in which Q_i is a reference momentum, which can be individually assigned for each K_i. Changing Q_i is a change of gauge.

For real momenta, we have

$$[p_i p_j] = \langle p_j p_i \rangle^\star \,, \tag{7}$$

which is a result we do not like, if we want to perform on-shell calculation. Let us consider forming an amplitude for three on-shell gluons

$$P_1 + P_2 + P_3 = 0 \,. \tag{8}$$

Then for real momenta

$$0 = P_1^2 = (P_2 + P_3)^2 = 2P_2 \cdot P_3 = -|\langle p_2 p_3 \rangle|^2 \,, \text{ etc.} \tag{9}$$

which means both

$$\langle p_i p_j \rangle = 0 \,, \quad \text{and} \quad [p_i p_j] = 0 \,, \qquad i, j = 1, 2, 3 \,. \tag{10}$$

This makes it impossible to define an on-shell tree level three-point gluon amplitude, which is the least demand to start a program.

On the other hand, for complex P's, $[p_j p_i]$ is no longer the complex conjugate of $\langle p_i p_j \rangle$ and for appropriate helicity arrangements the zero mass conditions can be satisfied by either

$$\langle p_i p_j \rangle = 0 \,, \tag{11}$$

then e.g.[9]

$$A(P_1^+, P_2^+, P_3^-) = -i \frac{[p_1 p_2]^4}{[p_1 p_2][p_2 p_3][p_3 p_1]} \,, \tag{12}$$

or

$$[p_i p_j] = 0\,, \tag{13}$$

then e.g.[9]

$$A(P_1^-, P_2^-, P_3^+) = i\frac{\langle p_1 p_2\rangle^4}{\langle p_1 p_2\rangle\langle p_2 p_3\rangle\langle p_3 p_1\rangle}\,. \tag{14}$$

3. Space-Cone Gauge

There are many different physical gauges to get rid of the unphysical degrees of freedom, but the one which is best for our purpose is the space-cone gauge. This is because for the specific analytic continuation into complex momenta, we use to arrive at recursion relations, we shall find that the vertices in this gauge are untouched. Let us be reminded that our aim is to factorize each term in an amplitude, which has both a numerator and a denominator, into something simpler, already known or done. If we do not have to touch the numerator in our manipulation to accomplish this, it will be just that much easier. In other words, we shall find that for this gauge the factorization is like what is needed in a scalar theory, where we shall be massaging products of propagators into something we can identify with a lower point on-shell amplitude.

Although we are free to have one reference vector (Q_i) for each emitted gluon, we shall use only two reference spinors for all gluons

$$|+\rangle\,, \qquad [-| \tag{15}$$

and normalize them to

$$\langle +-\rangle = [-+] = 1\,. \tag{16}$$

Any massless four-vector can be decomposed according to

$$P = p^+|-\rangle[-| + p^-|+\rangle[+| + p|-\rangle[+| + \bar{p}|+\rangle[-|\,, \tag{17}$$

and a gluon of momentum K has polarization vectors

$$\epsilon^+(K) = \frac{[-k]}{\langle +k\rangle}\,, \qquad \epsilon^-(K) = \frac{\langle +k\rangle}{[-k]}\,. \tag{18}$$

They satisfy

$$\epsilon^+(K)\epsilon^-(k) = 1 \tag{19}$$

which makes polarization sums very simple.

The space-cone gauge[10] is defined by the condition

$$N \cdot A = 0 \quad \text{or} \quad a = 0 \tag{20}$$

for each color index of the gauge field A. Here

$$N = |+\rangle[-| \tag{21}$$

is also a light-like vector. There is a constraint among the equations of motion, which can be used to express $\bar{a}$ in terms of $a^{\pm}$, and the resulting Lagrangian is

$$\begin{aligned} L = \mathrm{Tr}\Bigg[&\frac{1}{2} a^+ \partial_\mu \partial^\mu a^- - i\left(\frac{\partial^-}{\partial} a^+\right)[a^+, \partial a^-] \\ &- i\left(\frac{\partial^+}{\partial} a^-\right)[a^-, \partial a^+] + [a^+, \partial a^-] \frac{1}{\partial^2} [a^-, \partial a^+] \Bigg] . \end{aligned} \tag{22}$$

What is noteworthy is that in the interaction part of L we do not have the derivative component $\bar{\partial}$, which is very important for later discussion when we perform analytic continuation by shifting momenta, or derivatives. We shall find that only $\bar{\partial}$ will be affected, but this does not appear in the vertices, which means that the interaction will be unchanged. However, all components ∂, $\bar{\partial}$, $\partial^{\pm}$ appear in the Klein–Gordon operator $\partial_\mu \partial^\mu$, and therefore propagators will change when we do analytic continuation. It is as if we have a two-component scalar field theory.

The analysis is further simplified by color ordering.

4. The Largest Time Equation and Analytic Continuation

The causal nature of quantum field theory allows one to decompose a propagator into a positive frequency part and a negative frequency part. From this, identities for products of propagators and products in which some propagators are replaced by positive or negative parts can be written down. One easy way to arrive at them is to observe that if a system is driven from $t = -\infty$ to $t = +\infty$ by some external currents and then back to $t = -\infty$, the generating functional must be just unity. By equating the coefficients of various powers of external currents, one can obtain sets of identities. The physical outcome is that for every Feynman diagram in a scattering process, one can draw boundaries with inflowing energy lines on one side and outflowing energy lines on the other. The largest time equation by Veltman,[11] (which is closely associated with the closed time path cycle of Schwinger[12,13]), is to pick two out of possibly many space–time points in a diagram and time order them. This will relate a product of propagators with cut lines. The causal ordering is enforced by a parameter z, which is an integration variable

$$\theta(-\eta\cdot(x-y)) = \frac{1}{2\pi i}\int \frac{dz}{z-i\epsilon} e^{-iz\eta\cdot(x-y)} \tag{23}$$

with $\eta^\mu = (1,0,0,0)$.

If there are only two external lines, the equation yields the Lehman representation, and the parameter z can be rewritten as the invariant mass of an intermediate state.

For a scattering amplitude A, we have of course more than two external lines. It turns out that it can be analytically continued $\hat{A}(z)$ by making some of the momenta complex through complexifying z and η. One can find out the poles and cuts of A in its kinematical invariants, known as Mandlestam variables, by investigating the analyticity of $\hat{A}(z)$.[14] Furthermore, if there are only poles in $\hat{A}(z)$, one will obtain recursion relations, expressing A in terms of lower point on-shell scattering amplitudes, as a consequence of Cauchy's theorem.

To give an example, we look at one of the diagrams for the process $P_1^+P_2^+P_3^+P_4^-P_5^-$. We first write down a largest time equation

$$\begin{aligned}\Delta(x_1-x_2)\Delta(x_2-x_3) = {} & (\theta(-\eta\cdot(x_1-x_3))\Delta^+(x_1-x_2) \\ & +\theta(\eta\cdot(x_1-x_3))\Delta^-(x_1-x_2))\Delta(x_2-x_3) \\ & +(\theta(-\eta\cdot(x_1-x_3))\Delta^+(x_2-x_3) \\ & +\theta(\eta\cdot(x_1-x_3))\Delta^-(x_2-x_3))\Delta(x_1-x_2)\,. \end{aligned} \tag{24}$$

We let P_1 and P_2 go into x_1, P_3 into x_2, and P_4 and P_5 into x_3, put in the plane wave functions and integrate over all x's. Taking out the delta function which enforces energy–momentum conservation, we have corresponding to each term

$$\frac{1}{P_{12}^2}\frac{1}{P_{45}^2} = \frac{1}{P_{12}^2}\frac{1}{\hat{P}_{45}^2}\bigg|_{z=z_{12}} + \frac{1}{\hat{P}_{12}^2}\bigg|_{z=z_{45}}\frac{1}{P_{45}^2}\,, \tag{25}$$

where

$$\hat{P}_1 = P_1 + z\eta\,, \qquad \hat{P}_5 = P_5 - z\eta\,, \tag{26}$$

$$\hat{P}_{12} = \hat{P}_1 + P_2\,, \qquad \hat{P}_{45} = P_4 + \hat{P}_5\,, \tag{27}$$

and the on-shell conditions

$$\hat{P}_{12}^2 = 0 \rightarrow z_{12} = \frac{-P_{12}^2}{2\eta \cdot P_{12}}\,, \tag{28}$$

$$\hat{P}_{45}^2 = 0 \rightarrow z_{45} = \frac{P_{45}^2}{2\eta \cdot P_{45}}\,. \tag{29}$$

As every term is a rational function in η, we are allowed to maintain this equation when η is changed into N, the space-cone gauge vector. Let us accept the statement that the vertices do not change upon the shifts in momenta as described. We see that localizing z to different zeros of the kinematical invariants $\hat{P}_{12}^2$ and $\hat{P}_{45}^2$ is to factorize the amplitude into on-shell sub-amplitudes.

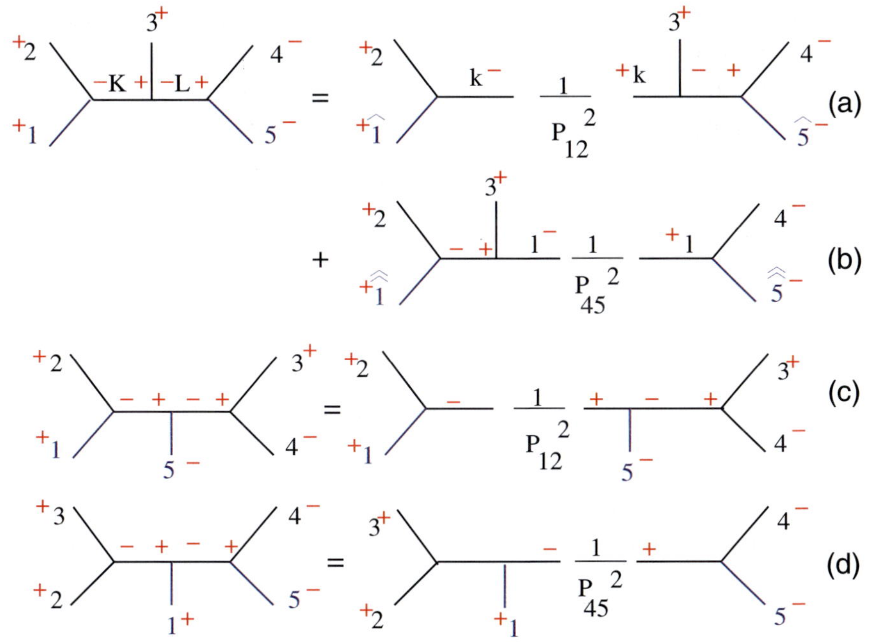

Fig. 1. Factorization of the five-point amplitude.

We have thus a recursion relation, which expresses a five-point amplitude into a sum of products of three- and four-point on-shell amplitudes.

Furthermore, we see that the above result is also easily obtained, if we define

$$\hat{A}(z) = \frac{1}{\hat{P}_{12}^2}\frac{1}{\hat{P}_{45}^2}\,, \tag{30}$$

and perform the integral

$$\oint \frac{dz}{z}\hat{A}(z) = 0 \tag{31}$$

over a closed contour in the complex a-plane. The physical amplitude is $\hat{A}(z = 0)$, which is the left-hand side, which is one of the poles in the integral. It is also given as the sum of the residues due to the other two poles of $\hat{A}(z)$.

5. Gauge Invariance

When we use the "on-shell" method to perform a calculation, we must be aware that the "effective vertices" are complexly continued lower point amplitudes. They are made on-shell, but they depend on the reference spinors $|+\rangle$, $[-|$. The final result, i.e. the physical amplitudes with real momenta, should be independent of any of such choice which is made for expediency. We also recall that fixing N is a choice of gauge and therefore the independency on N is tantamount to gauge invariance.

We explore this further in its infinitesimal form. Suppose we first make a choice

$$|+\rangle\,, \qquad [-|\,, \qquad [-+] = 1\,, \tag{32}$$

and then decide to make a small change

$$|+'\rangle = |+\rangle\,, \qquad |-'| = n([-| + \delta a[x|)\,, \tag{33}$$

in which n is a normalization factor and $[x|$ is at this point some spinor. However, when we normalize $[-'+'] = 1$, we find that $[x| = [+|$ is the only solution and therefore

$$[-'| = [-| + \delta a[+|\,. \tag{34}$$

The requirement of gauge invariance is that physical amplitudes with real external momenta should be independent of δa.

We learn from experience that gauge invariance imposes very stringent conditions on physical amplitudes. If there are several diagrams for a process, gauge invariance relates them in some mysterious way to cause tremendous amounts of cancellations to yield a simple result. Even though the "on-shell" method saves plenty of unnecessary labor, the cancellations are still incomplete.

Let us look at one example. We analyze a one-loop calculation of $P_1^- P_2^+ P_3^+ P_4^+$, with the choice

$$|+\rangle = |p_1\rangle\,, \qquad [-| = [p_3|\,. \tag{35}$$

We would like to make a remark with regard to complexificaton by shifting momenta. In order to reveal all the poles in the kinematical invariants that we are interested in, which transcribe into poles in the z-plane, we must choose shifts and reference spinors properly. For this example, one set of shifts is

$$[\hat{1}| = [1| + z[24][-|\,, \qquad |\hat{1}\rangle = |1\rangle\,, \tag{36}$$

$$[\hat{2}| = [2|\,, \qquad |\hat{2}\rangle - |2\rangle + z[4-]||1\rangle\,, \tag{37}$$

$$[\hat{3}| = [3|\,, \qquad |\hat{3}\rangle = |3\rangle\,, \tag{38}$$

$$[\hat{4}| = [4|\,, \qquad |\hat{4}\rangle = |4\rangle + z[-2]|1\rangle\,, \tag{39}$$

which preserve overall energy–momentum conservation. We find that there are three diagrams which make up this process, two of which are one particle reducible (1PR) and are easy to obtain and one is irreducible and requires some hard calculation.
From consideration of its collinear behavior, one can show that

$$A_4 = A_{4s}^{\mathrm{1PR}} f_s + A_{4u}^{\mathrm{1PR}} f_u\,. \tag{40}$$

When we make an infinitesimal gauge change, we have

$$0 = \delta A_4 = [(\delta A_{4s}^{\mathrm{1PR}}) f_s + A_{4s}^{\mathrm{1PR}}(\delta f_s)] + [(\delta A_{4u}^{\mathrm{1PR}}) f_u + A_{4u}^{\mathrm{1PR}}(\delta f_u)]\,. \tag{41}$$

The pole structure in $s = -(P_1 + P_2)^2$ and $u = -(P_1 + P_4)^2$ dictates that each pair of parentheses should vanish

$$\frac{\delta f_s}{f_s} = -\frac{\delta A_{4s}^{\mathrm{1PR}}}{A_{4s}^{\mathrm{1PR}}}\,, \qquad \frac{\delta f_u}{f_u} = -\frac{\delta A_{4u}^{\mathrm{1PR}}}{A_{4u}^{\mathrm{1PR}}}\,. \tag{42}$$

The right-hand sides are known and we can solve these equations to yield

$$f_s = \frac{-t}{u}\,, \qquad f_u = \frac{-t}{s}\,, \qquad t = -(s+u)\,. \tag{43}$$

We have a recursion relation, which connects four-point amplitudes to three-point amplitudes. f's are known as soft factors, and were postulated in Ref. 6.

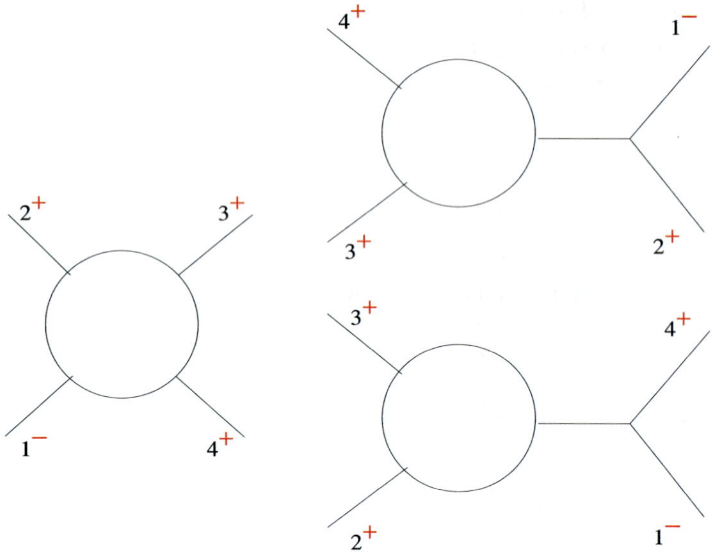

Fig. 2. The four-point one-loop $(+++-)$ amplitude.

This line of reasoning can be used for the evaluation of one-loop amplitudes for n gluons with all but one having the same helicity. Also, there are recursion relations for the one-loop gluon amplitudes with all $+$ or all $-$ helicity. They do not need any soft factors.[6,16]

6. Concluding Remarks

There is still much to be uncovered in non-Abelian gauge theories. We have used the freedom of choice in $|+\rangle$, $[-|$ and complex momentum shifts to explore the analyticity of its scattering amplitudes. The analysis is further simplified and augmented by the use of space-cone gauge. Much more can be and needs to be done.

There have been fruitful exchanges and inspiration between non-Abelian theories and higher dimensional conformal field theories and strings, particularly in calculational aspects of scattering amplitudes.[2–4,17,18] Through these infusions, one may even gain a better understanding of the strong coupling limit.

We have discussed those amplitudes which are rational functions of spinor products. For more complicated helicity arrangements, there is a lot of new developments, such as spinor/twistor integrations, generalized unitarity, loop integral evaluations, etc. They can only enrich the tool box.

All these are of interest to many of us, because of the relevance of non-Abelian fields to real physics, a tribute to Professor Yang's deep insight some five decades ago.

References

1. Z. Bern, L. J. Dixon, D. C. Dunbar and D. A. Kosower, *Nucl. Phys. B* **435**, 59 (1995).
2. E. Witten, *Commun. Math. Phys.* **252**, 189 (2004).
3. F. Cachazo, P. Svrcek and E. Witten, *JHEP* **0409**, 006 (2004).
4. R. Britto, F. Cachazo and B. Feng, *Nucl. Phys. B* **715**, 499 (2005).
5. R. Britto, F. Cachazo, B. Feng and E. Witten, *Phys. Rev. Lett.* **94**, 181602 (2005).
6. Z. Bern, L. J. Dixon and D. A. Kosower, *Phys. Rev. D* **71**, 105013 (2005).
7. F. A. Berends, R. Kleiss, P. De Causmaecker, R. Gastmans and T. T. Wu, *Phys. Lett. B* **103**, 124 (1981).
8. Z. Xu, D.-H. Zhang and L. Chang, *Nucl. Phys. B* **291**, 392 (1987).
9. S. Parke and T. Taylor, *Phys. Rev. Lett.* **56**, 2450 (1986).
10. G. Chalmers and W. Siegel, *Phys. Rev. D* **59**, 045013 (1999).
11. M. J. G. Veltman, *Physica* **29**, 186 (1963).
12. J. Schwinger, *J. Math. Phys.* **2**, 407 (1961).
13. K.-C. Chou, Z.-B. Su, B.-L. Hao and L. Yu, *Phys. Rep.* **118**, Nos. 1 and 2 (1988).
14. R. F. Streater and A. S. Wightman, *PCT, Spin & Statics and All That*, The Mathematical Physics Monograph Series (W. A. Benjamin, 1964).
15. D. Vaman and Y. P. Yao, *JHEP* **0604**, 030 (2006).
16. D. Vaman and Y. P. Yao, On-shell QCD recurrence relations and the space-cone gauge, in *42nd Rencontres de Moriond*, La Thuile, Italy, 10–24 March 2007, QCD Electronic Proc.: http://events.lal.in2p3.fr/Moriond/PDF_PROC_QCD/vaman%20-%20copie.pdf
17. J. M. Maldacena, *Adv. Theor. Math. Phys.* **2**, 231 (1998) [*Int. J. Theor. Phys.* **38**, 1113 (1999)].
18. L. F. Alday and J. M. Maldacena, *JHEP* **0706**, 064 (2007).

NEUTRINO OSCILLATION AND THE DAYA BAY θ_{13} EXPERIMENT

BING-LIN YOUNG

Department of Physics and Astronomy,
Iowa State University, Ames Iowa 50010, USA
and
Institute of Theoretical Physics,
Chinese Academy of Sciences, Beijing, China

A brief summary of the current status of neutrino oscillations will be given. Then the on-going construction of the Daya Bay Reaction Neutrino Experiment near the Daya Bay nuclear power plant is sketched. The Daya Bay experiment will measure the mixing angle θ_{13} to the level of $\sin^2 2\theta_{13} = 0.01$.

Keyword: Daya Bay neutrino experiment; neutrino oscillation; θ_{13} measurement.

1. Introduction

Neutrino oscillation is one of the most important discoveries and a very significant advancement in fundamental science in the last century. It all began in 1930 with the proposal of a ghostly particle by Pauli.[a] After the experimental observation[1] of neutrino in 1956, Pontecorvo proposed first the possibility of neutrino–antineutrino mixing[2,3] in 1957–1958, and then the neutrino flavor mixing[4] in 1968 after the observation of a second species of neutrino, i.e. the muon neutrino[5] in 1962. A more complete description of this early history of the neutrino oscillation can be found in Ref. 6. The early works are mostly theoretical development.[b] Mixing of states which cause the transition of one state to another is a common quantum mechanical

[a]On December 4 1930 Wolfang Pauli wrote his now famous letter to suggest the existence of a neutrino. The letter can be found in
"http://www.lapp.in2p3.fr/neutrinos/aplettre.html".

[b]Other earlier pivotal theoretical developments are the formal "official" proposal of the neutrino by Pauli at the 1933 7th Solvay Conference and Fermi's four-fermion theory of β-decay also related to the 1933 Solvay Conference. See Ref. 7 for a detailed description.

phenomenon. It requires that the masses of the states that mix are not degenerate. For neutrinos this provides a very important piece of physics, i.e. some or all species of neutrinos have nonvanishing mass.

The experimental establishment of neutrino oscillation has been made in a number of experiments with various neutrino sources during the final decade of the last century. A wealth of information on many aspects of neutrinos and neutrino oscillations can be found at the website of Neutrino Oscillation Industry.[c]

We can summarize the current status of neutrino oscillation as follows:

- Neutrino oscillation has been demonstrated in many experiments, of the various types, with both cosmic and terrestrial neutrino sources from the sun, atmosphere, reactors and accelerators. Several types of detector have been used. A concise review can be found in Ref. 8.
- The experimental data available to date, which are limited to the accuracy of the dominant mixing effect, can be understood in terms of theoretical frameworks of vacuum oscillations and adiabatic conversion in matter.[9]
- The large mixing angles (at least two of the three) and small masses of neutrinos are in stark contrast to the mixing and mass patterns of quarks, posting an interesting challenge to theoretical model building.
- The advancement of our knowledge of neutrino system have greatly expanded our tools of study in physics and astrophysics. It gives birth to the so-called neutrino astronomy. Neutrino detectors, located deep underground and in ocean, are used as neutrino telescopes to probe regions of stars and the cosmos that are not accessible to the electromagnetical radiation. Several experiments are in progress. A list of neutrino telescopes can be found on the internet[c] and some recent status is summarized in footnote d.
- Overall, even after the active development of more than a decade, this is still an experimentally driven area in which some important questions have to be answered with more precise experimentation.

The felicitous testimony of the importance of neutrino study is the fact that there have been to date three Nobel prizes to their credit, once every seven years since 1988: 1988, 1995 and 2002. Figure 1 shows the Nobel laureates and the citations for their prizes.

---

[c]The Neutrino Oscillation Industry: http://www.hep.anl.gov/ndk/hypertext/.

[d]*XII Int. Workshop on "Neutrino Telescopes"*, Venice, Italy, March 16–19 2007, http://neutrino.pd.infn.it/conference2007/talks.html.

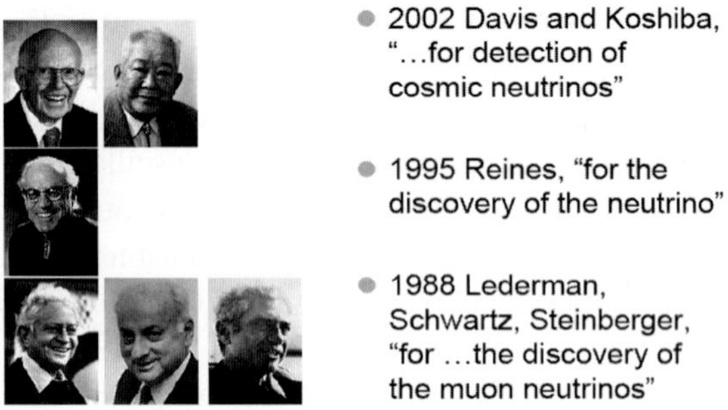

Fig. 1. Laureates of three sets of Nobel Prize, due to their works in neutrinos, together with the citations.

2. Description of the Neutrino System

All existing data can be understood in the framework of mixing of the three flavors of light neutrinos of the Standard Model, ν_e, ν_μ, ν_τ and their antiparticles. Since the quantum mechanical time evolution is defined in terms of particle states of definite masses, the dynamics of the system is describable by a 3×3 mass matrix which relates the flavor states ν_e, ν_μ and ν_τ to the mass eigenstates ν_1, ν_2 and ν_3. The most general case of neutrinos is that they are Majorana particles, which leads to a mass matrix consisting of nine parameters: three masses which are denoted as m_1, m_2 and m_3; three mixing angles θ_{12}, θ_{23} and θ_{13}; and three CP-violation phases δ_1, δ_2 and δ_3. Oscillation experiments, however, can only determine the following six parameters:

- Two mass-squared differences, e.g.

$$\Delta m^2_{21} \equiv m^2_2 - m^2_1 \,, \qquad \Delta m^2_{31} \equiv m^2_3 - m^2_1 \,. \tag{1}$$

  The third mass-squared difference is not independent, $\Delta m^2_{32} = m^2_3 - m^2_2 = \Delta m^2_{31} - \Delta m^2_{21}$.
- All the three mixing angles, θ_{12}, θ_{23} and θ_{13}.
- And one CP phase which is usually referred to as the Dirac CP phase which we denote as $\delta_{\rm D}$.

The 3×3 mixing matrix, denoted as $U = (U_{\alpha j})$ and generally referred to as the Pontecorvo–Maki–Nakagawa–Sakata mixing matrix, transforms the mass states to the flavor states:

$$\begin{pmatrix} \nu_e \\ \nu_\mu \\ \nu_\tau \end{pmatrix} = \begin{pmatrix} U_{e1} & U_{e2} & U_{e3} \\ U_{\mu 1} & U_{\mu 2} & U_{\mu 3} \\ U_{\tau 1} & U_{\tau 2} & U_{\tau 3} \end{pmatrix} \begin{pmatrix} \nu_1 \\ \nu_2 \\ \nu_3 \end{pmatrix} , \tag{2}$$

where U can be parametrized as, up to a factor consisting of Majorana phases,

$$U = \begin{pmatrix} \cos\theta_{12} & \sin\theta_{12} & 0 \\ -\sin\theta_{12} & \cos\theta_{12} & 0 \\ 0 & 0 & 1 \end{pmatrix} \times \begin{pmatrix} \cos\theta_{13} & 0 & \sin\theta_{13}e^{-i\delta_{\rm D}} \\ 0 & 1 & 0 \\ -\sin\theta_{13}e^{i\delta_{\rm D}} & 0 & \cos\theta_{13} \end{pmatrix}$$
$$\times \begin{pmatrix} 1 & 0 & 0 \\ 0 & \cos\theta_{23} & \sin\theta_{23} \\ 0 & -\sin\theta_{23} & \cos\theta_{23} \end{pmatrix}. \quad (3)$$

The neutrino masses themselves and two more Majorana phases cannot be determined in oscillation experiments. Information on neutrino masses, however, can be obtained from astrophysical observations and tritium beta-decay experiments. Information on Majorana phases can be extracted from neutrinoless double beta-decays.

In a two-neutrino approximation of flavor states ν_α and ν_β, and mass states ν_1 and ν_2, an oscillation experiment in the vacuum measures the oscillation probability of an appearance experiment, $\nu_\alpha \to \nu_\beta$,

$$P^{(\text{appearance})}_{\nu_\alpha \to \nu_\beta} = \sin^2 2\theta \sin^2\left(1.267\Delta m^2(\text{eV}^2)\frac{L(\text{km(m)})}{E_\nu(\text{GeV(MeV)})}\right), \quad (4)$$

or a survival experiment $\nu_\alpha \to \nu_\alpha$,

$$P^{(\text{survival})}_{\nu_\alpha \to \nu_\alpha} = 1 - \sin^2 2\theta \sin^2\left(1.267\Delta m^2(\text{eV}^2)\frac{L(\text{km(m)})}{E_\nu(\text{GeV(MeV)})}\right). \quad (5)$$

θ is the mixing angle, L is the baseline which is the distance that the neutrino travels between the points of its production and detection, E_ν is the neutrino energy and $\Delta m^2 = m_2^2 - m_1^2$ is the mass-squared difference of the mass states. As indicted L and E_ν are in units of km and GeV, or m and MeV. Δm^2 is in eV^2.

3. Summary of the Current Experimental Status

Observations of neutrino oscillations have been made in experiments with four kinds of neutrino source: atmospherical neutrinos (ν_μ, $\bar{\nu}_\mu$), solar neutrinos (ν_e), reactor neutrinos ($\bar{\nu}_e$) and accelerator neutrinos ($\nu_\mu, \bar{\nu}_\mu$). We briefly describe some of the key results of each neutrino source.

Atmospherical neutrinos ($\nu_\mu(\bar{\nu}_\mu) \to \nu_\tau(\bar{\nu}_\tau)$)

The atmospherical neutrino consists of ν_e, ν_μ and their antiparticles. From basic particle interactions, it can be predicted that the ratio of the total

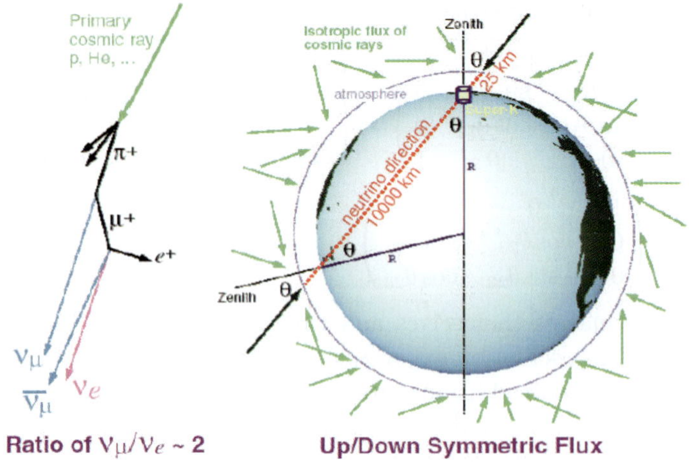

Fig. 2. Cartoon showing the production and detection of atmospheric neutrinos. The picture on the left shows the production of a ν_e together with a ν_μ and a $\bar{\nu}_\mu$. The picture on the right shows the different path lengths of the atmospheric neutrino that enters the Super-K detector at different zenith angles.

numbers of the muon (ν_μ plus $\bar{\nu}_\mu$) and the electron (ν_e or $\bar{\nu}_e$) flavored species is 2:1. A cartoon of the production of the neutrinos in the atmospheric and their detection in the Super-Kamiokande underground laboratory in Japan is shown in Fig. 2.[e] Earlier observations by IMB and Kamiokande collaborations showed that the detected neutrinos of the muon species are depleted in comparison with the electron species. This was referred to as the atmospherical neutrino anomaly. The key result was made by the Super-K collaboration and announced in 1998. The data showed convincingly that this anomaly is not due to experimental artifacts, but actually because of the disappearance of some of the muon flavors on their way from production in the atmosphere to the detector underground. It is the first strong evidence, called the smoking gun, of neutrino oscillation.[10]

Figure 3 shows the Super-K smoking gun. It compares data with the theoretical expectation of no-oscillation as a function of the neutrino azimuthal angle which can be converted into the neutrino baseline. While the

[e]Many figures such as Fig. 2 as a visual aid to the description of neutrino oscillation can be found online. Figure 2 is take from the document entitled "wyklad3-sources[1].pdf" which can be found in a University of Warsaw website:
http://neutrino.fuw.edu.pl/public/wyklad-From-neutrinos/. However I cannot decide if this is where Fig. 2 originated from. The documents in this website contains many such useful graphical illustrations.

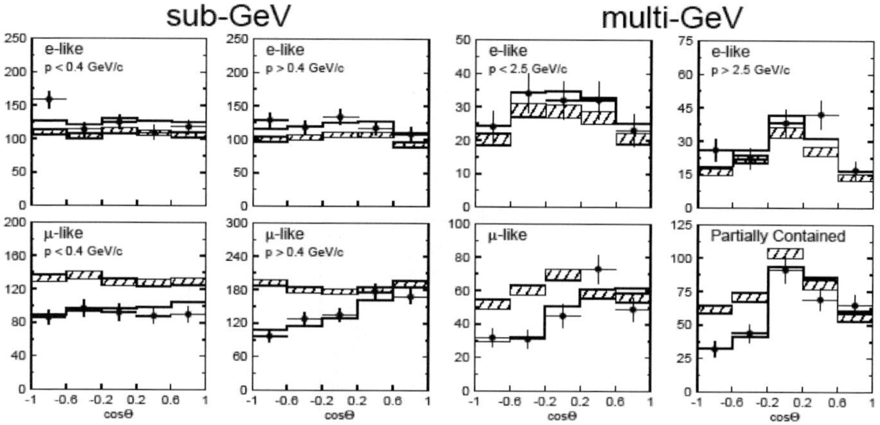

Fig. 3. The Super-K smoking gun data. The upper four panels are for the electron neutrino events and the lower ones are for the muon neutrino events. The data are shown in points with error bars and the theoretical nonoscillation expectation is shown as shaded rectangles. For the electron neutrino data, the observation and the prediction generally agree. But for the muon neutrino, the observation is depleted and the depletion generally increases with the muon neutrino baseline.

electron neutrino data agree with the theoretical expectation, the muon neutrino data are depleted. The extend of the depletion increases with the azimuthal angle, i.e. the baseline. This feature of the muon neutrino data is a characteristics of oscillation. The explanation of the depletion is the conversion of some of the muon neutrinos into some other species, i.e. the tau neutrino, which is not observable in the experiment. The atmospherical neutrino experiment determines the mixing angle θ_{23} and the mass-squared difference Δm^2_{31}.

Solar neutrino ($\nu_e \to \nu_\mu, \nu_\tau$)

The indication of oscillations in the solar neutrino appeared as early as 1968 in the chlorine experiment by Davis,[11] but its implication was not discerned at that time. Davis experiment was carried out in the Homestake mine.[12] A consistent result of solar neutrino experiments is the missing of a significant number of the detected solar neutrinos (ν_e) in comparison with the prediction of the standard solar model.[13] This is known as the solar neutrino puzzle. High statistics experiments were performed by the Super-K collaboration and the Sudbury Neutrino Observation (SNO). The SNO experiment in Canada is sensitive to both charged and neutral current events. The neutral current events which involve all flavors of Standard

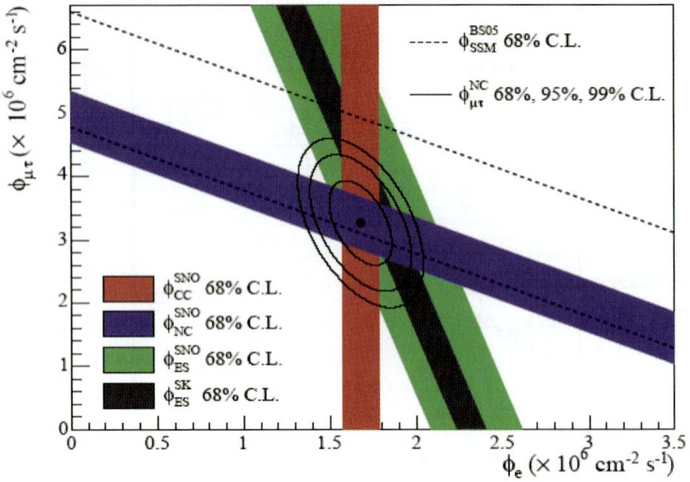

Fig. 4. The SNO and Super-K data summary plot for the ^8B solar neutrino flux. The horizontal axis is the ν_e flux and the vertical axis the sum of the flux of $\nu_\mu + \nu_\tau$. CC stands for *charge current*, NC for *neutral current*, ES for *elastic scatter* and SSM for *standard solar model*. The Super-K data is marked as $\phi_{\rm ES}^{\rm SK}$.

Model neutrinos account for the total solar neutrino events expected from the standard solar model. This verifies that the missing solar neutrinos are converted into different flavors of the Standard Model neutrinos, which is again an evidence of neutrino oscillations. The SNO data also confirm the so-called large mixing angle solution, which is one of the solutions of the Super-K data, of the solar neutrino puzzle. As an added important consequence, the SNO experiment also confirms the Standard Model of the sun. Figure 4 shows a collection of the SNO data.[14] The relevant Super-K data which are also shown in the plot can be found in Ref. 15. Solar neutrino data determine the mixing angle θ_{12} and mass-squared difference Δm_{21}^2.

Reactor experiments ($\bar{\nu}_e \to \bar{\nu}_\mu, \bar{\nu}_\tau$)

A number of reactor neutrino experiments have been performed. The beam consists of electron antineutrinos produced in a nuclear reactor and detected at either a short baseline of several tens of meter to a few km, or at a medium distance of (many) tens of km. We mention two short baseline experimental collaborations: Chooz[16,17] in France and Palo Verde[18] in USA. These experiments are suitable for the determination of the mixing angle θ_{13} by monitoring some tiny deletion of $\bar{\nu}_e$'s shown in the detector. Neither

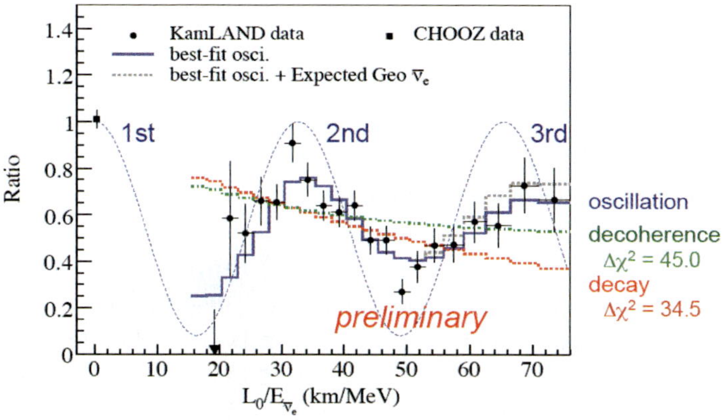

Fig. 5. A fit of reactor neutrino data to the vacuum oscillation.[20] The second and third maxima are covered by the KamLAND data.

of these collaborations has seen a signal. The Chooz collaboration gives an upper limit of the often quoted limit on θ_{13}: $\sin^2 2\theta_{13} \le 0.13$.[f] We will come back to these later when we describe the Daya Bay reactor experiment.

The medium baseline reactor experiment of KamLAND in Japan with the $\bar{\nu}_e$ beam is complementary to the solar neutrino experiment of very long baseline of ν_e beam. The KamLAND data[19,20] help further restrict the solar neutrino parameters and also found an oscillation signature. A similar oscillation signature has been found in the Super-K data.[21] A recent fit of the reactor data[20] from Chooz and KamLAND as a function of L/E_ν in vacuum oscillation is shown in Fig. 5. The Chooz data give the very short baseline beginning point and the KamLAND data provide a significant range of L/E_ν (see Eqs. (4) and (5)) which covers almost two complete cycles of oscillations including the second and third maxima. This is the best oscillation fit to date. As shown in Fig. 5 the oscillation fit rules out some alternative explanations of the solar neutrino puzzle: neutrino decay and decoherence.

Accelerator neutrino experiments $(\nu_\mu(\bar{\nu}_\mu) \to \nu_\tau(\bar{\nu}_\tau), \nu_e(\bar{\nu}_e))$

Accelerator neutrino experiments will be increasingly important in the future for precision determination of neutrino parameters. An accelerator

[f] The limit on $\sin^2 2\theta$ depends on the value of the relevant Δm^2 which, in the present case, is Δm^2_{31}. The limit quoted in Ref. 16 which presents the result of the final analysis of the Chooz data is $\sin^2 2\theta_{13} \le 0.10$.

neutrino beam consists mostly of ν_μ and $\bar{\nu}_\mu$ with a small admixture of ν_e and $\bar{\nu}_e$. The experiments that currently provide data for oscillation analysis are K2K[22] in Japan and MINOS[23] in USA. K2K is the first long baseline experiment at 250 km baseline. MINOS operates at a longer baseline of 730 km. K2K has finished its running. The MINOS data give the most recent contribution of accelerator data to the newer analysis. Data from both collaborations strengthen the results of the atmospheric neutrino experiment. An important accelerator program in the future is the verification of oscillation by the appearance experiment $\nu_\mu(\bar{\nu}_\mu) \to \nu_e(\bar{\nu}_e)$.

Another accelerator experiment is the MiniBooNE collaboration at Fermilab which has just completed its analysis.[24] The MiniBooNE experiment was designed to check the long controversial result of the LSND collaboration. The LSND data found a neutrino mass-squared difference of the order of $\Delta m^2 \sim 1$ eV^2. This much higher mass scale requires the existence of a heavier neutrino that is not contained in the Standard Model, generally referred to as the sterile neutrino. With much better statistics than the LSND experiment, the MiniBooNE has observed no LSND type of events. The LSND result was first published more than a decade ago[25] and had incited wide interests in the search for exotic neutrinos and the study of their possible theoretical structure. It should be noted that although the LSND-type sterile neutrino is now refuted, the MiniBooNE result does not rule out the existence of exotic neutrinos in general. Undoubtedly search of exotic neutrinos will be continued. Exotic, sterile neutrinos with energy-dependent mixing angles and masses, and possible those with other possible unconventional properties are compatible with the MiniBooNE data and they are very intriguing possibilities of new physics.[26]

Summary of information from oscillation experiments

Because of the different orders of magnitude of Δm^2_{31} and Δm^2_{21}, fits of various experiments for oscillation parameters are usually done in the framework of two-flavor approximation. From the muon flavored neutrinos in atmospheric and accelerator neutrino experiments, Δm^2_{31} and θ_{23} are extracted, while for the electron flavored in solar and reactor neutrino experiments, Δm^2_{21} and θ_{12} are obtained. Short baseline electron (anti)neutrino can lead directly to information on θ_{13}. However, global fits in the framework of three flavors that include different types of data have been performed and continuously updated in the past several years. We quote in Table 1 the results of a recent global fit given in Ref. 26. It should be noted that θ_{23} is nearly maximal, i.e. 45° and θ_{12} is not maximal. Δm^2_{21}

Table 1. Global fit of three-flavor neutrino oscillation parameters including atmospheric, solar, reactor, and accelerator data.[26]

Parameter	Best fit	2σ	3σ
$\Delta m^2_{21}(10^{-5}\text{ eV}^2)$	7.6	7.3–8.1	7.1–8.3
$\|\Delta m^2_{31}\|(10^{-3}\text{ eV}^2)$	2.4	2.1–2.7	2.0–2.8
$\sin^2\theta_{12}$	0.32	0.28–0.37	0.26–0.40
$\sin^2\theta_{23}$	0.50	0.38–0.63	0.34–0.67
$\sin^2\theta_{13}(\sin^2 2\theta_{23})$	0.007(0.028)	$\leq 0.033(0.13)$	$\leq 0.05(0.19)$

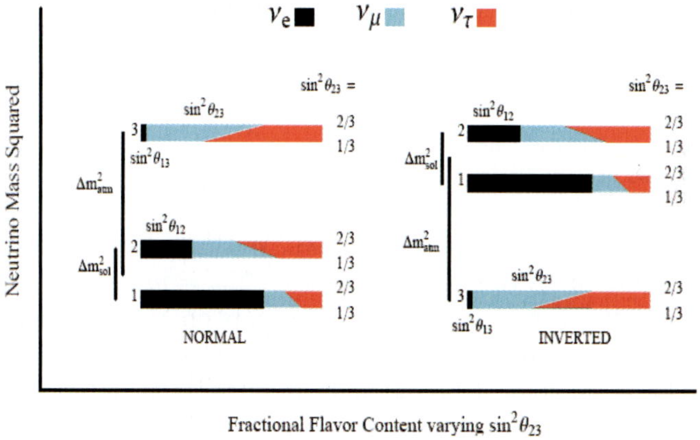

Fig. 6. Flavor contents of neutrino mass states shown with varying $\sin^2\theta_{23}$. The diagram on the left is for the normal spectrum and the diagram on the right is for the inverted spectrum.[27]

is determined for both its magnitude and sign, while for Δm^2_{31} only the magnitude is determined and the sign is not.

Presently, out of the six parameters, $\sin^2\theta_{12}$, $\sin^2\theta_{12}$, Δm^2_{21} and $|\Delta m^2_{31}|$ are known to within 10–20%, and θ_{13} has an upper limit, $\sin^2 2\theta_{13} \leq 0.13$, as given by the Chooz limit. The unknowns are the sign of Δm^2_{31}, the CP violation phase δ_{D} and the direct observation of the matter effect.

Because the sign of Δm^2_{31} is not known, there are two possibilities for the spectrum of the three neutrinos: the normal spectrum for $\Delta m^2_{31} > 0$ and the inverted spectrum for $\Delta m^2_{31} < 0$. In Fig. 6 the normal and inverted spectra are shown together with the flavor compositions of the neutrino mass eigenstates.[27]

It is interesting to note that although the oscillation experiments are not sensitive to individual masses of the neutrinos, cosmological observations

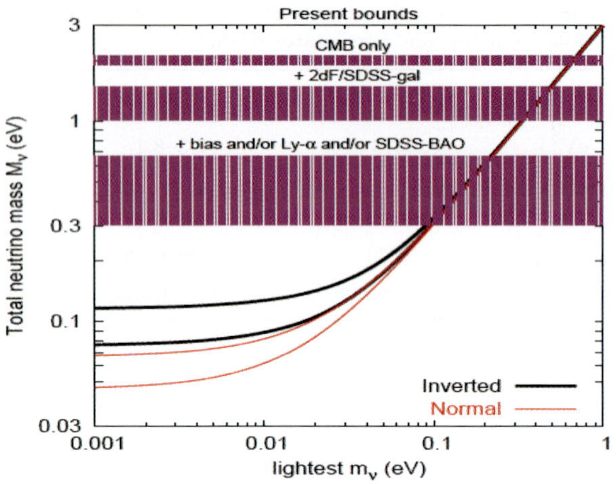

Fig. 7. Bounds on the total neutrino masses $m_1 + m_2 + m_3$ from cosmological observations which also provide limits on the lightest neutrino mass.[29]

can help to set bounds on neutrino masses. With the present observational accuracy the sum of all the neutrino masses, $\Sigma_m = m_1 + m_2 + m_3$, can be bounded.[28,29] The current cosmological bounds on Σ_m is shown in Fig. 7. Σ_m is also shown in this figure as a function of the mass of the lightest neutrino. The cosmological bound indicates that the individual neutrino masses are likely to be in the sub-eV region.

4. Implications and Future Study of Massive Neutrinos

Massive neutrinos have potentially far reaching implications in physics and astrophysics. To elucidate the detailed structure of the neutrino systems, which has just been revealed, requires more detailed study. We list some of the obvious implications below:

- Broadly speaking, the massive neutrino is the first concrete evidence of physics beyond the Standard Model.
- The large mixing angles (two of the three mixing angles) are in vivid contrast to the mixing pattern of the quark system, providing a constraining guide to theoretical model building of a grand unifying theory, providing both a challenge and an opportunity.
- The smallness of the neutrino masses, of the order eV or sub-eV, forms a distinctive group of fundamental particles of the Standard Model away from the rest of massive particles. We know now the Standard Model spectrum extending to no less than 13 orders of magnitude.

Given the electroweak symmetry breaking scale of $\Lambda_{\rm SM} \sim 100$ GeV, the very small neutrino mass implies the existence of a high energy scale $\Lambda_{\rm NEW} \equiv \Lambda_{\rm SM}^2/m_\nu \sim 10^{14}$ GeV. This will open a window to physics at this new high energy scale.

The 2004 Theory of Neutrino Report[30] of the American Physical Society (APS) posted a number of key questions and comments concerning the neutrino system. We quote in parallel phrasing some of the questions and comments:

- What are implications of massive neutrinos for such long standing ideas as grand unification, supersymmetry, extra dimension, Majorana versus Dirac fermions, etc.?
- What are implications of the possible existence of additional neutrino species for physics and cosmology?
- Whether neutrinos have anything to do with the observed matter–antimatter asymmetry in the universe, if so, whether there is any way to determine it via low energy experiments.
- Knowing neutrino properties in detail may also play a crucial role in clarifying the blue print of new physics laws beyond those embodied in the Standard Model.

Concerning the programs of future neutrino study, the APS Neutrino Study Group (NSG)[31] and the U.S. Neutrino Scientific Assessment Group (NuSAG)[32] suggest the precision measurement of all neutrino oscillation parameters through a series of experiments, including:

- Measure θ_{13} to 3°, i.e. $\sin^2 2\theta_{13}$ to 0.01.
- Determine the sign of Δm^2_{31} and the matter effect, and measure the CP phase $\delta_{\rm D}$.
- Determine all other parameters to the accuracy of a few %.
- Search for exotic properties and exotic species of neutrinos.
- Study neutrinoless double beta decay and the individual masses of neutrinos.
- Improve cosmological bound on neutrino masses.

5. A Near Term Neutrino Program

For future neutrino programs, NSG and NuSAG recommended, in particular, as a near term priority:[31,32]

An expeditiously deployed multidetector reactor experiment with sensitivity to $\bar{\nu}_e$ *down to* $\sin^2 2\theta_{13} \approx 0.01$*, an order of magnitude below present limits.*

Motivations of measuring θ_{13}

There are clear reasons for measuring θ_{13} to the above suggested accuracy as an initial step to implement the long-term strategy of the neutrino program:

- θ_{13} is referred to as the gateway to the lepton CP violation. In oscillation experiment, the Dirac CP phase $\delta_{\rm D}$ appears in the so-called Jarlskog invariant $J(\delta_{\rm D})$, with the mixing matrix parametrized as in Eq. (3):

$$J(\delta_{\rm D}) = \frac{1}{8} \cos\theta_{13} \sin 2\theta_{13} \sin 2\theta_{12} \sin 2\theta_{23} \sin \delta_{\rm D} \,. \tag{6}$$

  Since θ_{12} is sizable and θ_{23} is near maximal, the effect of CP violation depends crucially on the size of θ_{13}. A very small θ_{13} will make lepton CP-violation measurement difficult and, under such a situation, it calls for new experimental technology and new neutrino beam facilities, and the future neutrino program will be more challenging.
- The value of θ_{13} has interesting theoretical implications. A vanishing or very small θ_{13} may imply a special new symmetry in the lepton sector.
- θ_{13} plays an important role in model building for the construction of a neutrino mass matrix. A recent tabulation of predictions on $\sin^2 2\theta_{13}$ from 63 neutrino mass models is given in Ref. 33. Its graphic summary of the various model predictions on $\sin^2\theta_{13}$ is reproduced here in Fig. 8. As can be seen from Fig. 8, a determination of the value of θ_{13} or even a tighter bound, say, the limit of $\sin^2 2\theta_{13} = 0.01$, as suggested by NSG and NuSAG, can rule out a significant number of the existing models.

What is known about θ_{13}*?*

As stated earlier, the present knowledge on $\sin^2 2\theta_{13}$ is the Chooz limit.[16,17] In addition, there are global fits of oscillation parameters which also give the likelihood value of $\sin^2\theta_{13}$. Recent global fits of $\sin^2\theta_{13}$ is shown in Fig. 9. The directly measurable quantity in a short baseline reactor experiment is $\sin^2 2\theta_{13} \approx 4\sin^2\theta_{13}$, which can be read off from Fig. 9. Let us remark that the global fit for $\sin^2\theta_{13}$ has started a few years back and the likelihood value has been changing. Needless to say, a direct experimental determination is necessary.

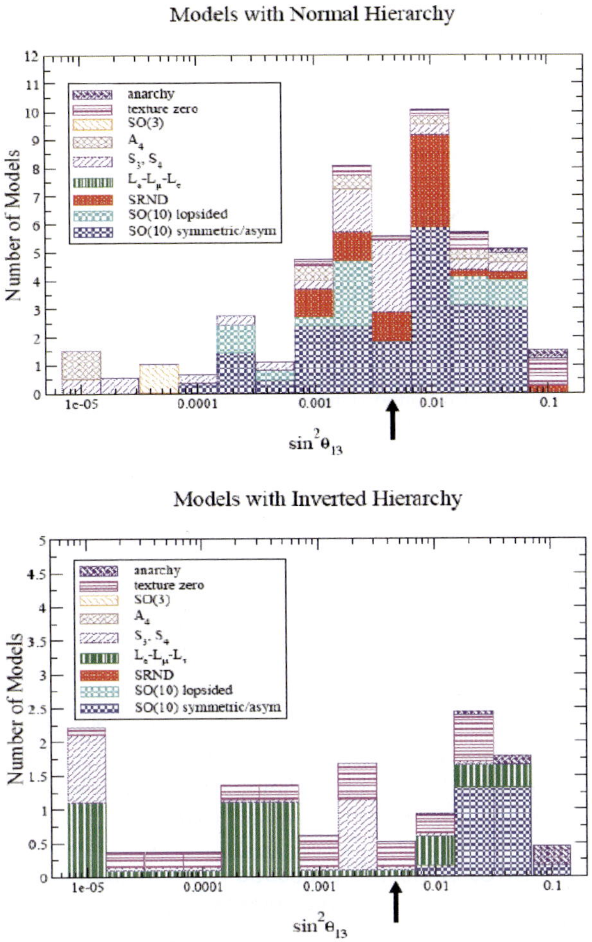

Fig. 8. Model predictions of $\sin^2\theta_{13}$.[33]

Reactor θ_{13} experiments worldwide

The APS and NuSAG recommendation of measuring $\sin^2 2\theta_{13}$ to 0.01 is a high precision experiment, which has to satisfy a set of stringent requirements on the selection of the experimental site and the design of the detector(s), although the complications of the construction of a accelerator is avoided. The experimental site has to be near a high power reactor to provide intensive beam of neutrinos. The site should also be near a mountain range, under which an underground laboratory can be constructed to have sufficient shield from cosmic ray so as to reduce the muon background.

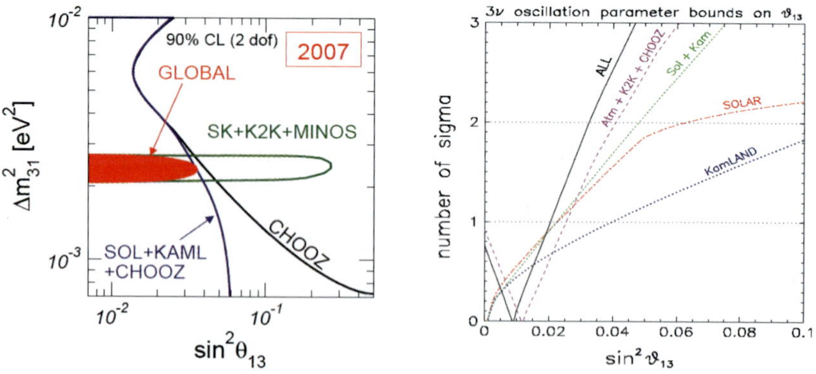

Fig. 9. (Left): Experimental bound (Chooz) and global fits of $\sin^2\theta_{13}$ vs Δm^2_{31} as given in Ref. 26. The maximal value of the global fit is somewhat smaller than the Chooz bound. (Right): Global fits of $\sin^2\theta_{13}$ and the number of σ with fixed $\Delta m^2_{31} = 2.4 \times 10^{-3}$ eV^2 as given in Ref. 34. The minimum of the "All" fit is $\sin^2 2\theta_{13} = 0.04$.

To reach the required accuracy, two sets of identical detector, one near and one far, are necessary, because of the insufficient knowledge of the neutrino beam intensity of a nuclear reactor. The near detector is sufficiently close to the neutrino source to monitor the reactor neutrino intensity, and the far detector located at a suitable distance from the neutrino source measures the possible change in the neutrino beam it receives. Several possible sites over the globe were proposed in the past decade. Now four sites have been actively pursued (see Fig. 10): Double Chooz in France, Daya Bay in China,

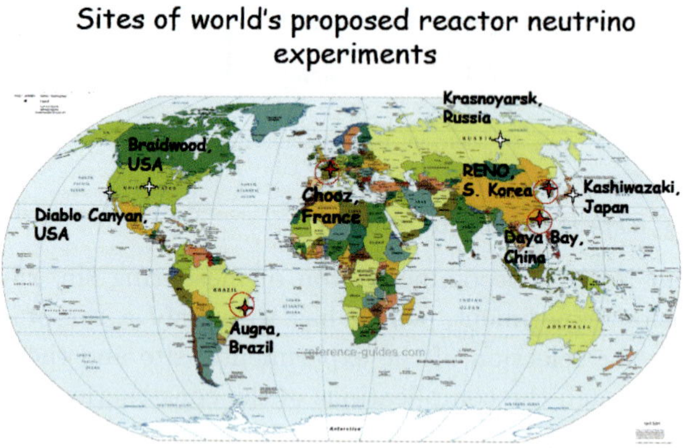

Fig. 10. World's proposed sites for reactor neutrino experiment to measure θ_{13} in the past decade. There were totally eight sites. The four active sites, Chooz, Daya Bay, RENO and Angra (denoted by circled solid stars) are under various stages of development.

RENO in South Korea and Angra in Brazil. The detectors of Double Chooz and Daya Bay are under construction. RENO is in the R&D phase. The detector in Angra is a proposal.

6. The Daya Bay Reactor Neutrino Experiment

The goal of the Daya Bay experiment is to measure $\sin^2 2\theta_{13}$ to the accuracy of 1% in a multidetector deployment. The physics and technical details, including the detector requirement and design, events and background simulations, the geological and topographical structure of the experimental site, physical parameters of laboratory, tunnel and laboratory construction, etc., are given in Ref. 35. The experimental setup can also be used as an early warning system for supernova events with the unique feature of concidence among its three sets of detectors.[36] In the following we present a sketch of some of the salient points.[g]

Antineutrino source

The experimental site is near a nuclear power plant complex on the Dapeng peninsula in south China's Guangdong province, about 50 km northeast of Hong Kong. The power plant complex consists of two nuclear power plants (NPP's), one called the Daya Bay NPP and the other Ling Ao NPP. Each NPP has two reactor cores of 2.9 $\mathrm{GW_{th}}$ thermal power. So the total thermal power of the complex is 11.6 $\mathrm{GW_{th}}$. This makes the Daya Bay–Ling Ao complex the 12th most powerful NPP in the world. Note that a 1 $\mathrm{GW_{th}}$ thermal power output produces about 2×10^{20} $\bar{\nu}_e$/s. The total neutrino intensity of the complex is 2.3×10^{21} $\bar{\nu}_e$/s. A third NPP, called Ling Ao-II, is under construction and is to be completed in 2011. The total thermal power of the enlarged complex will be 17.4 $\mathrm{GW_{th}}$, making it the world's 5th most powerful reactor complex with a total neutrino intensity of the order of 3.5×10^{21} $\bar{\nu}_e$/s.

Experimental site

A photo of the Daya Bay–Ling Ao complex is shown in Fig. 11. The adjacent mountains, which are mostly hard rocks, are suitable for the construction of underground tunnels and laboratories with sufficient overburden to cut down significantly the cosmic rays background.

[g]Figures which appear below in this section are taken from documents and presentations of the Daya Bay Collaboration.

Fig. 11. The Daya Bay–Ling Ao nuclear power plant complex. Daya Bay is shown on the left with two (yellow) buildings. Ling Ao, visible in this photo, shown with some light colored buildings, is to the right of the near mountain range on the left and in front of the mountain further back. Ling Ao-II under construction is further beyond and not visible in this photo.

Since the Daya Bay–Ling Ao complex consists of multiple reactor cores, the detector and laboratory layout is more elaborate. There are two near detector sets, one called the Daya Bay near detector set and the other the Ling Ao near detector set (including Ling Ao-II). The near detector sets are close to their respective reactors for monitoring primarily the neutrino fluxes from them. The far detector set receives antineutrinos from

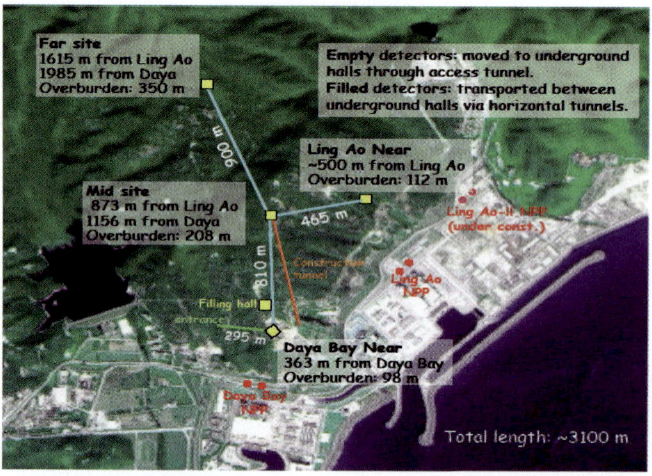

Fig. 12. The experimental halls and tunnel layout. The Daya Bay, Ling Ao and Ling Ao-II reactor cores are denoted as (red) dots, and the detector halls as (yellow) squares. The relevant dimensions noted are: lengths of tunnel sections; distances of the three detector halls to the NPP's; and overburdens of the three detector halls. The total length of the tunnel is about 3100 m. The dark (blue) region on the lower right is the ocean. The dark (blue) area on the left is a reservoir.

both Daya Bay and Ling Ao. Under the given topographical configuration, the optimal detector positions are determined by detailed simulations. Figure 12, which is constructed from an aerial photograph of the NNP complex and nearby mountains, shows the detector positions relative to the NPP's. The baseline and overburden parameters are also shown.

Suppression of the cosmic-ray muons

A crucial factor in reaching the accuracy of the Daya Bay experiment is the control of background. Hence it is important to reduce the cosmic muon events reaching the detectors. This requires significant overburdens at all three sets of detector. The near detectors which have a much higher neutrino flux can tolerate more cosmic events than the far detectors. This fact reflects in the requirement of smaller overburden on the near detectors than on the far detectors. The overburdens of the three detector sites are given in Fig. 13. The overburden figures shown on the left of Fig. 13 are in meters of rocks. The chart shown on the right, where the horizontal axis is the overburden in meters of water equivalent (MWE), compares the overburdens of the Daya Bay detectors with underground laboratories worldwide.

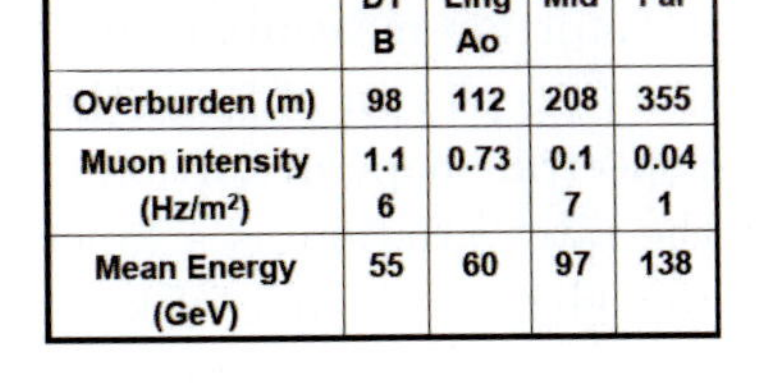

	DYB	Ling Ao	Mid	Far
Overburden (m)	98	112	208	355
Muon intensity (Hz/m²)	1.16	0.73	0.17	0.041
Mean Energy (GeV)	55	60	97	138

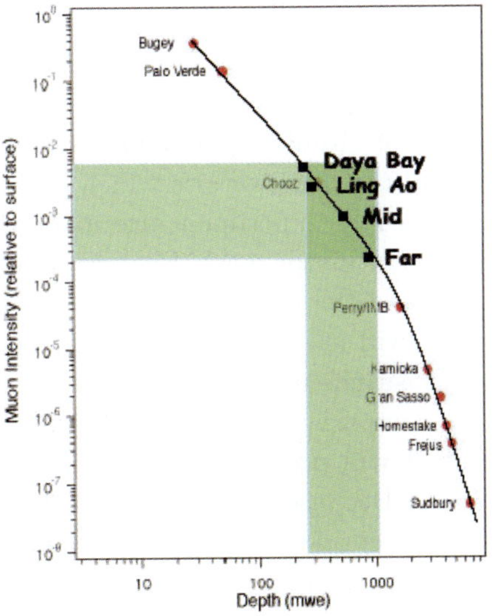

Fig. 13. Details of the overburdens of the three sets of detector of the Daya Bay experiment. Note that the overburden figures on the left are in meters of rock, while the overburden in the chart on the right, shown with other underground laboratories in the world, are listed in meters of water equivalent (MWE).

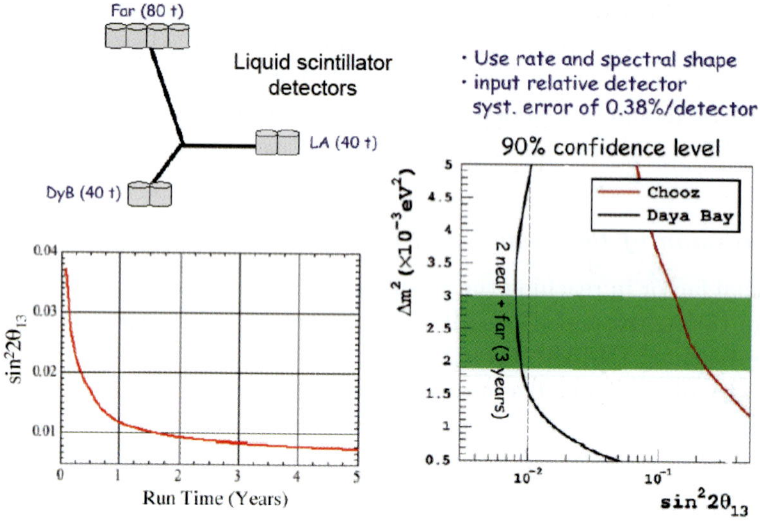

Fig. 14. The limit of $\sin^2 2\theta_{13}$ that is expected to be reached and the running time for the Daya Bay experiment. As a comparison the three year limit for the Double Chooz experiment is also shown in the chart on the right.

Sensitivity of $\sin^2 2\theta_{13}$

Given the thermal power of the NPP complex, the sensitivity of the measurement depends on several factors. In addition to the background suppression of cosmic-ray muons stressed earlier, it depends on the running time and the detector size which determine the statistics. The detectors of the Daya Bay experiment are made of liquid scintillator modules, 20 ton each in scintillator mass. Each near site contains two such modules of total 40 ton, and the far site consists of four modules of total 80 ton. The data analysis will use both the rate and spectral distortion measurements. The rate measurement is to determine if there is any depletion of $\bar{\nu}_e$ at the far site in comparison with the near sites. The spectral shape distortion measurement will provide more details about the rate variation as a function of L/E_ν. Because of the very short baseline which is a couple of km, the matter effect on the neutrino beam is completely negligible. Figure 14 shows the detector configuration (upper left) and the expected limit of the value of $\sin^2 2\theta_{13}$ to be reached as a function of the run time (lower left). The chart on the right shows the expected limits of $\sin^2 2\theta_{13}$ versus the relevant mass-squared difference, which is Δm^2_{31} in the present case, in three years running for both the Daya Bay and the Double Chooz experiments.

The collaboration

The Daya Bay experiment is an international collaboration, currently involving 18 institutions and universities in Asia, 14 in North America and 3 in Europe. The total number of collaborators in the latest count is 215. A list of participating institutions and an incomplete list of collaborators can be found in Ref. 35.

Plans and milestones

The current status, milestone and future plans:

- Passed scientific reviews: China, April 2006 and U.S., October 2006.
- Approval of project in China: Chinese Academy of Sciences, April 2006 and Ministry of Science and Technology, January 2007.
- Passed U.S. CD-1 review: April 2007.
- Passed China's nuclear safety review: April 2007.
- Began three-year project funding from China: April 2007.
- Began civil construction: October 2007.
- Ground-breaking ceremony for underground laboratory: October 13, 2007.
- U.S. CD 2/3a: January 2008.
- Data taking with two detectors at Daya Bay Near Hall: July 2009.
- Data taking with all eight detectors: September 2010.

Acknowledgments

It is a great honor and pleasure for me to join so many distinguished colleagues to celebrate Prof. C. N. Yang's 85th birthday. The subject of my talk concerns neutrinos. Their development and elucidation have been fundamentally influenced by Prof. Yang's work. I would like to take this opportunity to wish Prof. Yang a very happy birthday and many happy returns.

I would like to thank the organizers of the conference, in particularly Professor K. K. Phua, for inviting me to the conference and for the kind hospitality extended to me.

References

1. C. L. Cowan, F. Reines, F. B. Harrison, H. W. Kruse and A. A. McGuire, *Science* **124**, 103 (1956).
2. B. Pontecorvo, *J. Exp. Theor. Phys.* **33**, 549 (1957) [*Sov. Phys. JETP* **6**, 429 (1958)].

3. B. Pontecorvo, *J. Exp. Theor. Phys.* **34**, 247 (1958) [*Sov. Phys. JETP* **7**, 172 (1958)].
4. B. Pontecorvo, *J. Exp. Theor. Phys.* **53**, 1717 (1977) [*Sov. Phys. JETP* **26**, 984 (1968)].
5. G. Danby, J. M. Gaillard, K. Goulianos, L. M. Lederman, N. Mistry, M. Schwartz and J. Steinberger, *Phys. Rev. Lett.* **9**, 36 (1962).
6. S. M. Bilenky, *Phys. Scripta T* **121**, 17 (2005), arXiv:hep-ph/0410090.
7. L. M. Brown and H. Rechenberg, *The Origin of the Concept of Nuclear Forces* (Inst. of Physics Pub. Inc., 1998).
8. M. D. Messier, *Proc. 4th Flavor Physics and CP Violation Conference* (*FPCP2006*), Vancouvor, British Columbia, Canada, 9–12 April 2006, *eConf* **C060409**, 018 (2006), arXiv:hep-ex/0606013.
9. A. Yu. Smirnov, *Proc. IPM School and Conf. on Lepton and Hadron Physics* (*IPM-LHP06*), Tehran, Iran, 15–20 May 2006, ed. Y. Farzan, *eConf* **C0605151** (2006), arXiv:hep-ph/0702061.
10. Super-Kamiokande (Y. Fukuda *et al.*), *Phys. Rev. Lett.* **81**, 1562 (1998), arXiv:hep-ex/9807003.
11. Homestake-Chlorine (R. Davis, D. S. Harmer and K. C. Hoffman), *Phys. Rev. Lett.* **20**, 1205 (1968).
12. Homestake-Chlorine (B. T. Cleveland *et al.*), *Astrophys. J.* **496**, 505 (1998).
13. J. N. Bahcall, A. M. Serenelli and S. Basu, *Astrophys. J.* **621**, L85 (2005), arXiv:astro-ph/0412440.
14. SNO (B. Aharmin *et al.*), *Phys. Rev. C* **72**, 055502 (2005), arXiv:nucl-ex/0502021.
15. Super-Kamiokande (S. Fukuda *et al.*), *Phys. Lett. B* **539**, 179 (2002), arXiv:hep-ex/0205075.
16. Chooz (M. Apollonio *et al.*), *Phys. Lett. B* **420**, 397 (1998), arXiv:hep-ex/9711002.
17. Chooz (M. Apollonio *et al.*), *Eur. Phys. J. C* **27**, 331 (2003), arXiv:hep-ex/0301017.
18. Palo Verde (F. Boehm *et al.*), *Phys. Rev. D* **64**, 112001 (2001), arXiv:hep-ex/0107009.
19. KamLAND (T. Araki *et al.*), *Phys. Rev. Lett.* **94**, 081801 (2005), arXiv:hep-ex/0406035.
20. KamLAND (I. Shimuzu), Talk at the *10th Int. Conf. on Topics in Astroparticle and Underground Physics* (*TAUP2007*), http://www.awa.tohoku.ac.jp/taup2007/.
21. Super-Kamiokande (Y. Ashie *et al.*), *Phys. Rev. Lett.* **93**, 101801 (2004), arXiv:hep-ex/0404034.
22. K2K (M. H. Ahn *et al.*), *Phys. Rev. D* **74**, 072003 (2006), arXiv:hep-ex/0606032.
23. MINOS (P. Adamson *et al.*), A study of muon neutrino disappearance using the Fermilab main injector neutrino beam, arXiv:0711.0769.
24. MiniBooNE (A. A. Aguilar-Arevalo *et al.*), *Phys. Rev. Lett.* **98**, 231801 (2007), arXiv:0704.1500.
25. LSND Collab. (C. Anhanassopoulos *et al.*), *Phys. Rev. Lett.* **75**, 2650 (1995).

26. Schwetz, LSND versus MiniBooNE: Sterile neutrinos with energy dependent masses and mixing angles, arXiv:0710.2985.
27. O. Mena and S. J. Parke, *Phys. Rev. D* **69**, 117301 (2004), arXiv:hep-ph/0312131.
28. W. Hu, E. J. Eisenstein and M. Tegmar, *Phys. Rev. Lett.* **80**, 5255 (1998), arXiv:astro-ph/0712057.
29. J. Lesgourgues and S. Pastor, *Phys. Rep.* **429**, 307 (2006), arXiv:astro-ph/0603494.
30. R. N. Mohapatra *et al.*, *Rep. Prog. Phys.* **70**, 1757 (2007), arXiv:hep-ph/0510213.
31. APS Multidivisional Neutrino Study, The neutino matrix, arXiv:physics/0411216.
32. NuSAG Final Report: Recommendations to the Department of Energy and the National Science Foundation on a Future US Program in Neutrino Oscillations, July 13, 2007, http://www.er.doe.gov/hep/hepap_reports.shtm.
33. C. H. Albright and M.-C. Chen, *Phys. Rev. D* **74**, 113006 (2006), arXiv:hep-ph/0608137.
34. G. L. Fogli, E. Lisi, A. Marrone and A. Palazzo, *Prog. Part. Nucl. Phys.* **57**, 742 (2006), arXiv:hep-ph/0506083.
35. X.-H. Guo *et al.*, A precision measurement of the neutrino mixing angle θ_{13} using reactor antineutrinos at Daya Bay, arXiv:hep-ex/0701029.
36. X.-H. Guo and B.-L. Young, *Phys. Rev. D* **73**, 093003 (2006), arXiv:hep-ph/0605122.

SCATTERING AND PRODUCTION AT HIGH ENERGIES

TAI TSUN WU

Harvard University
Cambridge, MA 02138, U.S.A.
and
CERN, Geneva, Switzerland

A summary is given of some of the developments in the elastic scattering and production processes at high energies. First, Professor Yang's geometrical model of hadronic and nuclear collisions is reviewed. When there was some preliminary experimental evidence that the total cross section for proton–proton collision might not approach a constant at very high energies, theoretical studies were initiated to determine the high-energy asymptotic behavior in relativistic quantum gauge theory. With this starting point, the successive stages of development are:

I. Theoretical prediction of increasing total cross sections;
II. Development of phenomenology and quantitative predictions that were verified experimentally afterwards; and
III. Theory and phenomenology of production processes, especially that of the Higgs particle, at the Large Hadron Collider.

Since the Large Hadron Collider (LHC) is still being built, the last topic is at an early stage of development, and unexpected results may be forthcoming.

Keyword: Scattering, production, high energy, cross section, Yang–Mills non-Abelian gauge theory, phenomenology, Large Hadron Collider, Higgs particle.

1. Introduction

Professor Yang has opened up numerous important fields in physics. The best known one is:[1,2]

Yang–Mills non-Abelian gauge theory.

This is the basis of our present understanding of just about everything in particle physics, including both the strong and electroweak interactions.

In Ref. 2, the authors stated explicitly that they had "not been able to conclude anything about the mass of the b quanta." This turned out to be prophetic: now we know that the Yang–Mills particle (the gluon) for the strong interactions is massless, while those (the W and Z) for the electroweak interactions are massive.

During the Symposium in 1999 on the occasion of Professor Yang's retirement from the State University of New York at Stony Brook, I gave a talk on the Yang–Mills theory.[3] On this happy occasion of the 85th birthday celebration of Professor Yang, I would like to discuss another field opened up by Professor Yang:

High-energy hadron–hadron collisions.

2. Geometrical Model of Hadronic Collisions

Starting in 1965, Professor Yang has published over twenty papers on this topic. Some of his collaborators are: J. Benecke, N. Byers, A. W. Chao, T. T. Chou, T. T. Wu, and E. Yen.

The first paper,[4] published in 1965, is actually on high-energy large momentum transfer processes. The starting points for this paper are the following experimental facts:

(i) The total proton–proton cross section remains essentially constant at high energies;
(ii) above 300 MeV of excitation energy the nucleon has many excited states; and
(iii) the large-angle elastic proton–proton cross section drops down spectacularly with energy.

These facts suggest picturing the nucleon as an extended object, and that the difficulty in making large momentum transfers is due to the *difficulty in accelerating the various parts of the nucleon without breaking it up*.

Shortly thereafter, Professor Yang realized that this picture of the nucleon should have a small-momentum aspect in addition to the large-momentum aspect emphasized in this paper. For this reason, Professor Yang has considered Ref. 4 to be a precursor which launched a geometrical model of hadronic and nuclear collisions, p. 60 of Ref. 5.

This geometrical picture by Professor Yang has led to numerous predictions. In particular, it has been found[6] that, on the basis of this geometrical picture, for very-high-energy collisions the lab system (L) and the projectile system (P, where the incoming projectile is at rest) are to be preferred over the center-of-mass system, because in these systems, some of the outgoing particles approach limiting distributions.

As an example, the following limit exists:

$$\lim_{E\to\infty}\left(\begin{array}{l}\text{partial cross section that a particle of mass } m_1 \text{ and momentum}\\ \vec{p}_1, \text{and a particle of mass } m_2 \text{ and momentum } \vec{p}_2 \text{ are emitted,}\\ \text{together with any number of other particles.}\end{array}\right)$$

$$= \rho_2(\vec{p}_1, \vec{p}_2)\, d^3p_1\, d^3p_2, \tag{1}$$

where E is the center-of-mass energy, and $\rho_2 > 0$.

In particular, for the total cross section and elastic differential cross section,

$$\sigma_0 = \lim_{E\to\infty} \sigma_{\text{tot}} \tag{2}$$

and

$$\Sigma(t) = \lim_{E\to\infty} \left.\frac{d\sigma}{dt}\right|_{\text{elastic}}, \tag{3}$$

both exist in the limit of very high energies.

3. Do the Limiting Values Exist?

Let us consider Eqs. (2) and (3) in more detail. Of these two, Eq. (2) is more important. Because the total cross section is the sum of the contributions from all the sub-processes, this total cross section is one of the best candidates for having a limiting value. Furthermore, Eq. (2) is closely related to the special case $t = 0$ of Eq. (3).

Shortly after Professor Yang developed the geometrical picture of hadron–hadron collisions, the question was asked: What are the experimental evidences for or against the existence of a limiting value for the total cross section? At that time, above the center-of-mass energy of about 6 GeV, the measured proton–proton total cross section was barely changing, decreasing slightly. The proton–antiproton total cross section was decreasing more rapidly, but approaching that of proton–proton. The π^+p total cross section is also nearly constant, but that for K^+p is increasing significantly.

In particle physics, it is often useful to question long-held beliefs. Such inquiry is especially likely to lead to progress when there are some, even preliminary, experimental data giving counter-indications to the belief.

Is there any other quantity that is sensitive to the issue whether the total cross sections approach limiting values or not? One of the most promising quantities is the ratio of the real and the imaginary parts of the forward scattering amplitude. The reasoning is as follows. The forward scattering amplitude is expected to satisfy a dispersion relation. This means that the real part of the forward scattering amplitude can be written as an integral over the imaginary part, which is essentially the total cross section. To avoid questions of normalization, it is usual to consider the ratio

$$\rho = \frac{\text{Real part of } pp \to pp \text{ in the forward direction}}{\text{Imaginary part of } pp \to pp \text{ in the forward direction}} \tag{4}$$

in the case of proton–proton collisions.

If the proton–proton total cross section does approach a finite limit, then this ratio ρ must approach zero at high energies. At that time, the experimentally measured values of ρ were negative and increasing toward zero.

However, the rate of increase seemed to be a little too fast, i.e., this ρ seemed to have a tendency to overshoot to become positive. This is a preliminary indication; if this indication is taken seriously, then it is desirable to study whether a high-energy limit does exist for the proton–proton total cross section.

4. Relativistic Quantum Gauge Theory

How could we study such a problem theoretically? Clearly we need a model. How do we choose a model for this purpose?

Interactions between elementary particles share the following basic features:

- relativistic kinematics,
- unitarity, and
- particle production,

in addition to the space–time dimension being $3 + 1$.

Even though each of these three basic features may be considered to be "trivial," it is nevertheless not easy to have a model with all these features. The simplest way to have these features is to have a relativistic quantum field theory. It remains to make a choice which relativistic quantum theory to study.

As already mentioned in Sec. 1, Professor Yang has taught us that all interactions in particle physics are described by gauge theories, Abelian or Yang–Mills (non-Abelian).[1,2] Therefore, in order to study the high-energy behavior of various quantities, including especially the total cross sections, we chose the model of four-dimensional relativistic quantum field theory with a gauge invariance of the second kind.

There is no known example of such a theory that is exactly solvable. For this reason, it is necessary to study the perturbation series of such a theory at high energies, renormalized when needed.

On the basis of this model, the problem is to calculate asymptotically, order by order, each term of the perturbation series. Here "asymptotically" means the leading term when the ratio

$$\frac{E}{m} \quad \text{is large,} \tag{5}$$

where m is the mass of the incident particles, typically that of a proton. After these asymptotic terms are obtained for each order, these leading terms are summed over the order. This procedure will be illustrated in the next section.

The summation of these leading terms, carried out by Cheng and me,[7,8] turned out to require a rather complicated and lengthy calculation. In the context of the high-energy behavior of the relativistic quantum gauge theory, the argument given at the end of Sec. 3 above is supplemented by the following observation and question. In any relativistic quantum field theory, there is not only direct channel (s-channel) unitarity, but also cross channel (t-channel) unitarity. How does the cross channel unitarity affect the high-energy behavior of, for example, the total cross section?

5. Theoretical Prediction of Increasing Total Cross Sections

For definiteness, consider fermion–fermion elastic scattering in a relativistic quantum gauge theory. As a first step in trying to understand the high-energy behavior of such a theory, consider the exchange of Yang–Mills gauge particles directly between the two fermions. The Feynman diagrams for second order, fourth order, and sixth order are shown in Figs. 1(a), 1(b) and 1(c), respectively, while the general diagram is given schematically in Fig. 1(d). Actually, there is another set of diagrams where the two outgoing fermions are exchanged. In each of this second set of diagrams, the Yang–Mills gauge particles have to carry a large momentum, and hence the contribution at high energies is negligible.

For high energies and fixed momentum transfers, the diagrams of Fig. 1 have been evaluated. When the gauge particle is massless, then there is the usual divergence in the total cross section due to the one-"photon" exchange of Fig. 1(a). If the gauge particle is not massless, for example through the Higgs mechanism,[9] then this divergence disappears. Such trivial divergences are ignored in the following discussions.

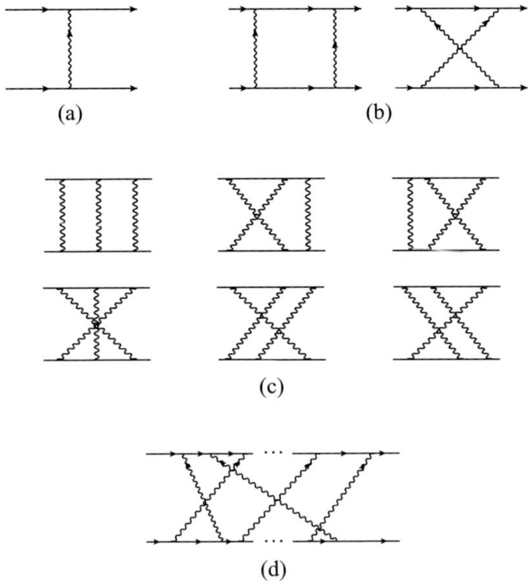

Fig. 1. Feynman diagrams for the direct exchange of Yang–Mills gauge particles between two fermions. (a) second order; (b) fourth order; (c) sixth order; (d) general.

By the way, the name Higgs mechanism should more accurately be called the Englert–Brout–Higgs (or EBH) mechanism, because the paper of Englert and Brout[10] was actually submitted for publication four weeks earlier than that of Higgs.[9] In the rest of this paper, to be fair to Englert and Brout, the term EBH or EBH particle shall be used.

All results from the diagrams of Fig. 1 are in perfect agreement with the geometrical picture of Professor Yang. In particular, the existence of the high-energy limit is verified for $\sigma_{\rm tot}$ and $(d\sigma/dt)|_{\rm elastic}$, as given by Eqs. (2) and (3).

In this verification of the geometrical picture of Professor Yang, two of the three basic features listed in Sec. 4 on the interaction between elementary particles have played central roles: they are relativistic kinematics and unitarity. The question to be raised at this stage is thus: What is the role of particle production in the total cross section at very high energies?

The high-energy behaviors of a large number of production processes have been investigated. It turned out that the simple diagram of Fig. 2 plays a crucial role; it involves the production of a single Yang–Mills gauge particle. Because of the importance of this process, it may be of interest to discuss it in some detail.

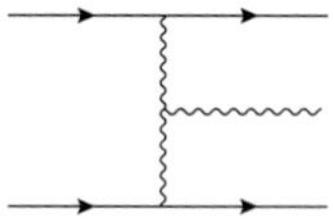

Fig. 2. A Feynman diagram for the production of a single Yang–Mills gauge particle.

In a high-energy collision of elementary particles, additional particles may be produced with a wide range of energies and momenta. For example, a produced particle may have a relatively low momentum in the projectile system; see Sec. 2. At high energies, such a particle necessarily has a high momentum in the so-called lab system, i.e., where the other incoming particle is at rest. Besides the produced particles that have relatively low momenta in the projectile or lab system, there may be others that have relatively high momenta in both these systems; for example, they may have relatively not high momenta in the center-of-mass system. In cosmic ray events, particles satisfying this description have been seen; they are mostly pions and are therefore referred to as pionization products. In this language, the Yang–Mills gauge particle produced as shown in Fig. 2 is a pionization product. It is known[8] that the integrated cross section for the production process shown in Fig. 2 is proportional to the available rapidity range, which is of the order of $\ln(E/m)$.

This is the first clear evidence of an increasing partial cross section, and hence an increasing total cross section. This does not mean that the total cross section should increase as $\ln(E/m)$; the actual situation is much more interesting.

Since the appearance of one pionization product leads to an extra factor of $\ln(E/m)$ in the integrated cross section, the appearance of two independent pionization products is expected to lead to $[\ln(E/m)]^2$. This is indeed the case, and the relevant diagram is shown in Fig. 3(a). Similarly, the diagram of Fig. 3(b) gives an integrated cross section that increases as $[\ln(E/m)]^3$.

But such an increase is impossible because it violates the bound of Froissart[11] and Martin.[12] Therefore, factors such as $[\ln(E/m)]^3$, or more precisely a positive power of E/m when these various powers of $\ln(E/m)$ from the diagrams of Figs. 2 and 3 are summed, must be regarded as representing a strongly absorptive potential. This shows the importance of particle production, the third basic feature listed at the beginning of Sec. 4.

The incident plane wave is a superposition of waves with the impact distance b ranging from zero to infinity. If b is small, the wave is prone

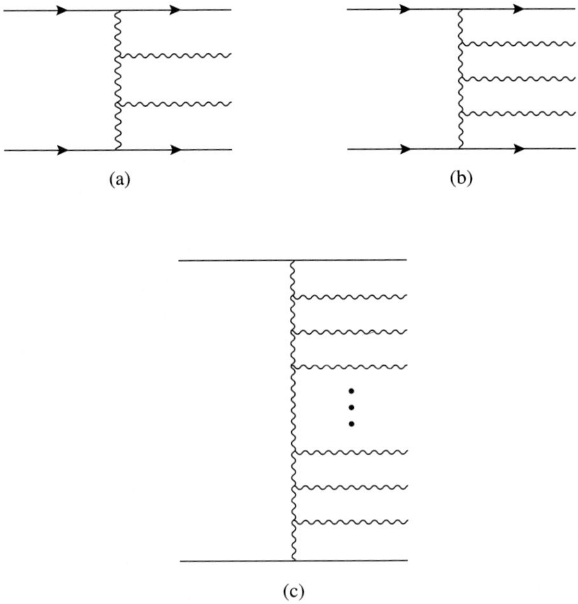

Fig. 3. Feynman diagrams for the production of (a) two, (b) three, and more generally (c) n Yang–Mills gauge particles.

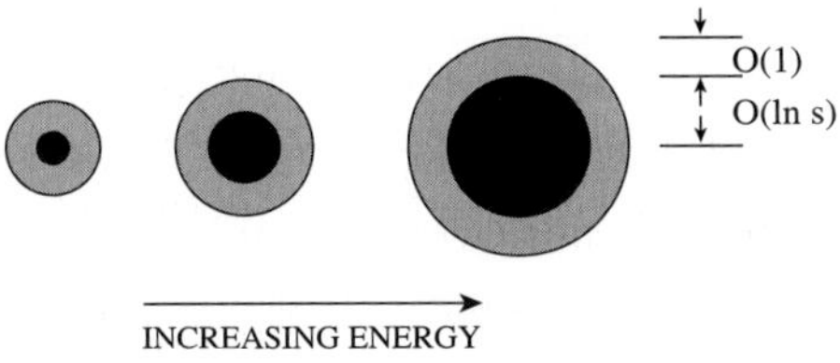

Fig. 4. Schematic representation of the appearance of a high-energy particle [$s = E^2$].

to create pionization products and is then lost to the beam. The resulting physical picture is shown schematically in Fig. 4, where the black core of nearly complete absorption and increasing radius is surrounded by a gray fringe of partial absorption.

6. Phenomenology and Predictions

Physics is basically an experimental science. This theoretical result of increasing cross sections *must be followed by quantitative predictions that can be verified or refuted by future experiments.*

In order to obtain such quantitative predictions, it is necessary to develop a phenomenological model which satisfies the following: (a) it must possess the properties obtained from relativistic quantum gauge theory for very, very high energies; and (b) it must be sufficiently realistic that direct comparison with experimental data is possible.

At very high energies, the elastic scattering amplitude is given approximately by the impact-distance representation

$$\mathcal{M}(s, \vec{\Delta}) \sim \frac{is}{2\pi} \int d\vec{x}_\perp \, e^{-i\vec{\Delta}\cdot\vec{x}_\perp}[1 - e^{-\Omega(s, x_\perp^2)}], \tag{6}$$

where $\vec{\Delta}$ is the momentum transfer and all spin variables have been omitted. In view of the increasing cross sections as discussed in Sec. 5, the absorption $\Omega(s, x_\perp^2)$, or more precisely the real part of $\Omega(s, x_\perp^2)$, should increase with increasing energy. A natural and simple choice for this $\Omega(s, x_\perp^2)$ is

$$\Omega(s, x_\perp^2) = \mathcal{S}(s)F(x_\perp^2). \tag{7}$$

As discussed in Sec. 4, relativistic quantum gauge theory is used to arrive at the conclusion that cross sections increase at high energies. It is therefore essential to choose this $\mathcal{S}(s)$ to be crossing symmetric. The particular choice used is

$$\mathcal{S}(s) = \frac{s^c}{(\ln s)^{c'}} + \frac{u^c}{(\ln u)^{c'}}, \tag{8}$$

where u is the third Mandelstam variable and c and c' are two constants to be determined. Note that this $\mathcal{S}(s)$ is complex.

This choice (8) is far from being straightforward and has a rather deep basis. A competing possibility is for the $\mathcal{S}(s)$ of Eq. (7) to be replaced by the Fourier transform of

$$\bar{\mathcal{S}}(s, t) = s^{\alpha(t)}, \tag{9}$$

or its crossing symmetric version. This form (9) is similar to the usual Regge pole, except that the singularity is above 1. In the Regge terminology, (9) is a moving Regge pole while (8) is a fixed cut. In order to choose between these two possibilities, among others, we appeal once more to relativistic quantum gauge field theory. On the basis of the high-energy behavior of such theory, it has been found that Eq. (8) holds but not Eq. (9), with or without any dependence on t. The underlying reason for Eq. (8) is that such gauge field theory in four dimensions is renormalizable but not super-renormalizable.

Table 1. Parameters for the Phenomenological Model.

Parameter	1978 value[13]	1984 value[14]
c	0.151	0.167
c'	0.756	0.748
m_1 (GeV)	0.619	0.586
m_2 (GeV)	1.587	1.704
a (GeV)	2.257	1.953
f (GeV^{-2})	8.125	7.115

For Abelian gauge theory, the value of the c' of Eq. (8) has been found to be

$$c' = \frac{3}{2} \tag{10}$$

to leading order; see Appendix B of Ref. 8. The corresponding value of c' has not been successfully calculated for Yang–Mills theory yet. It has been conjectured that, to leading order,

$$c' = \frac{3}{4} \tag{11}$$

for SU(3) Yang–Mills non-Abelian gauge field theory—quantum chromodynamics. See below.

We now turn to the choice of the $F(x_\perp^2)$ for phenomenology. In the first realistic phenomenology for pp and $p\bar{p}$ elastic scattering,[13] developed in 1978 and published in 1979 in collaboration with Bourrely and Soffer, the choice is that $F(x_\perp^2)$ is the Fourier transform of

$$\tilde{F}(t) = f[G(t)]^2 \frac{a^2 + t}{a^2 - t} \tag{12}$$

with

$$G(t) = \frac{1}{\left(1 - \dfrac{t}{m_1^2}\right)\left(1 - \dfrac{t}{m_2^2}\right)}. \tag{13}$$

Thus there are six parameters: c, c', m_1, m_2, a and f; these are determined by an overall fit using existing data at that time.

Six years later, when there were significantly more experimental data at high energies, Bourrely, Soffer and I repeated the overall fit.[14] The values of these parameters are listed in Table 1. That these values do not change much implies that this phenomenological model is quite robust.

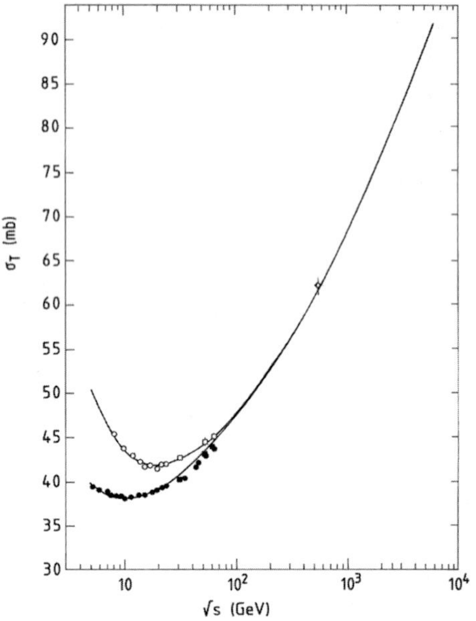

Fig. 5. Comparison of the result from the phenomenological model with the measured total cross sections for pp (black points) and $p\bar{p}$ (open points).

Of these six parameters, perhaps the most interesting one is the second one, c'. That this value, determined by an overall fit of the experimental data, is close to 0.75 is the basis for the conjecture (11).

Since the phenomenological formulas are very simple, numerous computations have been carried out using the 1984 values. For the lower part of the energy range, Regge backgrounds are also included. A few examples are given in the next three figures.

Figure 5 shows a comparison of the experimental data with the result of the phenomenological model for the pp and $p\bar{p}$ total cross sections. This figure is taken from Ref. 14, the same paper that gives the 1984 values of the parameter values in Table 1. The difference between the pp and $p\bar{p}$ total cross sections is due entirely to the Regge background. The predicted total pp cross section for the Large Hadron Collider, being built at CERN and to be described in the next section, is

$$\sigma_{\text{tot}} = 103.6 \text{ mb}. \tag{14}$$

This is more than twice the total pp cross section measured at the highest energy in the mid-sixties.

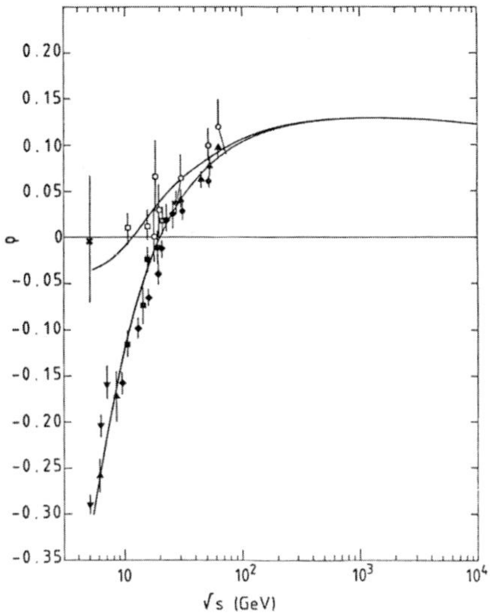

Fig. 6. The ratio ρ for pp (black points) and $p\bar{p}$ (open points) elastic scattering.

Figure 6 shows a similar comparison for the ratio of the real to the imaginary part of the proton–proton forward scattering amplitude; see the definition (4). In the mid-sixties, only the data below $\sqrt{s} \sim 10$ GeV were available. As seen from Fig. 6, these data points gave a negative value for ρ, but were increasing quite rapidly for increasing $\sqrt{s}$, with a tendency to become positive. This is the preliminary indication discussed in the last two paragraphs of Sec. 3.

For the Large Hadron Collider, the predicted value is

$$\rho = 0.122. \tag{15}$$

The phenomenology is in good agreement with proton–proton experimental data, not only for the total cross section but also for the elastic differential cross section. This pp elastic differential cross section has a dip somewhat above a momentum transfer of 1 GeV/c. Since the dip is due to the cancellation of various contributions, it gives a severe test to the phenomenology. Figure 7 shows such a comparison. Note the good agreement in the dip region. The depth of the dip in the phenomenology is intimately related to the fact that the crossing symmetric $\mathcal{S}(s)$, as defined by Eq. (8), is not real.

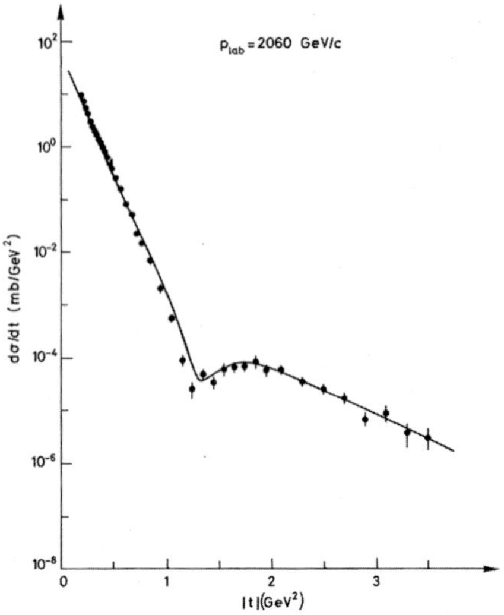

Fig. 7. Proton–proton differential cross section at $p_{\rm lab} = 2060$ GeV/c. Note the good agreement in the dip region.

7. Production Processes at the LHC

In the last two sections, we have presented results that are reasonably well understood. A more detailed summary for these results may be found in Ref. 15. In this section, some ongoing work is to be discussed.

So far we have been concerned almost exclusively with elastic scattering and directly related quantities such as the total cross section. That the total cross section is increasing with energy must have far-reaching consequences beyond elastic scattering, production processes being an obvious candidate. It is, however, believed that, for the effect of Sec. 5 to be important, even higher energies are required for production processes than for elastic scattering. Such higher energies have not been available so far from particle accelerators; this situation is expected to change in 2008 when the construction of the Large Hadron Collider (LHC) will be completed at CERN, Geneva, Switzerland.

The LHC is a proton–proton collider with two 7 TeV beams, i.e.,

$$\sqrt{s} = 14 \text{ TeV} = 14{,}000 \text{ GeV}. \tag{16}$$

The values of the total cross section and ρ in Eqs. (14) and (15) are both specifically for this energy. At LHC, there will be four experiments: ATLAS,

CMS, ALICE, and LHC-B. The ATLAS detector is the largest of the four, and CMS the second largest.

There are many exciting experiments to be carried out at LHC. One of them is on the production and decay of the EBH particle already mentioned in Sec. 5. Experiments at LEP, which is an electron–positron colliding accelerator at CERN, have given first possible experimental evidence[16,17] for the EBH particle, whose mass M is about

$$M \sim 115 \text{ GeV}/c^2. \tag{17}$$

Gastmans and I have been studying the production of the EBH particle at LHC on the basis of the high-energy behavior of relativistic quantum gauge field theory. Compared with the work presented in Sec. 5, there are the following two major differences:

(1) While the problem of Sec. 5 is to determine the asymptotic behavior of a four-point function, here the problem is to treat that for a five-point function; and

(2) While there is one large parameter, as specified by (5), for the problem of Sec. 5, there are here two large parameters, namely,

$$\frac{E}{M} \quad \text{and} \quad \frac{M}{m} \quad \text{are both large.} \tag{18}$$

As is well known, asymptotic calculations with two independent large parameters are very much more difficult than those with one large parameter. We are now only beginning to know how to deal with the two independent large parameters of (18) for the present problem. Using the LEP preliminary result of (17) and taking m to be the mass of the proton, then it follows from the LHC energy (16) that

$$\frac{E}{M} \sim \frac{M}{m} \sim 122. \tag{19}$$

That these two independent large parameters happen to be nearly equal is, of course, accidental.

It has been seen from Sec. 5 and especially Fig. 3(c) that the increasing total cross section at high energies is intimately related to the production through pionization of Yang–Mills gauge particles. In Fig. 8(a) is shown the simplest elastic scattering through the exchange of one Yang–Mills gauge particle, a process that leads to a constant total cross section at high energies. From Fig. 3, this simplest scattering, when modified by the production of additional Yang–Mills gauge particles as shown in Fig. 8(b), leads to the increasing total cross section discussed in Sec. 5.

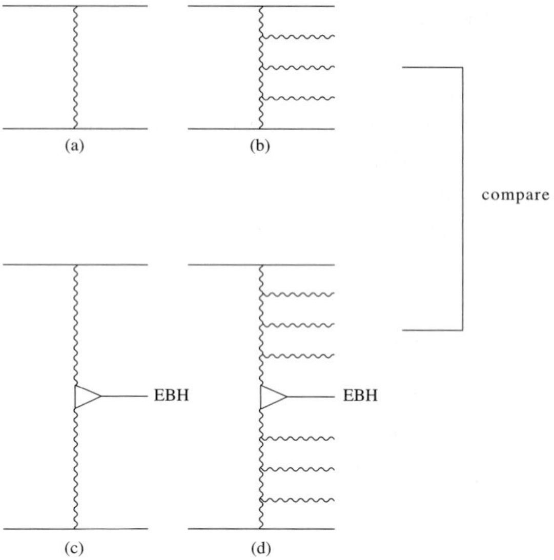

Fig. 8. Schematic comparison of increasing cross sections for elastic scattering and EBH production at high energies.

What is the corresponding situation for the production of the EBH particle at LHC? The simplest production process is shown in Fig. 8(c). In this process, each of the incident fermions (quarks) emits a Yang–Mills gauge particle (gluon) and the two gluons combine to produce the EBH particle. This process is commonly referred to as gluon fusion and is believed to be the most important EBH production process. In this case, each of the two originally emitted Yang–Mills gauge particles can produce additional Yang–Mills gauge particles. Very similar to Fig. 8(a) leading to Fig. 8(b), this gluon fusion process Fig. 8(c) leads to the diagram of Fig. 8(d). [In both Fig. 8(b) and Fig. 8(d), each group of three particles stands symbolically for an arbitrary number of particles.]

As seen from Fig. 8, especially from 8(b) and 8(d), there is a great deal of similarity between the diagram for elastic scattering and *half* of the diagram for EBH production. There are also major differences: for elastic scattering there are of course two incident particles, but, for half of EBH production, there is only one incident particle unless somehow the triangle due to a heavy fermion can serve as the second incident particle. This is one of the major problems being studied. It should also be re-emphasized that the diagrams of Fig. 8(b) are evaluated for one large parameter as given by (5), while those of Fig. 8(d) are for two large parameters (18).

Next let us turn our attention briefly to phenomenology. That for elastic scattering, as discussed in Sec. 6, depends critically on experimental data for pp and $p\bar{p}$ scattering. In particular, these data are used to determine the values of the parameters given in Table 1. But for the production of the EBH particle, there are no experimental data at all. Nevertheless, Gastmans and I are optimistic that this major obstacle to phenomenology can be overcome. For this and other reasons, this work on LHC physics is an ongoing project.

8. Conclusion and Discussions

Already for some time, the increasing total cross section, together with its immediate consequences on elastic cross sections, is well established, both theoretically and experimentally. Indeed, Professor Yang has incorporated this new phenomenon into his geometrical picture of scattering.[18]

It is also clear that the increasing cross section has direct consequences on inelastic processes. One way to see this point is as follows. As already discussed in some detail in Sec. 5, an integrated cross section that increases as $[\ln(E/m)]^3$, given by the diagram of Fig. 3(b), violates the bound of Froissart[11] and Martin[12] and is hence impossible. Therefore, such results from the diagrams of Fig. 3 must be regarded as representing a strongly absorptive potential. Nevertheless, the problem still needs to be understood: What are the high-energy behaviors of the integrated cross sections for pionization?

One way to put this issue succinctly is as follows. Since pionization leads to increasing cross sections, the increasing cross section must in turn affect pionization cross sections. Precisely how does this come about and what are the resulting pionization cross sections? This and many related questions about production processes have not yet been answered satisfactorily.

Of the numerous processes that can be and will be studied at the Large Hadron Collider—which is scheduled to begin operation at CERN in the summer of 2008—the production and decay processes of the EBH particle are surely among the most exciting. [This EBH particle is commonly called the Higgs particle; see Sec. 5.] The remainder of this section is to be devoted to some comments on these production and decay processes.

Consider the diagram of Fig. 8(d): the upper part is similar to the pionization diagram 8(b), and so is the lower part. If both parts are removed, what remains is the diagram of Fig. 9, EBH production by gluon fusion. In the standard model of Glashow, Weinberg, and Salam,[19] by far the most important contribution to this gluon fusion is from the top triangle.

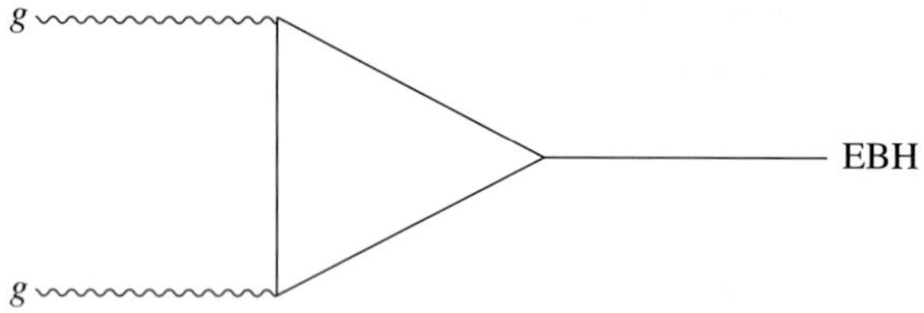

Fig. 9. EBH production by gluon fusion.

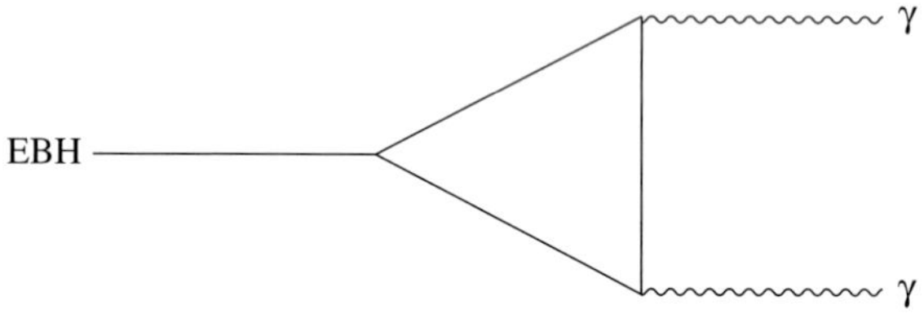

Fig. 10. The decay EBH $\to \gamma\gamma$.

If this diagram of Fig. 9 is turned around and the gluons are reinterpreted as photons, then the result is that of Fig. 10, the decay of the EBH particle into two photons. Once again, in the standard model, the most important contribution to this two-photon decay is from the top triangle.

The situation is much more interesting if the standard model is not the whole story. In this case, there may well be additional heavy particles that contribute significantly to these triangles. In general, the list of new particles that give such significant contributions may not be the same for the diagrams of Figs. 9 and 10: for the triangle of Fig. 9, the particle must have color so that it couples to the gluons; and for the triangle of Fig. 10, the particle must have charge so that it couples to the photon. If the contributing particle is of spin-0, then the triangle is, of course, not the only diagram.

This situation bears some resemblance to that of the LEP Collider in 1989. The first major experimental result from LEP was the determination of the number of generations with light neutrinos from the width of the Z. Thus the number of light neutrinos, and hence the number of such generations, was found to be 3. This was accomplished without observing experimentally any neutrino at all.

In the present case, from the production cross section of the EBH particle, or more precisely from the decay of the EBH into two photons, information can be obtained about the possible existence of new heavy particles. In the case of the decay into two photons, the rate may be normalized to any decay process that does not depend on the triangular loop, one example being the leptonic decay

$$\text{EBH} \to l^+ l^- . \tag{20}$$

Similar to the LEP case, information about these heavy particles can be obtained without observing them experimentally. Because of this possibility, the triangular loop in Figs. 9 and 10 has been sometimes referred to as the magic triangle.

Finally, the following even more exciting possible scenario may be mentioned briefly: there is more than one physical EBH particle, leading to various magic triangles. This possibility follows from noting that, in the standard model, the presence of only one EBH doublet is an assumption based on simplicity.

Acknowledgments

A deep sense of gratitude is owed to Professor Chen Ning Yang, who introduced me to this subject and with whom I had numerous subsequent discussions. I am indebted to Professors Claude Bourrely, Hung Cheng, Raymond Gastmans, and Jacques Soffer for many years of collaboration. I thank Professor Kok Khoo Phua and President Choon Fong Shih for inviting me to this exciting conference.

References

1. C. N. Yang and R. L. Mills, *Phys. Rev.* **95**, 631 (1954).
2. C. N. Yang and R. L. Mills, *Phys. Rev.* **96**, 191 (1954).
3. T. T. Wu, Remarks on Yang–Mills Theory, in *Symmetry & Modern Physics: Yang Retirement Symposium*, eds., A. Goldhaber, R. Shrock, J. Smith, G. Sterman, P. van Nieuwenhuizen, W. Weisberger (World Scientific, Singapore, 2003).
4. T. T. Wu and C. N. Yang, *Phys. Rev.* **137**, B708 (1965).
5. C. N. Yang, *Selected Papers 1945–1980 with Commentary* (W. H. Freeman and Co., San Francisco, CA, 1983).
6. J. Benecke, T. T. Chou, C. N. Yang, and E. Yen, *Phys. Rev.* **188**, 2159 (1969).
7. H. Cheng and T. T. Wu, *Phys. Rev. Lett.* **24**, 1456 (1970).
8. H. Cheng and T. T. Wu, *Expanding Protons: Scattering at High Energies* (M.I.T. Press, Cambridge, MA, 1987).
9. P. W. Higgs, *Phys. Lett.* **12**, 132 (1964).
10. F. Englert and R. Brout, *Phys. Rev. Lett.* **13**, 321 (1964).
11. M. Froissart, *Phys. Rev.* **123**, 1053 (1961).
12. A. Martin, *Il Nuovo Cimento* **42A**, 930 (1966); **44A**, 1219 (1966).

13. C. Bourrely, J. Soffer, and T. T. Wu, *Phys. Rev. D* **19**, 3249 (1979).
14. C. Bourrely, J. Soffer, and T. T. Wu, *Nucl. Phys. B* **247**, 15 (1984).
15. T. T. Wu, Chapter 4.3.4 in *Scattering—Scattering and Inverse Scattering in Pure and Applied Science*, Vol. 2, eds., R. Pike and P. Sabatier (Academic Press, London, 2002).
16. ALEPH Collaboration, R. Barate *et al.*, *Phys. Lett. B* **495**, 1 (2000); DELPHI Collaboration, P. Abreu *et al.*, *Phys. Lett. B* **499**, 23 (2001); OPAL Collaboration, G. Abbiendi *et al.*, *Phys. Lett. B* **499**, 38 (2001); L3 Collaboration, P. Achard *et al.*, *Phys. Lett. B* **517**, 319 (2001).
17. P. A. McNamara III and S. L. Wu, *Report on Progress in Physics* **65**, 465 (2002).
18. T. T. Chou and C. N. Yang, *Phys. Rev. Lett.* **46**, 764 (1981).
19. S. L. Glashow, *Nucl. Phys.* **22**, 579 (1961); S. Weinberg, *Phys. Rev. Lett.* **19**, 1264 (1967); A. Salam, *Proc. Eighth Nobel Symp., May 1968*, ed., N. Svartholm (Wiley, New York, 1968).

GRAVITY AND ITS MYSTERIES: SOME THOUGHTS AND SPECULATIONS

A. ZEE

Department of Physics, University of California, Santa Barbara, CA 93106, USA

Kavli Institute for Theoretical Physics, University of California, Santa Barbara, CA 93106, USA

I gave a rambling talk about gravity and its many mysteries at Chen-Ning Yang's 85th Birthday Celebration held in November 2007. I do not have any answers.

1. Introduction

It is an honor for me to be giving this talk on the occasion of Professor Chen-Ning Yang's 85th birthday. Like many ethnic Chinese physicists of my generation, I was inspired to go into physics by accounts of the work of Lee and Yang on parity violation. When the organizers invited me, I clearly understood that I was not to talk about "what I did last month" as is appropriate at a standard physics conference, but to give a broader perspective on some facet of theoretical physics. I am going to talk about how the mysteries of gravity have puzzled and fascinated me. Some of the following will reflect my own confusion and lack of understanding. I also confess to ignorance of entire chunks of the literature.

2. The Graviton Knows About Everything

Gravity knows about everything, whatever its origin, luminous or dark, even the energy contained in fluctuating quantum fields.

As is well known, this leads us to one of the gravest puzzles of theoretical physics. Consider the Feynman diagram with the graviton coupling to a matter field (for example an electron field) loop. If we claim to understand the physics of the electron field up to an energy scale of M, then the graviton sees an energy density given schematically by $\Lambda \sim M^4 + M^2 m_e^2 \log\left(\frac{M}{m_e}\right) + m_e^4 \log\left(\frac{M}{m_e}\right) + \cdots$. Just about any reasonable choice of M leads to a humongous energy density!!! In fact, even if the first two terms were to be mysteriously deleted, there is still an energy density of order m_e^4, that is, an energy density corresponding to one electron mass in a volume the size of the Compton wavelength of the electron, filling all of space, which is clearly unacceptable.

Apparently, this disastrous prediction of quantum field theory has nothing to do with quantum gravity. Indeed, the quantum field theory we need for the matter field is merely free field theory: we are just adding up zero point energy of harmonic oscillators.

The cosmological constant paradox may be summarized as follows. In some suitable units, the cosmological constant was expected to have the value $\sim 10^{123}$. This was so huge that it was decreed to be equal to $= 0$ identically, while the measured value turned out to be ~ 1.

I am presuming that the observed dark energy is the fabled cosmological constant. The evidence seems increasingly to favor this simplest of hypotheses. Even if this were not the case, much of the paradox still remains.

I define Λ by writing the Einstein–Hilbert action as $\int d^4x \sqrt{g}\left(\frac{1}{G}R + \Lambda\right)$. It is useful to define the mass scale of the cosmological constant according to $\Lambda \equiv {M_\Lambda}^4$. Since observationally the cosmological constant almost closes the universe we could write the Einstein–Friedmann equation $\left(\frac{1}{a}\frac{da}{dt}\right)^2 = G\rho$ as $L_{\text{universe}}^{-2} \sim G\Lambda$ with L_{universe} the size of the universe, say the Hubble radius. Let us define $M_U \equiv 1/L_{\text{universe}}$ as some sort of Compton mass of the universe. Then we have $M_U^2 \sim {M_\Lambda}^4/M_{\text{Pl}}^2$ so that

$$M_\Lambda \sim \sqrt{M_{\text{Pl}} M_U}\,. \tag{1}$$

With $M_{\text{Pl}} \sim 10^{19}$ GeV and $M_U \sim 2 \times 10^{-33}$ eV we find that $M_\Lambda \sim 4 \times 10^{-3}$ eV.

Neutrino masses, while possibly quite different from family to family, appear to have generic values, very roughly, of order 10^{-3} eV. Is this just a coincidence? In any case, there might be some physics we have yet to understand at a mass scale of $\sim 10^{-3}$ eV.

Instead of thinking about the cosmological constant as an energy density we could regard it as a sort of "curvature" by moving a left parenthesis and writing the Einstein–Hilbert action as $\int d^4x \sqrt{g}\frac{1}{G}(R + \lambda)$. Then λ has the dimension of an inverse square of a length, which we define as l_Λ. Again, observationally, we know that the two terms in the action have comparable weight, and hence the length scale associated with the cosmological constant is of the order of the size of the universe. In other words, $l_\Lambda = M_{\rm Pl}/M_\Lambda^2 \sim 1/M_U \sim L_{\rm universe}$.

Incidentally, while Λ was decreed to be identically zero by theorists, it was never banished by observational cosmologists, who needed it to reconcile various discrepancies in the data (for example, a universe younger than the earth due to an erroneous value of the Hubble constant in the 1930's and the clustering of the redshift data of quasars in the 1960's).

Gravity, knowing about everything, is the only interaction sensitive to a shift of the Lagrangian by an additive constant. In classical physics, additive constants do not affect the equation of motion. In quantum mechanics, experiments typically measure only energy differences ΔE and not the energies themselves. The Casimir effect measures the change in vacuum energy ΔE before and after the mirrors are introduced, not the vacuum energy itself (as is sometimes erroneously stated). But gravity knows about the vacuum energy $\sum \frac{1}{2}\hbar\omega$.

Is the zero point energy $\frac{1}{2}\hbar\omega$ real? I should think so, since it comes directly from the uncertainty principle. The textbook demonstration of reality is of course the liquidity of helium at zero temperature, but in fact, during the early days of quantum mechanics, many of the greats were skeptical. At the 1913 Solvay Congress Einstein declared that he did not believe in zero point energy, writing to Ehrenfest that the concept was "dead as a door nail." Pauli also had his doubts, but the experiment $\gamma + H_2 \to H + H$ convinced him. He was apparently the first to worry about the gravitational effect of the zero point energy filling space. He used for M the classical radius of the electron and concluded that the resulting universe "could not even reach to the moon!" With the passage of time people found "better" things to worry about and the issue was forgotten until Zel'dovich raised it again in the late sixties.

3. The Proton Lifetime as an Analogy

I would like to argue by analogy: this is a time-honored tradition in physics, historically often helpful and suggestive. Let us try to think of a physical quantity once expected to be huge, later decreed to be zero, then measured to be small but not zero. What I came up with was an alternative history of proton decay. It did not happen exactly this way in our civilization, but it could have easily happened in some other civilization somewhere else. Suppose that in the early 1950's, a bright young theorist decided to estimate the rate Γ for the decay $p \to e^+ + \pi^0$. He wrote down the effective Lagrangian $\mathcal{L} \sim f\pi e p$, and comparing with the pion nucleon interaction $\mathcal{L} \sim g\pi n p$

he "naturally" expected $f \sim \alpha g$ with a factor of α thrown in for isospin violation. Obviously, this naturally expected rate Γ came out way too large by many many orders of magnitude. The rate was decreed to be identically zero, by Wigner I think, and the decree came with some nice sounding words like "baryon number conservation," in a typical example of proof by authority. Even though in our own world only an upper bound on Γ exists, we could easily imagine that in our alternative world Γ was later measured to be tiny compared to the natural expectation, but definitely nonzero.

We now know the resolution of this huge paradox. It did not come from thinking about a theory of proton decay, or what the right mechanism for it might be, but from a "totally unexpected" direction, namely baryon spectroscopy.

The Lagrangian $\mathcal{L} \sim f\pi e p$ with scaling dimension

$$\frac{3}{2} + \frac{3}{2} + 1 = 4 \tag{2}$$

got transmuted into $\mathcal{L} \sim \frac{1}{M^2} qqql$ with scaling dimension

$$\frac{3}{2} + \frac{3}{2} + \frac{3}{2} + \frac{3}{2} = 6\,. \tag{3}$$

Here M denotes the mass scale of the physics responsible for proton decay. Instead of the proton p and the pion π fields we write the quark field q, and l is just a fancier way of writing e. Note that Lorentz invariance requires 4 fermion fields rather than 2.

Remarkably, this boost in scaling dimension from 4 to 6 is enough to solve an enormous paradox!!! The reason is that it appears in the exponential. Thus, now Γ is proportional to $(1/M^2)^2 = 1/M^4$. We also need the matrix element $\langle p|qqql|\pi e\rangle$, which is set by low mass scale physics and so should be $\sim m_p$ the proton mass. Hence by "high school dimensional analysis, we obtain $\Gamma \sim (m_p/M)^4 m_p$, which is enough to account for the "absurdly" small value of Γ!

We could even imagine that in this alternative civilization a bright young theorist could have argued that the long lifetime of the proton pointed to quarks.

Do we learn something from this story? Anything? Nothing?

If this story is somehow relevant to the cosmological constant paradox, we might ask whether we could in some way promote $\mathcal{L} \sim \Lambda\sqrt{g}$ with scaling dimension 0 to an operator $\mathcal{O}$ with dimension p so that the effective Lagrangian for the energy density becomes $\mathcal{L} \sim \frac{1}{M^{p-4}}\mathcal{O}$ with M some new high mass scale? We would then obtain $\Lambda \sim \frac{1}{M^{p-4}}\langle\mathcal{O}\rangle = (m/M)^p M^4$ with $\langle\mathcal{O}\rangle = m^p$ and m some new low mass scale.

With m small enough and M large enough and p big enough, we could get the suppression we want.

I certainly do not have a detailed theory of how this could happen. One question: how could we promote $\sqrt{g}$ without also promoting the Einstein–Hilbert term $\frac{1}{G}\sqrt{g}R$? Interestingly, the same question arises in my historical analogy and it might be instructive to watch how the question was resolved: πep was promoted to $qqql$ with dimension jumping from 4 to 6 but πnp was changed to qqA (here A denotes the gluon field) with dimension remaining at 4.

4. Could Gravity be Part of a Larger Structure?

Could Einstein–Hilbert be replaced by something more fundamental which could lead to $\frac{1}{G}\sqrt{g}R$ effectively at low energy much as quantum chromodynamics leads to the Yukawa theory?

I am not necessarily suggesting here that the graviton is composite. Indeed, there is a seemingly convincing argument against the graviton being composite. Consider the same Feynman diagram mentioned earlier, but at the point where the graviton couples to the electron we insert a form factor with some energy scale. The trouble is that the momentum q carried by the graviton is in what I would call the "extreme ultra infrared" with $q \sim 1/L_{\text{cosmological}} \sim 0$ where $L_{\text{cosmological}}$ denotes a cosmological distance scale. In other words, the universe could care less if the graviton is composite at an energy scale of say 1 TeV.

The alternative may be that gravitation is part of a larger structure (perhaps along the line I sketched in *Phys. Rev. Lett.* **55** 2379 (1985)). We now understand the electromagnetic field as part of a larger structure. Gerard 't Hooft has given an elegant expression for the Maxwell field $\mathcal{F}_{\mu\nu}$ in terms of the Yang–Mills field $F^a{}_{\mu\nu}$. Is there an analog for gravity? Being part of a larger structure, even if the structure is not seen at low energies, does lead to physical consequences. Thus, electric charge is quantized if the larger structure is a grand unified theory based on a simple group, and we understand why $Q_{\text{electron}} = -Q_{\text{proton}}$ exactly, a fact of cosmological significance.

Alternatively, we could argue that quantum field theory such as quantum electrodynamics only makes sense when formulated on a lattice, and then the electromagnetic U(1) is "necessarily" compact which leads to charge quantization. In either case, what we see and know is part of a larger structure.

The question, stated in the format of an IQ test question, is then "What is to gravity as Yang–Mills is to electromagnetism?"

5. The Horizon

Many have made careers out of worrying about quantum gravity. But classical gravity is already plenty puzzling. When we first studied physics, we were told that physics should be local, that something happening here could only affect something happening nearby, and for a physical effect to propagate across space–time a field is needed. (The mysteries of quantum mechanics have however also led to entanglement and the Aharonov–Bohm effect.) Already in classical, nonquantum, gravity we have black holes, and the horizon around a black hole is a strikingly nonlocal concept. Nothing happens locally. Observers falling in do not notice anything. The hand-wringing over the horizon only affects the mythical observer stationed way off at infinity. To me that is a basic puzzle of physics. In technical terms, the Riemann curvature is nice and smooth at the horizon and could be made arbitrarily small for massive black holes. But somehow the other fields know about the metric $g_{\mu\nu}$ directly, not Riemann curvature. For a nice pedagogical treatment of how the horizon appears as a black hole forms, see the not terribly well-known work by R. Adler, J. Bjorken, P. S. Chen and J. S. Liu.

The horizon is an inherently nonlocal concept. But confusingly, while we cannot perform local measurements to detect the presence of a horizon directly, we could do so indirectly. By measuring whether light rays tend to converge or diverge, we could detect the presence of a "trapped surface" (or apparent horizon). A sequence of highly plausible theorems (each of which nevertheless involves some technical assumptions) by Penrose, Ellis and others, combined with the unproven cosmic censorship conjecture, states that the presence of a trapped surface implies the presence of a horizon.

More physically, the horizon is nonlocal in the following sense. By drawing a Penrose diagram we can see that we could be sitting peacefully with an incoming shell of matter far away threatening to form a black hole soon and we might be inside the horizon even before the black hole forms.

In the standard Schwarzschild coordinates, $g_{00} = 0$ and $g_{rr} = \infty$ at the horizon. Time and space then exchange roles. It would appear that to have a proper formulation of quantum mechanics and quantum field theory we need to have a well-behaved time variable to evolve unitarily with. As is well known, there are textbook formulations of quantum field theory in curved space–time and standard treatments lead to Hawking radiation. Are these treatments correct? Is there a modification of Einstein's theory such that metric singularities such as $g_{00} = 0$ and $g_{rr} = \infty$ are somehow forbidden. Of course, any student knows that these are but artifacts due to a poor coordinate choice. We could transform to coordinates in which

g_{00} and g_{rr} are perfectly well behaved at the horizon, but as far as I know, these coordinates only cover the black hole exterior. Various mathematical maneuvers, so on and so forth, have to be performed to continue into the interior. I do not find the treatment completely satisfactory yet.

Historically, the horizon was a source of great confusion and Kruskal's contribution cannot be overestimated. For example, on page 203 of Bergmann's standard text *Introduction to the Theory of Relativity* (with a foreword by A. Einstein) he quoted Robertson as concluding that "at least part of the singular character" of the metric at $r = 2GM$ must be attributed to the choice of coordinates. Curiously, people at the time did not follow the modern expedient of simply noting the smoothness of the Riemann curvature tensor, which Schwarzschild himself, at the very least, must have calculated. (Bergmann then went on and cited Einstein's 1939 work showing that in a toy model of a spherical cluster of noninteracting particles the Schwarzschild singularity could not form.)

Could we possibly modify general relativity so as to avoid having an horizon? Once again, apparently not because black hole is a low energy phenomenon. Naively, we might also think addition of local terms would not remove a nonlocal phenomenon like a horizon.

When we turn on quantum mechanics the black hole emits Hawking radiation and eventually disappears, so that the horizon is not only not local in space, but also not local in time.

Quantum field theory in curved space–time is a well developed subject and leads to Hawking radiation for example, but again I still have lingering doubts. In calculating a loop diagram for a process, say the electron's magnetic moment, at the horizon, are there subtleties involving virtual particles propagating inside the horizon and then out again? Presumably it is okay over a distance scale of the order the Compton wavelength of the particles. More generally, in doing a quantum gravity path integral sum over all gravitational field configurations, are we to include configurations containing black holes or not? I imagine that there are experts walking around who are sure of the answers to these questions and more.

6. The Gravitational Field is Not Just Another Field

According to an apparently appealing philosophy due to many eminent physicists (…, Gupta, Kraichnan, Feynman, Weinberg, Boulware, Deser, …), we should regard the gravitational field as just another field. As Feynman showed in his posthumous book on gravity, we could pretend that we never heard of general relativity and Riemannian geometry, and simply develop the field theory of a massless spin-2 particle called the graviton. The program worked: general relativity and Riemannian geometry emerge from playing with Feynman diagrams, but most people hate this antigeometric approach. (By the way, Kraichnan did his work as an 18-year old undergraduate at MIT. According to his recollection, Einstein was appalled by this approach. Partly as a result, Kraichnan delayed publication for eight years and so ended up publishing after Gupta. Feynman was apparently unaware of the work of Kraichnan and Gupta.)

Nevertheless, this view is somehow fundamentally wrong. The way I like to put it is that we are in some avant-garde theater. Unique among the actors, the graviton is not just another actor on the stage. The actor is himself the stage. It provides the arena in which the other fields work and play.

The founders of quantum field theory wrote profound equations such as

$$A_\mu = 0 + A_\mu \tag{4}$$

and

$$\varphi = 0 + \varphi\,. \tag{5}$$

Fields execute quantum fluctuation around vanishing classical values. But then physicists became more sophisticated in the 1960's and wrote fancier equations like

$$\varphi = v + h \tag{6}$$

with $v = \langle\varphi\rangle$. A great deal of money has been, and is being spent, to see if this idea is correct.

The basic equation for the graviton field has the same form

$$g_{\mu\nu} = \eta_{\mu\nu} + h_{\mu\nu}\,. \tag{7}$$

This naturally suggests that $\eta_{\mu\nu} = \langle g_{\mu\nu}\rangle$ and perhaps some sort of spontaneous symmetry breaking. But gravity exhibits a fundamentally new feature: $g_{\mu\nu}$ is a matrix, and hence has a signature. Large fluctuations of $h_{\mu\nu}$ could change the signature of $g_{\mu\nu}$ and there could be regions with two times. An obvious thing to write down would be a potential for $g_{\mu\nu}$ (which breaks general coordinate invariance) of the form $V(g) = \lambda(g_{\mu\nu} - \eta_{\mu\nu})^2$, or more generally a potential with a deep well pinning $g_{\mu\nu}$ to values close to $\eta_{\mu\nu}$. This induces a graviton mass of order $m_g^2 \sim \lambda M_{\rm Pl}^2$ so that $\lambda^{\frac{1}{2}}$ is given by the ratio of the largest mass and possibly the smallest mass known to physics.

This line of thought raises the possibility that the potential V might have minima elsewhere. Perhaps there is a phase with $g_{\mu\nu} = 0$. That could be the ultimate terrorist plot, to unleash a $g_{\mu\nu} = 0$ bomb that would annihilate space–time in the victimized country.

An intriguing idea is that of emergent gravity developed by X. G. Wen and others. This line of development emerged from the days of the chiral spin liquid, in which gauge fields readily emerged from systems that consist solely of electron spins. (See for example, A. Zee, in the *M. A. B. Bég Memorial Volume*, eds. A. Ali and P. Hoodbhoy (1990).) Besides gravity, fermions also puzzle me. (Jordan's manuscript languished in Born's pocket for a whole year.) Sometimes I feel that the world ought to contain only bose fields. Perhaps half integral spin could also be emergent. (See for example, A. Zee, in *Quantum Coherence: 30 Years of Aharonov–Bohm Effect*, ed. J. Anandan (1989).) I am also intrigued by the effect discovered by 't Hooft *et al.*, that binding a boson to magnetic monopole produces a fermion.

7. Unimodular Gravity

The notion of unimodular gravity goes back to Einstein in some sense, and was developed later by van der Bij, van Dam, Y. J. Ng, Wilczek, Zee, Dolgov, Weinberg, and many others. Suppose $g \equiv \det g_{\mu\nu}$ is fixed to be equal to 1, then the cosmological constant term in the action $S = \int d^4x \sqrt{g}\Lambda + \cdots$ becomes impotent, and hence irrelevant. But in fact, it comes back. Since the constraint $\delta \det g_{\mu\nu} = 0$ is equivalent to $g^{\mu\nu}\delta g_{\mu\nu} = 0$ we only get the traceless part of Einstein's equation

$$R_{\mu\nu} - \frac{1}{4} g_{\mu\nu} R = T_{\mu\nu} - \frac{1}{4} g_{\mu\nu} T \,. \tag{8}$$

Writing $-\frac{1}{4}$ on the left-hand side as $-\frac{1}{2} + \frac{1}{4}$ and taking the covariant derivative, we obtain $\partial_\mu R = -\partial_\mu T$, which could be solved to give $R = -T + C$. The integration constant C reappears as the cosmological constant when this equation is inserted back into the traceless part of Einstein's equation.

Thus, unimodular gravity does not solve the problem but makes some people "feel more comfortable" because in theoretical physics, supposedly, we have the license to set integration constants to whatever we want.

8. Equivalence Principle

Let us go back to the Feynman diagram described at the beginning of this article, with the graviton coupling to a matter field, say the electron field, loop. Ultimately, it is this graph that causes all our hand-wringing over the cosmological constant. Suppose one were to work long and hard and come up with a rule or theory that cleverly deletes this graph, thus solving the cosmological constant paradox. As emphasized by J. Polchinski, any such rule or theory would always be doomed to fail because of the equivalence principle.

The argument is as follows. Connect the graph by some photon lines to the propagator of some atomic nucleus, say aluminum or iron. This graph thus contributes to the gravitational mass of the nucleus. On the other hand, consider the same graph with the atomic nucleus but with the graviton removed, a graph that presumably has nothing to do with gravity. But this graph contributes to the inertial mass of the nucleus. Thus, with the enormous accuracy to which the equivalence principle has been tested, we already know that the graph with the graviton attached could not be deleted. But we are claiming that, in order to resolve the cosmological constant paradox, we have some "rule" to delete this graph.

The trouble is once again that physics as we understand it should be local: at the point the graviton couples to the electron, how could the graviton "know" what the electron loop is going to do? It could not know whether the electron is just going to loop back upon itself, or that before looping back, the electron is "planning" to emit two photons which subsequently will be absorbed by a nucleus.

9. The Extreme Ultra Infrared

The local nature of Feynman diagrams, plus the constraint from the experimental verification of the equivalence principle, make it difficult to imagine how any "rule" could be invented to delete one Feynman diagram and not another. Perhaps one loophole is offered by the phrase "nothing to do with gravity"; perhaps even a graph without the graviton is subject to the requirements of some ultimate theory of gravity.

Another way out is suggested by the fact that, upon closer inspection, we see that there is evidently a huge difference between the graph responsible for the cosmological constant paradox and the same graph attached by two photon lines to a nucleus. The momentum carried by the graviton, called it q, has a value $q \sim 1/L_{\text{cosmological}}$ in the former, but a vastly larger value $q \sim 1/L_{\text{laboratory}}$ or $q \sim 1/L_{\text{terrestrial}}$ in the latter.

In particle physics we always profess ignorance about physics at high energies, about the ultraviolet regime, but truth be told, we know almost nothing about the "extreme ultra infrared." Thus, we could always modify the left-hand side of Einstein's equation by acting with some operator $f(L^2D^2)$ where D denotes the covariant derivative and L is some cosmological length scale. The left-hand side is effectively multiplied by $f(L^2/L^2_{\text{phenomenon}})$ where $L_{\text{phenomenon}}$ denotes the length scale of the phenomenon under study. All we require in order to distinguish between the two graphs is for f to have the properties $f(\sim\infty) = 1$ and $f(\sim 0) = 0$. Needless to say, such a momentum dependent function implies that the theory is highly nonlocal.

10. The Universe is Secretly Acausal but only the Universe Knows About It

One realization of this sort of idea is due to Arkani-Hamed *et al.* (2002) who proposed modifying Einstein's equation to

$$M_{\rm Pl}^2\left(R_{\mu\nu} - \frac{1}{2}g_{\mu\nu}R\right) - \frac{1}{4}\bar{M}^2 g_{\mu\nu}\bar{R} = T_{\mu\nu}\,, \tag{9}$$

where $\bar{R}$ denotes the space–time averaged scalar curvature $\bar{R} \equiv \frac{\int d^4x\sqrt{g}R}{\int d^4x\sqrt{g}}$. This equation is manifestly nonlocal and acausal: physics now depends on physics in the far future. But by construction the modification to Einstein's equation takes effect only if the future is de Sitter with constant scalar curvature determined by the cosmological constant $\bar{R} = -\frac{4\Lambda}{M_{\rm Pl}^2+\bar{M}^2}$. To account for observation, the new mass scale $\bar{M}$ has to be huge, taking values ranging from $\sim 10^{48}$ GeV to $\sim 10^{80}$ GeV depending on the assumed value of the cosmological constant one wishes to "neutralize." Unhappily, another enormous mass scale has to be introduced into physics.

In this approach, the modification is clearly designed not to matter for any situation other than cosmological. For the solar system for example, $\bar{R}$ would come out to be practically zero. The universe is secretly acausal but only the universe knows about it! I must say that in recent years, theoretical physicists have become increasingly adept at hiding new physics from experimentalists.

Arkani-Hamed *et al.* argued that any mechanism to "neutralize" the cosmological constant must be acausal: when a vacuum energy density "turns on," the alleged mechanism must "wait" for a cosmological time period to "find out" if the energy density is indeed a cosmological constant. I am very much troubled by the thought that physics may be ultimately nonlocal, but the argument appears to be plausible.

11. Induced Gravity

At one time induced gravity appeared to offer a way out of our problems with gravity and thus enjoyed a following. Consider the path integral

$$\int d\phi\, d\psi\, dA\, e^{iS(g,\phi,\psi,A)} = e^{i\int d^4x\sqrt{g}(\Lambda+R/G+\cdots)}\,. \tag{10}$$

There is no question that integration over the matter fields ϕ, ψ, and A would generate the Einstein–Hilbert term. The difficulty is that Λ comes out naturally large, but this is of course just the cosmological constant problem again.

One fundamental question is whether we need to integrate next over $\mathcal{D}g$. If not, that is, if we do not integrate over the metric, then the classical equation of motion of the gravitational field would not emerge automatically as Planck's constant approaches zero but has to be imposed by hand.

This leads us to the perennial question of whether gravity has to be quantized. If not, as was first proposed by Møller (1962) and Rosenfeld (1963), then we have the equation

$$R_{\mu\nu} - \frac{1}{2} g_{\mu\nu} R = \langle T_{\mu\nu} \rangle \,. \tag{11}$$

Once again, this produces a huge cosmological constant on the right-hand side. But let us leave that aside. The objection to this equation is that it violates the uncertainty principle. If gravity is not quantized, then it acts as a classical probe, and we could use a massive ball attached to a torsion balance to measure the position and momentum of a passing electron. In 1981 Page and Geiliker experimentally demonstrated the difficulty one runs into. Consider a Cavendish experiment in which the heavy ball is moved from one position "here" to another position "there" as determined by some radioactive decay. This amounts to a Schrödinger cat experiment with the quantum state in the preceding equation given by $|\rangle = \frac{1}{\sqrt{2}}(|\text{here}\rangle + |\text{there}\rangle)$. The torsion pendulum would then point to a "phantom ball" situated halfway between here and there.

There are those (for example Dyson) who would raise the question of whether gravity has to be quantized on phenomenological grounds, since no conceivable experiment could detect a single graviton.

12. Ever More Speculative Ways Out

Over the years, many physicists have had many ("crazy") thoughts about gravity. I listed some of them in a talk almost a quarter of a century ago on an occasion similar to this one, dedicated to Paul Dirac. One possibility, considered highly speculative at the time, was to entertain a decaying cosmological constant $\frac{d\Lambda}{dt} \neq 0$, but these days, with a multitude of scalar fields around, this possibility would be considered commonplace rather than outrageous. (See *High Energy Physics in Honor of P. A. M. Dirac in His 80th Year*, ed. S. Mintz *et al.* (1983).)

The cosmological constant paradox suggests to some people that we might have to break free of local field theory entirely. This line of thought led Steve Hsu and I to propose adding terms not of the form $\int d^4x(\cdots)$ to the action, in a vaguely "Landau–Ginzburg" sort of approach to the action. We obtained

$$M_\Lambda \sim \sqrt{M_{\text{Pl}} M_U}\,, \tag{12}$$

where M_U was defined earlier as the Compton mass of the universe. This relation, regardless of high shakily it is derived, has the pleasing form of giving the mass scale of the cosmological constant (or dark energy) M_Λ as the geometric mean of perhaps the largest and smallest mass scales in physics M_{Pl} and M_U. As explained earlier, it goes back to Einstein since it amounts to the statement that the observed dark energy is just about enough to close the universe.

13. Reversal of Fortune

We have witnessed a remarkable shift in attitude towards quantum field theory over the last 30 years. An operator in the action is classified according to whether its mass dimension is < 4, $= 4$, or > 4, operators known respectively as "superrenormalizable," "renormalizable," and "nonrenormalizable." Textbooks taught that superrenormalizable interactions are nice, renormalizable interactions are what we want, while nonrenormalizable interactions should fill us with fear and loathing.

This traditional doctrine was replaced by a new attitude which regards quantum field theory as a low energy effective theory. In an astonishing reversal of fortune, the nonrenormalizable terms are now welcomed and well-liked as terms that are inevitably here with us. They are regarded as innocuous since they are suppressed by powers of some higher mass scale $\frac{1}{M^p}$, while the renormalizable terms are uniquely fixed by the gauge principle, etc. In contrast, our "friends" the superrenormalizable terms are now regarded as nasty guys.

Since these nasty guys have nominal mass dimension < 4, there are fortunately only a finite number of them. They represent the challenges confronting fundamental physics today, and are in turn known as the Higgs mass term, the Einstein–Hilbert term, and the cosmological constant term. The Higgs mass term has dimension 2. The Einstein–Hilbert term has nominal dimension 2 which after rescaling by the Planck mass becomes dimension $4 + 5 + 6 + \cdots$. The cosmological constant term has nominal dimension 0 which after rescaling becomes dimension $0 + 1 + 2 + \cdots$.

Perhaps there is something seriously wrong with this picture.

Our understanding of physics is based on this notion of effective field theory, to which all we know could be reduced. Yet there are many questions, many doubts, but no clear answer. Field theory itself, and Einstein gravity as an effective field theory, could fail at truly long distances. Ultraviolet regularization has been understood for long time, but as I have said, not the extreme infrared.

Quantum field theory is very much based on the momentum–distance relation, also known as the uncertainty relation, as expressed in the Fourier relation $e^{\frac{ipx}{\hbar}}$. This connection could fail and be modified. (Indeed, this is what happens in string theory.) From the discussion of the cosmological constant paradox it is clear that some kind of connection between ultraviolet and infrared (such as that offered by the anthropic principle: Λ is ultraviolet while we humans are infrared) is needed.

Black hole offers a well known "violation" of the standard momentum–distance relation: the more massive the black hole, the larger its size $R = GM$. Clearly, the exception is due to the existence of a fundamental mass scale $l_{\mathrm{Pl}} = \frac{1}{\sqrt{G}}$.

Another possibility is the breakdown of quantum mechanics when the splitting between energy levels ΔE is less than the inverse of some cosmological time scale, such as the age of universe.

Meanwhile, Bern, Kosower, and many others, using the twistor formalism, have discovered amazing cancellations and simplifications in complicated Feynman diagram calculations. I find particularly intriguing the hint from the explicit calculation performed by Bern *et al.* that amplitudes in Einstein gravity could be regarded as the square, or sum of squares, of appropriately "color stripped" amplitudes in Yang–Mills theory, a hidden relation originally suggested in a string theory context by Kawai, Lewellen and Tye. Recent work by Arkani-Hamed and others give tantalizing evidence that superficially more complicated theory like Einstein gravity and Yang–Mills theory may have better ultraviolet behavior than a simple scalar field theory.

I have always been bothered by the liberal and indiscriminate use of scalar fields in particle theory and cosmology. Quantum field theory textbooks start with scalar fields precisely because they are "without qualities." If Nature wanted to show us an elementary scalar field, would not she have shown us one long ago? We have encountered elementary spin 1 fields, an elementary spin-2 field, and in a mysterious twist, even elementary spin-1/2 fields. We know about meson fields, but they are clearly composite. When and if the Higgs field is discovered, an interesting questions might be whether or not it could be regarded as composite. I have speculated elsewhere that perhaps quantum field theory somehow forbids elementary scalar fields. In a new formulation of quantum field theory, and one appears to be suggested by recent work, might elementary scalar fields not be allowed?

Note that in our historical analogy, when the pseudoscalar π field was banished in favor of two quark fields the scaling dimension of the relevant operator goes up by 2 and physicists have one less "naturalness" paradox to contend with.

14. Hierarchy Problems

It seems to me that the discovery of a small nonvanishing cosmological constant may have liberated us from having to worry about the various hierarchy problems of particle physics. The small cosmological constant, if indeed a cosmological constant, would be a living exception to the 't Hooft "naturalness doctrine" regarding the occurrence of small dimensionless numbers in physics. In practical terms, one of the arguments in favor of the rather unlikely and contrived idea of low energy supersymmetry might have evaporated.

15. The Coincidence Problem

No discussion of the cosmological constant paradox is complete without mentioning the cosmic coincidence problem. The energy density ρ in matter varies with the scale factor a of the expanding universe like $1/a^3$, while the energy density in curvature varies like $1/a^2$, and the energy density in the cosmological constant varies like $1/a^0$. It is remarkable that they are comparable now. Why now?!

The only plausible "explanation" is the "anthropic lack of principle." In some sense, the smallness of Λ was predicted by Weinberg using a very weak version of the anthropic principle. This very weak version of the anthropic principle should be acceptable to most theoretical physicists: it merely correlates two observations, namely that galaxies formed and the smallness of Λ.

16. Closing Remarks

I was recently reading about the history of special relativity. Young Einstein was able to accomplish what Lorentz and Poincaré were not able to accomplish, even though the two established giants had most of it worked out, at least mathematically. After all, Lorentz had the Lorentz transformation in all its glory. The two older physicists were not able to abandon the perfectly sensible notion that if there is a wave *something* must be waving. So they had the ether as a dynamical variable. Einstein simply trashed the ether and asserted that *nothing* could also wave.

Nowadays, any student is able to accept, without blinking twice, that an electromagnetic wave consists of A_μ waving, yes, just a mathematical symbol known as a field waving. Of course, there is energy and momentum densities associated with the wave, and so it is real in that sense. But what is a field? After spending years writing a textbook on quantum field theory, I could understand a field as only something that does what a field does. No more, no less.

To move forward, physics had to abandon an apparently ironclad piece of commonsense that "where there is a wave *something* must be waving." I would not be at all surprised if it turns out that to move forward, we have to abandon an equally ironclad piece of commonsense. I leave it to the reader to identify that piece.

We conclude with a rather dark motto about dark energy I learned after giving a related talk in Bologna: "Per obscura ad obscuriora."

Acknowledgments

Over the decades I have benefited from conversations about gravity with many colleagues far too numerous to name here, starting with John Wheeler who guided me through my very first research project. During this past year leading up to this talk, and in preparing for this talk, I was enlightened on various occasions by Nima Arkani-Hamed, Steve Hsu, Joe Polchinski, Xiao-Gang Wen and Frank Wilczek. I thank Joe Polchinski and Rafael Porto for reading this manuscript. I am also grateful to Richard Neher and Rafael Porto for technical help in preparing this article.

Some versions of this talk were also given at the Lorentz Institute, Leiden, the Netherlands, the National Taiwan University, Taipei, Republic of China, and the Institute for Theoretical Physics, Sao Paulo, Brazil.

I am supported in part by NSF under Grant No. 04-56556.

GEOMETRIC PHASE AND CHIRAL ANOMALY; THEIR BASIC DIFFERENCES

KAZUO FUJIKAWA

Institute of Quantum Science, College of Science and Technology, Nihon University, Chiyoda-ku, Tokyo, Japan

All the geometric phases are shown to be topologically trivial by using the second quantized formulation. The exact hidden local symmetry in the Schrödinger equation, which was hitherto unrecognized, controls the holonomy associated with both of the adiabatic and non-adiabatic geometric phases. The second quantized formulation is located in between the first quantized formulation and the field theory, and thus it is convenient to compare the geometric phase with the chiral anomaly in field theory. It is shown that these two notions are completely different.

1. Introduction

Phases are intriguing notions, as was emphasized by C.N. Yang on various occasions. Here we discuss two phases, and the first phase is the geometric phase in quantum mechanics[1–9] for which we present the recent developments on the basis of the second quantized formulation of all the geometric phases.[10–13] The second phase is the chiral anomaly in field theory,[14–17] which is by now well understood.[18] The second quantized formulation is located in between the first quantized formulation and the field theory, and thus it is convenient to compare the geometric phase with the chiral anomaly in field theory.[19–21]

We then show

(i) A unified treatment of adiabatic and non-adiabatic geometric phases is possible in the second quantized formulation by using the exact hidden local (*i.e.*, time-dependent) symmetry in the Schrödinger equation.
(ii) The topology of all the geometric phases is trivial by using an exactly solvable example.
(iii) Geometric phases in the Schrödinger problem and the chiral anomaly in field theory are completely different.

2. Second Quantized Formulation

We start with defining an *arbitrary* complete basis set

$$\int d^3x v_n^\star(t,\vec{x}) v_m(t,\vec{x}) = \delta_{nm} \tag{1}$$

and expand the field operator $\hat{\psi}(t,\vec{x})$ as

$$\hat{\psi}(t,\vec{x}) = \sum_n \hat{b}_n(t) v(t,\vec{x}). \tag{2}$$

The action

$$S = \int_0^T dt d^3x [\hat{\psi}^\star(t,\vec{x}) i\hbar \frac{\partial}{\partial t} \hat{\psi}(t,\vec{x}) - \hat{\psi}^\star(t,\vec{x}) \hat{H}(t) \hat{\psi}(t,\vec{x})] \tag{3}$$

which gives rise to the field equation

$$i\hbar \frac{\partial}{\partial t} \hat{\psi}(t,\vec{x}) = \hat{H}(t) \hat{\psi}(t,\vec{x}) \tag{4}$$

then becomes

$$S = \int_0^T dt \{ \sum_n \hat{b}_n^\dagger(t) i\hbar \partial_t \hat{b}_n(t) - \hat{H}_{eff} \}. \tag{5}$$

The effective Hamiltonian is given by

$$\begin{aligned} \hat{H}_{eff}(t) = \sum_{n,m} \hat{b}_n^\dagger(t) [\int d^3x v_n^\star(t,\vec{x}) \hat{H}(t) v_m(t,\vec{x}) \\ - \int d^3x v_n^\star(t,\vec{x}) i\hbar \frac{\partial}{\partial t} v_m(t,\vec{x})] \hat{b}_m(t) \end{aligned} \tag{6}$$

and the canonical commutation relations $[\hat{b}_n(t), \hat{b}_m^\dagger(t)]_\mp = \delta_{n,m}$, but statistics (fermions or bosons) is not important in our application.

The Schrödinger picture $\hat{\mathcal{H}}_{eff}(t)$ is obtained by replacing $\hat{b}_n(t)$ with $\hat{b}_n(0)$ in $\hat{H}_{eff}(t)$ (6). Then the evolution operator is given by[11]

$$\begin{aligned} &\langle m | T^\star \exp\{ -\frac{i}{\hbar} \int_0^t \hat{\mathcal{H}}_{eff}(t) dt \} | n \rangle \\ &= \langle m(t) | T^\star \exp\{ -\frac{i}{\hbar} \int_0^t \hat{H}(\hat{\vec{p}}, \hat{\vec{x}}, X(t)) dt \} | n(0) \rangle \end{aligned} \tag{7}$$

with time ordering symbol $T^\star$. In the second quantized formulation on the left-hand side we have $|n\rangle = \hat{b}_n^\dagger(0)|0\rangle$, and in the first quantized formulation on the right-hand side we have $\langle \vec{x} | n(t) \rangle = v_n(t,\vec{x})$.

The exact Schrödinger probability amplitude which satisfies

$$i\hbar\partial_t\psi_n(t,\vec{x}) = \hat{H}(t)\psi_n(t,\vec{x})$$

with $\psi_n(0,\vec{x}) = v_n(0,\vec{x})$ is given by

$$\begin{aligned}\psi_n(t,\vec{x}) &= \langle 0|\hat{\psi}(t,\vec{x})\hat{b}_n^\dagger(0)|0\rangle \\ &= \sum_m v_m(t,\vec{x})\langle 0|\hat{b}_m(t)\hat{b}_n^\dagger(0)|0\rangle \\ &= \sum_m v_m(t,\vec{x})\langle m|T^\star \exp\{-\frac{i}{\hbar}\int_0^t \hat{\mathcal{H}}_{eff}(t)dt\}|n\rangle \qquad (8)\end{aligned}$$

which is confirmed by using the relation $i\hbar\partial_t\hat{\psi}(t,\vec{x}) = \hat{H}(t)\hat{\psi}(t,\vec{x})$ in (4). We note that the general geometric terms automatically appear as the second terms in the *exact* $\hat{\mathcal{H}}_{eff}(t)$ in (8). See $\hat{H}_{eff}(t)$ in (6).

2.1. *Hidden local symmetry*

Since the basic field variable is written as $\hat{\psi}(t,\vec{x}) = \sum_n \hat{b}_n(t)v_n(t,\vec{x})$, we have an exact *hidden* local (i.e., time dependent) symmetry[11]

$$\begin{aligned}&v_n(t,\vec{x}) \to v_n'(t,\vec{x}) = e^{i\alpha_n(t)}v_n(t,\vec{x}), \\ &\hat{b}_n(t) \to \hat{b}_n'(t) = e^{-i\alpha_n(t)}\hat{b}_n(t), \qquad n = 1,2,3,... \qquad (9)\end{aligned}$$

which keeps $\hat{\psi}(t,\vec{x})$ invariant. This symmetry means arbitrariness in the choice of the coordinates in the functional space. The Schrödinger amplitude $\psi_n(t,\vec{x}) = \langle 0|\hat{\psi}(t,\vec{x})\hat{b}_n^\dagger(0)|0\rangle$ is then transformed as

$$\psi_n'(t,\vec{x}) = e^{i\alpha_n(0)}\psi_n(t,\vec{x}) \qquad (10)$$

under the hidden symmetry for any t. Namely, it gives the ray representation with a constant phase. We thus have the enormous hidden local symmetry behind the ray representation, which was not recognized in the past. The product $\psi_n(0,\vec{x})^\star\psi_n(T,\vec{x})$ is then manifestly gauge invariant for a periodic system.

If one chooses a specific basis

$$\hat{H}(X(t))v(\vec{x};X(t)) = \mathcal{E}_n(X(t))v(\vec{x};X(t)) \qquad (11)$$

in (1) for a periodic Hamiltonian $\hat{H}(X(0)) = \hat{H}(X(T))$ and assumes "diagonal dominance" in the effective Hamiltonian, we have from (8)

$$\psi_n(t,\vec{x}) \simeq v_n(\vec{x};X(t))\exp\{-\frac{i}{\hbar}\int_0^t [\mathcal{E}_n(X(t)) - \langle n|i\hbar\frac{\partial}{\partial t}|n\rangle]dt\} \qquad (12)$$

which reproduces the result of the conventional adiabatic approximation.

This shows that

Adiabatic approximation = Approximate diagonalization of H_{eff}

and thus the geometric phases are *dynamical*, i.e., a part of the Hamiltonian. In fact, it has been recently shown that the second quantized formulation nicely resolves some of the subtle problems in the conventional adiabatic approximation.[22]

In the adiabatic approximation (12), we have a gauge invariant quantity (for a general choice of the hidden local symmetry)

$$\psi_n(0,\vec{x})^\star\psi_n(T,\vec{x}) = v_n(0,\vec{x};X(0))^\star v_n(T,\vec{x};X(T)) \times \exp\{-\frac{i}{\hbar}\int_0^T [\mathcal{E}_n(X(t)) - \langle n|i\hbar\frac{\partial}{\partial t}|n\rangle]dt\}. \quad (13)$$

If one chooses a specific hidden local gauge such that $v_n(T,\vec{x};X(T)) = v_n(0,\vec{x};X(0))$, the pre-factor $v_n(0,\vec{x};X(0))^\star v_n(T,\vec{x};X(T))$ becomes real and positive and thus the factor on the exponential in (13) represents the entire gauge invariant phase. This unique gauge invariant quantity reproduces the conventional adiabatic phase.[3,4]

2.2. *Parallel transport and holonomy*

The parallel transport of $v_n(t,\vec{x})$ is defined by

$$\int d^3x v_n^\dagger(t,\vec{x})\frac{\partial}{\partial t}v_n(t,\vec{x}) = 0 \quad (14)$$

which is derived from the conditions

$$\int d^3x v_n^\dagger(t,\vec{x})v_n(t+\delta t,\vec{x}) = \text{real and positive} \quad (15)$$

and

$$\int d^3x v_n^\dagger(t+\delta t,\vec{x})v_n(t+\delta t,\vec{x}) = \int d^3x v_n^\dagger(t,\vec{x})v_n(t,\vec{x}). \quad (16)$$

By using the hidden local gauge $\bar{v}_n(t,\vec{x}) = e^{i\alpha_n(t)}v_n(t,\vec{x})$ for a general $v_n(t,\vec{x})$, which may not satisfy the condition (14), the parallel transport condition

$$\int d^3x \bar{v}_n^\dagger(t,\vec{x})\frac{\partial}{\partial t}\bar{v}_n(t,\vec{x}) = 0 \quad (17)$$

gives

$$\bar{v}_n(t,\vec{x}) = \exp[i\int_0^t dt'\int d^3x v_n^\dagger(t',\vec{x})i\partial_{t'}v_n(t',\vec{x})]v_n(t,\vec{x}). \quad (18)$$

Since $\bar{v}_n(t,\vec{x})$ satisfies the parallel transport condition, the **holonomy**, i.e., the phase change after one cycle, is given by[13]

$$\begin{aligned}&\bar{v}_n^\dagger(0,\vec{x})\bar{v}_n(T,\vec{x})\\&= v_n^\dagger(0,\vec{x})v_n(T,\vec{x})\exp[i\int_0^T dt'\int d^3x v_n^\dagger(t',\vec{x})i\partial_{t'}v_n(t',\vec{x})]. \end{aligned}\quad (19)$$

This holonomy of *basis vectors*, not of the Schrödinger amplitude, associated with the hidden local symmetry determines *all* the geometric phases. In fact, the adiabatic phase in (13) is an example.

2.3. *Non-adiabatic phase: Cyclic evolution*

The cyclic evolution is defined by[6]

$$\begin{aligned}&\int d^3x\psi^\dagger(t,\vec{x})\psi(t,\vec{x}) = 1,\\&\psi(t,\vec{x}) = e^{i\phi(t)}\tilde{\psi}(t,\vec{x}),\quad \tilde{\psi}(T,\vec{x}) = \tilde{\psi}(0,\vec{x}).\end{aligned}\quad (20)$$

namely, $\psi(T,\vec{x}) = e^{i\phi}\psi(0,\vec{x})$ with $\phi(T) = \phi,\quad \phi(0) = 0$.

If one chooses the first element of the arbitrary basis set $\{v_n(t,\vec{x})\}$ in (1) such that $v_1(t,\vec{x}) = \tilde{\psi}(t,\vec{x})$, one can confirm that the exact Schrödinger amplitude (8) is written as

$$\begin{aligned}\psi(t,\vec{x}) = v_1(t,\vec{x})\exp\{-\frac{i}{\hbar}[&\int_0^t dt\int d^3x v_1^\star(t,\vec{x})\hat{H}v_1(t,\vec{x})\\&-\int_0^t dt\int d^3x v_1^\star(t,\vec{x})i\hbar\partial_t v_1(t,\vec{x})]\}.\end{aligned}\quad (21)$$

Under the hidden local symmetry of basis vectors, we have

$$\psi(t,\vec{x}) \to e^{i\alpha_1(0)}\psi(t,\vec{x}) \quad (22)$$

and the gauge invariant quantity is given by

$$\begin{aligned}&\psi^\dagger(0,\vec{x})\psi(T,\vec{x})\\&= v_1^\star(0,\vec{x})v_1(T,\vec{x})\exp\{-\frac{i}{\hbar}\int_0^T dt\int d^3x[v_1^\star(t,\vec{x})\hat{H}v_1(t,\vec{x})\\&\qquad -v_1^\star(t,\vec{x})i\hbar\partial_t v_1(t,\vec{x})]\}.\end{aligned}\quad (23)$$

If one chooses the specific hidden local symmetry $v_1(0,\vec{x}) = v_1(T,\vec{x})$, $v_1^\star(0,\vec{x})v_1(T,\vec{x})$ becomes real and positive, and the factor

$$\beta = \oint dt\int d^3x v_1^\star(t,\vec{x})i\frac{\partial}{\partial t}v_1(t,\vec{x}) \quad (24)$$

gives the unique *non-adiabatic phase.*[6] Eq. (23) gives another example of the holonomy (19), namely, the holonomy of the basis vector, not of the Schrödinger amplitude, determines the non-adiabatic phase in our formulation.[12]

Note that the so-called "projective Hilbert space" and the transformation of the Schrödinger amplitude[6]

$$\psi(t,\vec{x}) \to e^{i\omega(t)}\psi(t,\vec{x}), \tag{25}$$

which is not the symmetry of the Schrödinger equation, is not used in our formulation. We note that the consistency of the "projective Hilbert space" (25) with the superposition principle is not obvious.[12] More about this will be discussed later.

2.4. *Non-adiabatic phase: Non-cyclic evolution*

Any exact Schrödinger amplitude is written in the form

$$\begin{aligned}\psi_k(\vec{x},t) &= v_k(\vec{x},t)\exp\{-\frac{i}{\hbar}\int_0^t\int d^3x[v_k^{\dagger}(\vec{x},t)\hat{H}(t)v_k(\vec{x},t)\\ &\qquad -v_k^{\dagger}(\vec{x},t)i\hbar\frac{\partial}{\partial t}v_k(\vec{x},t)]\}\end{aligned} \tag{26}$$

if one chooses $\{v_k(\vec{x},t)\}$ suitably.[13] Note, however, the periodicity

$$v_k(T,\vec{x}) = v_k(0,\vec{x}) \tag{27}$$

is lost in general, and thus *non-cyclic.*

In this case, the quantity

$$\begin{aligned}\int d^3x\psi_k^{\dagger}(0,\vec{x})\psi_k(T,\vec{x}) &= \int d^3x v_k^{\dagger}(0,\vec{x})v_k(T,\vec{x})\\ \times\exp\{\frac{-i}{\hbar}\int_0^T dtd^3x[v_k^{\dagger}(t,\vec{x})\hat{H}(t)v_k(t,\vec{x})\\ -v_k^{\dagger}(t,\vec{x})i\hbar\partial_t v_k(t,\vec{x})]\}\end{aligned} \tag{28}$$

is manifestly invariant under the hidden local symmetry (9). By choosing a suitable hidden symmetry $v_k(t,\vec{x}) \to e^{i\alpha_k(t)}v_k(t,\vec{x})$, one can make the pre-factor

$$\int d^3x v_k^{\dagger}(0,\vec{x})v_k(T,\vec{x}) \tag{29}$$

real and positive. It is important that we can make only the integrated pre-factor (29) real and positive in the present non-cyclic case, since one cannot make $v_k^{\dagger}(0,\vec{x})v_k(T,\vec{x})$ real and positive by a time dependent gauge transformation for all $\vec{x}$ for the non-cyclic case.[11] Then the exponential factor in (28) defines the unique non-cyclic and non-adiabatic phase.[7] We have a structure similar to (19) in the present non-cyclic case also, though it may not be called holonomy in a rigorous sense. We emphasize that we do not use the projective Hilbert space defined by (25) in the present formulation of non-adiabatic and non-cyclic geometric phase.[13]

2.5. *Geometric phase for mixed states*

We start with a given hermitian Hamiltonian $\hat{H}(t)$ and given $\mathcal{U}(t) = T^{\star}\exp[-\frac{i}{\hbar}\int_0^t \hat{H}(t)dt]$. We employ a diagonal form of the density matrix

$$\rho(0) = \sum_k \omega_k \psi_k(0,\vec{x})\psi_k^{\dagger}(0,\vec{x}), \tag{30}$$

where the exact Schrödinger amplitudes are defined by

$$\psi_k(t,\vec{x}) = \langle\vec{x}|\mathcal{U}(t)|k\rangle = \int d^3y \langle\vec{x}|\mathcal{U}(t)|\vec{y}\rangle v_k(0,\vec{y}). \tag{31}$$

We define the total phases for pure states $\psi_k(t,\vec{x})$ by

$$\phi_k(t) = \arg\int d^3x \psi_k^{\dagger}(0,\vec{x})\psi_k(t,\vec{x}) \tag{32}$$

and the complete set of basis vectors in (1) by

$$v_k(t,\vec{x}) = e^{-i\phi_k(t)}\psi_k(t,\vec{x}), \quad \int d^3x v_k^{\dagger}(t,\vec{x})v_l(t,\vec{x}) = \delta_{k,l}. \tag{33}$$

One can then confirm that the exact Schrödinger amplitudes are written as

$$\begin{aligned}\psi_k(\vec{x},t) &= v_k(\vec{x},t)\\ &\times \exp\{-\frac{i}{\hbar}\int_0^t [\int d^3x v_k^{\dagger}(\vec{x},t)\hat{H}(t)v_k(\vec{x},t) - \langle k|i\hbar\frac{\partial}{\partial t}|k\rangle]\}\end{aligned} \tag{34}$$

with

$$\langle k|i\hbar\frac{\partial}{\partial t}|k\rangle \equiv \int d^3x v_k^{\dagger}(\vec{x},t)i\hbar\frac{\partial}{\partial t}v_k(\vec{x},t). \tag{35}$$

The Schrödinger amplitude $\psi_k(t,\vec{x})$ is transformed under the hidden local symmetry as $\psi_k(t,\vec{x}) \to e^{i\alpha_k(0)}\psi_k(t,\vec{x})$ independently of t and thus the Schrödinger equation is invariant under the hidden local symmetry.

The quantity $\mathrm{Tr}\mathcal{U}(T)\rho(0)$ is then written as

$$\mathrm{Tr}\mathcal{U}(T)\rho(0) = \sum_k \omega_k \psi_k^\dagger(0,\vec{x})\psi_k(T,\vec{x})$$
$$= \sum_k \omega_k v_k^\dagger(0,\vec{x}) v_k(T,\vec{x}) \exp\{\frac{i}{\hbar}\int_0^T dt d^3x [v_k^\dagger(t,\vec{x}) i\hbar\partial_t v_k(t,\vec{x})$$
$$- v_k^\dagger(t,\vec{x})\hat{H}(t) v_k(t,\vec{x})]\} \qquad (36)$$

without integration over $\vec{x}$. If all the pure states perform cyclic evolution with the same period T, one can choose the hidden local gauge such that

$$v_k^\dagger(0,\vec{x}) v_k(T,\vec{x}) = \text{real and positive} \qquad (37)$$

for all k, and the exponential factor in (36) exhibits the entire geometrical phase together with the "dynamical phase"

$$(1/\hbar)\int_0^T dt d^3x v_k^\dagger(t,\vec{x})\hat{H}(t) v_k(t,\vec{x})$$

of each pure state. In practice, the cyclic evolution of all the pure states $\psi_k(t)$ with a period T may be rather exceptional. For a generic case, we need to define the phase for non-cyclic evolution[7] as the phase of (see (28))

$$\mathrm{Tr}\mathcal{U}(T)\rho(0) = \sum_k \omega_k \int d^3x \psi_k^\dagger(0,\vec{x})\psi_k(T,\vec{x}) \qquad (38)$$
$$= \sum_k \omega_k \int d^3x v_k^\dagger(0,\vec{x}) v_k(T,\vec{x})$$
$$\times \exp\{\frac{i}{\hbar}\int_0^T dt d^3x [v_k^\dagger(t,\vec{x}) i\hbar\partial_t v_k(t,\vec{x}) - v_k^\dagger(t,\vec{x})\hat{H}(t) v_k(t,\vec{x})]\}$$

These quantities (36) and (38) are manifestly invariant under the hidden local symmetry,[13] and thus not only the total phase $\arg\mathrm{Tr}\mathcal{U}(T)\rho(0)$ but also the visibility $|\mathrm{Tr}\mathcal{U}(T)\rho(0)|$ in the interference pattern[8]

$$I \propto 1 + |\mathrm{Tr}\mathcal{U}(T)\rho(0)| \cos[\chi - \arg\mathrm{Tr}\mathcal{U}(T)\rho(0)] \qquad (39)$$

which are experimentally observable are manifestly gauge invariant. Here χ stands for the variable $U(1)$ phase (difference) in the interference beams. We note that the gauge invariance of the interference pattern (39) does not hold in the sense of the projective Hilbert space (25) in the conventional formulation,[8,9] which is related to the fact that the projective Hilbert space defined by (25) is not consistent with the superposition principle to describe interference.[12]

3. Exactly Solvable Example

We discuss the model

$$\hat{H} = -\mu\hbar\vec{B}(t)\vec{\sigma},$$
$$\vec{B}(t) = B(\sin\theta\cos\varphi(t), \sin\theta\sin\varphi(t), \cos\theta) \tag{40}$$

with $\varphi(t) = \omega t$ and constant ω, B and θ. This model has been analyzed in the past by various authors by using the adiabatic approximation.[3] It has been recently shown that this model is exactly treated in the framework of the second quantized formulation.[13,22]

The exact effective Hamiltonian (6) is given by

$$\hat{H}_{eff}(t) = [-\mu\hbar B - \frac{(1+\cos\theta)}{2}\hbar\omega]\hat{b}_+^\dagger\hat{b}_+$$
$$+[\mu\hbar B - \frac{1-\cos\theta}{2}\hbar\omega]\hat{b}_-^\dagger\hat{b}_- - \frac{\sin\theta}{2}\hbar\omega[\hat{b}_+^\dagger\hat{b}_- + \hat{b}_-^\dagger\hat{b}_+] \tag{41}$$

if one uses the instantaneous eigenstates

$$\hat{H}(t)v_\pm(t) = \mp\mu\hbar B v_\pm(t) \tag{42}$$

as the complete basis set in (1) and the expansion $\hat{\psi}(t) = \sum \hat{b}_n v_n(t)$. This effective H_{eff} is not diagonal, but it is *diagonalized* if one performs a unitary transformation

$$\begin{pmatrix} \hat{b}_+(t) \\ \hat{b}_-(t) \end{pmatrix} = \begin{pmatrix} \cos\frac{1}{2}\alpha & -\sin\frac{1}{2}\alpha \\ \sin\frac{1}{2}\alpha & \cos\frac{1}{2}\alpha \end{pmatrix} \begin{pmatrix} \hat{c}_+(t) \\ \hat{c}_-(t) \end{pmatrix} \tag{43}$$

with a constant α satisfying the parameter equation

$$\tan\alpha = \frac{\hbar\omega\sin\theta}{2\mu\hbar B + \hbar\omega\cos\theta}. \tag{44}$$

The corresponding new basis vectors are then explicitly given by

$$w_+(t) = \begin{pmatrix} \cos\frac{1}{2}(\theta-\alpha)e^{-i\varphi(t)} \\ \sin\frac{1}{2}(\theta-\alpha) \end{pmatrix}, w_-(t) = \begin{pmatrix} \sin\frac{1}{2}(\theta-\alpha)e^{-i\varphi(t)} \\ -\cos\frac{1}{2}(\theta-\alpha) \end{pmatrix} \tag{45}$$

which satisfies $\hat{\psi}(t) = \sum \hat{b}_n v_n(t) = \sum \hat{c}_n w_n(t)$. These new basis vectors are periodic $w_\pm(0) = w_\pm(T)$ with $T = \frac{2\pi}{\omega}$, and one can confirm

$$w_\pm^\dagger(t)\hat{H}w_\pm(t) = \mp\mu\hbar B\cos\alpha$$
$$w_\pm^\dagger(t)i\hbar\partial_t w_\pm(t) = \frac{\hbar\omega}{2}(1\pm\cos(\theta-\alpha)). \tag{46}$$

The effective Hamiltonian H_{eff} (41) is now diagonalized in terms of $w_{\pm}(t)$, and thus the *exact* solution of the Schrödinger eq., $i\hbar\partial_t\psi(t) = \hat{H}\psi(t)$, is given by

$$\psi_{\pm}(t) = w_{\pm}(t)\exp\{-\frac{i}{\hbar}\int_0^t dt'[w_{\pm}^{\dagger}(t')\hat{H}w_{\pm}(t') \\ -w_{\pm}^{\dagger}(t')i\hbar\partial_{t'}w_{\pm}(t')]\} \tag{47}$$

if one uses the formula (8). This amplitude may be regarded either as an exact version of the adiabatic phase or as a non-adiabatic cyclic phase in our formulation in (21).

We examine the two extreme limits of this formula:

(i) For the *adiabatic limit* $\hbar\omega/(\hbar\mu B) \ll 1$, the parameter equation (44) gives

$$\alpha \simeq [\hbar\omega/2\hbar\mu B]\sin\theta, \tag{48}$$

and if one sets $\alpha = 0$ in the exact solution (47), one recovers the ordinary Berry phase[3,4]

$$\psi_{\pm}(T) \simeq \exp\{i\pi(1 \pm \cos\theta)\} \\ \times \exp\{\pm\frac{i}{\hbar}\int_0^T dt\mu\hbar B\}v_{\pm}(T) \tag{49}$$

where the first exponential factor stands for the "monopole-like phase" and

$$v_{+}(t) = \begin{pmatrix} \cos\frac{1}{2}\theta e^{-i\varphi(t)} \\ \sin\frac{1}{2}\theta \end{pmatrix}, v_{-}(t) = \begin{pmatrix} \sin\frac{1}{2}\theta e^{-i\varphi(t)} \\ -\cos\frac{1}{2}\theta \end{pmatrix}. \tag{50}$$

(ii) For the *non-adiabatic* limit $\hbar\mu B/(\hbar\omega) \ll 1$, the parameter equation (44) gives

$$\theta - \alpha \simeq [2\hbar\mu B/\hbar\omega]\sin\theta \tag{51}$$

and if one sets $\alpha = \theta$ in the exact solution (47), one obtains the trivial phase

$$\psi_{\pm}(T) \simeq w_{\pm}(T)\exp\{\pm\frac{i}{\hbar}\int_0^T dt[\mu\hbar B\cos\theta]\} \tag{52}$$

with

$$w_{+}(t) = \begin{pmatrix} e^{-i\varphi(t)} \\ 0 \end{pmatrix}, \\ w_{-}(t) = \begin{pmatrix} 0 \\ -1 \end{pmatrix}. \tag{53}$$

This shows that the "monopole-like singularity" is smoothly connected to a trivial phase in the exact solution, and thus the geometric phase is *topologically trivial.*[22]

The adiabatic and non-adiabatic phases are treated in a unified manner in the present second quantized formulation, and thus this example shows that all the geometric phases are topologically trivial.

4. Chiral Anomaly

We consider the evolution operator

$$\int \mathcal{D}\bar{\psi}\mathcal{D}\psi \exp\{i\int d^4x[\bar{\psi}i\gamma^\mu(\partial_\mu - igA_\mu)\psi]\} \tag{54}$$

for the Dirac fermion $\psi(t,\vec{x})$ inside the background gauge field $A_\mu(t,\vec{x})$. The chiral anomaly in gauge field theory is understood in path integrals as arising from the non-trivial Jacobian under the chiral transformation. For an infinitesimal chiral transformation of field variables

$$\psi(x) \to e^{i\omega(x)\gamma_5}\psi(x), \qquad \bar{\psi}(x) \to \bar{\psi}(x)e^{i\omega(x)\gamma_5} \tag{55}$$

we have a non-trivial Jacobian

$$\mathcal{D}\bar{\psi}\mathcal{D}\psi \to \exp\{-i\int d^4x\omega(x)\frac{g^2}{16\pi^2}\epsilon^{\mu\nu\alpha\beta}F_{\mu\nu}F_{\alpha\beta}\}\mathcal{D}\bar{\psi}\mathcal{D}\psi \tag{56}$$

which is valid for a general class of regularization including the lattice gauge theory.[18] The Jacobian factor is identified with the chiral anomaly, and the integrated or summed Jacobian is called the Wess-Zumino term.[16]

Some of the known essential and general properties of the quantum anomalies are:[18]

(i) The anomalies are not recognized by a naive manipulation of the classical Lagrangian or action (or by a naive canonical manipulation in operator formulation), which leads to the naive Nöther's theorem.
(ii) The quantum anomaly is related to the quantum breaking of classical symmetries (and the failure of the naive Nöther's theorem). For example, the Gauss law operator (or BRST charge) becomes time-dependent and thus it cannot be used to specify physical states in anomalous gauge theory.
(iii) The quantum anomalies are generally associated with an infinite number of degrees of freedom. The anomalies in the practical calculation are thus closely related to the regularization, though the anomalies by themselves are perfectly finite.

(iv) In the path integral formulation, the anomalies are recognized as non-trivial Jacobians for the change of path integral variables associated with classical symmetries, as is explained above.

None of these essential properties are shared with the geometric phases discussed in Sections 2 and 3. One rather recognizes the following basic differences between the geometric phases and chiral anomaly:[21]

(i) The Wess-Zumino term, which is obtained by a sum of the infinitesimal Jacobian such as in (56), is added to the classical action in path integrals, whereas the geometric term appears *inside* the classical action sandwiched by field variables as in (6)

$$\hat{H}_{eff}(t) = \sum_{n,m} \hat{b}_n^\dagger(t)[\int d^3x v_n^\star(t,\vec{x})\hat{H}(t)v_m(t,\vec{x}) - \int d^3x v_n^\star(t,\vec{x}) i\hbar\frac{\partial}{\partial t}v_m(t,\vec{x})]\hat{b}_m(t). \tag{57}$$

The geometric phase thus depends on each state in the Fock space generated by $\hat{b}_n^\dagger$, whereas the chiral anomaly is state-independent.

(ii) The topology of chiral anomaly, which is provided by given gauge field, is exact, whereas the topology of the adiabatic geometric phase, which is valid only approximately in the adiabatic limit, is trivial as we have shown in Section 3.

(iii) The geometric phases are basically different from the topologically exact objects such as the Aharonov-Bohm phase or chiral anomaly. For example, the Aharonov-Bohm phase is identical for adiabatic or non-adiabatic motion of the electron.

(iv) *Similarity* between the geometric phase and a special class of chiral anomaly was noted by M. Stone on the basis of a model[20]

$$\mathcal{H}(t) = \frac{\vec{L}^2}{2I} - \psi^\dagger \mu \mathbf{n}(t)\cdot\vec{\sigma}\psi \tag{58}$$

where $\mathbf{n}(t)$ plays a role of the magnetic field in (40) which acts on the spin $\vec{\sigma}$, and $\vec{L}$ induces the rotation of $\mathbf{n}(t)$. But it is obvious from our analysis of topological properties in Section 3 that these two notions are fundamentally different.

(v) The topology of Berry's phase is valid only when the adiabatic approximation is strictly valid, whereas the anomaly appears in field theory only when the adiabatic approximation *fails* in a version of the Hamiltonian analysis.[19] Thus these notions cannot be compatible.

5. Conclusion

We have illustrate the advantages of the second quantized formulation of all the geometric phases. The second quantized formulation is located in between the first quantization and field theory, and thus it is convenient to compare the geometric phase with other phases such as chiral anomaly. We clarified the basic differences between these two notions.

In the early literature on the geometric phase, the similarity between the geometric phase and other phases such as the chiral anomaly and the Aharonov-Bohm phase, was often emphasized. But in view of the wide use of the loosely defined terminology "geometric phase" in various fields in physics today, it is our opinion that a more precise distinction of "identical phenomena" from "similar phenomena" is important. To be precise, what we are suggesting is to call chiral anomaly as chiral anomaly, Wess-Zumino term as Wess-Zumino term, and Aharovov-Bohm phase as Aharonov-Bohm phase, etc., since those terminologies convey very clear messages and well-defined physical contents which the majority in physics community can readily recognize. Even in this sharp definition of terminology, one should still be able to clearly identify the geometric phase and its physical characteristics, which are intrinsic to the geometric phase and cannot be described by other notions.

References

1. H. Longuet-Higgins, Proc. Roy. Soc. A**344**(1975) 147.
2. C. Mead and D. Truhlar, J. Chem. Phys. **70** (1978) 2284.
3. M.V. Berry, Proc. Roy. Soc. A**392**(1984) 45.
4. B. Simon, Phys. Rev. Lett. **51**(1983) 2167.
5. F. Wilczek and A. Zee, Phys. Rev. Lett. **52**, 2111 (1984).
6. Y. Aharonov and J. Anandan, Phys. Rev. Lett. **58**(1987) 1593.
7. J. Samuel and R. Bhandari, Phys. Rev. Lett. **60**(1988) 2339.
8. E. Sjöqvist, et al., Phys. Rev. Lett. **85** (2000) 2845.
9. K. Singh, D.M. Tong, K. Basu, J.L. Cheng and J.F. Du, Phys. Rev. A**67** (2003) 032106.
10. K. Fujikawa, Mod. Phys. Lett. A**20**, 335 (2005),
    S. Deguchi and K. Fujikawa, Phys. Rev. A**72**(2005) 012111.
11. K. Fujikawa, Phys. Rev. D**72**(2005) 025009.
12. K. Fujikawa, Int. J. Mod. Phys. A**21** (2006) 5333.
13. K. Fujikawa, Ann. of Phys. 322 (2007) 1500.
14. J.S. Bell and R. Jackiw, Nuovo Cim. **60A** (1969) 47.
15. S.L. Adler, Phys. Rev. **177** (1969) 2426.
16. J. Wess and B. Zumino, Phys. Lett. B**37** (1971) 95.
17. K. Fujikawa, Phys. Rev. Lett. **42**, 1195 (1979); Phys. Rev. D**21**, 2848 (1980).

18. K. Fujikawa and H. Suzuki, *Path Integrals and Quantum Anomalies* (Oxford University Press, Oxford, 2004).
19. P. Nelson and L. Alvarez-Gaume, Comm. Math. Phys. **99** (1985) 103.
20. M. Stone, Phys. Rev. **D 33** (1986) 1191.
21. K. Fujikawa, Phys. Rev. D **73** (2006) 025017.
22. K. Fujikawa, Phys. Rev. D **77** (2008) 045006.

CONSEQUENCES OF A MINIMAL LENGTH

LAY NAM CHANG

College of Science, Virginia Tech,
Blacksburg, Virginia 24060, USA

Many approaches at incorporating quantum effects in gravity result in a fundamental minimal length. There have been several ways of manifesting this length, notably those of Yang and Snyder. In this paper, I describe the consequences of such a possibility within the context of several simple quantum mechanical systems, and point out how these can be used to constrain the value of the minimal length. I also discuss some implications for relativistic dynamics.

The idea of a minimal length has been around for as long as local quantum field theory. The idea initially was to utilize such a length to ameliorate the divergences in such descriptions. Later on, with the introduction of renomalization, the idea of such a length was shown not to be critical. Indeed, successful applications of local field theory have incorporated the essence of scale invariance, doing away with any need for a minimal length.

The situation however changes once we attempt to incorparate gravity. There is a fundamental length present in most quantum descriptions of gravity, with the emergence of the Planck length. Unfortunately, this length is very small compared to the distances presently accesible and so whether there is in fact such a length is not such a pressing issue. Nonetheless, it is instructive to examine the manner a minimal length would affect some well-know physical systems. This paper explores some of the nature of these consequences.

Lately, there have been suggestions, based upon considerations involving higher dimensions,[1] that such a length may in fact be larger than thought, and so it becomes tantalizing to entertain a minimal length again. And to ask if there are ways such a length might show an effect at much larger scales.

The context then would be to suppose that we have a description in which points are still relevant, but that there is a length below which the point description breaks down. One possible way to do so would be to imagine that position coordinates no longer commute, but that the remaining dynamical structure requires no further modifications. Just as the breakdown in commutation between position and momentum would imply a minimum action, below which phase space is discrete, such a breakdown would imply a minimum length in the system, below which space itself may be discrete.

In this setting, one of the more significant earlier suggestions was due to Snyder[2] and Yang.[3] They asked the question if space-time can be discrete, and if so, what would happen to the kinematic symmetries of Lorentz transformations and translations. What I describe below is an elaboration of what they proposed then.

I begin by displaying the algebra considered by Yang:

$$\begin{aligned} [x_i, x_j] &= ia^2 L_{ij} \,, \\ [p_i, p_j] &= iL_{ij}/R^2 \,, \\ [x_i, p_j] &= i\xi\delta_{ij} \,. \end{aligned}$$

Here ξ is an operator with the commutation relations:

$$\begin{aligned} [\xi, x_i] &= ia^2 p_i \,, \\ [\xi, p_i] &= ix_i/R^2 \,. \end{aligned}$$

This algebra is characterized by two lengths, and describes a De Sitter geometry. In order to gain a better appreciation of what consequences such lengths entail, it is instructive to consider a contracted version of the algebra, obtained by going into the limit $R \to \infty$. In this limit, the operators p_i commute with each other, and may be thought of as generators of translations.

In 1-D we may imagine that the operator ξ can be expressed as a function of p. In which case, we can express the commutation relation between x and p as

$$\begin{aligned} [x, p] &= i\hbar\left\{1 + \beta p^2\right\} \,, \\ \beta &= -a^2/2 \,. \end{aligned}$$

We may consider this form for ξ as a perturbative expansion of a more general function of p. It gives rise immediately to a relation between the position and momentum uncertainties:

$$\Delta x \geq \frac{\hbar}{2}\left(\frac{1}{\Delta p} + \beta \Delta p\right),$$
$$\Delta x \geq \hbar\sqrt{\beta}$$

making manifest the significance of the parameter β as the minimal length. The relation also indicates a UV/IR mixing characteristic of many quantum gravity descriptions, or indeed any system with such a length.

In D-dimensions, we may posit the following algebra among the operators x_i to incorporate a minimal length:

$$[x_i, x_j] = i\mathcal{A}_{ij}\,.$$

We shall assume the simplest structure to bring closure of the algebra. In particular we suppose that the underlying geometry has rotational symmetry, with generators L_{ij} satisfying the usual $O(D)$ algebra. The operators x_i and p_i will be assumed to transform as vectors under this $O(D)$ algebra. The simplest form of this closure can then be taken to be:

$$\mathcal{A}_{ij} = if\ L_{ij}\,,$$
$$[x_i, f] = i\gamma p_i$$

with f regarded simply as a function in p^2. Using rotational covariance,

$$[x_i, p_j] = i\hbar\left\{f_1\delta_{ij} + f_2 p_i p_j\right\}\,.$$

Jacobi identity may then be used to show that the two functions $f_{1,2}$ are not independent:

$$f_2 = \frac{f_1 + 2f_1\dot{f}_1}{f_1 - 2p^2\dot{f}_1}\,.$$

Here $\dot{f}_{1,2}$ refers to differentiation relative to the argument p^2.

To proceed, we will imagine that we may develop a perturbative expansion for the functions $f_{1,2}$. To second order,

$$f_1 = 1 + \beta p^2\,,$$
$$f_2 = \beta'\,.$$

The complete algebra can then be obtained:

$$[x_i, p_j] = i\hbar\left\{\delta_{ij}\left(1 + \beta p^2\right) + \beta' p_i p_j\right\}\,,$$
$$[p_i, p_j] = 0\,,$$

$$[x_i, x_j] = i\hbar \left\{(\beta' - 2\beta) - (2\beta + \beta')\beta p^2\right\} L_{ji}\,,$$
$$L_{ij} = \frac{x_i p_j - x_j p_i}{(1 + \beta p^2)}\,.$$

This algebra can be realized as Poisson brackets:

$$\{F, G\} = \left(\frac{\partial F}{\partial x_i}\frac{\partial G}{\partial p_j} - \frac{\partial F}{\partial p_i}\frac{\partial G}{\partial x_j}\right)\{x_i, p_j\} + \frac{\partial F}{\partial x_i}\frac{\partial G}{\partial x_j}\{x_i, x_j\}\,.$$

As such, we may investigate the effects of a minimal length will have on classical equations of motion.[4] We will be invoking a form of the correspondence principle, so that we have a Hamiltonian H having the same form as normal dynamics. The equations of motion are then taken to be

$$H = p^2/2M + V\,,$$
$$\dot{x}_i = \{x_i, H\} = \{x_i, p_j\}\frac{\partial H}{\partial p_j} + \{x_i, x_j\}\frac{\partial H}{\partial x_j}\,,$$
$$\dot{p}_i = \{p_i, H\} = -\{x_i, p_j\}\frac{\partial H}{\partial x_j}\,.$$

Newton's equation now reads, for $\beta' = 0$:

$$m\ddot{x}_i = (1 + \beta p^2)(1 + 3\beta p^2)F_i\,,$$
$$F_i = -\frac{\partial V}{\partial x_i}\,.$$

Note the dependence on momentum in the equation for acceleration. Single point particle motion in gravity now exhibits a breakdown in equivalence principle. That this is so is a facet of the UV/IR mixing alluded earlier. Minimal lengths endow point particles with a structure. The resultant breakdown in the equivalence principle can be traced to tidal effects. However, there could well be a modified form of this principle expressed as general covariance on the underlying noncommutative geometry.

We list here some consequences for the specific case of motion in central potentials.

1. Because of the underlying rotational symmetry, all motion continues to occur in a plane, characterized by angular momentum, which is conserved.

2. As a result of the momentum dependence in Newton's equation, orbits are generally not closed, even for the special cases of a simple harmonic oscillator, and for Kepler motion.[4] For the case of simple harmonic motion, the "perihelion" precession will advance for the special case of $\beta \neq 0$ but with $\beta' = 0$, but switches sign in the complementary case of $\beta = 0$, $\beta' \neq 0$. For the important case of Kepler motion, the precession lags.
3. One may compute modifications to Kepler's third law by regarding the minimal length as a perturbation.[5] In which case, if a is the semi-major axis of the orbit, M the mass of the Sun, and m the mass of the planet, T the period, then

$$\begin{aligned}
\frac{a^3}{T^2} &= \frac{GM}{4\pi} f(\epsilon) \,, \\
f(\epsilon) &= \frac{1}{2} \left\{ \left(1 + 4\epsilon - \epsilon^2\right) + (1+\epsilon)\sqrt{1 + 6\epsilon + \epsilon^2} \right\} \,, \\
\epsilon &= \beta m^2 \frac{GM}{a} \,.
\end{aligned}$$

There is no dependence on β'. Deviations from the third law have been reported in the literature. However we have not taken into account the effects from other forces on planets and satellites using the deformed dynamics, and comparisons have yet to be performed.

We now turn our attention to how minimal lengths affect quantum mechanical systems. In preparation for doing so, it is important to see how we may quantify the density of states. As a first step, we will need an expression for the volume in phase space which remains invariant under dynamical evolution. This analog of the classical Liouville volume can be obtained by straightforward computation using the form of the equations of motion exhibited above.[6] The result is

$$\frac{d^D \mathbf{x} d^D \mathbf{p}}{\left[1 + \beta p^2\right]^{D-1} \left[1 + (\beta + \beta') p^2\right]^{1 - \beta'/2(\beta+\beta')}} \,.$$

As remarked earlier, the existence of a minimal length could provide a UV cut-off to phase space integrals. As an example of how this happens, let us now consider how an estimate of the cosmological constant Λ can be effected through use of the Liouville volume. For concreteness, we shall set β' to zero, and integrate over the spatial volume. The cosmological constant is the energy density contributed by each state within the volume. For this estimate, we will suppose that the volume is populated by free particles, each of which has the elementary energy of $\sqrt{p^2+m^2}$. In three dimensions,

$$\begin{aligned}\Lambda(m) &= \int \frac{d^3p}{(1+\beta p^2)^3}\left[\frac{1}{2}\sqrt{p^2+m^2}\right] \\ &= 2\pi\int_0^\infty \frac{p^2\,dp}{(1+\beta p^2)^3}\sqrt{p^2+m^2} \\ &= \frac{\pi}{2\beta^2} f\left(\beta m^2\right),\end{aligned}$$

where

$$f(x) = \left\{\begin{array}{l} 1+\dfrac{x}{2(1-x)}+\dfrac{x^2}{4(1-x)^{3/2}}\ln\left(\dfrac{1-\sqrt{1-x}}{1+\sqrt{1-x}}\right) \\ 1-\dfrac{x}{2(x-1)}+\dfrac{x^2}{2(x-1)^{3/2}}\tan^{-1}\sqrt{x-1} \end{array}\right\}.$$

This function is monotonically increasing, and has these limiting values:

$$\begin{aligned} f(x) &= \sqrt{x} \text{ for large } x \\ &= (1+x)^{0.42} \quad 0<x<1\,. \end{aligned}$$

For massless excitations,

$$\Lambda(0) = \frac{\pi}{2\beta^2}\,.$$

In all instances, as expected, the minimal length $\sqrt{\beta}$ provides a UV cut-off to render the integrals finite. Unfortunately, the expression doesn't really solve the current dark energy conundrum; the value of $1/\sqrt{\beta}$ is expected to be at the Planck scale and is therefore too high to be of significance.

The presence of the weight factor in the phase space measure, expressed for photons as

$$\left(\frac{1}{1+\beta\nu^2}\right)^3$$

has an effect on blackbody radiation. Inclusion of this weight gives a modification to the spectral distribution function:

$$\begin{aligned} u_\beta &= \frac{1}{\left[1+\left(y/a_\beta\right)^2\right]^3} u_0\,, \\ u_0 &= \frac{8\pi h{\nu_{\mathrm{B}}}^3}{c^3}\left\{\frac{y^3}{\exp y - 1}\right\}, \end{aligned}$$

$$
\begin{aligned}
y &= h\nu / k_{\rm B}T\,, \\
a_\beta &= T_\beta / T\,, \\
T_\beta &= \frac{c}{k_{\rm B}}\frac{1}{\sqrt{\beta}}\,, \\
\nu_{\rm B} &= \frac{k_{\rm B}T}{h}\,.
\end{aligned}
$$

Here ν is the frequency of the photon, T is the temperature, and $k_{\rm B}$ is Boltzmann's constant. The effects of the weight is therefore noticeable only for temperatures close to T_β. For values of the minimal length near the Planck scale, these are not significant in the cosmic microwave background (CMB), given that decoupling in the CMB is at the level of MeV.[7] That will remain the case even if there were higher dimensions that would lower the Planck scale.[1]

We now examine a system within a quantum mechanical context. To get an initial sense of how a minimal length would affect the description, we begin with the well-known example of a simple harmonic oscillator in D-dimensions:

$$
\begin{aligned}
H &= \frac{\hat{p}^2}{2M} + \frac{1}{2}M\omega^2\hat{x}^2\,, \\
\hat{x}_i &= i\hbar\left\{\left(1+\beta p^2\right)\frac{\partial}{\partial p_i} + \beta' p_i p_j \frac{\partial}{\partial p_j} + \gamma p_i\right\}\,, \\
\hat{p}_i &= p_i\,.
\end{aligned}
$$

Here $\hat{p}_i$ and $\hat{x}_i$ are operators that satisfy the deformed algebra involving a minimal length $\sqrt{\beta}$. γ is a central extension that is allowed by the algebra. Its value does determine the inner product

$$
\begin{aligned}
\langle f|g\rangle &= \int \frac{dp}{\left(1+\beta p^2\right)^{1-\alpha}} f^*(p)g(p)\,, \\
\alpha &= \frac{\gamma}{\beta}\,.
\end{aligned}
$$

Relative to this inner product, $\hat{x}_i$ is hermitian. It will turn out that the energy eigenvalues are independent of γ.

Solving the resultant differential equation from the Hamiltonian is straightforward, and details are presented in Ref. 8. The energy eigenvalues are given by:

$$
\epsilon = m\hbar\omega\,,
$$

$$E_{n\ell} = \hbar\omega \left\{ \left(n + \frac{D}{2}\right) \sqrt{1 + \left(\beta^2 L^2 + \frac{(D\beta + \beta')^2}{4}\right)\epsilon^2} \right\}$$
$$+ \hbar\omega \left\{ (\beta + \beta') \left(n + \frac{D}{2}\right)^2 \right\}$$
$$+ \hbar\omega \left\{ (\beta + \beta') \left(L^2 + \frac{D^2}{4}\right) + \beta' \frac{D}{2} \right\} \frac{\epsilon}{2} .$$

Here L is the angular momentum of the eigenstate, given by

$$L^2 = \ell\,(\ell + D - 2)\,, \qquad \ell = 0, 1, 2, \ldots .$$

These are no longer equally spaced, a result that is not unexpected since the classical orbits as we have seen are no longer closed. Note in addition that the dependence on the quantum numbers is not linear. The bottom of the harmonic potential is now flatter, giving rise to a system that is more reminiscent of a square-well potential, and thence, to a quadratic dependence on the quantum number. This result agrees with previous ones, particularly those obtained by Kempf.[9] Those results however were obtained on the basis of perturbation theory to first order in β and β'.

One place to look for such departures would be through careful measurements of electron orbits in a Penning trap, which is in essence a 1-D simple harmonic oscillator. Treating the resultant single parameter $\beta + \beta'$ as a perturbation in β alone, the first order departure from linear spacing is

$$\frac{\Delta E_n}{\hbar\omega} = \left(\frac{\beta m \hbar\omega}{2}\right) n^2 ,$$

which grows rapidly with n. The key parameter in the expression $\beta m \hbar\omega$ turns out to be independent of the electron mass, since the frequency is simply in this case the cyclotron frequency $\omega = \omega_c = eB/m_e$:

$$\beta m \hbar\omega_c = \beta\, e\hbar B .$$

For a field strength of 6 teslas,

$$e\hbar\, B = 1.0 \times 10^{-52} \mathrm{km}^2\, \mathrm{m}^2/\mathrm{s}$$

and the absence of such a deviation would mean

$$\frac{\beta\, e\hbar B}{2} n^2 < 1$$

or

$$\frac{1}{\sqrt{\beta}} > (13\ \mathrm{eV}/c)\, n .$$

On the other hand, in order to ensure that the non-relativistic approximation remains valid, we cannot tolerate too large a value for n:

$$\frac{n\hbar\omega}{m_e\, c^2} \ll 1$$

which translates to

$$n \ll 10^9 \, .$$

This limit will also allow the electron orbit to be well within the geometry of the trap. Allowing for a 10% relativistic correction at most, we will require that $n < 10^8$, thereby allowing for $1\sqrt{\beta} > 1$ GeV c, or

$$\hbar\sqrt{\beta} < 10^{-16}\ \mathrm{m}\, .$$

We next turn to the more interesting case of the hydrogen atom. (The problem was first solved in the limit of $\beta' = 2\beta$ in perturbation theory by Brau.[10]) This system presents a challenge in that the potential is singular in the limit $R \to 0$, where R is the spatial distance from the proton to the electron. In a system with a minimal length, R takes on discrete values, and the singularity is smeared out. Using the representation for $\hat{x}_i$ given above, we can solve for the eigenfunctions and eigenvalues of the operator $R^2 = \hat{x}_i\hat{x}^i$. These are parametrized by an angular momentum quantum number ℓ and a principal quantum number n, just as in the case with no minimal length. The eigenvalues are given by $r_{n\ell}^2 = \hbar^2(\beta + \beta')\rho_{n\ell}^2$, while the eigenfunctions are given by $R_{n\ell}$:

$$\rho_{n\ell}^2 = (2n + a + b + 1)^2 - (1 - \eta)^2 \left(L^2 + \frac{(D-1)^2}{4} \right)$$

$$R_{n\ell}(z) \propto (\beta + \beta')^{D/4} \left(\frac{1-z}{2} \right)^{\lambda/2} \left(\frac{1+z}{2} \right)^{\ell/2} P_n^{(a,b)}(z)\, .$$

Here $P_n^{(a,b)}$ are Jacobi polynomials.

We may expand the eigenfunctions of the Hamiltonian in terms of those for R^2, and convert the Schrödinger equation into a recursion relation for the expansion coefficients. Applying standard techniques for termination of the series, we obtain the relevant eigenvalues, and the associated eigenfunctions. The process can be completed numerically, and the results are presented in Ref. 11.

In the same reference, we also perform a perturbative analysis of the same problem, and the results turn out to be in excellent agreement with the numerical answers. The perturbation is carried out as an expansion in β and $\beta\prime$. The results give a better appreciation of the numerical results.

Since the unperturbed potential is singular at the origin, the method fails for S-waves, which are not shielded by any centrifugal barriers. In order to check against the numerical answers, which as pointed above do not suffer from this shortcoming, we truncated the approximation at a small distance from the origin, and attempt to do a matching around the origin. The results show that both methods yield substantially the same answers.

We have attempted at an estimate on a bound on the inverse minimal length, based upon current values of the 1S-2S Lamb shift, and find that the constraint lies somewhere between 1.75 GeV and 6.87 GeV. The bounds were obtained by attributing the entire discrepancy between observations and and standard QED calculations to the contributions obtained by these methods. However we have not made an effort to incorporate corrections to QED resulting from the minimal length deformed algebra. Such corrections lie outside of the scope of the present discussion.

Acknowledgments

I am indebted to my collaborators in the preparation of this presentation. In addition, I have had very helpful conversations on the subject with F. Brau, O. W. Greenberg, M. Koike, J. Slawny, and Y. P. Yao. The research is supported in part by the US Department of Energy Grant DE-FG05-92ER40709.

References

1. L. Randall, R. Sundrum, Phys. Rev. Lett. **83**, 3370 (1997); N. Arkani-Hamed, S. Dimopoulos, G. Dvali, Phys. Lett. **B429**, 263 (1998); L. N. Chang, O. Lebedev, W. Loinaz, T. Takeuchi, Phys. Rev. Lett. **85**, 3765 (2000).
2. H. Snyder, Phys. Rev. **71**, 38 (1947).
3. C. N. Yang, Phys. Rev. **72**, 874 (1947).
4. S. Benczik, L. N. Chang, D. Minic, N. Okamura, S. Rayyan, T. Takeuchi, Phys. Rev. **D66** 026003 (2002).
5. S. Benczik, L. N. Chang, D. Minic, N. Okamura, S. Rayyan, T. Takeuchi, hep-th/0209119 (2002).
6. L. N. Chang, D. Minic, N. Okamura, T. Takeuchi, Phys. Rev. **D65** 125028 (2002).
7. G. F. Smoot, D. Scott, Eur. Phys. J. C. **15**, 145 (2000).
8. L. N. Chang, D. Minic, N. Okamura, T. Takeuchi, Phys. Rev. **D65** 125027 (2002).
9. A. Kempf, J. Phys. **A30**, 2093 (1997).
10. F. Brau, J. Phys. **32**, 7691 (1999).
11. S. Benczik, L. N. Chang, D. Minic, T. Takeuchi, Phys. Rev. **A72** 012104 (2005).

MODEL KINETIC EQUATIONS WITH NON-BOLTZMANN PROPERTIES

BRUCE H. J. MCKELLAR*, IVONA OKUNIEWICZ† and JAMES QUACH

School of Physics, University of Melbourne, Australia 3010
**E-mail: bhjm@unimelb.edu.au*

We reconsider the question of the relative importance of single particle effects and correlations in the solvable interacting neutrino models introduced by Friedland and Lunardini and by Bell, Rawlinson and Sawyer. We show, by an exact calculation, that the two particle correlations are not "small", and that they dominate the time evolution in these models, in spite of indications to the contrary from the rate of equilibration. The failure of the Boltzmann single particle approximation in this model is tentatively attributed to the simplicity of the model, in particular the restriction to two flavor mixing, and the neglect of the position dependence of the interaction.

Keyword: Neutrino Kinetic Equation, Boltzmann Equation.

1. Introduction

This paper summarizes work in progress. It is based on earlier work of BMcK and IO in collaboration with Alexander Friedland.[1] Preliminary accounts of some of the results have appeared in the PhD thesis of IO[2] and the BSc honours report of JQ.[3] Detailed calculations are omitted here, and the interested reader is referred to the thesis and the report.

This work has its origin in the recent discussion of the validity of neutrino kinetic equations of the Boltzmann type, which describe the evolution of the single particle density of a neutrino species. About 20 years ago it was recognized that the neutral current interaction between neutrinos should be taken into account in discussing the preparation of neutrinos through a dense neutrino background,[4–11] and these equations have been

†Present address: WA Office of Energy, 9th floor, 197 St Georges Tce, Perth WA Australia 6000

used in studies of neutrino phenomena in the early universe, (*e.g.*[12,13]) and in supernovae, (*e.g.*[14]).

In 2003, Friedland and Lunardini,[15] and Bell, Rawlinson and Sawyer,[16] re-examined the validity of the single-particle approximation inherent in these kinetic equations. Part of the motivation was a concern that the entanglement induced by the interactions could make correlations so important that the single particle approximation could be invalidated. They studied a simplified model which allowed exact solutions of the many body problem, and examined the rate of equilibration of the system. This discussion was extended by Friedland and Lunardini[17] and by Friedland, McKellar and Okuniewicz.[1] The crucial point in the discussion is the observation that the incoherent equilibration time and the coherent equilibration times scale with the number of particles in very different ways:

$$t_{\rm eq}^{\rm inc} \propto 1/(g\sqrt{N}) \tag{1}$$

$$t_{\rm eq}^{\rm coh} \propto 1/(gN). \tag{2}$$

Here g is a suitably normalized interaction strength. In a large system coherent equilibration is much faster than incoherent equilibration, and thus the equilibration timescale of the manybody system can be used as a proxy to determine the validity or otherwise of the single particle approximation, in as much as one does not have coherent equilibration in such an approximation. Bell, Rawlinson and Sawyer found a coherent timescale in their numerical modeling, but Friedland and Lunardini and Friedland, McKellar and Okuniewicz found incoherent timescales. While these three papers studied systems that differed in detail, they lead to the conclusion that, at least in many circumstances, the single particle approximation is valid.

In this paper we demonstrate that that is a false conclusion in this particular model. The flaw in the logic is that while coherent equilibration timescales are an indication that correlations are important, incoherent equilbration timescales do not necessarily imply that correlations are insignificant.

Because the model of Friedland, McKellar and Okuniewicz is exactly solvable, we can calculate the two body correlations in this model, and we find that they are significant. Moreover we show that the time evolution of the system is determined completely and correctly by the two body correlations through the BBGKY equations, and demonstrate that the one particle approximation, *ie* the quantum Boltzmann equation, leads to the erroneous conclusion that the single particle density matrix is constant in time.

2. The Model of Friedland, McKellar and Okuniewicz

Friedland, McKellar and Okuniewicz used the model of Friedland and Lunardini, and generalized it to obtain the solution to a system of neutrinos with unequal numbers of flavours. So that the system could be solved exactly it is restricted to two neutrino types and the spatial or momentum degrees of freedom are ignored. In principle one could extend the analysis to more than two neutrino flavors, but the SU(N) analogue of the 6-j coefficients of SU(2) are not so well studied, making the analysis more difficult to complete. The advantage of the model is that is allows an exact investigation of the dynamics of flavor conversion.

With two neutrino flavors we could regard the neutrino system as equivalent to the up-down quark system (or the proton neutron system) and use the language of isospin in talking about the system. For consistency with the earlier papers, and to avoid confusion with the weak isospin of the neutrinos, we instead use the language of spin.

Neglecting the spatial degrees of freedom (and thus the Dirac matrices), the weak interaction of a pair of neutrinos inducing flavor mixing is

$$H_{\mathrm{W}} = g\,(\Psi_i^*\delta_{ij}\Psi_j\Phi_k^*\delta_{kl}\Phi_l + \Psi_i^*\delta_{il}\Psi_l\Phi_k^*\delta_{kj}\Phi_j) \tag{3}$$

where g is a coupling constant. In the two flavour case this can be rewritten as

$$\begin{aligned} H_{\mathrm{W}}(ij) &= g\left(\frac{3}{2} + \frac{\vec{\sigma_1}\cdot\vec{\sigma_2}}{2}\right) \\ &= g\left(\frac{\vec{\sigma}_i}{2} + \frac{\vec{\sigma}_j}{2}\right)^2 . \end{aligned} \tag{4}$$

The interaction Hamiltonian for N neutrinos is then

$$\begin{aligned} H &= g\sum_{i=1}^{N-1}\sum_{j=i+1}^{N}\left(\frac{\vec{\sigma_1}}{2} + \frac{\vec{\sigma_2}}{2}\right)^2 \\ &= g\left(J^2 + \frac{3}{4}N(N-2)\right), \end{aligned} \tag{5}$$

with eigenvalues

$$E_J = [gJ(J+1) + g\frac{3}{4}N(N-2)] \tag{6}$$

and eigenstates Ψ_{JM}. J is the total angular momentum, and M is the projection of the angular momentum on the quantization axis. Note that, because we are ignoring the position (or momentum) variables, we are unable to generate states which obey Fermi statistics any two spin up particles are necessarily identical and thus antisymmetrizing them produces a null state, as required by the Pauli principle.

It is then a straightforward procedure to calculate the complete density matrix and then the one body density matrix. This was the basic tool used by Friedland, McKellar and Okuniewicz to calculate the time evolution of single particle probabilities. They found that, although the time evolution of these probabilities was periodic, the period was unrealistically large, and on normal time scales the probabilities decayed to a quasi-equilibrium value on an incoherent timescale, and thus concluded that the single particle approximation on which the neutrino kinetic equations are based is a valid approximation.

If this is indeed the case, then one would expect two body correlations to be small in this model. Because the model is exactly solvable, we can calculate the two body correlations and test this hypothesis.

3. Two Body Correlations

The technques which enabled the calculaion of the one body density matrix[1] can be readily extended to give the two body density matrices. Details of the calculation are given by Okuniewicz.[2]

When both spins are initially up the result is

$$\begin{aligned}\rho_{12}^{(2)}(t) = \sum_{\substack{J,J',\\ m,m_{12},m'_{12}}} & \exp\{-itg[J(J+1) - J'(J'+1)]\} \\ & (2j_{12}+1)(2j+1) \\ & \langle j_U j_D; JM | m_U m_D\rangle \langle j_U j_D; J'M | m_U m_D\rangle \\ & \langle j_{12} j; JM | m_{12} m\rangle \langle j_{12} j; J'M | m'_{12} m\rangle \\ & \begin{Bmatrix} 1 & j_U - 1 & j_U \\ j_D & J & j \end{Bmatrix} \begin{Bmatrix} 1 & j_U - 1 & j_U \\ j_D & J' & j \end{Bmatrix} \\ & |j_{12} m_{12}\rangle \langle j_{12} m'_{12}| \,. \end{aligned} \tag{7}$$

When one spin is initially up and the other down the two body density matrix involves 9-j symbols:

$$\begin{aligned}\rho_{12}(t) = \sum_{\substack{J,J',j,j_{12},j'_{12}\\ m,m_{12},m'_{12}}} & \exp\{-itg[J(J+1)-J'(J'+1)]\}\\ & (2j_U+1)(2j_D+1)(2j+1)[(2j_{12}+1)(2j'_{12}+1)]^{\frac{1}{2}}\\ & \langle j_U j_D; JM|m_U m_D\rangle \langle j_U j_D; J'M|m_U m_D\rangle\\ & \langle j_{12} j; m_{12} m|JM\rangle \langle j'_{12} j; m'_{12} m|J'M\rangle\\ & \begin{Bmatrix} j_1 & k_U & j_U\\ j_2 & k_D & j_D\\ j_{12} & j & J\end{Bmatrix} \begin{Bmatrix} j_1 & k_U & j_U\\ j_2 & k_D & j_D\\ j_{12} & j & J'\end{Bmatrix}\\ & |j_{12} m_{12}\rangle \langle j'_{12} m'_{12}| \,. \end{aligned} \tag{8}$$

Using these results we calculate the two body correlation function which is defined by

$$\Gamma = \langle s_1 s_2\rangle - \langle s_1\rangle\langle s_2\rangle, \tag{9}$$

where $\langle s_i\rangle$ is the expectation value of the angular momentum s_i, where $i = 1, 2$ refers to particle i, and $\langle s_1 s_2\rangle$ is the expectation value of the product of the spins. It should be emphasised that in this model the correlation function is not a function of the separation distance of the spins, as the position of the spins does not enter the model. Γ simply measures the correlation between spins, but its magnitude and time dependence are of interest.

A plot of the correlation function, Γ, for the initial conditon in which particle 1 and particle 2 are in the state up is given in figure 1. In these plots the initial number of up spins, $N = 101$, and the number of down spins, M is varied such that $N \geq M$. The time is scaled as in Friedland *et al*,[1] $\tau = gt(N+M)$. Note that the correlation function is positive for all of these cases. Other results are similar.

The correlation function shows the same features as those found by Friedland, McKellar and Okuniewicz for the probability of one of the initial spin up particles remaining in the spin up state. Namely,

- When $N \sim M$, the system comes to equilibrium after some time. This features can be seen for the case N=101, M=101. Just as in the study of the equilibration of the single spin by Friedland, McKellar and Okuniewicz, the equilibration time is incoherent.
- When $N - M$ is large the correlation function exhibits oscillations but the amplitude is small.

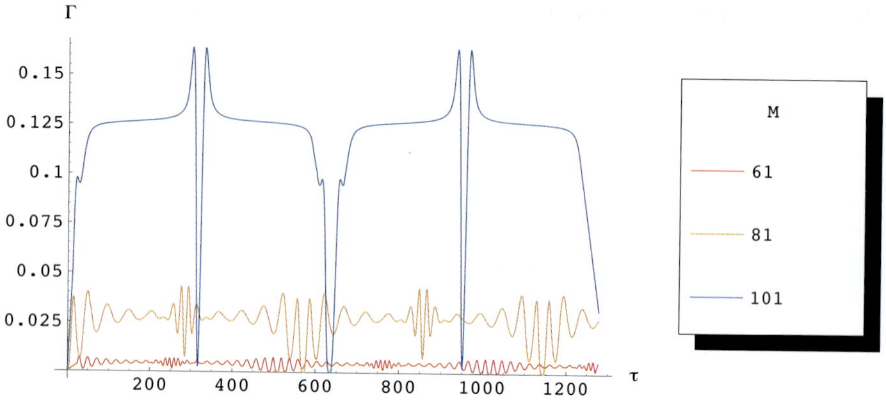

Fig. 1. A plot of the correlation function, G, for the initial condition in which particle 1 and particle 2 are in the state up, as described in the text. For the case $N = M$ the glitches in the plot are not real, rather they are an artifact of mathematica graphics.

- The correlation function is periodic on large time scales, with the same period as before.

Friedland *et al*[1] found that is was for the case that $N \sim M$ that the incoherent equilibration was particularly marked. Yet it precisely this case that the the correlation function is its greatest, more than 10% most of the time. It must be a cause of concern whether or not a correlation of 10% is indeed *"small"*, leading us to ask the question

Are the correlations big enough to induce significant corrections to the single particle kinetic equations, in spite of the incoherent time evolution?

Since we are working with a solvable model, and thus we know the time evolution of the complete many particle density matrix, we can proceed to answer this question.

4. The BBGKY Equations and the Evolution of the Single Particle Density Matrix

The time evolution of the density matrix, is described by the Von Neumann equation,

$$i\frac{\partial \rho}{\partial t} = [H, \rho] \tag{10}$$

We can split the Hamiltonian into its single particle and interaction constituents,

$$H = \sum_{i=1}^{N} K_i + \frac{1}{2}\sum_{i\neq j} V_{ij} \tag{11}$$

where K is the single particle Hamiltonian of the ith particle and V_{ij} is the interaction Hamiltonian between the ith and jth particle.

The rate of change of a single particle density matrix is found by tracing over all the other particles,

$$\begin{aligned} i\frac{\partial \rho_1^{(1)}}{\partial t} &= [K_1, \rho_1^{(1)}] + \frac{1}{2}\sum_{i\neq j} \mathrm{Tr}_{2..N}[V_{ij}, \rho_{1..N}] \\ &= [K_1, \rho_1^{(1)}] + \frac{1}{2}\sum_{j} \mathrm{Tr}_j[V_{1j}, \rho_{1j}^{(2)}] + \frac{1}{2}\sum_{i} \mathrm{Tr}_i[V_{i1}, \rho_{1i}^{(2)}] \\ &\quad + \frac{1}{2}\sum_{i\neq j} \mathrm{Tr}_{ij}[V_{ij}, \rho_{1ij}^{(3)}] \\ &= [K_1, \rho_1^{(1)}] + \sum_{j} \mathrm{Tr}_j[V_{1j}, \rho_{1j}] + \frac{1}{2}\sum_{i\neq j} \mathrm{Tr}_{ij}[V_{ij}, \rho_{1ij}^{(3)}]. \end{aligned} \tag{12}$$

In the last line of Eq. (12) we have employed the symmetry of our interaction Hamiltonian. Eq. (12) is the first quantum Bogolubov-Born-Green-Kirkwood-Yvon (BBGKY) equation.[18]

In the model of Friedland *et al* there is no single particle Hamiltonian. In this general discussion we retain such a Hamiltonian, which is responsible for the free space neutrino oscillations, and omit it when discussing the consequences of the model.

For the interaction Hamiltonian of our model it can readily be shown that the last term goes to zero, and the three particle density matrix does not contribute to the time dependence of the single particle density matrix. This is a significant simplification in the dynamics of the model. To demonstrate this represent the one, two and three particle operators on the spin space on the basis of direct products of the extended Pauli matrices σ_μ, $\mu \in \{0, 1, 2, 3\}$, with the convention that $\sigma_0 = 1$, and $\sigma_j, j \in \{1, 2, 3\}$ are the usual Pauli matrices. The superscript $^{[i]}$ indicates that the Pauli martix or function refers to particle i, and similarly for multiple bracketed superscripts. Explicitly

$$K_i = \frac{1}{2}\mathcal{K}^{[i]}_{\mu}\sigma^{[i]}_{\mu} \tag{13}$$

$$V_{ij} = \frac{1}{4}v^{[ij]}_{\mu\nu}\sigma^{[i]}_{\mu} \otimes \sigma^{[j]}_{\nu} \tag{14}$$

$$\rho^{(1)}_i = \frac{1}{2}\mathcal{P}^{[i]}_{\mu}\sigma^{[i]}_{\mu} \tag{15}$$

$$\rho^{(2)}_{ij} = \frac{1}{4}p^{[ij]}_{\mu\nu}\sigma^{[i]}_{\mu} \otimes \sigma^{[j]}_{\nu} \tag{16}$$

$$\rho^{(3)}_{ijk} = \frac{1}{8}\pi^{[ijk]}_{\kappa\mu\nu}\sigma^{[i]}_{\kappa} \otimes \sigma^{[j]}_{\mu} \otimes \sigma^{[k]}_{\nu}. \tag{17}$$

Here the superscript $^{[i]}$ refers to particle i, and we have employed a summation convention over repeated Greek indices. The few particle density matrices satisfy

$$\mathrm{Tr}(\rho^{(1)})_i = 1 \tag{18}$$

$$\mathrm{Tr}_j(\rho^{(2)}_{ij}) = \rho^{(1)}_i. \tag{19}$$

Note that Eq. (19) is valid for all $N-1$ possible values of j. From these follow the relationships

$$\mathcal{P}^{[i]}_0 = 1 \tag{20}$$

$$p^{[ij]}_{00} = 1 \tag{21}$$

$$p^{[ij]}_{\mu 0} = \mathcal{P}^{[i]}_{\mu} \tag{22}$$

$$p^{[ij]}_{0\nu} = \mathcal{P}^{[j]}_{\nu}. \tag{23}$$

The algebra of the extended Pauli matrices is

$$\sigma^{[i]}_{\mu}\sigma^{[i]}_{\nu} = \delta_{\mu\nu} + \delta_{\mu,0}\sigma^{[i]}_{\nu} + \delta_{\nu,0}\sigma^{[i]}_{\mu} + i\epsilon_{\kappa\mu\nu}\sigma^{[i]}_{\kappa}, \tag{24}$$

with the additional convention that

$$\epsilon_{0\mu\nu} = 0. \tag{25}$$

Using this algebra, the components of the final term in Eq. (12) are readily computed and shown to vanish

$$\mathrm{Tr}_{ij}[V_{ij}, \rho^{(3)}_{1ij}] = 0. \tag{26}$$

The BBGKY equation for $\dot{\rho}^{(1)}$ now depends only on $\rho^{(1)}$ and $\rho^{(2)}$:

$$i\frac{\partial \rho^{(1)}_1}{\partial t} = [K_1, \rho^{(1)}_1] + \sum_j \mathrm{Tr}_j[V_{1j}, \rho^{(2)}_{1j}] \tag{27}$$

In terms of the Pauli matrix representation of the density matrices, Eq. (27) is

$$\dot{\mathcal{P}}_{\mu}^{[i]} = \epsilon_{\alpha\beta\mu}\left(\mathcal{K}_{\alpha}^{[i]}\mathcal{P}_{\beta}^{[i]} + \sum_{j} v_{\alpha\lambda}^{[ij]} p_{\beta\lambda}^{[ij]}\right). \qquad (28)$$

The first term of Eq. (28) gives the precession of the vector $\vec{\mathcal{P}}^{[i]}$ about the vector $\vec{\mathcal{K}}^{[i]}$, which is the usual neutrino oscillation phenomenon.[8] Note that, in the general case, $\mathcal{P}_0^{[i]}$ is constant, as it must be, and that $(\vec{\mathcal{P}}^{[i]})^2$ is not constant. The fact that $(\vec{\mathcal{P}}^{[i]})^2$ is not constant implies that the single particle density matrix describes a mixed state rather than a pure state. We have the one and two body density matrices, in the case that $K_j = 0$, from the previous section, and it is a good check on our calculations to verify that the Eq. (27) is indeed satisfied. This has been done, and as the calculation is rather involved, the interested reader is referred to the Honours thesis of JQ[3] for the details.

5. The Boltzmann Equation and its Failure in the Model

While it is reassuring that the exact BBGKY equation is satisfied in our exactly solvable model, we really need to test the analogue in the model of the neutrino kinetic equations. As shown in detail by Thomson[4] and McKellar and Thomson[8] there are two key approximations in deriving the neutrino kinetic equations.

(1) Two particle correlations can be neglected.
(2) The time of a collision is negligible compared to the time between collisions.

In the present model, as we have no spatial dependence in the interactions we cannot model the second approximation. The first approximation can be implemented in the model by approximating the two body density matrix as the direct product of single particle density matrices,

$$\rho_{ij}^{(2)} \approx \rho_i^{(1)} \times \rho_j^{(1)} \qquad (29)$$

and substituting this approximation in Eq. (27). In this way we obtain a version of the quantum Boltzmann equation

$$\dot{\mathcal{P}}_{\mu}^{[i]} = \epsilon_{\alpha\beta\mu}\left(\mathcal{K}_{\alpha}^{[i]}\mathcal{P}_{\beta}^{[i]} + \sum_{j} v_{\alpha\lambda}^{[ij]}\mathcal{P}_{\lambda}^{[j]}\mathcal{P}_{\beta}^{[i]}\right). \qquad (30)$$

Before we specify the interaction appropriate to the model, we see that in the Boltzmann equation Eq. (30) the interactions now give a modified oscillatory behaviour, as $\vec{\mathcal{P}}^{[i]}$ precesses about the vector with components $\mathcal{K}_k^{[i]} + \sum_j v_{k\lambda}^{[ij]}\mathcal{P}_\lambda^{[j]}$, and $\left(\vec{\mathcal{P}}^{[i]}\right)^2$ is constant. and thus can be fixed to 1. With the general result that $\mathcal{P}_0^{[i]} = 1$ this implies that $(\rho^{(1)})^2 = \rho^{(1)}$. In the Boltzmann approximation the single particle density matrix represents a pure state.

Now we note that in the model, the Pauli matrix representation of the interaction is

$$v_{\alpha\beta} = \phi_\alpha \delta_{\alpha\beta} \quad \text{with} \quad \phi_0 = 6g \quad \text{and} \quad \phi_k = 2g, \tag{31}$$

and the single particle Hamiltonian vanishes, so the model Boltzmann equation is

$$\dot{\mathcal{P}}_k^{[i]} = 2g\epsilon_{\ell mk}\dot{\mathcal{P}}_\ell^{[j]}\dot{\mathcal{P}}_m^{[i]}. \tag{32}$$

From this it follows that:

(1) $\left(\vec{\mathcal{P}}^{[i]} + \vec{\mathcal{P}}^{[j]}\right)$ is a constant vector.

(2) The vector $\left(\vec{\mathcal{P}}^{[i]} \times \vec{\mathcal{P}}^{[j]}\right)$ precesses about the vector $\left(\vec{\mathcal{P}}^{[i]} + \vec{\mathcal{P}}^{[j]}\right)$. Thus, if at any time $\left(\vec{\mathcal{P}}^{[i]} \times \vec{\mathcal{P}}^{[j]}\right) = 0$, then it vanishes at all times.

For the the body density matrices constructed explicitly in section 3, initially both spins are up, or one is up and one is down. In these cases, at $t = 0$,

(1) Both spins up:

$$\vec{\mathcal{P}}^{[i]} = \vec{\mathcal{P}}^{[j]} = (0, 0, 1). \tag{33}$$

(2) One spin, i, up and the other spin down:

$$\vec{\mathcal{P}}^{[i]} = -\vec{\mathcal{P}}^{[j]} = (0, 0, 1). \tag{34}$$

In each case $\left(\vec{\mathcal{P}}^{[i]} \times \vec{\mathcal{P}}^{[j]}\right) = 0$ at $t = 0$, and so vanishes for all times. When this occurs, it follows from the model Boltzmann equation (32), that $\vec{\mathcal{P}}^{[i]}$, and thus the single body density matrix $\rho^{(1)}$ are constant.

This is in complete contradiction with the results of Friedland, McKellar and Okuniewicz,[1] in which the exact non-trivial time development of $\rho^{(1)}$ was computed. The Boltzmann equation is not satisfied in the model in the most dramatic way possible — **all of the time dependence of $\rho^{(1)}$ is**

generated by the correlations — even thought the correlation function, as we saw, has a magnitude of at most about 10%.

Because we have explicit representations of the relevant density matrices in the angular momentum formalism of Friedland *et al*,[1] and in section 3, it is possible to verify this result by detailed calculation of each side of Eq. (30). Details are in the BSc Honours report of JQ,[3] and of course the result is that the left hand side of Eq. (30) is non-vanishing and the right hand side is zero — clearly the equation is not even approximately valid.

6. Discussion

It is tempting to suggest that the breakdown of the Boltzmann equation in this exactly solvable model is closely connected to the structure of the model, and is not a general effect. In particular, because we are modelling just two neutrino flavours, or equivalently describing a system of spin half particles, we have only two choices for the initial state of a pair of particles, and thus are forced to the situation in which the single particle density matrix is constant for the interactions of the model in the Boltzmann approximation.

It is appropriate to direct further investigations to

(1) Noting that result is specific to the model interaction, we could try to generalize the interaction, maintaing the solvability of the model, but restoring a time dependence to the single particle density matrix in the Boltzmann approximation.
(2) Investigating the situation with more than two flavours. This would require working with the Clebsch-Gordan algebra for SU(N), or the SU(N) Gell-Mann matrices, or both.

Of these approaches the second seems the most straightforward, and we hope to report some results using it in the near future.

Regrettably, we have found that, while a coherent time scale may indicate the importance of correlations, an incoherent timescale is not a reliable indicator of the absence of significant correlations. As a consequence this particular solvable model is not a good guide to the validity or otherwise of the neutrino kinetic equations in either the early universe or supernovae.

Acknowledgments

This work has been supported in part by the Australian Research Council.

It is an honour to dedicate this paper to Professor C. N. Yang, in celebration of his 85th Birthday.

References

1. A. Friedland, B. H. J. McKellar and I. Okuniewicz, Phys. Rev. D **73**, 093002 (2006).
2. I. Okuniewicz, "Construction and Analysis of a Simplied Many-Body Neutrino Model", PhD thesis, University of Melbourne , 2006.
3. J. Quach, "A Theoretical Investigation Into The Validity Of The Boltzmann Property In The FMO Model", University of Melbourne BSC(Hons) Report, 2006.
4. M. J. Thomson, "Neutrino Oscillations in the Early Universe", 'Ph D thesis, University of Melbourne, 1990
5. M. J. Thomson and B. H. J. McKellar, "Thermal Excitation of Sterile Neutrinos in the Early Universe", University of Melbourne report UM-P-90-44, 1990
6. M. J. Thomson and B. H. J. McKellar, "The non-linear MSW equation and neutrino oscillations in the early universe", paper contributed to Neutrino '90 Conf., Geneva, Switzerland, 1990
7. B. H. J. McKellar and M. J. Thomson, "Master equations for oscillating doublet neutrinos", Franklin Symposium, April 30 — May 1, 1992. (published in *Discovery of the Neutrino*, (edited by C. E. Lane and R. I. Steinberg, World Scientific, Singapore, 1993) pp 169 — 174).
8. B. H. J. McKellar and M. J. Thomson, Phys. Rev. D **49**, 2710 (1994).
9. J. T. Pantaleone, Phys. Lett. **B287**, 128 (1992)
10. J. T. Pantaleone, Phys. Rev. D **46**, 510 (1992.
11. G. Sigl, and G. Raffelt, Nucl. Phys. **B406**, 423 (1993)
12. N. F. Bell, R. R. Volkas and Y. Y. Y. Wong, Phys. Rev. D **59**, 113001 (1999)
13. A. D. Dolgov, Phys. Rept. **370**, 333 (2002)
14. B. Dasgupta and A. Dighe, arXiv:0712.3798 [hep-ph].
15. A. Friedland and C. Lunardini, Phys. Rev., **D68**, 013007, (2003)
16. N. F .Bell, A. A. Rawlinson and R. F. Sawyer, Phys. Lett., **B573**, 86, (2003).
17. A. Friedland and C. Lunardini, JHEP, **10**, 043, (2003)
18. K. Huang, "Statistical Mechanics", John Wiley and Sons Inc, Hoboken, NJ, USA. Second edition, 1987.

RIGID LIMIT IN $\mathcal{N} = 2$ SUPERGRAVITY AND WEAK-GRAVITY CONJECTURE

TOHRU EGUCHI

Yukawa Institute for Theoretical Physics, Kyoto University, Kyoto, 606-8502, Japan

YUJI TACHIKAWA

Institute for Advanced Study, Princeton, NJ, 08540, USA

We analyze the coupled $\mathcal{N} = 2$ supergravity and Yang–Mills system using holomorphy, near the rigid limit where the former decouples from the latter. We find that there appears generically a new mass scale around $gM_{\rm pl}$ where g is the gauge coupling constant and $M_{\rm pl}$ is the Planck scale. This is in accord with the weak-gravity conjecture proposed recently.

Keyword: $\mathcal{N} = 2$ supergravity; string theory; Calabi–Yau manifold; holomorphy; rigid limit.

1. Introduction

Quantization of general relativity has been one of the most serious challenges for theoretical physics for a long time. Its coupling constant is dimensionful, which makes the theory apparently nonrenormalizable. Thus, we need to complete the theory in the ultraviolet (UV) to make it into a consistent quantum theory. The prime candidate for quantized gravity is the superstring theory, and the progress we made during the last decade makes us confident that there exist many consistent four-dimensional theories with a high degree of supersymmetry containing quantized graviton in their spectrum. These low energy field theories coupled to gravity have a consistent UV completion and are obtained via compactification of superstring theory on suitable internal manifolds.

When we come to theories with a smaller number of supersymmetries the situation becomes somewhat delicate. Recent developments suggest that there exists an enormous number of $\mathcal{N} = 1$ supersymmetric four-dimensional models with negative cosmological constant (for a review, see e.g. Ref. 1). This *landscape* of superstring vacua, if taken at face value, predicts a disturbingly huge number, 10^{200} or larger, of solutions with varying gauge groups and matter contents. Then it is natural to ask which theory is realized as a low-energy effective description of a consistent theory with quantized gravity.[2] Several criteria have already been proposed in Refs. 3 and 4 which characterize models in the *swampland* which cannot be UV completed to a consistent theory of quantum gravity.

The criterion we will focus in this paper is *the weak-gravity conjecture* proposed in Ref. 3; one way to state the conjecture is that if a consistent theory coupled to gravity with the Planck scale $M_{\rm pl}$ contains a gauge field with the coupling constant g, then there should necessarily be a new physics around the mass scale $gM_{\rm pl}$. We refer the reader to the original paper for the arguments which led to this proposal.[3] Our objective in this paper is to show how this conjecture will generically hold within the framework of $\mathcal{N} = 2$ supersymmetric Yang–Mills coupled to $\mathcal{N} = 2$ supergravity.

The system of $\mathcal{N} = 2$ supersymmetry is well suited to the analysis of the effects of quantum gravity on the gauge theory. One advantage is that the dynamics of $\mathcal{N} = 2$ supersymmetric Yang–Mills theories has been studied in great detail since the pioneering work of Ref. 5. Another advantage is that the limit where the $\mathcal{N} = 2$ Yang–Mills theory decouples from the $\mathcal{N} = 2$ supergravity is fairly well understood in the context of the string compactification on Calabi–Yau (CY) manifold with a fiber of ADE singularities. This limit is known as the *rigid limit* or *decoupling limit* since supersymmetry becomes rigid and gravity decouples from the gauge theory in the limit. It is also called the *geometric-engineering limit*,[6–8] since non-Abelian gauge symmetry is generated by ADE singularities.

In this paper we consider a type II string theory on CY manifolds which possess $K3$ fibration over $\mathbb{CP}^1$ and thus has a dual heterotic string description. At the geometric engineering limit $\epsilon \to 0$ when the $K3$ surface develops ADE singularity, such a CY manifold acquires periods which behave as a power and logarithm of ϵ. We shall show that the ratio of these periods leads to the hierarchy of gauge and gravity mass scales which has exactly the form of the weak-gravity conjecture. Since the geometric engineering limit is the only way to generate non-Abelian gauge symmetry in type II theory, the weak-gravity conjecture seems to hold generically in $\mathcal{N} = 2$ gauge theory coupled to $\mathcal{N} = 2$ supergravity. Actually as is well known, $M_{\rm het}=gM_{\rm pl}$ is the mass scale of heterotic string theory and thus the weak-gravity conjecture seems to fit very nicely with the type II–heterotic duality.

In our analysis the holomorphy and the special geometry of $\mathcal{N}=2$ theories play the basic role. Holomorphic functions are determined by their behavior at the singularities, in particular by the monodromy properties around the singular locus.

The organization of the paper is as follows. In Sec. 2 we discuss an example of a type II string theory compactified on a CY manifold with a $K3$ fibration. We shall show how a hierarchy of mass scales is generated in the rigid limit $\epsilon \to 0$ which fits exactly to the weak-gravity conjecture. We also point out that the presence of a logarithmic period $\log \epsilon$ predicts a kinetic term for a field S:

$$\frac{\partial_\mu S \partial_\mu S}{(\mathrm{Im}\, S)^2}\,. \tag{1}$$

S corresponds to the gauge coupling constant $S = \theta/(2\pi)+4\pi i/g^2$ and maps to the heterotic dilaton under the type II/heterotic duality. We discuss in Sec. 3 the mechanism of how the logarithmic periods necessarily appear in a CY manifold with a $K3$ fibration. We conclude this paper with some discussions in Sec. 4.

2. An Example

2.1. *A Calabi–Yau and the rigid limit*

Let us start with an example from the string theory. As is well known, in the type IIA superstring theory, an $\mathcal{N}=2$ supergravity system in four dimensions can be obtained by compactification on a CY manifold M. It is also known that the SU(n) $\mathcal{N}=2$ gauge symmetry arises if M has a sphere of A_{n-1} type singularities. In the simplest case of A_1 singularity such a CY manifold has at least two Kähler parameters: one for the size of the sphere of the singularities and the other for the size of resolution of singularities. One explicit example is given by a CY manifold X_8 which is a degree 8 hypersurface in the weighted projective space $\mathbb{WCP}^4_{1,1,2,2,2}$ with Hodge numbers $h_{11}=2$, $h_{21}=86$.

Our analysis is facilitated by going to the mirror type IIB theory where worldsheet instanton corrections in IIA theory are summed up by mirror transformation. Mirror pair of X_8 and X_8^* has been extensively studied in the literature (e.g. Refs. 9 and 10). We first briefly review their properties.

Defining equation of the mirror X_8^* is given by

$$X_8^* : \quad W = \frac{B}{8}x_1^8 + \frac{B}{8}x_2^8 + \frac{1}{4}x_3^4 + \frac{1}{4}x_4^4 + \frac{1}{4}x_5^4 - \psi_0 x_1 x_2 x_3 x_4 x_5 - \frac{1}{4}\psi_2 (x_1 x_2)^4 = 0 \tag{2}$$

in an orbifold of $\mathbb{WCP}^4_{1,1,2,2,2}$. $[B : \psi_0 : \psi_2]$ parametrizes the complex structure moduli of X_8^*. We first note that this hypersurface has a structure of a $K3$ fibration over $\mathbb{CP}^1$: by a change of variables $x_0 = x_1 x_2$, $\zeta = x_1/x_2$, W is rewritten as

$$W = \frac{B'}{4}x_0^4 + \frac{1}{4}x_3^4 + \frac{1}{4}x_4^4 + \frac{1}{4}x_5^4 - \psi_0 x_0 x_3 x_4 x_5 = 0\,, \tag{3}$$

$$B' = \frac{B}{2}\left(\zeta + \frac{1}{\zeta}\right) - \psi_2\,. \tag{4}$$

ζ parametrizes the base of the $K3$ fibration. $K3$ surface (3) (with fixed ζ) has singularities at

$$B' = 0\,; \qquad \text{large complex structure limit}\,, \tag{5}$$

$$B' = \psi_0^4\,; \qquad \text{conifold singularity}\,. \tag{6}$$

These are located by imposing equations $W = 0$, $\partial W/\partial x_i = 0$, $i = 0, 3, 4, 5$ simultaneously. If we solve (5), (6) for ζ, we find

$$B' = 0 \Rightarrow \zeta = e_0^{\pm}\,, \qquad \text{where } e_0^{\pm} = \frac{\psi_2}{B} \pm \sqrt{\left(\frac{\psi_2}{B}\right)^2 - 1}\,, \tag{7}$$

$$B' = \psi_0^4 \Rightarrow \zeta = e_1^{\pm}\,, \quad \text{where } e_1^{\pm} = \frac{(\psi_2 + \psi_0^4)}{B} \pm \sqrt{\left(\frac{\psi_2 + \psi_0^4}{B}\right)^2 - 1}\,. \tag{8}$$

Singularities of the total space X_8^* are located by further imposing $\partial B'/\partial\zeta = 0$:

$$\frac{\partial B'}{\partial \zeta} = 0 \Rightarrow B = 0 \quad \text{or} \quad \zeta = \pm 1\,. \tag{9}$$

Substituting $\zeta = \pm 1$ into (5) and (6), we find singular loci in the moduli space of X_8^*:

$$B = \pm\psi_2\,, \qquad B = \pm(\psi_2 + \psi_0^4)\,. \tag{10}$$

These coincide with the locations where $e_0^{\pm}$, $e_1^{\pm}$ become degenerate.

Thus the discriminant of the mirror CY manifold is given by

$$\Delta = B^2(B^2 - \psi_2^2)(B^2 - (\psi_2 + \psi_0^4)^2)\,. \tag{11}$$

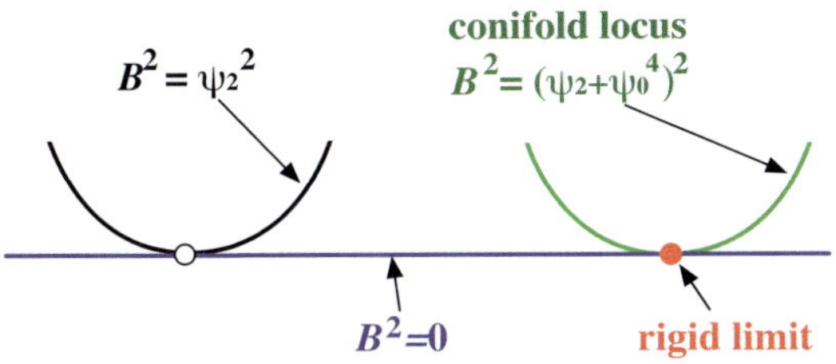

Fig. 1. Discriminant loci of the moduli of the CY X_8^*, before the blowup.

Three components of the discriminant loci are depicted in Fig. 1. The first and the second factors intersect tangentially at the large complex structure point and the third factor is the conifold locus. The conifold locus and the locus $B^2=0$ also meet tangentially at the rigid limit,[a] so that the moduli space needs to be blown up at these points.

We now concentrate on the region near the rigid limit. The blowing up introduces an exceptional curve which is a $\mathbb{CP}^1$ parametrized by $[\Lambda^2 : u]$ via the relation

$$\epsilon\Lambda^2 = B\,, \qquad \epsilon u = \psi_2 + \psi_0^4\,. \tag{12}$$

The exceptional curve is at $\epsilon = 0$. The discriminant loci after the blowup are shown in Fig. 2.

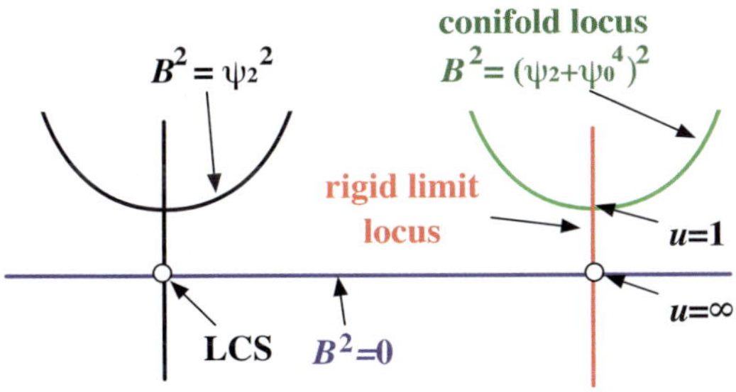

Fig. 2. Discriminant loci of the moduli of the CY X_8^* after the blowup. LCS stands for the large complex structure point.

The defining polynomial W in the limit $\epsilon \to 0$ is given by

$$W = \frac{\epsilon}{2}\left[\frac{1}{2}\left(w + \frac{\Lambda^4}{w}\right) + x^2 + y^2 + z^2 - u\right] + O(\epsilon^2) \tag{13}$$

[a]The parameter sets (B, ψ_0, ψ_2) and $(-B, \psi_0, \psi_2)$ describe the same complex structure and so the natural coordinate of the moduli is B^2 rather than B.

after a suitable redefinition of the coordinates. This is a fibration of A_1 singularity over $\mathbb{CP}^1$ parametrized by w. It is in fact the Seiberg–Witten geometry of the $\mathcal{N} = 2$ supersymmetric pure SU(2) Yang–Mills theory with the modulus $u = \langle \mathrm{tr}\, \phi^2 \rangle$ and the dynamical mass scale Λ. Thus, the exceptional curve we have introduced is identified as the u-plane of SU(2) gauge theory: the u-plane is naturally compactified at $u = \infty$ into a sphere. We call this sphere the rigid limit locus.

Note that before taking the rigid limit $\epsilon \to 0$, the theory contains $h_{11} + 1 = 3$ gauge fields: they are the graviphoton, the gauge partner of the scalar field S and the U(1) (Cartan-subalgebra) part of SU(2) gauge field. Here S denotes the scalar field which corresponds to the gauge coupling constant in field theory,

$$S = \frac{\theta}{2\pi} + \frac{4\pi i}{g^2}\,. \tag{14}$$

We recall that when CY manifold M possesses a $K3$ fibration on $\mathbb{CP}^1$, there exists a duality between type IIA on M and heterotic theory on $K3 \times T^2$.[12] The field S corresponds to the size of the base $\mathbb{CP}^1$ of $K3$ fibration in type IIA theory and becomes the heterotic dilaton under this duality. In the decoupling limit $\epsilon \to 0$, two of the gauge fields, the graviphoton and the partner of S, disappear and we are only left with the (Cartan part of) SU(2) gauge field.

2.2. *Behavior of the Kähler potential*

Let us next quickly recall the structure of vector multiplet scalars in the $\mathcal{N} = 2$ theories. First, in the case of field theories of rigid $\mathcal{N} = 2$ supersymmetry with the gauge group $U(1)^n$, there exist n complex scalar fields ϕ^i, $(i = 1, \ldots, n)$. Their Kähler potential is given by

$$K = \mathrm{Im} \sum_i (a_i^D)^* a^i\,, \tag{15}$$

where a^i and a_i^D are holomorphic functions of the VEV's of ϕ^i. a^i and a_i^D are called the special coordinates or the periods of the theory. Dual periods are related to each other as

$$a_i^D = \frac{\partial F_{\mathrm{gauge}}}{\partial a^i}\,, \qquad i = 1, \ldots, n\,, \tag{16}$$

where F_{gauge} denotes the prepotential of the gauge theory.

Second, in the case of $\mathcal{N}=2$ supergravity with N vector multiplets, there exist $2(N+1)$ periods X^a, F_a, $a=1,\dots,N+1$. The Kähler potential is given by

$$e^{-K}=\mathrm{Im}\sum_a F_a^* X^a\,. \tag{17}$$

The periods X^a, F_a are holomorphic functions of scalars Φ^i ($i=1,\dots,N$). Under the Kähler transformation $K\to K-f-f^*$ periods are transformed as $X^a\to e^f X^a$, $F_a\to e^f F_a$. The mass squared of a BPS-saturated soliton with charges (q_a,m^a) is then given by

$$m^2=e^K\left|\sum_a(q_aX^a+m^aF_a)\right|^2, \tag{18}$$

which is invariant under the Kähler transformation. An important property of the supergravity periods is the transversality condition:

$$\sum_a X^a\frac{\partial F_a}{\partial\Phi^i}-\sum_a\frac{\partial X^a}{\partial\Phi^i}F_a=0\,, \tag{19}$$

which guarantees the existence of the prepotential. Prepotential of $\mathcal{N}=2$ supergravity is a homogeneous function of degree 2 in X_a.

In the case of CY compactification of type IIB string theory, the periods are given by

$$X^a=\int_{A^a}\Omega\,,\qquad F_a=\int_{B_a}\Omega\,, \tag{20}$$

where Ω is the $(3,0)$-form of the CY and A^a, B_a are the canonical basis of $H_3(M^*,\mathbb{Z})$ of CY manifold. In this case the condition (19) comes from the Griffiths transversality $\int\Omega\wedge\partial_{\Phi^i}\Omega=0$.

Now let us go back to the example of the previous section, type IIB string theory compactified on X_8. In the field theory limit we have only one gauge field ($n=1$) and two periods a and a^D of SU(2) Seiberg–Witten theory. At the level of supergravity there exist three gauge fields (two vector multiplets, $N=2$) and six periods X^a, F_a, $a=1,2,3$. Behavior of these periods near the decoupling limit and in particular their monodromy properties around rigid limit locus have been discussed in great detail in Ref. 11.

It turns out that two of the periods, say X^1 and F_1, are converted to the gauge theory periods in the rigid limit. They behave as

$$X^1=\epsilon^{1/2}a+O(\epsilon)\,,\qquad F_1=\epsilon^{1/2}a^D+O(\epsilon)\,. \tag{21}$$

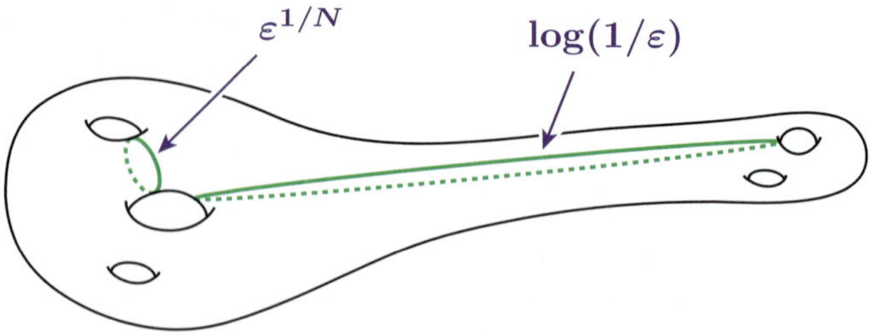

Fig. 3. Behavior of periods near decoupling limit.

Remaining four periods behave as

$$X^2,\ X^3 = 1 + O(\epsilon^{1/2}), \qquad F_2,\ F_3 = \frac{1}{2\pi i}\log\epsilon + O(1). \tag{22}$$

The origin of logarithmic behaviors in F_2, F_3 will be discussed in Sec. 3: they come from the geometry of $K3$ fibration of the CY manifold.

Behavior of these periods near decoupling limit may be pictorially represented as in Fig. 3. (N is put to 2 for X_8 (SU(2)) case).

Then using (17) we find that e^K behaves as $\log 1/|\epsilon|$. Therefore the supergravity Kähler potential is expanded as

$$K = \log(\log 1/|\epsilon|) + \frac{|\epsilon|}{\log 1/|\epsilon|}\,\mathrm{Im}(a^D)^* a + \cdots \tag{23}$$

as $\epsilon \to 0$. Note that $\mathrm{Im}(a^D)^* a$ is the Kähler potential of the field theory (15). Thus we can clearly see that SU(2) super-Yang–Mills theory decouples from gravity.

The factor $|\epsilon|$ in front of the Kähler potential of the field theory determines the hierarchy between the Planck scale and the scale of the gauge theory: it is basically in accord with the expectation[8] with $|\epsilon|^{1/2}$ being identified with the dynamical mass scale Λ_{gauge} of the gauge theory. The existence of an extra factor of $\log 1/|\epsilon|$ in the denominator was first recognized by the authors of Ref. 11. We will see in the following that this factor implies the weak-gravity conjecture in the present context.

Let us now consider the weak coupling region of gauge theory for the sake of simplicity. There the periods a and a^D behave as

$$a \approx \sqrt{2u}, \qquad a^D \approx \frac{i}{\pi}\sqrt{2u}\log u. \tag{24}$$

Using the relation of periods to the low-energy gauge coupling constant τ:

$$\tau = \frac{\theta}{2\pi} + \frac{4\pi i}{g^2(m_W)} = \frac{\partial a^D}{\partial a}, \tag{25}$$

we find

$$e^{-2\pi^2/g^2(m_W)} = u^{-1/2}. \tag{26}$$

The coupling constant g in the above equation is to be evaluated at the scale of the mass m_W of the massive gauge boson where the coupling stops running. m_W is, in turn, given by the formula (18):

$$m_W^2 = e^K|X^1|^2 = \frac{|\epsilon|}{\log 1/|\epsilon|}u. \tag{27}$$

From (26) and (27), we find the dynamical scale of the gauge theory

$$\Lambda_{\text{gauge}} = m_W e^{-2\pi^2/g^2(m_W)} = \frac{|\epsilon|^{1/2}}{(\log 1/|\epsilon|)^{1/2}} M_{\text{pl}}, \tag{28}$$

where we reinstated the Planck scale to recover the correct mass dimension.

Let us next introduce a chiral superfield $S = \theta/2\pi + 4\pi i/g^2$ via the relation

$$S = \frac{1}{\pi i}\log\epsilon. \tag{29}$$

Then, the monodromy around $\epsilon = 0$ is generated by the shift $S \to S + 2$. $\operatorname{Im} S$, which is the partner of the dynamical theta angle, is the natural bare gauge coupling constant in the supergravity. Furthermore, S coincides with the heterotic dilaton which we have discussed at the end of Subsec. 2.1. There will be subleading corrections to (29) if one goes outside the region of weak coupling or small ϵ. Another notable fact is that, because of the Kähler potential (23), the field S in fact has the standard kinetic term for the dilaton,

$$g_{SS^*}\partial_\mu S \partial_\mu S^* = \frac{\partial_\mu S \partial_\mu S^*}{(\operatorname{Im} S)^2}. \tag{30}$$

Using the field $S = \theta/2\pi + 4\pi i/g^2$, the relation (28) now becomes

$$\Lambda_{\text{gauge}} = e^{-2\pi^2/g^2} \cdot g M_{\text{pl}}. \tag{31}$$

There exists an extra factor of g in front of M_{pl} in the above equation, which means that the ultraviolet gauge coupling g is defined not at the Planck scale M_{pl} but at a lower energy scale gM_{pl}. The running of the gauge coupling from the value at low energy $\operatorname{Im}\tau$ to the one at high energy $\operatorname{Im} S$ is schematically depicted in Fig. 4. The existence of the new scale gM_{pl} is what the weak-gravity conjecture has predicted. Thus the analysis of the $\mathcal{N} = 2$ SU(2) gauge theory coupled to supergravity supports the weak-gravity conjecture.

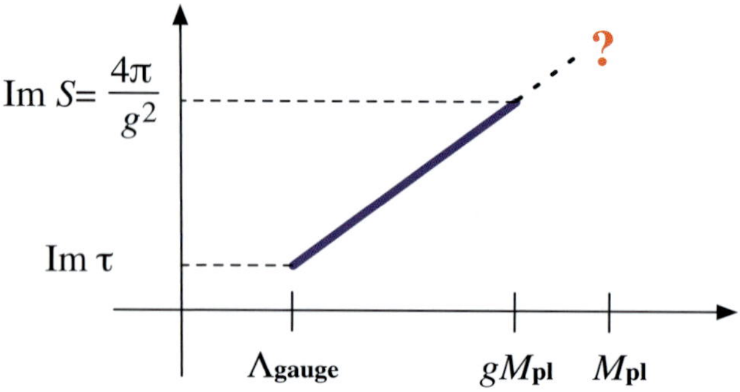

Fig. 4. Running of the coupling in the gauge theory coupled to supergravity.

3. Explicit Description of Logarithmic Periods

For a CY which is a $K3$ fibration over $\mathbb{CP}^1$, 3-cycles can be constructed explicitly. We follow the approach of Ref. 11 and Appendix in Ref. 14. Consider a CY with a defining equation

$$w + \frac{\mu^2}{w} + W_{K3}(x, y, z; t_\ell) = 0\,, \tag{32}$$

where t_ℓ denote the moduli of the $K3$. The holomorphic 3-form is given by

$$\Omega = \frac{dw}{w} \wedge \Omega_{K3}\,, \qquad \Omega_{K3} = \frac{dx \wedge dy}{\partial_z W_{K3}}\,. \tag{33}$$

3-cycles of CY are made of the product of a 1-cycle of the $\mathbb{CP}^1$ base and a 2-cycle of $K3$.

2-cycles of $K3$ to be used here are those which are not holomorphically embedded into $K3$, since holomorphic cycles have the representative which are of the $(1,1)$-form so that their integrals with the $(2,0)$-form Ω_{K3} must vanish. Holomorphic cycles of $K3$ form the Picard lattice Λ^{Pic}

$$\Lambda^{\mathrm{Pic}} = H^{1,1}(K3) \cap H^2(K3, \mathbb{Z}) \tag{34}$$

and its dimension is called the Picard number $\rho(K3)$. Cycles which are not holomorphically embedded are called transcendental and the lattice Λ of the 2nd homology of $K3$ has an orthogonal decomposition into Picard and transcendental lattices

$$\Lambda = \Lambda^{\mathrm{Pic}} \oplus \Lambda^{\mathrm{tr}}\,. \tag{35}$$

It is well known that the lattice Λ has a signature of $(3, 19)$. In the case of projective $K3$, the Kähler form becomes algebraic and the Picard lattice has a signature $(1, \rho(K3)-1)$. Then the signature of Λ^{tr} becomes $(2, 20-\rho(K3))$.

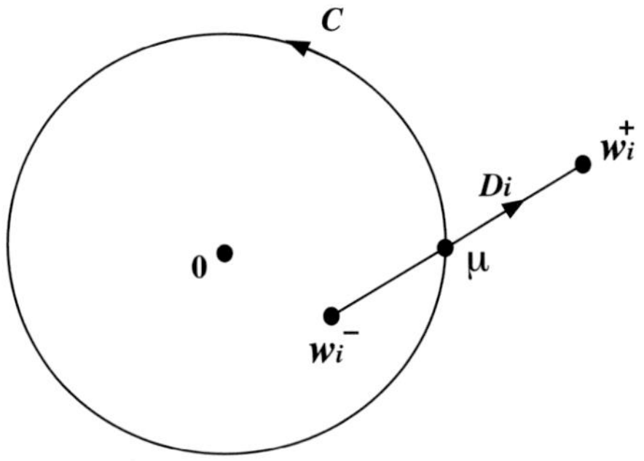

Fig. 5. Cuts in the base $\mathbb{CP}^1$.

In the case of the quartic $K3$ surfaces which featured in our example X_8, the Picard number is $\rho(K3) = 19$ and thus there are three transcendental cycles with signature $(2,1)$. The 2-cycle with a negative signature, i.e. a negative self-intersection number is the vanishing cycle of A_1 singularity. Two 2-cycles of the positive signature generate periods which have logarithmic behavior in ϵ as we see below. Reference 11 discusses another example of CY manifold X_{24} which also possesses a $K3$ fibration and produces the SU(3) gauge theory in the decoupling limit. In this case there exist four transcendental cycles with a signature $(2,2)$. Two 2-cycles with the negative signature describe the vanishing cycles of A_2 singularity. In the case of general A_r singularity there will be $2+r$ transcendental cycles with the signature $(2,r)$. As we shall see below, two transcendental cycles of $K3$ with the positive signature will generate logarithmic cycles of CY manifold.

The CY (32) can be thought of as a one-parameter family of $K3$, whose moduli depend on w. Suppose a transcendental two-cycle S_i degenerates at $w + \mu^2/w = k_i$. For a small μ, this happens at $w_i^+ \sim k_i$ and $w_i^- \sim \mu^2/k_i$, see Fig. 5. Let C be the circle around the origin $|w| = |\mu|$, and D_i denote the path connecting $w_i^{\pm}$. Then $C \times S_i$ and $D_i \times S_i$ are closed 3-cycles of CY manifold.

In general, Yang–Mills gauge theories are geometrically engineered by fine-tuning the parameters $\{t_\ell\}$ of $K3$ so that the $K3$ develops ADE singularities: see Ref. 7 for SU(n), Ref. 15 for SO(n) and Refs. 16 and 17 for E_n groups. Suppose we have a singularity of type G with $\operatorname{rank} G = r$ around $x = y = z = 0$. The moduli $\{t_\ell\}$ of $K3$ are decomposed into two sets of parameters

$$\{u_2, \ldots, u_h\}\,, \qquad \{v_1, v_2, \ldots\}\,, \tag{36}$$

Table 1. Data of ADE singularities.

	Defining equations	h	d_x	d_y	d_z	Degree of Casimirs
A_{n-1}	$0 = x^2 + y^2 + z^n$	n	$n/2$	$n/2$	1	$2, 3, \ldots, n$
D_{n+1}	$0 = x^2 + y^2 z + z^n$	$2n$	n	$n-1$	2	$2, 4, \ldots, 2n$
E_6	$0 = x^2 + y^3 + z^4$	12	6	4	3	$2, 5, 6, 8, 9, 12$
E_7	$0 = x^2 + y^3 + yz^3$	18	9	6	4	$2, 6, 8, 10, 12, 14, 18$
E_8	$0 = x^2 + y^3 + z^5$	30	15	10	6	$2, 8, 12, 14, 18, 20, 24, 30$

where u_i corresponds to the degree i Casimir invariant of the group G. u_i are tuned to vanish as $\epsilon^{i/h}$ in the geometric engineering limit and we rescale them as $\epsilon^{i/h} \cdot u_i$. Here h is the dual Coxeter number of G. v_j are the moduli which remain finite in the engineering limit. We also introduce the rescaled coordinates as

$$w = \epsilon \tilde{w} , \quad x = \epsilon^{d_x/h} \tilde{x} , \quad y = \epsilon^{d_y/h} \tilde{y} , \quad z = \epsilon^{d_z/h} \tilde{z} . \tag{37}$$

$d_{x,y,z}$ are the degrees of x, y, z (see Table 1). We also set $\mu = \epsilon \Lambda^h$. Then the defining equation (32) of the CY becomes

$$\epsilon \left(\tilde{w} + \frac{\Lambda^{2h}}{\tilde{w}} + W_{\mathrm{ADE}}(\tilde{x}, \tilde{y}, \tilde{z}; u_i) + O(\epsilon^{1/h}) \right) = 0 . \tag{38}$$

The holomorphic 3-form is given by

$$\begin{aligned} \Omega &= \frac{dw}{w} \wedge \frac{dx \wedge dy}{\partial_z W_{K3}} \\ &= \epsilon^{(d_x + d_y + d_z)/h - 1} \frac{d\tilde{w}}{\tilde{w}} \wedge \frac{d\tilde{x} \wedge d\tilde{y}}{\partial_{\tilde{z}} W_{\mathrm{ADE}}} \\ &= \epsilon^{1/h} \frac{d\tilde{w}}{\tilde{w}} \wedge \Omega_{\mathrm{ADE}} , \end{aligned} \tag{39}$$

where we used the fact $d_x + d_y + d_z = h + 1$.

There are r independent two-cycles S_i of $K3$ which vanish simultaneously in the engineering limit. These give rise to $2r$ 3-cycles $\bar{A}_i = C \times S_i$ and $\bar{B}_i = D_i \times S_i$, $i = 1, \ldots, r$ of the CY as explained above. We can take their linear combinations, A^i and B_i, so that they have the canonical intersection form, $(A^i, A^j) = (B_i, B_j) = 0$, $(A^i, B_j) = \delta^i_j$. Then

$$a^i = \int_{A^i} \frac{d\tilde{w}}{\tilde{w}} \wedge \Omega_{\mathrm{ADE}} , \qquad a^D_i = \int_{B_i} \frac{d\tilde{w}}{\tilde{w}} \wedge \Omega_{\mathrm{ADE}} , \tag{40}$$

are identified with the special coordinates of Seiberg–Witten theory. Corresponding supergravity periods behave as

$$\begin{aligned} X^i &= \int_{A^i} \Omega = \epsilon^{1/h} a^i + O(\epsilon^{2/h}) \,, \\ F_i &= \int_{B_i} \Omega = \epsilon^{1/h} a_i^D + O(\epsilon^{2/h}) \,. \end{aligned} \tag{41}$$

$K3$ surface has two extra 2-cycles which have a positive signature, as we have noted above. We call them T_a, $a = 1, 2$ and arrange them so that they do not intersect S_i and stay at finite values of x, y and z. Now the defining equation of CY near the 2-cycles T_a is given by

$$w + \frac{\epsilon^2 \Lambda^{2h}}{w} + W_{K3}(x, y, z; 0, v_j) = 0 \,. \tag{42}$$

Thus from the cycle $U_a = C \times T_a$ we obtain the period

$$X^a \equiv \int_{U_a} \Omega = \oint_C \frac{dw}{w} \int_{T_a} \Omega_{K3} = 2\pi i c_a \approx O(1) \,,$$

where

$$c_a = \int_{T_a} \Omega_{K3}(u_i = 0; v_j) \,. \tag{43}$$

In the case of the cycles $V_a = D_a \times T_a$, the end points of the w integration become

$$w_a^- \sim \frac{\epsilon^2 \Lambda^{2h}}{k_a} \,, \qquad w_a^+ \sim k_a \,, \tag{44}$$

where $k_a = w + \epsilon^2 \Lambda^{2h}/w$ is the value at which the 2-cycle T_a degenerates. Then we find the logarithmic behavior

$$F^a \equiv \int_{V_a} \Omega = \int_{w_a^-}^{w_a^+} \frac{dw}{w} \int_{T_a} \Omega_{K3} \approx -2c_a \log \epsilon \,. \tag{45}$$

The analysis of the monodromy under the phase rotation of ϵ suggests

$$F^a \approx -\frac{1}{\pi i} \log \epsilon \cdot X^a + O(\epsilon^{1/h}) \,, \tag{46}$$

although the precise form of this expression will depend on the intersection form of T_a. Thus we have established the existence of periods behaving logarithmically near the engineering limit.

4. Discussion

In this paper, we have seen how the holomorphy inherent in $\mathcal{N}=2$ supersymmetry can be effectively used to study the effect of gravity upon the running of gauge theory. More specifically, we showed how the monodromy of the periods around the locus of the rigid limit translates to the hierarchical separation of the dynamical scale of gauge theory and the Planck scale. We have argued that, as compared to the naive relation

$$\Lambda_{\text{gauge}} \approx e^{-4\pi^2/hg^2} M_{\text{pl}} \,, \tag{47}$$

there is generically an extra factor of the gauge coupling constant g on the right-hand side,

$$\Lambda_{\text{gauge}} \approx e^{-4\pi^2/hg^2} \cdot g M_{\text{pl}} \,, \tag{48}$$

supporting the weak-gravity conjecture.

The result presented here is only a small step in utilizing the holomorphy to understand the dynamics of the coupled $\mathcal{N}=2$ supergravity-gauge systems. We believe many more properties can be learned in a similar manner. It would also be interesting to make a comparison with the result in Ref. 19 where the authors calculated the one-loop effect of gravity to the beta function of the gauge theory. It was argued in Ref. 20 that the beta function in Ref. 19 alone leads to the weak-gravity conjecture. We will have to supersymmetrize the result of Ref. 19 to carry out the comparison to our case.

It will be very important to see if it is possible to extend our results to the realm of $\mathcal{N}=1$ supersymmetric theories. In the case when $\mathcal{N}=1$ theories are obtained from those of $\mathcal{N}=2$ by introducing fluxes, branes, etc. many of the structures of the latter survive. Hopefully we will have enough control over mass scales of these theories to derive the characterization of consistent $\mathcal{N}=1$ field theories coupled to gravity.

Acknowledgments

T. Eguchi would like to thank Prof. K. K. Phua for his invitation to the conference celebrating Prof. C. N. Yang's 85th birthday. This talk is based on Ref. 21. Research of T. Eguchi is supported in part by a Grant-in-Aid from the Japan Ministry of Education and Science. Research of Y. Tachikawa is supported by DOE grant DE-FG02-90ER40542.

References

1. M. R. Douglas and S. Kachru, Flux compactification, arXiv:hep-th/0610102.
2. C. Vafa, The string landscape and the swampland, arXiv:hep-th/0509212.
3. N. Arkani-Hamed, L. Motl, A. Nicolis and C. Vafa, The string landscape, black holes and gravity as the weakest force, arXiv:hep-th/0601001.
4. H. Ooguri and C. Vafa, *Nucl. Phys. B* **766**, 21 (2007), arXiv:hep-th/0605264.
5. N. Seiberg and E. Witten, *Nucl. Phys. B* **426**, 19 (1994) [Erratum: *ibid.* **430**, 485 (1994)], arXiv:hep-th/9407087.
6. S. Kachru, A. Klemm, W. Lerche, P. Mayr and C. Vafa, *Nucl. Phys. B* **459**, 537 (1996), arXiv:hep-th/9508155.
7. A. Klemm, W. Lerche, P. Mayr, C. Vafa and N. P. Warner, *Nucl. Phys. B* **477**, 746 (1996), arXiv:hep-th/9604034.
8. S. H. Katz, A. Klemm and C. Vafa, *Nucl. Phys. B* **497**, 173 (1997), arXiv:hep-th/9609239.
9. P. Candelas, X. de la Ossa, A. Font, S. H. Katz and D. R. Morrison, *Nucl. Phys. B* **416**, 481 (1994), arXiv:hep-th/9308083.
10. S. Hosono, A. Klemm, S. Theisen and S. T. Yau, *Commun. Math. Phys.* **167**, 301 (1995), arXiv:hep-th/9308122.
11. M. Billó, F. Denef, P. Frè, I. Pesando, W. Troost, A. Van Proeyen and D. Zanon, *Class. Quantum Grav.* **15**, 2083 (1998), arXiv:hep-th/9803228.
12. P. S. Aspinwall and J. Louis, *Phys. Lett. B* **369**, 233 (1996), arXiv:hep-th/9510234.
13. T. Eguchi and Y. Tachikawa, *J. High Energy Phys.* **0601**, 100 (2006), arXiv:hep-th/0510061.
14. F. Denef, *Nucl. Phys. B* **547**, 201 (1999), arXiv:hep-th/9812049.
15. M. Aganagic and M. Gremm, *Nucl. Phys. B* **524**, 207 (1998), arXiv:hep-th/9712011.
16. J. H. Brodie, *Nucl. Phys. B* **506**, 183 (1997), arXiv:hep-th/9705068.
17. J. Hashiba and S. Terashima, *J. High Energy Phys.* **9909**, 020 (1999), arXiv:hep-th/9909032.
18. N. Dorey, V. V. Khoze and M. P. Mattis, *Phys. Lett. B* **390**, 205 (1997), arXiv:hep-th/9606199.
19. S. P. Robinson and F. Wilczek, *Phys. Rev. Lett.* **96**, 231601 (2006), arXiv:hep-th/0509050.
20. Q. G. Huang, *J. High Energy Phys.* **0703**, 053 (2007), arXiv:hep-th/0703039.
21. T. Eguchi and Y. Tachikawa, *J. High Energy Phys.* **0708**, 068 (2007), arXiv:0706.2114.

FIVE DECADES AFTER THE REVOLUTION: HOW MUCH DO WE KNOW ABOUT THE NEUTRINO?

NGEE-PONG CHANG
Physics Department, City College of CUNY, New York, NY, 10031, USA

It is a pleasure to dedicate this talk to Prof Yang on the occasion of his 85th birthday, especially here in my birthplace, Singapore. I still recall the year 1957 when the local Chinese newspapers broke with the story of breakdown of parity, I didn't understand the phrase 宇称不守恒, and when I checked into my high school physics text, I couldn't find any explanation. The Chinese newspapers 南洋商报, 星洲日报 call this Lee–Yang breakthrough a revolution, but it was hardly noticed by the English newspapers at the time. It wasn't until I came to the U.S. in the Fall of 1957, that a true revolution occurred. I think I can speak for a whole generation of ethnic Chinese physicists who grew up in the U.S. that we personally benefitted far more from this *revolution* than the *revelation* of parity non-conservation.

Red Moon over U.S.

Sputnik launch, October 4, 1957.

In line with the festive nature of this occasion, let me collect together some of the headlines of those exciting days as much as I can gather them. I regret that I could not locate the archives of 南洋商报，星洲日报 so as to give you a flavor of the respect and pride that the Chinese community had for the news. Instead I have been able to assess the archives of the New York Times and the APS.

Five Decades after the Revelation

New York Times, Wednesday, Jan 16, 1957.

Basic Concept in Physics Is Reported Upset in Tests

Conservation of Parity Law in Nuclear Theory Challenged by Scientists at Columbia and Princeton Institute

By HAROLD M. SCHMECK Jr.

Physical Review **104**, 254 (1956)

Question of Parity Conservation in Weak Interactions*

T. D. LEE, *Columbia University, New York, New York*

AND

C. N. YANG,† *Brookhaven National Laboratory, Upton, New York*

(Received June 22, 1956)

The question of parity conservation in β decays and in hyperon and meson decays is examined. Possible experiments are suggested which might test parity conservation in these interactions.

T.D. Lee and C.N. Yang at Institute for Advanced Study.

Physical Review **105**, 1413 (1957)

Experimental Test of Parity Conservation in Beta Decay*

C. S. Wu, *Columbia University, New York, New York*

AND

E. Ambler, R. W. Hayward, D. D. Hoppes, and R. P. Hudson,
National Bureau of Standards, Washington, D. C.

(Received January 15, 1957)

IN a recent paper[1] on the question of parity in weak interactions, Lee and Yang critically surveyed the experimental information concerning this question and reached the conclusion that there is no existing evidence either to support or to refute parity conservation in weak interactions. They proposed a number of experiments on

C.S. Wu (1912–1997).

The precision experiment by C.S. Wu and collaborators established the maximum breakdown of mirror symmetry in low temperature polarized Co 60 radioactive β decays. It was quickly confirmed by many experiments on $\pi \to \mu\nu$ decays. And in October 1957 came the announcement of the Nobel Prize award to **Chen-Ning Yang** and **Tsung-Dao Lee**.

The Nobel Prize in Physics 1957

"for their penetrating investigation of the so-called parity laws which has led to important discoveries regarding the elementary particles"

Lee and Yang at OCPA 2000, Chinese University of Hong Kong.

During the hey-day of 1957 and the 60's, it quickly became clear that the neutrino is left-handed, and Lee and Yang observed that while C and P are separately maximally violated, CP appears to be largely conserved, but encouraged experimentalists to look for CP non-conservation. And it wasn't until 1964 when Cronin and Fitch in a small BNL experiment discovered a very small CP violation in $K_L - K_S$ decay.

Today, we know a lot about the neutrino. We know there are three types of neutrinos, ν_e, ν_μ, and ν_τ, and since the Kamiokande discovery in 1998, we now know that the three types of neutrino can oscillate into one another. Such an oscillation can come only from a mass-like term in the Lagrangian. There is much information on the so-called neutrino-mixing matrix parameters. But what is the mass of the electron neutrino?

The neutrino was proposed by W. Pauli in a postcard dated Dec 4, 1930 to Lise Meitner *et al.* (Radioactive Ladies and Gentlemen), and Pauli gave a formal talk on this idea at the Solvay Congress, Oct 22–29, 1933. Fermi was at the Congress, and on Dec 31, 1933, he published in Ricerca Scientifica the first exposition on the 4-fermion theory of weak interaction. It was he who gave the name 'little neutral one' to this invention by Pauli. Interestingly, this first paper had been rejected by Nature. Aspiring young theorists, take heart!

While the rest of the particle physics world are engaged in big accelerator experiment measurements of these oscillation parameters, there is a team of low energy physicists doing painstaking measurements on the electron neutrino.

$$\text{Tritium Decay:} \quad {}^3H \longrightarrow {}^2He + e^- + \bar{\nu}_e$$

The determination of the electron neutrino mass requires a precision measurement of the shape of the electron decay energy near the end-point, where the electron energy is maximum. Early generations of measurements gave puzzling results. In 1997, second generation experiments were done at Mainz and, separately, at Troitsk, with the surprising results

$$m_{\nu_e}^2 = (-1.6 \pm 2.5 \pm 2.1)eV^2$$

(Mainz)

$$m_{\nu_e}^2 = (-2.3 \pm 2.5 \pm 2.0)eV^2$$

(Troitsk)

Currently, there is an ongoing KATRIN experiment at Karlsruhe which seeks to measure the endpoint to a sensitivity of 0.2 eV.

Notice the unusual tachyonic nature of the quoted result. In early days of the Particle Data Group, the results were presented as is (i.e. negative mass-squared). Now the Particle Data Group suppresses it in favor of their own interpretation that the result is consistent with a small physical neutrino mass.

I will end this brief talk with the tantalizing possibility that the mysterious neutrino indeed is a tachyonic neutral particle. Chodos, Hauser and Kostelecky in 1985 proposed an ad-hoc equation for such a tachyon:

$$\gamma \cdot \partial\psi = -m\gamma_5\psi$$

Can such a tachyonic field theory be quantized? It is not my intention to use this brief time to go into a complex subject. I just want to tickle your curiosity and remind one and all that we must always look to the experiments rather than our own prejudices.

INTERACTING MULTI-COMPONENT FERMIONS AND THE YANG-BAXTER EQUATION: FUTURE PROSPECTS

M. T. BATCHELOR

Theoretical Physics, RSPSE and Mathematical Sciences Institute, The Australian National University, Canberra ACT 0200, Australia
E-mail: Murray.Batchelor@anu.edu.au

The remarkable exact solution of 1D interacting fermions heralded a new era in the study of quantum many-body problems. The mathematical structure of the underlying Yang-Baxter equation led to profound advances in other fields. Recently there has been a significant revival of interest in 1D quantum many-body problems due to the striking and ongoing experimental developments in the study of cold quantum matter. In particular, through the trapping and cooling of interacting systems of atomic bosons and fermions in low dimensions. There is, e.g., already rather spectacular agreement between theory and experiment for 1D interacting bosons. This includes an experimental test of Yang-Yang thermodynamics for bosons on a chip.

In this talk I will describe recent calculations for 1D interacting multi-component fermions. We find that two-component ultracold fermions undergo a phase transition between a BCS pairing state and a normal Fermi liquid [1]. However, for three-component ultracold fermions Zeeman splittings may trigger novel quantum phase transitions: from a trionic state to a BCS pairing state; from a trionic Luttinger liquid to a normal Fermi liquid, and from a BCS pairing state to a normal Fermi liquid [3]. The exact results for the ground state energy, phase diagrams, critical fields and magnetizations [1, 2] are obtained analytically from the Yang-Yang-type thermodynamic Bethe ansatz equations [3, 4]. More generally, such models of multi-component interacting fermions and their experimental realizations with cold atoms may shed light on spin-neutral Fermi-liquids and colour superconductivity as well as the dynamics of quarks and baryons.

[1] X.-W. Guan, M. T. Batchelor, C. Lee and M. Bortz, *Phys. Rev. B* **76**, 085120 (2007).
[2] X.-W. Guan, M. T. Batchelor, C. Lee and H.-Q Zhou, *Phys. Rev. Lett.* **100**, 200401 (2008).
[3] C. N. Yang and C. P. Yang, *J. Math. Phys.* **10**, 1115 (1969).
[4] M. Takahashi, *Prog. Theor. Phys.* **44**, 899 (1970).

FREE ELECTRON LASER DEVELOPMENTS IN CHINA

ZHENTANG ZHAO

Shanghai Institute of Applied Physics,
Chinese Academy of Sciences
E-mail: zhaozt@ssrc.ac.cn

As a development step towards constructing a hard X-ray FEL in China, a soft X-ray FEL test facility (SXFEL) was proposed under C N Yangs initial suggestion. This test facility will be built on the campus of the Shanghai Synchrotron Radiation Facility and it can be converted into a user facility or a part of hard X-ray FEL in the future. This presentation describes the FEL developments in China and reports the preliminary design and status of this Chinese soft X-ray Test facility.

PROSPECT OF PARTICLE PHYSICS IN CHINA

HESHENG CHEN

Institute of High Energy Physics, Chinese Academy of Sciences,
Beijing, 100049, China
E-mail: chenhs@ihep.ac.cn

The nuclear physics and particle physics researches in China have a long tradition. BEPC is a milestone of the particle physics in China. The many interesting physics results were obtained from BEPC. Beijing Synchrotron Radiation Facility based on BEPC became the major synchrotron radiation facility in China. The upgrade of BEPC, which will increase the luminosity by two orders of magnitude, is going smoothly. The non-accelerator based experiments were promoted also. Professor Yang made very important contribution to Chinese physics, especially to promote the large science facilities for the multiple discipline researches. The particle physics faces the great challenges, and meet the great opportunities in the 21st century. The medium term plan of the particle physics in China was discussed.

Keyword: Particle physics in China; C.N.Yang; BEPC; Beijing Synchrotron Radiation Facility.

1. Particle Physics in China

The nuclear physics and particle physics researches in China have a long tradition. Actually, Professor Zhongyao Zhao discovered the Positron at first. But he did not publish the result. During the Second World War, Professor Ganchang Wang proposed an idea of the experiment to search for neutrino.

Chinese Academy of Sciences established Nov. 1949. As one of its first institutes, the Institute of Modern Physics, which is the mother institute of IHEP, established 1950 with emphases of the nuclear physics research. China jointed the Joint Institute of Nuclear Research (JINR) at Dubna 1956. Chinese physicists at JINR obtained some interesting results. An experimental group led by Professor Ganchang Wang discovered the anti-Σ particle in a bubble chamber experiment. However, China withdrew from JINR at the beginning of 1960's. The Chinese Government decided to keep the money used to be contributed to JINR for the Chinese own particle physics facility.

Institute for High Energy Physics (IHEP) became an independent institute for the particle physics in Feb. 1973. Institute of High Energy Physics, under the Chinese Academy of Sciences, is the comprehensive and largest fundamental research center in China. The construction of Beijing Electron Positron Collider (BEPC) is a milestone of the Chinese particle physics. The Chinese particle physics community also provides big scientific platforms, such as the synchrotron radiation light sources in Beijing, Hefei and Shanghai, as well as the Chinese Spallation Neutron Source, for the multiple discipline researches. The major research fields of IHEP include:

- Particle physics: Charm physics @ BEPC, LHC experiments, cosmic ray measurements, particle astrophysics and neutrino physics
- Accelerator technology and applications
- Synchrotron radiation technologies and applications

There are 1030 employees in IHEP, two third of them are physicists and engineers. In addition, there are about 400 PhD Students and postdoctors. There are strong experimental particle physics groups in Peking University, Tsinghua University, University of Science and Technology of China (USTC), Shandong University, Huazong Normal University... etc.. In addition, there are many active groups of the particle and nuclear physics theory in dozens of universities.

The major activities of the particle physics experiments in China include both accelerator-based experiments and non-accelerator-based experiments. BEPC and its upgrade BEPCII are the major activities of the domestic experimental base of the high energy accelerator.

The non-accelerator based experiments in China include:

- Yangbajing cosmic-ray international observatory at Tibet
- L3 cosmic measurement (finished)
- The hard X-ray modulation telescope
- The Daya Bay reactor neutrino experiment.

Chinese particle physicists joint many international collaborations of the particle physics experiments in World:

- Mark-J at PETRA of DESY (IHEP and USTC)
- L3 (IHEP and USTC)and ALEPH (IHEP) at LEP of CERN
- ATLAS, CMS, LHCb and Alice at LHC of CERN
- Alpha Magnetic Spectrometer (AMS)

- BELLE at KEKB of KEK
- Kamland and SuperK at Kamioka
- Star and Phenix at RHIC of BNL
- R&D of ILC.

2. Beijing Electron Positron Collider

BEPC was constructed successfully during 1984-1988, on the schedule, within the budget, and reached the design specification soon after commissioning. Beijing Spectrometer (BES) is the general purpose magnetic spectrometer in BEPC. The physics window of BEPC is very interesting. The production cross sections of J/Ψ and Ψ' are very high, and the backgrounds are very low compared with the hadron collision experiments. The decays of J/Ψ and Ψ' are gluon rich with larger phase space to hadrons of 1-3 GeV. They are very good channels to study the light hadron spectroscopy and to search for new particles and new phenomena. The BES international collaboration included more than 30 universities and institutes from China, US, Japan and UK. BES collected 66M J/Ψ events, the largest data sample of J/Ψ in the world.

BES obtained many important physics results. Actually more than 400 results from BES are quoted by PDG 2006. The precision measurement of the τ lepton mass changed its world average value by 3σ, and improved its accuracy by a factor of 10. The result together with the measurements of the leptonic decay branch ratio and the lifetime of τ approved the lepton universality of τ particle. The precision measurement of the hadronic cross section R at 2-5GeV reduced the uncertainty ΔR/R from 15-20% at PDG to 6.6%. The result has the large impact to the Higgs mass prediction from Standard Model and to the theoretical calculation of the g-2 experiment. The result also improves the calculation of the fine structure constant $\alpha(M_z^2)$ from 128.890 ± 0.090 to 128.936 ± 0.046. The BES experiment also made the systematic study of Ψ(2S) and J/Ψ decays. BES observed the resonance X(1835) in $J/\Psi \to \gamma p\bar{p}$ and $J/\Psi \to \gamma\eta'\pi^+\pi^-$. A possible explanation to the resonance is the bound state of $p\bar{p}$. From the measurements of the $D\bar{D}$ cross section and the R value of Ψ(3770), BESII found for the first time a significant decay branch fraction of non-$D\bar{D}$, which was confirmed later by CLEOc.

3. Non-accelerator Based Experiment

Yangbajing Cosmic Ray Observatory in Tibet located 90km north of Lasha, 4300m above sea level. It is the best site in World for the high altitude cosmic ray observatory. There is a plain of about 100km by 20 Km, no firn even in winter. The nearby large terrestrial heat power station provides all the necessary infrastructure. There two experiments in the Yangbajing: China-Japan Air Shower array since 1990 and IHEP-INFN Argo RPC carpet since 2000. Some interesting results were reported from Yangbajing, including the observation of the new anisotropy component and the co-rotation of the galaxy cosmic ray.

Chinese institutions jointed the Alpha Magnetic Spectrometer (AMS) experiment. The permanent magnet and the main structure of the AMS01 detector were designed, constructed and space qualified by the institute of electric engineering and IHEP of CAS as well as the Chinese aerospace. The magnet became the first large magnet made by mankind in the space, as the payload of the AMS test flight by the space shuttle Discovery June 1998. Chinese institutions have large activities in AMS02. IHEP and Chinese Aerospace work with LAPP of IN2P3 and Pisa of INFN for the electromagnetic calorimeter of 700Kg.

The gamma ray burst detector made by IHEP was flown in the Shenzhou-II spacecraft 2001. The Chinese moon project, ChangEr-1, was launched 24 Oct. 2007. Its payload included: the optical system, the X ray spectrometer, the γ ray spectrometer, the laser altimeter, the Solar wind detector. All of them were made by Chinese Academy of Sciences. The X ray spectrometer to identify the elements of Moon was made by IHEP. The instruments work well in the Moon orbit.

The Hard X ray modulation telescope (HXMT) satellite is designed based on the special algorithm of the modulation method proposed by Chinese physicists. HXMT provides much higher sensitivity to scan the X-point sources for the full sky. The scientific goals of HXMT are searching for highly obscured supermassive black holes, and new types of high energy objects. HXMT has been approved by the Chinese Space Agency. It should be launched by 2011.

4. BEPCII: High Luminosity Double-Ring Collider

BEPC II uses the double ring design. The second ring was built in the existing BEPC tunnel. Two rings cross over at the south and the north interaction points to form the two equal rings for electrons and positrons. There are 93 bunches per ring. The total beam current is more than 0.9A in each ring. The beams collide at the south interaction region with large horizontal cross-angle with ± 11 mr. The collision spacing is 8 ns. The design luminosity is $10^{33} cm^{-2} s^{-1}$ at the C.M. energy of 3.78GeV. The upgrade of Linac will provide the positron ejection rate of 50mA/min. with the full energy injection up to 1.89GeV. The synchrotron radiation performance will be also improved to 250mA at 2.5 GeV. The hard X-ray flux will be increased by one order of magnitude. A new detector BES III at BEPCII is aimed to adapt to the high event rate of 3KHz at the luminosity of $10^{33}cm^{-2}s^{-1}$ and the bunch spacing of 8ns. The systematic errors will be reduced significantly to match with the high statistics of the event samples, especially the photon measurement and the particle identification. The acceptance will be increased and give the space for the Superconducting quads.

Physics topics at BEPCII/BESIII are very interesting:

- The precision measurements of the CKM matrix elements from the precision measurements of the decay branching rations of the charm mesons: V_{cd}/V_{cs} by the Leptonic and the semi-leptonic decays, and V_{cb} by the hadronic decays. The measurements of f_D and f_{Ds} from the Leptonic decays and the form factors of the semi-leptonic decays together with the measurements at B-factories will improve the accuracies of V_{td}/V_{ts} and V_{ub} significantly. The unitarity of the CKM matrix will be also tested.
- The precision test of Standard Model, the test of the Lepton universality
- QCD and the hadron production, the precision measurement of hadronic R value
- The light hadron spectroscopy, the Baryon spectroscopy, the Charmonium spectroscopy
- The Charmonium physics
- The Search for new particles: Glueball, non-$q\bar{q}$ states...
- The Search for new physics: the $D\bar{D}$ mixing, the CP violation, the rare decays, FCNC, the Lepton number violation...
- ...

The construction of BEPC is carried out in three stages, interleaving with the synchrotron radiation running. The construction started from Jan. 2004, and expected to be finished by the end of 2008. The project goes smoothly. The major task of the first Stage, the Linac upgrade, reached the design goals on the schedule. The tasks of the second stage are the installation of two storage rings, the commissioning of the Collider, and the construction of the BESIII detector. The second stage is going smoothly. The electron beam current in the storage ring reaches 500 mA, the positron beam current reaches 200mA. The measurement of the storage ring parameters are in good agreement with the prediction. The luminosity is quite good.

BESIII Detector consists of the Beryllium beam pipe, the drift chamber, the time of flight counters, the electromagnetic calorimeter, the muon detector and the magnetic yoke. The main drift chamber uses the small cell and the Helium-based gas. It has 7000 gold-plated tungsten signal wires of diameter 25 μm and 22000 gold-plated Aluminum field wires of diameter 110μm. The momentum resolution is expected 0.5% at 1GeV, and the dE/dX resolution about 6%. The wiring was completed with good quality. The inner chamber and outer chamber assembled. The construction of MDC has been finished.The cosmic-ray test finished with the single wire resolution better than 120μm. The design goals of the CsI(Tl) crystal electromagnetic calorimeter are the energy resolution of 2.5% and the spatial resolution of 0.6cm both at 1 GeV. The length of the crystal is 28 cm. Each crystal is read by 2 photodiodes, 2 Preamp and 1 Amplifier. The preamplifier noise is less than 1100 e (220keV), and the shaping time of amplifier is 1μs. There are 5280 crystals in the barrel with the weight of 21564 kg and 960 crystals in the endcaps with the weight of 4051 kg, in total 6240 crystals with the weight of 25.6 T. The muon detector uses 9 layers of RPC in total 2000 m^2. The special bakelite plates without the linseed oil were used. The total number of channels is 10000. The noise rate in the RPC is less than 0.1 Hz/cm^2. The assembling of the BESIII detector is going smoothly. The BESIII will be moved into the interaction region next April. The physics running will be started by next summer.

5. Professor C. N. Yang's Contribution to the Particle Physics in China

Professor Yang made his pioneer visits to China at July 1971. Chairman Mao met Professor Yang July 19,1973. Afterwards he frequently visits China: met the Chinese government leaders, and urged to reconstruct the education system, and promoted the sciences and technology in China. He visited many universities and institutes, and gave seminars and lectures to introduce Chinese physicists the latest developments in particle physics and theoretical physics. Professor Yang also introduced western scientists about the development in China, and promoted the scientific exchanges between China and Western countries. With his help, many Chinese scholars and PhD students studied at US. Many of them became the backbone of Chinese physics research and Chinese universities.

Professor Yang gave the Chinese high energy community the foreseen advice about developing large accelerator-based science facilities for the multiple discipline research since middle of 1980's. He encouraged the synchrotron radiation facility and its applications in Beijing Electron Positron Collider. Following his advice, Beijing Synchrotron Radiation Facility based on Beijing Electron Positron Collider became the major hard X-ray light source in China and produces many first class results.

Professor Yang made great efforts to promote the Chinese hard X ray FEL based on HGHG and its test facility. He wrote the first letter to the state councilor Song Jian and the president of Chinese Academy of Sciences Zhou Guangzhao to propose the Chinese XFEL May 20, 1997. He visited IHEP several times during 2005-2006 to discuss the test facility of CXFEL. We also had long discussions in his office and house at the Tsinghua University. March 2, 2005 Professor Yang wrote the letter to the state councilor Chen Zhili about the HGHG test facility of CXFEL. This is his 7th letter about CXFEL. With his encouragement the conceptual design and the proposal of the test facility of CXFEL was finished in rather short period. The test facility will be built at the Shanghai Institute of Applied Physics.

Chinese spallation neutron source (CSNS) consists of: RCS H- ion source, RFQ, DTL linac of 81 MeV(upgradeable to 230MeV superconducting linac) and rapid-cycling synchrotron of 1.6 GeV at 25 Hz. In the phase I of CSNS, the beam current is 83 mA, the beam power at the target is 120kW. The first target system allows 18 spectrometers. With the superconducting linac in the phase II, the beam power at the target will be 500kW, and the second target system could be built. The Chinese Government approved the CSNS project in principle. IHEP is in charge of the CSNS project. The site of CSNS is in Dongguan, Guangdong province, as a branch of IHEP. The total budget of CSNS is 1.4B RMB. The local governments will provide the free land and the additional fund for the infrastructure. We expect the first beam after 5.5 year from the ground breaking. CSNS will be the major project for the machine team and the detector team after BEPCII/BESIII finished. We received many helps to CSNS from Professor Yang.

6. Particle Physics in 21st Century

The particle physics and the particle astrophysics face the great challenges in the 21st century. The symmetry breaking mechanism is a big open question: by Higgs or by SUSY particles, or some thing unexpected? The recent discoveries in the Neutrino physics provided the hits beyond of the Standard Model. The origin of the CP violation is still a mystery. The latest astronomy observations indicate that the dark matter and the dark energy are 23% and 73% of the total matter in Universe respectively. The Standard

Model cannot explain them at all. The biggest challenges are the search to the dark matter and to understand the dark energy. The great challenges to the particle physics also mean the great opportunities in 21st century. We expect some great discoveries in near future.

Major frontiers of the particle physics in the 21st century include: the accelerator based experiments and the non-accelerator based experiments. Most of the experiments are based on the big facilities. The accelerator based experiments have two frontiers: 1) the high energy frontier: LHC and ILC. 2) the high precision frontier: the factories, such as BEPCII, KEKB, PEPII and Daϕne. The non-accelerator based experiments include the particle astrophysics experiments (cosmic ray measurements, space-based observations) and the Neutrino physics experiments.

7. Chinese Particle Physics in 21st Century

As the Chinese economy grows quickly and steadily, the Chinese government increases the supports to sciences and technology significantly and constantly. The Chinese government realizes that only sciences and technology could meet the great challenges, such as energy, resources and environments, which China is facing. The supports to the particle physics and the large science facilities are increasing quickly. With construction of BEPCII/BESIII, the Shanghai light source and CSNS, the new generation of Chinese accelerator and detector teams are shaping. They are young and growing fast. They could catch the future opportunity in the particle physics. The outline of the medium plan of the Chinese particle physics includes:

- Charm physics @ BEPCII
- International collaborations: LHC experiments, ILC R&D...
- The particle astrophysics experiments at Space
  - (a) The hard X-ray modulation telescope satellite
  - (b) Polar @ Chinese Spacelab: to measure the polarization of the γ burst
- Cosmic ray measurements:
  - (a) Yangbajing Cosmic ray Observatory
  - (b) Cosmic ray neutrino telescope
- Neutrino experiments:
  - (a) Daya Bay Reactor neutrino to measure $sin^2 2\theta_{13}$
  - (b) Very LBL oscillation: J-Prac $\rightarrow$ Beijing (under discussion)
- National underground Lab. (under discussion)

There are strong demands on the large scientific facilities based on accelerators, as well as the application of accelerator and detector technology.

- The high power proton accelerator:
  (a) Chinese spallation neutron source
  (b) Accelerator driven subcritical system for the transmute of the nuclear waste and possible clean nuclear energy.
- The hard X-ray FEL

One expected that BEPC will be converted into a dedicated synchrotron radiation source after BEPCII finished physics running. Actually, IHEP is extending its research fields into the protein structure, the nano-science, the material science ... etc., and will become a multiple discipline research center for the long term.

Statistical Physics, Condensed Matter and Biophysics

NEARSIGHTEDNESS OF ELECTRONIC MATTER

WALTER KOHN

University of California, Santa Barbara
E-mail: kohn@physics.ucsb.edu

We use the term "electronic matter" for electrons (charged or, hypothetically, uncharged) in their ground-state under the action of a fixed external potential (Born-Oppenheimer approximation). For a given chemical potential mu, the properties (e.g. the density)of the electrons anear a given point, say $r = 0$, depend primarily on the positions and nuclear charges of the nearby nuclei. This so-called "nearsightednes", its validity and limitations, will be analyzed and exemplified.

COMPLEX COOPERATIVE BEHAVIOUR IN RANGE-FREE FRUSTRATED MANY-BODY SYSTEMS

DAVID SHERRINGTON

Rudolf Peierls Centre for Theoretical Physics,
University of Oxford, 1 Keble Road, Oxford OX1 3NP, UK

A brief introduction and overview is given of the complexity that is possible and the challenges its study poses in many-body systems in which spatial dimension is irrelevant and naively one might have expected trivial behaviour.

1. Introduction

This paper is concerned with many body systems and their cooperative behaviour; in particular when that behaviour is complex and hard to anticipate from the microscopics, even qualitatively and even when the systems are made up of simple individual units with simple inter-unit interactions.

'Range-free' (or 'infinite-ranged') refers to situations where the interactions are not dependent on the physical separations of individual units, and hence neither on the dimensionality nor on the structure of the embedding space. Such systems are also often referred to as 'mean-field', since one can often show (and usually believes) that their behaviour in the thermodynamic limit ($N \to \infty$ units) is identical to that of an appropriate mean-field approximation to a short-range system.

'Frustration' refers to incompatability between different microscopic ordering tendencies.

Self-consistent mean-field theories do have the ability to describe spontaneous symmetry breaking and phase transitions and they have played an important role in statistical physics. However as pure systems, without quenched Hamiltonian disorder or out-of-equilibrium self-induced disorder, they do not exhibit the interesting non-simple dimension-dependent but details-independent (universal) critical behaviour whose study drove much of the interest of statistical mechanics in the seventies and eighties [1]. For this reason 'mean-field' used to be interpreted as fairly trivial.

On the other hand, with quenched disorder and frustration in their interactions range-free many-body systems can, and regularly do, exhibit behaviour that is complex and rich. This paper represents a brief introduction to and partial overview of such systems.

2. General Structure and Features

The general class of systems we consider can be summarized as characterised by schematic 'control functions' of the form $H(\{J_{ij...k}\}, \{S_i\}, X)$ where

(i) in thermodynamics (statics) the $\{S\}$ are the variables and the $\{J\}$ are quenched (frozen) parameters, or vice-versa,
(ii) in dynamics the $\{S\}$ are the 'fast' variables and the $\{J\}$ are 'slow' variables, or vice-versa, where 'fast' and 'slow' refer to the characteristic microscopic time-scales,
(iii) in both cases, the X are intensive control parameters, influencing the system deterministically, quenched-randomly or stochastically,
and
(iv) we shall be particularly interested in typical behaviour in situations in which any quenched disorder is drawn independently from identical intensive distributions, enabling (at least in principle) useful thermodynamic-limit measures of the macroscopic behaviour.

The interest arises when the effects of different interactions are 'frustrated', in competition with one another. In such cases with detailed balance, at low enough noise the macrostate structure/space is typically fractured (or clustered), in a manner often envisaged in terms of a 'rugged landscape' paradigm in which the dynamics is imagined as motion in a very high dimensional landscape of exponentially many hills and valleys, often hierarchically structured, with concomitant confinements, slow dynamics and history dependence. In dynamical systems without detailed balance, strictly there is no such simple Lyapunov 'landscape' but the 'motion' is analogously complexly hindered, with many effective macroscopic time-scales.

First studied (in physics) in the context of magnetic alloys, such systems are now recognised in many different contexts; in inanimate physical systems, computer science, and information science; in animate biology, economics and social science. In these different systems 'controllers' of the 'control functions' vary; including the laws of physics, devisors of computer algorithms, human behaviour, governmentally-devised laws etc.

3. The Sherrington-Kirkpatrick Model

A simply-formulated but richly-behaved canonical model is that of Sherrington and Kirkpatrick (SK) [2], originally introduced as a potentially soluble model corresponding to a novel mean-field theory introduced by Edwards and Anderson (EA) [3] to capture the essential physics of some unusual magnetic alloys, known as spin glasses [4, 5].

The SK model is characterized by a Hamiltonian

$$H = -\sum_{(ij)} J_{ij}\sigma_i\sigma_j ; \sigma = \pm 1 \tag{1}$$

where the i, j label spins σ, taken for simplicity as Ising, and the interactions $\{J_{ij}\}$ are chosen randomly and independently from a distribution $P_{exch}(J_{ij})$. Dynamically the system can be considered to follow any standard single-spin-flip dynamics corresponding to a temperature T. Were normal equilibration to occur it would be characterized by Boltzmann-Gibbs statistics, $p(\sigma) \sim \exp(-H\{\sigma\}/T)$.

However, if the distribution $P_{exch}(J)$ has sufficient variance compared with its mean and the temperature is sufficiently low, normal equilibration does not occur and complex macro-behaviour results beneath a transition temperature. The interesting regime, known as the 'spin glass phase', occurs at intensive T if the variance of $P_{exch}(J)$ scales with N as J^2/N, the mean as J_0/N. As Parisi showed, in a series of papers (*e.g.* [6–8]) which involved amazing insight and highly original conceptualization and methodology, this glassy state is characterized by a hierarchy of 'metastable' macrostates, differences between restricted and Gibbsian thermodynamic averages, as well as non-self-averaging; see also *e.g.* [9].

These features can be characterized by the macrostate overlap distribution functions $P(q) = \sum_{S,S'} \delta(q- \mid N^{-1}\sum_i \langle\sigma_i\rangle_S \langle\sigma_i\rangle_{S'} \mid)$ where $\langle O\rangle_S$ denotes a thermodynamic average of O over the macrostate S. For a conventional system, with a single macrostate, $P(q)$ has a single delta function, while for a system with entropically extensively many macrostates $P(q)$ has more structure. When the state structure is continuously hierarchical, as it is for SK, there is a continuum of weight in the disorder-averaged overlap distribution function $\bar{P}(q) = \int DJ[\prod_{(ij)} P_{exch}(J_{ij})] P_{\{J_{ij}\}}(q)$. Non-self-averaging arises in different $P_{\{J_{ij}\}}(q)$ for different realizations of quenched disorder even when that disorder is chosen *i.i.d.* from an intensive $P_{exch}(J)$. Ultrametricity [8] is a feature of the hierarchical order.

Later studies [10] have further exposed the existence of slow dynamics and aging and a remarkable non-trivial quantitative relationship to the thermodynamics [11, 12]. One feature of this is a modification of the normal fluctuation-dissipation relationship to $-dR/dC = \beta X(C)$ where R is a response function and C a related correlation function, parenthetically connected through the measurement time t in the limits of large initial waiting time/field application time t_w and t itself, and $X(Q)$ is related to the average overlap distribution $\bar{P}(q)$ by $X(Q) = \int_0^Q \bar{P}(q)dq$. In the normal fluctuation-dissipation theorem $X(C)$ is replaced by unity.

The original exposure of the subtleties of the SK model utilised unusual and non-rigorous mathematics and ansätze, together with unconventional physical conceptualization, going far beyond the conventional realms of rigorous mathematical physics and probability theory. The predictions have however long been shown to be in accord with computer simulations of the model (*e.g.* [13]) and consequently were believed by physicists. Their rigorous demonstration, though, has been a non-trivial challenge which has required deep analysis and led to new rigorous mathematical methodologies in recent years [14–17], finally completely vindicating Parisi's theory [18]. Thus there is no doubt that there is deep complexity in the SK model.

On the other hand, while it is generally believed that real spin glass transitions do occur also in the short-ranged EA model for dimensions three or more and also in the experimental magnetic alloys that first stimulated its study, it remains controversial as to whether or to what extent all the subtle predictions of the SK model apply to these systems (see *e.g.* [11] [19]).

Hence it becomes appropriate to ask whether range-free frustrated and disordered systems have a wider relevance beyond as over-idealized models of real magnetic alloys and as challenges for mathematicians. Possibly remarkably, it turns out that the answer is a resounding "Yes"; they turn out to be rather ubiquitous in many areas of science.

4. Beyond Magnetic Alloys

Range-free frustrated and disordered many-body problems occur in many scenarios outside of physics, for example in many of the hard optimization problems studied by computer scientists, and in situations in which correlation between individuals occurs through the transfer of information available to all, irrespective of physical separation, as epitomised by modern interaction through the internet and telephones, or commonly available through the world-wide web, newspapers, radio and television. These systems are usually different in detail from the SK model but share some of the same conceptual and technical challenges, as well as providing further challenges of their own.

4.1. *The SK model as an optimization exercise*

A simple illustration of the the possibilities of extension comes from viewing the SK model as an optimization problem which is describable in everyday terms as "The Dean's Problem" [20]. One imagines a University Dean who has to place N students in two dormitories, but with the challenge that every pair of students (i, j) either likes or dislikes one another to an extent J_{ij}[a]. His problem is to choose to which dorm to allocate each student so as to ensure the greatest satisfaction overall. Labelling the dorm choices by $\sigma_i = \pm 1$ the SK Hamiltonian becomes the 'cost function' that the Dean should minimise, with the $\{\sigma_i\}$ in its ground state the optimal choice[b]. Allowing the Dean a degree of uncertainty in his decision-making provides an analogue of 'temperature'.

4.2. *Simulated annealling*

There are many other combinatorial optimization problems that can be viewed as finding the minimum of a cost function of the form $H(\{J_{ij...k}\}, \{S_i\}, X)$.

In a 'simulated annealing' [21] 'spin-flip' computer algorithm to minimise the cost function, an annealing 'temperature' T_A is introduced artificially via a stochastic probability measure determined by $\exp(-\delta H/T_A)$, where δH is the change in the cost function engendered by the flip, so that equilibration at T_A would yield the corresponding Boltzmann distribution, and T_A is gradually reduced to the value of interest (zero for a minimum of H, or T if that is the stochastic noise of actual uncertainty). Correspondingly, T_A can be introduced into an effective equilibrium statistical mechanics, with Boltzmann weighting $\exp(-H\{S\}/T_A)$, examined analytically and the minimum found from $H_{min} = \lim_{T_A \to 0}\{-T_A \ln \sum_{\{S\}} \exp(-H\{S\}/T_A)\}$.

4.3. *p-spin spin glass, satisfiability and error-correction*

In the p-spin glass model one replaces the binary interaction of SK by one involving p spins;

$$H = -\sum_{(i_1 i_2 .. i_p)} J_{i_1 i_2 .. i_p} \sigma_{i_1} \sigma_{i_2} ... \sigma_{i_p}, \tag{2}$$

[a] $J > 0$ corresponding to 'like' and $J < 0$ to dislike.

[b] If the Dean also has to put an equal number of students in each dorm there is an extra constraint $\sum \sigma_i = 1$.

with the J again drawn randomly and independently, all from the same distribution, and then quenched[c]. This apparently innocuous extension of the SK model yields new behaviour in several ways. Firstly, instead of a continuous onset of a hierarchy of different levels of metastability, with a growing continuous range of state overlaps, there is a discontinuous onset of many orthogonal but otherwise equivalent metastates of finite overlap order parameter[d] [25, 26]. Secondly, the dynamical transition is no longer at the same temperature as the thermodynamic transition but is higher [30]. Thirdly, there is another lower temperature (continuous) thermodynamic transition to a state with a continuous range of overlap distributions [31].

This type of behaviour, of a dynamical transition pre-empting a thermodynamic one, both with discontinuous onset of overlap, turns out to be rather common in frustrated many-body systems[e]. So too is the lower temperature thermodynamic transition to a continuous range of overlap distributions [f]. Consequently the determination of the minimum achievable cost function often has the difficulties associated with the full hierarchical character of the SK model at $T = 0$.

In fact, many of the problems of interest in computer science are effectively range-free but on random graphs of finite connectivity, in contrast to the full connectivity of the SK- and p- spin models of Eqns. (1) and (2) above. The conceptual ideas extend, albeit made more complicated (and currently incompletely solved) by the need for higher order overlap functions within the full (replica) theory of [2, 3, 6] [g]. Dilution, however, has also brought to the fore an alternative and highly successful computational methodology in the form of 'survey propagation' [24].

[c]In this case variance scaling as $N^{-(p-1)}$ yields an intensive transition temperaure.

[d]These differences can be seen in the character of the onset of structure in $P(q)$ at $q > 0$ in addition to the main delta-function peak at $q = 0$; note that this is in contrast to a conventional ferromagnet (or antiferromagnet) for which the whole delta-function peak, which in the paramagnetic phase is at $q = 0$, would move to a finite value of q. For a continuous transition $P(q)$ develops extra finite weight growing continuously from $q = 0$, whereas in a discontinuous transition $P(q)$ develops by acquiring weight directly at finite $q = q_1 \sim \mathcal{O}(1)$ but with its weight growing continuously. In the case of $p = 2$ (SK) the (continuous) onset also has finite weight for a continuous range of q within $0 \leq q \leq q_1$. For SK q_1 grows continuously with $(T - T_c)$ where T_c is the transition temperature. For $p \geq 3$ the initial onset is discontinuous at a single finite q_1.

[e]In general, the symmetry of definiteness of SK and EA seems to be more the exception than the norm.

[f]Exceptions to this lower transition occur for so-called spherical spins (individual S_i unbounded but with $\sum_{i=1,..N}(S_i)^2 = N$) or for $p = \infty$.

[g]They also no longer require inverse N-scaling to achieve finite transition temperatures; the relevant criterion is scaling as z^{-1} where z is the graph coordination number.

One example of the conceptual transfer of these ideas is to random satisfiability problems in computer science, both in explaining the existence of satisfiable-unsatisfiable (SAT-UNSAT) phase transitions [32] and in leading to the recognition that for random $K(>2)$-SAT, where K refers to the length of the individual clauses to be satisfied simultaneously, there should be a region of 'HARD-SAT' separating practically satisfiable SAT problems from UNSAT as the constraint density, the ratio of the number of constrained clauses to the number of variables, is increased [28]; this ratio can be considered as playing a role reminiscent of that of the inverse of temperature in the p-spin model with the transitions analogues of the p-spin dynamical and thermodynamic transitions[h]. In fact, on closer examination, random K-SAT exhibits an even richer sequence of phase transions; see *e.g.* [29].

It is possible to interpolate between the type of behaviour of the $p \geq 3$ model and that of the $p = 2$ SK model. One way is to add a magnetic field h to the p-spin glass. This leads to a sequence of behaviours as h is increased; for small h it is qualitatively as described above for the zero-field p-spin model, followed at a first critical field by the coming together of the dynamical and thermodynamic spin glass transitions and replacement of the discontinuous onset of non-trivial $P(q)$ by a continuous one [30], but still with a single delta function onset at non-zero q, in addition to that at $q = 0$, and then at a higher critical field by a transition to a continuously distributed hierarchy of metastable states and overlaps[i]. This suggests a possible utility in adding an extra 'effective field' in the computer algorithmic optimization, to avoid the dynamical pre-emption of a thermodynamic transition.

4.4. *Interacting agents*

Another interesting class of range-free problems is of systems where many 'agents', each with individual characteristics but with no direct interactions between them, behave in a cooperatively complex fashion by all reacting to common 'information'. This common information acts as an effectuator for correlation between the agents. Frustration and complexity arise when the goals are such that not all can 'win'.

[h] Random K-SAT maps to an Ising spin glass model with terms of several $p \leq K$.
[i] In fact, this sequence of events was first recognised in a Potts spin glass [33, 34], still with two-body interactions but with the (symmetric) Ising interaction $\sigma_i \sigma_j$ replaced by a non-symmetric Potts interaction $\delta_{s_i,s_j}; s_i = 1, 2, ...p$ where p is the Potts dimensionality. In this case the sequence is from SK at $p = 2$ to pure p-spin-glass-like at $p = 4$.

4.4.1. *The Minority Game*

A minimalist model that illustrates this class is the so-called 'Minority Game'(MG) [35, 36], introduced to emulate some features of a stockmarket in which players make profits by buying when the price is low and selling when the price is high. In a simple version of this model N agents at each time-step t simultaneously make one of two choices, which we shall denote ± 1. Their 'objectives' are to make the minority choice. They make their choices on the basis of (i) some 'information' $\mathbf{I}(\mathbf{t})$ commonly available to all, (ii) the operation on that information by each agent i of one of a pair of individual strategy operators $\hat{S}_i^\alpha; \alpha = +, -$, with the output determining the 'choice' made, (iii) individual 'point-scores' $p_i(t)$ that enable the agents to 'decide' which of their two strategies to employ at each step. The strategy pairs are chosen randomly and independently at the outset and thereafter fixed. The information $\mathbf{I}(\mathbf{t})$ varies at each time-step and hence so does the outcome of the strategies acting upon it . The space of the strategies spans the two possible outputs equally. In the simplest deterministic version of the game the strategy $\hat{S}_i^\alpha$ employed by agent i at time t is that labelled by the same sign as $p_i(t)$. The points are updated according to

$$p_i(t+1) = p_i(t) - [\hat{S}_i^{\mathrm{sign}(p_i(t))}(\mathbf{I}(\mathbf{t}))]A(t) \tag{3}$$

where $[\hat{S}_i^\alpha(\mathbf{I})] = \pm 1$ is the action choice of the strategy $\hat{S}_i^\alpha$ acting on the information $\mathbf{I}$ and and $A(t)$ is the average 'choice' over the stategies actually employed,

$$A(t) = N^{-1} \sum_j [\hat{S}_j^{\mathrm{sign}(p_j(t))}(\mathbf{I}(\mathbf{t}))]; \tag{4}$$

i.e. by increasing the point-score bias for strategies leading to minority behaviour. In the original formulation [37] the information used was the Booolean string indicating the minority choice in the previous m time-steps of play and the $\hat{S}$ were Boolean operators. However, essentially similar behaviour is obtained for a system in which $\mathbf{I}(t)$ is randomly generated at each time t, equally probably from the whole space of m binaries [38].

The most obviously relevant macroscopic measure in the MG is the volatility, the variance of the choices. Computer simulations demonstrated that it has scaling behaviour, the volatility per agent versus the the information dimension per agent $d = D/N = 2^m/N$ approaching independence of N as the latter is increased, and also has a cusp-like minimum at a critical d_c with behaviour ergodic for $d > d_c$ but non-ergodic for

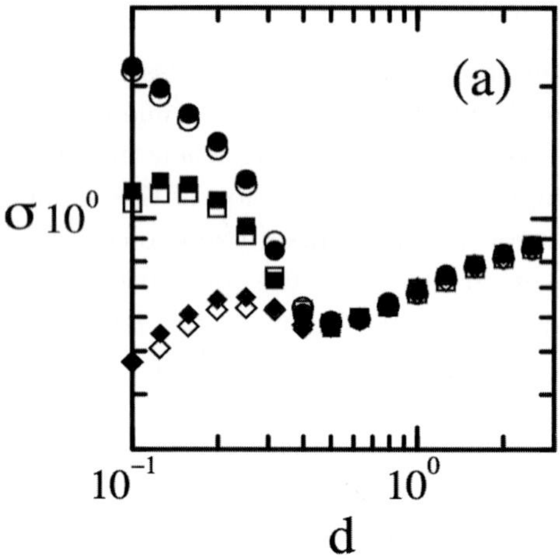

Fig. 1. Volatilities in Minority Games with 2 strategies per agent; Shown are (i) different biases of initial point asymmetries between each agent's 2 strategies: $p_i(0) = 0.0$ (circles), 0.5 (squares) and 1.0 (diamonds), (ii) a comparison betweeen the results of simulation of the deterministic many-agent dynamics (open symbols) and the numerical evaluation of the analytically-derived stochastic single-agent ensemble dynamics. From.[39]

$d < d_c$ [j]. Fig 1 shows this behaviour for a slightly different variant of the model in which the strategies are taken as $D = dN$-dimensional binary strings $\mathbf{S}_i^\alpha = \{S_i^{\alpha,1}, S_i^{\alpha,2}, ...S_i^{\alpha,D}\}; i = 1, ..N, \alpha = \pm$, with each component $S_i^{\alpha,\mu}; \mu = 1, ..D$ chosen randomly and independently at the outset and thereafter fixed (quenched), and the stochastic 'information' consists in randomly choosing $\mu(t)$ at each time-step and then using the corresponding strategy elements. This is reminiscent of the behaviour of the susceptibility of the SK spin glass, shown in Fig 2, if one compares the volatility with the inverse susceptibility and the information dimension with the temperature. Hence one is tempted to analyze the MG using methodolgy developed for spin glasses.

Updating the point-score only after M steps where $M \geq \mathcal{O}(N)$ leads to an averaging over the random information to produce an effective interaction between the agents and yield the so-called 'batch' game (with

[j]In the case shown in Fig. 1, of uncorrelated strategies, the cusp-like behaviour is most pronounced for a *tabula rasa* start. For anti-correlated strategies [39] there is no cusp for *tabua rasa* start but the non-ergodicity-onset is clear.

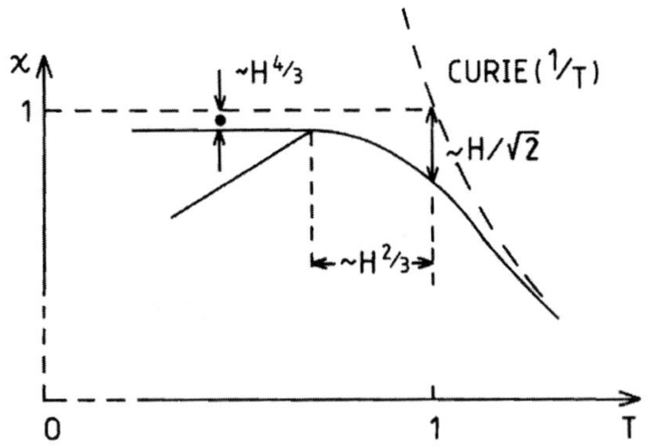

Fig. 2. Schematic susceptibility of the SK spin glass in an applied field H, as predicted by Parisi theory. The upper curve shows the full Gibbs average, obtained from the full $q(x)$ and interpreted as the field-cooled (FC) susceptibility. The lower curve shows the result of restricting to one thermodynamic state, as obtained from $q(1)$ and interpreted as the zero-field-cooled susceptibility. From [40].

temporally-rescaled update dynamics)

$$p_i(t+1) = p_i(t) - \sum_j J_{ij}\mathrm{sgn}(p_j(t)) - h_i \equiv p_i(t) - \partial H/\partial s_i \,|_{s_i=\mathrm{sgn}(p_i(t))}, \quad (5)$$

where H is an effective 'Hamiltonian'

$$H = \sum_{(ij)} J_{ij}s_i s_j + h_i s_i \quad (6)$$

and J_{ij} and h_i are effective 'exchange' and 'field' terms given by

$$J_{ij} = N^{-1}\sum_{\mu=1}^{D} \xi_i^\mu \xi_j^\mu, \ h_i = N^{-1/2}\sum_{\mu=1}^{D} \omega_i^\mu \xi_i^\mu, \quad (7)$$

where $\omega_i = (\mathbf{S}_i^1 + \mathbf{S}_i^2)/2$, $\xi_i = (\mathbf{S}_i^1 - \mathbf{S}_i^2)/2$. Since the $\mathbf{S}$ are random so are the exchange and field terms. Hence H is a disordered and frustrated control function. The expression for the $\{J_{ij}\}$ is very reminiscent of the Hebbian-inspired synapses of the Hopfield neural network model [41], where the $\{\xi_i^\mu\}$ are the stored memories, but crucially with the opposite sign ensuring that here the $\{\xi_i^\mu\}$ are now repellors rather attractors[k].

[k]Note also that the usual spin glass or neural network dynamics is different in detail from that of the minority game, *e.g.* random sequential rather than parallel.

Methodologies

There are two main methodologies employed to study statics, the replica procedure and the cavity method (see *e.g.* [9]). The most common method for the cooperative dynamics is the generating functional method [42, 43].

In the replica method one studies the disorder-averaged free energy

$$\int D\{J\}P_{exch}(\{J\})(-T\ln \mathrm{Tr}_{\{\sigma\}}\exp(-H_{\{J\}}(\{\sigma\})/T)), \tag{8}$$

using the identity $\ln Z = \mathrm{Lim}_{n\to 0}\{Z^n - 1\}/n$, identifying the power n as describing n replicas, $\alpha = 1,..n$; with n eventually taken to 0. Macroscopic order parameters are introduced through multiplication by unity of the form

$$1 = \int \prod_{(\alpha\beta)} Dq^{\alpha\beta}\delta(q^{\alpha\beta} - N^{-1}\sum \langle \sigma_i^\alpha \sigma_i^\beta \rangle_{H_{eff}}), \tag{9}$$

where the α, β label replicas and H_{eff} is the effective Hamiltonian after disorder averaging. The microscopic variables $\{\sigma_i^\alpha\}$ are integrated out and the dominant extremum[1] with respect to the $q^{\alpha\beta}$ is taken in the limit $N \to \infty$. In the most natural asantz, replica symmetry among $q^{\alpha\beta}; \alpha \neq \beta$ was assumed [2, 3], but this proved to be too naive. The correct solution for the SK model requires Parisi's much more subtle ansatz of replica symmetry breaking [6]. This ansatz introduces a hierarchy of spontaneous replica symmetry breaking (RSB) with a sequence of $q_i, x_i; i = 1,..K$ that in the limit of $K \to \infty$ yields a continuous order function $q(x) : 0 \leq x \leq 1$, later shown [7] to be related to the average overlap distribution through $\bar{P}(q) = \int dx\delta(q - q(x))$.

The dynamical functional method for the SK model is discussed in [10, 19]. Here we describe instead its use for the Minority Game [36]. A generating functional can be defined by

$$Z = \int \prod_t d\mathbf{p}(t)W(\mathbf{p}(t+1) \mid \mathbf{p}(t))P_0(\mathbf{p}(0)), \tag{10}$$

where $\mathbf{p}(t) = (p_1(t), \ldots, p_N(t))$, $W(\mathbf{p}(t+1) \mid \mathbf{p}(\mathbf{t}))$ denotes the transformation operatation of eqn. (5) and $P_0(\mathbf{p}(0))$ denotes the probability distribution of the initial score differences.

---

[1]The correct extremum is actually the maximum [9, 15, 16].

Averaging over the specific choices of quenched strategies, introducing macroscopic two-time correlation order functions via

$$1 = \int \prod_{t,t'} DC(t,t')\delta(C(t,t') - N^{-1} \sum_i \mathrm{sign} p_i(t) \mathrm{sign} p_i(t')) \qquad (11)$$

and similar expressions for response functions $G(t,t')$ and derivative-variable correlators $K(t,t')$, and integrating out the microscopic variables, the averaged generating functional may then be transformed exactly into a form

$$Z = \int D\mathbf{C}D\tilde{\mathbf{C}}D\mathbf{G}D\tilde{\mathbf{G}}D\mathbf{K}D\tilde{\mathbf{K}} \exp\left(N\Phi(\mathbf{C},\tilde{\mathbf{C}},\mathbf{G},\tilde{\mathbf{G}},\mathbf{K},\tilde{\mathbf{K}}\right), \qquad (12)$$

where Φ is N-independent, the bold-face notation denotes matrices in time and the tilded variables are complementary ones introduced to exponentiate the delta functions in eqn. (11) and its partners. Being extremally dominated, in the large-N limit this yields the effective single agent stochastic dynamics

$$p(t+1) = p(t) - \alpha \sum_{t' \leq t} (\mathbf{1} + \mathbf{G})^{-1}_{tt'} \mathrm{sgn} p(t') + \sqrt{\alpha}\eta(t), \qquad (13)$$

where $\eta(t)$ is coloured noise determined self-consistently over the corresponding ensemble by

$$\langle \eta(t)\eta(t') \rangle = [(\mathbf{1} + \mathbf{G})^{-1}(\mathbf{1} + \mathbf{C})(\mathbf{1} + \mathbf{G}^{\mathbf{T}})^{-1}]_{tt'}. \qquad (14)$$

Fig. 1 demonstrates the veracity of this result in a comparison of the results of computer simulation of the original deterministic many-body problem eqn. (5) and the numerical evaluation of the self-consistently noisy single-agent ensemble of eqn. (13).

The analogous equations for the p-spin spherical spin glass formed the basis for recognition of the dynamical transitions mentioned earlier and the existence of aging solutions and modifications to conventional fluctuation-dissipation relations.

5. Critical Behaviour and Correlation Length

Having commented earlier that standard non-frustrated non-disordered infinite-ranged systems do not have interesting critical behaviour, it is relevant to note that again frustrated disordered systems are different [46, 47, 49, 50], having interesting critical behaviour at low temperature and applied magnetic field, even though mean-field.

Parisi replica symmetry breaking involves an infinite sequence of hierarchies. K-RSB has K step-breaks in the order function $q(x) : 0 \leq x \leq 1$ [m]. The exact free energy is formally obtained by finding the supremum with respect to the break and plateau values q_i, x_i and taking $K \to \infty$ [6, 18]. The continuum limit was given as a set of implicit equations already in Parisi's early work[n]. Most (but not all) of the subsequent analysis has been perturbative near to the transition temperature for spin glass onset[o]. Numerical evaluations have until a couple of years ago been restricted to just the first few steps of RSB, but very recently very high accuracy numerical extremizations for high orders of RSB have been performed at zero and low temperatures and have shown interesting features [46–49].

At low temperatures the steps x_i scale as $x_i \sim a_i T$ with the a_i having non-zero limits as $T \to 0$ and exposing critical points at both $a = 0$ and $a = \infty$. As $T \to \infty$ the K-step approximation of q_i against a_i approaches a fixed-point function $q^*(a)$ of form close to $q^*(a) = (\sqrt{\pi}/2a)\mathrm{erf}(\xi/a)$ with ξ a 'correlation function' in a-space given by $\xi \approx 2/\sqrt{\pi}$ [p]. The degree of RSB can be viewed as an effective one-dimensional lattice of size K, with $K \to \infty$ the analogue of the infinite-length lattice (or thermodynamic) limit. Similarly, finite-K approximation yields an analogue of finite-size effects, including finite-size scaling. Note however that this new type of finite-size scaling is for a mean-field problem in the thermodynamic limit and is in a space of degree of approximation. There are also finite K-size scalings when the system is perturbed away from the $T = 0$ critical point (at $a = 0$) and for finite applied field h near $a = \infty$ [49]. Correspondingly there are further 'correlation lengths' in temperature-deviation and in field-deviation, which of course also determine the extent of RSB needed to get a good approximation as temperature or field become non-zero.

Conclusions

In this short paper it has only been possible to present a brief and non-detailed vignette of the complexity that can and does exist in disordered and frustrated many-body systems, even within a dimension-free mean-field situation. The puzzles, intrigues and challenges have developed and been a source of intense study for over 30 years. Finite-range systems have

[m] $q(x)$ is related to the averaged overlap distribution $\bar{P}(q)$ by $\bar{P}(q) = \int dx \delta(q - q(x))$.
[n] See also [51].
[o] For the most complete perturbative study the reader is referred to [52].
[p] Strictly the behaviour is found to deviate slightly but subtly - see Refs. [47, 49].

also been a great source of interest, again with significant progress but still subject to some controversy [44, 45].

The case of systems with variables having different fundamental timescales, such as fast neurons and slow synapses or evolutionary models with different timescales for phenotypes and genotypes, have not been discussed. Nor has the problem of dynamical sticking in effectively self-determined disordered states of some systems without quenched disorder in their control functions but started far from equilibrium.

Also, in this brief review, only some of the simplest models have been described. It is however clear that many extensions and more realistic/complete scenarios exist that are still effectively range-free, yet complex, interesting and challenging.

Acknowledgements

The author would like to thank his numerous collaborators, students, colleagues and friends, too many to name all individually, for their parts in helping his understanding and appreciation of the subject of this paper. He also acknowledges, with gratitude, the financial support of the EPSRC (and its predecessors), the EC and the ESF.

References

1. J Cardy, *Scaling and Renormalization in Statistical Physics* (Cambridge University Press, Cambridge, 1996)
2. D. Sherrington and S. Kirkpatrick, *Phys. Rev. Lett.***35** 1972 (1976)
3. S.F. Edwards and P. W. Anderson, *J. Phys. F* **5**, 965 (1975)
4. J. A. Mydosh, *Spin glasses; an experimental introduction*, Taylor and Francis, London (1993)
5. D. Sherrington, in *Spin Glasses*, eds. E. Bolthausen and A. Bovier, (Springer, Berlin, 2007)
6. G. Parisi, *J. Phys. A* **13**, 1101 (1980)
7. G. Parisi, *Phys. Rev. Lett.* **50**, 1946 (1983)
8. M. Mézard, G. Parisi, N. Sourlas, G. Toulouse and M.A. Virasoro, *J.Physique* **45**, 843 (1984)
9. M. Mézard, G. Parisi and M.A. Virasoro, *Spin GlassTheory and Beyond* (World-Scientific, Singapore, 1987)
10. L. F. Cugliandolo and J. Kurchan, *J.Phys.A* **27**, 5749 (1993)
11. A.P. Young A P (ed.) *Spin Glasses and Random Fields* (World Scientific, Singapore, 1997)
12. G. Parisi, in *Stealing the Gold: a Celebration of the Pioneering Physics of Sam Edwards*, eds. P. M. Goldbart, N. Goldenfeld and D. Sherrington (Oxford University Press, Oxford, 2004)

13. A. P. Young, *Phys. Rev. Lett.* **51**, 1206 (1983)
14. M.Talagrand, *The Sherrington-Kirkpatrick model: a challenge for mathematicians, Probab. Theor. Rel.* **110**, 109 (1998)
15. M. Talagrand *Spin Glasses: a Challenge for Mathematicians* (Springer, Berlin, 2003)
16. F. Guerra, *cond-mat/*057581 (2005)
17. E. Bolthausen and A. Bovier eds., *Spin Glasses* (Springer, Berlin, 2007)
18. M. Talagrand, *Ann. Math.***163**, 221 (2006)
19. L. F. Cugliandolo and J. Kurchan, *J. Phys. A***41**, 324018 (2008)
20. A variant was posed to describe one of the Clay Millenium Prize Problems; see http://www.claymath.org/millennium/P_vs_NP. That the problem of finding the ground state of a spin glass in three and more dimensions is NP-complete has been known since at least the early 1980s.
21. S. Kirkpatrick, C. D. Gelatt and M. P. Vecchi, *Science* **220**, 672 (1983)
22. D. J. Gross, I. Kanter and H. Sompolinsky, *Phys.Rev.Lett.***55**, 304 (185)
23. T. Kirkpatrick and P. G. Wolynes, *Phys.Rev.***B** 36, 8552 (1987)
24. M. Mézard and R. Zecchina, *Phys.Rev. E***66**, 056126 (2002)
25. D. J. Gross and M. Mézard, *Nuc. Phys. B***240**, 431 (1984)
26. A. Crisanti and H-J Sommers, *Z. Phys. B***87**, 341 (1992)
27. P. Gillin, H. Nishimori and D. Sherrington, *J. Phys. A***34**, 2949 (2001)
28. M. Mézard, G. Parisi and R. Zecchina, *Science* **297**, 812 (2002)
29. F. Kzakala, A. Montanari. F. Ricci-Tersenghi, G. Semerjian and L. Zdeborova, *Proc. Nat. Acad. Sci.* **104**, 10318 (2007)
30. A. Crisanti, H. Horner and H-J. Sommers, *Z. Phys. B*,**92**, 257 (1993)
31. E. Gardner, *Nuc. Phys. B* **257**, 747 (1985)
32. S. Kirkpatrick amd B. Selman, *Science* **264**, 1297 (1994)
33. D. Elderfield and D. Sherrington, *J. Phys. C***16**, L497 (1983)
34. D. J. Gross, I Kanter and H. Sompolinsky, *Phys. Rev. Lett.* **55**, 304 (1985)
35. D. Challet, M. Marsili and Y-C Zhang, *Minority Games* (Oxford University Press, Oxford 2005)
36. A.C.C. Coolen, *The Mathematical Theory of Minority Games* (Oxford University Press, Oxford 2005)
37. D. Challet and Y-C Zhang, *Physica A***246**, 407 (1997)
38. A. Cavagna, *Phys. Rev E***59**, R3783 (1998)
39. T.Galla and D. Sherrington *Physica A* 324, 25 (2003)
40. D. Sherrington, in *Heidelberg Symposium on Glassy Dynamics*, 2, (Springer-Verlag, Berlin 1987)
41. J. J. Hopfield, *Proc.Nat.Acad.USA***79**, 2554 (1982)
42. C. de Dominicis, . *J. Physique C***1**, 247 (1976)
43. H.Janssen,*Z. Phys. B***23**, 377 (1976)
44. A. P. Young, *J.Phys.A*41,324016 (2008)
45. G. Parisi, *J. Phys. A*.**41**, 324002 (2008)
46. R. Oppermann and D. Sherrington, *Phys. Rev. Lett.* **95**, 197203 (2005)
47. R. Oppermann, M. J. Schmidt and D. Sherrington, *Phys. Rev. Lett.***98**,127201 (2007)
48. R.Oppermann and M. J. Schmidt, *arXiv:* o801.1756 (2008)

49. R. Oppermann and M. J. Schmidt, *arXiv:*0803.3918 (2008)
50. S. Pankov, *Phys. Rev. Lett.* **96**, 197204 (2006)
51. H-J. Sommers and W. Dupont, *J. Phys. F* **17**, 5785 (1984)
52. A. Crisanti and T. Rizzo, *Phys. Rev. E* **65**, 046137 (2002)

ASYMMETRIC HEAT CONDUCTION IN NONLINEAR SYSTEMS

BAMBI HU

Department of Physics,
Hong Kong Baptist University
E-mail: bhu@hkbu.edu.hk

Heat conduction is an old yet important problem. Since Fourier introduced the law bearing his name two hundred years ago, a first-principle derivation of this law from statistical mechanics is still lacking. Worse still, the validity of this law in low dimensions, and the necessary and sufficient conditions for its validity are still far from clear. In this talk I'll give a review of recent works done on this subject. I'll also report our latest work on asymmetric heat conduction in nonlinear systems. The study of heat condution is not only of theoretical interest but also of practical interest. The study of electric conduction has led to the invention of such important electric devices such as electric diodes and transistors. The study of heat conduction may also lead to the invention of thermal diodes and transistors in the future.

THE SPIN-CHARGE GAUGE APPROACH TO THE THEORY OF DOPED MOTT INSULATORS

LU YU

Institute of Theoretical Physics, Academia Sinica
E-mail: lyu@aphy.iphy.ac.cn

We briefly review the spin-charge gauge approach to the 2D t-J model (prototypic doped Mott insulators) in the limit t≫J, introducing a U(1) field gauging the global charge symmetry and an SU(2) field gauging the global spin-rotational symmetry. We show that this approach can naturally explain many experimental features of transport properties for High Tc cuprates, in particular the metal-insulator crossover phenomena in the "pseudogap phase" (PG) and 1/T behavior of conductivities in the "strange metal phase" (SM) at higher T or doping concentration. Furthermore, it is able to reproduce the universality and the quadratic in T behavior (above the crossover) of in-plane resistivity in PG. A composite particle formed by binding the charge carrier (holon) and spin excitation (spinon) via the slave-particle gauge field is invoked to interpret the obtained results.

See the review: P.A. Marchetti, Z.B. Su and L. Yu, J. Phys.: Condens. Matt. 19 (2007) 125212 and references therein.

DIRECT AND NON-DEMOLITION OPTICAL MEASUREMENT OF PURE SPIN CURRENTS IN SEMICONDUCTORS

JING WANG and BANG-FEN ZHU*

Department of Physics, Tsinghua University, Beijing, 100084, China
**E-mail: bfz@tsinghua.edu.cn*

REN-BAO LIU

Department of Physics, The Chinese University of Hong Kong,
Shatin, N.T., Hong Kong, China
E-mail: rbliu@phy.cuhk.edu.hk

The photon helicity may be mapped to a spin-1/2, whereby we put forward an effective interaction (a scalar) between a light beam and an electron spin current through virtual optical transitions in a direct-gap semiconductor such as GaAs. Such an effective interaction is possible since the pure spin current and the photon spin current, both keeping the time-reversal symmetry but breaking the space-inversion symmetry of the system, are of the same tensor type, namely, the rank-2 pseudo-tensor. The optical effects due to the effective coupling induces the circular birefringence, which is similar to the Faraday rotation in magneto-optics but nevertheless involve no net magnetization. Such optical birefringence effect of a pure spin current originate from the intrinsic spin-orbit coupling in valence bands but involves neither the Rashba effect from structure inversion asymmetry nor the Dresselhaus effect due to bulk inversion asymmetry of the material. This novel optical birefringence effect may be exploited for direct, non-demolition measurement of a pure spin current.

Keyword: Pure spin current; direct measurement; direct-gap semiconductors; circular birefringence.

1. Introduction

With size of the modern integrate circuit (IC) decreased, the power dissipation for IC becomes a more and more urgent problem. In principle, spin current, which conserves the time-reversal symmetry, is dissipationless. Thus the field of spintronics is very promising in the application and is attracting a lot of research interest. Central issues in this field are how to inject, detect, and manipulate the spin degree of freedom of carriers in semiconductor spintronics devices.[1] Among them, the measurement of spin currents is a central topic in the field of spintronics.[1] Till now, the direct and unambiguous measurement of the pure spin current is still a challenge. Spin polarized currents have been detected by the Faraday or Kerr rotation[2–4] or through ferromagnetic filters.[5,6] In a few pioneering experiments, the pure spin currents have been observed by converting spin current into voltage signal[7] via the spin Hall effect,[8–11] or by terminating the spin current to accumulate spin-polarized electrons or excitons which cause Faraday rotation,[12] polarized light emission,[13] or polarization-selective absorption.[14,15] These methods, however, involve either the Rashba effect[16] or the Dresselhaus effect[17] in inversion asymmetric systems, or even the destruction of the spin current, and yet are not the direct measurement.

Thus a question is naturally raised, "Can we have a direct and non-demolition method to detect the pure spin current?" Illuminated by the electro-magnetics experiments, in which to measure a charge current, we may use another charge current, or a small magnetic needle which can detect the induced magnetic filed accompanying the charge current, the spin current may be measured by using another spin current. In fact, this is a basic principle: for a current breaking the symmetry of a system, there should be a force that couples to the current to recover its original symmetry. Regarding the fact that, as a rank-2 pseudo-tensor, a spin current conserves the time-reversal symmetry but breaks the space-inversion symmetry; what we are looking for is a physical quantity of the same type which is coupled with the spin current to form a scalar effective Hamiltonian. Thus, an obvious solution is to find another "spin current". A polarized light beam may play the role of such a "spin current".[18] Here we propose a direct, non-demolition measurement scheme by optical means,[19] which may be applied to detect pure spin currents in direct-gap semiconductors.

The paper is organized as follows. In Sec. 2, we present the physical model, and in Sec. 3, we discuss the optical effects of the pure spin current. Finally in Sec. 4, we give our conclusions.

2. Physical Model

In fact, as a spin-1 massless boson, photon has two physical helicity states, resembles a spin-1/2 in the Jones vector representation of light polarization,[20] such as

$$\cos\frac{\theta}{2}e^{i\phi/2}\mathbf{n}_+ + \sin\frac{\theta}{2}e^{-i\phi/2}\mathbf{n}_-,$$

in which the right (left) circular polarization $\mathbf{n}_+$ ($\mathbf{n}_-$) is mapped to a spin-1/2 state parallel to the light propagating direction. The "spin current" tensor for a light with electric field

$$\mathbf{F}(\mathbf{r},t) = \left(F_+\mathbf{n}_+ + F_-\mathbf{n}_-\right) e^{i\mathbf{q}\cdot\mathbf{r}-i\omega_q t} + \text{c.c.}$$

can then be defined by

$$\mathbb{I} \equiv q\left(I_x\mathbf{xz} + I_y\mathbf{yz} + I_z\mathbf{zz}\right), \tag{1a}$$

$$I_j = \frac{1}{2}\sum_{\mu,\nu=\pm} \sigma^j_{\mu\nu} F^*_\mu F_\nu, \tag{1b}$$

where σ^j ($j = x, y, z$) is the Pauli matrix, and the unit axis vectors $\mathbf{x}$, $\mathbf{y}$, and $\mathbf{z}$ are defined through $\mathbf{n}_\pm \equiv (\mp\mathbf{x} - i\mathbf{y})/\sqrt{2}$ and $\mathbf{z} \equiv \mathbf{q}/q$.

We take an n-doped bulk III-V compound semiconductor with a direct band-gap such as GaAs for consideration. The doped electrons in the conduction band (CB) is in a steady non-equilibrium distribution $\hat{\rho}$, with a spin current

$$\mathbb{J} = e\sum_{\mu,\nu,\mathbf{k}} \frac{1}{2}\boldsymbol{\sigma}_{\mu,\nu} f_{\mu\nu,\mathbf{k}} \hbar^{-1}\nabla_{\mathbf{k}} E_{e\mathbf{k}}, \tag{2}$$

where $\hat{e}_{\mu\mathbf{k}}$ is the electron annihilation operator,

$$f_{\mu\nu,\mathbf{k}} \equiv \text{Tr}\left[\hat{\rho}\hat{e}^\dagger_{\mu\mathbf{k}}\hat{e}_{\nu\mathbf{k}}\right]$$

is the electron population, and $E_{e\mathbf{k}}$ is the electron energy.

The derivation of the effective Hamiltonian is straightforward. In the following we outline the key steps and make the results plausible.

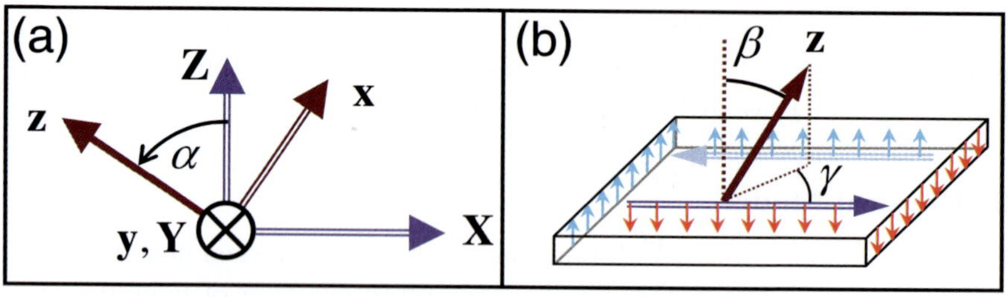

Fig. 1. (Color online) (a) The spin current direction (**Z**), the light beam direction (**z**), and the associated coordinate systems. (b) Geometry configuration for direct measurement of a transverse pure spin current.

Under the assumption that the incident light is tuned to be little lower than the Fermi energy, we consider optical transitions between the conduction band and the heavy-hole(HH) and light-hole (LH) band, but neglect the split-off (SO) band (see Fig. 2 (a) for the energy band diagram), because in semiconductors like GaAs the band edge separation between the HH or LH band and the SO band is large compared to other quantities of interest. Here we would like to emphasize that since the system we investigate is the bulk semiconductors, there should not be the Rashba effect.[16] Furthermore, the spin splitting due to the Dresselhauss effect, originating from the intrinsic spatial asymmetry in polar semiconductors, is estimated around 0.01 meV in GaAs at the doping density of 10^{16}cm^{-3},[21] which is much less than the detuning of the light from the lowest transition available. So we can safely neglect it.

The pure spin current in bulk GaAs given by Eqn. (2) can also be cast in the tensor form as

$$\mathbb{J} = J_X \mathbf{XZ} + J_Y \mathbf{YZ} + J_Z \mathbf{ZZ} \equiv \mathbf{JZ}, \tag{3}$$

provided $\mathbf{Z}$ is the current direction and $\mathbf{X}$ and $\mathbf{Y}$ are the transverse directions (as shown in Fig. 1 (a)). Note that in the dyadic form of $\mathbb{I}$ and $\mathbb{J}$, the polarization components $I_x(I_y, I_z)$ and $J_X(J_Y, J_Z))$ are pseudo-scalars.

The Luttinger-Kohn Hamiltonian H_{LK} for the valence bands near the band edge is[22]

$$H_{\text{LK}} = \frac{\hbar^2}{2m}\left[\left(\gamma_1 + \frac{5}{2}\gamma_2\right)\boldsymbol{\nabla}^2 - 2\gamma_2\left(\boldsymbol{\nabla}\cdot\mathbf{K}\right)^2\right], \tag{4}$$

where $\mathbf{K}$ is a spin-3/2 for the total angular momentum of a hole in the valence band. Here we have for simplicity neglected the anisotropy of the valence bands which would not change the essential results in this paper. The energy dispersion for a hole with magnetic quantum number K_j quantized along the wavevector $\mathbf{p}$ is

$$E_{K_j}(\mathbf{p}) = \frac{\hbar^2 p^2}{2m}\left[\gamma_1 + \left(\frac{5}{2} - 2K_j^2\right)\gamma_2\right].$$

Thus the non-interacting Hamiltonian of the electrons and holes reads

$$\hat{H}_0 = \sum_{\mu,\mathbf{p}}(E_{e\mathbf{p}}\hat{e}^\dagger_{\mu\mathbf{p}}\hat{e}_{\mu\mathbf{p}} + E_{h\mathbf{p}}\hat{h}^\dagger_{\mu\mathbf{p}}\hat{h}_{\mu\mathbf{p}} + E_{l\mathbf{p}}\hat{l}^\dagger_{\mu\mathbf{p}}\hat{l}_{\mu\mathbf{p}}), \tag{5}$$

where $\mu = \pm$ denotes the spin moment quantized along $\mathbf{p}$-direction, $\hat{h}_{\pm\mathbf{p}}$ and $\hat{l}_{\pm,\mathbf{p}}$ are the annihilation operators for the hole with $K_j = \pm 3/2$ and $\pm 1/2$, respectively. $E_{h\mathbf{p}} = E_{\pm 3/2}(\mathbf{p})$, and $E_{l\mathbf{p}} = E_{\pm 1/2}(\mathbf{p})$.

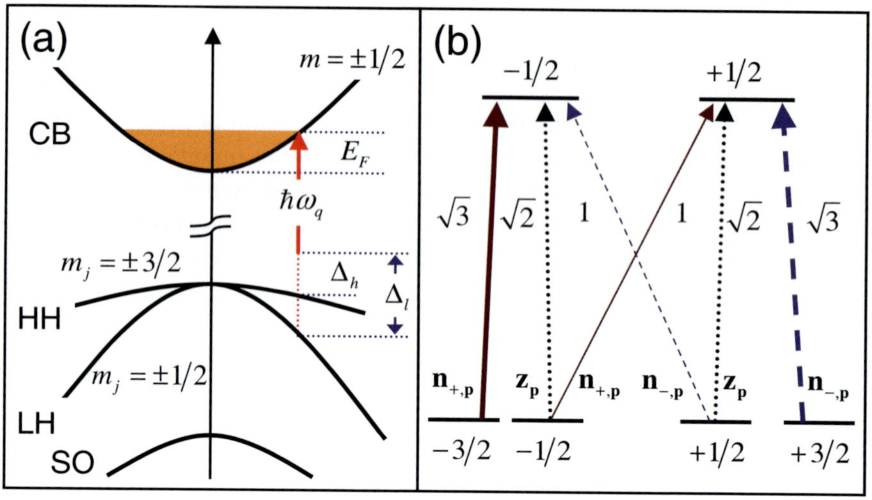

Fig. 2. (color online) (a) Schematic band structure near the Γ point of an n-doped III-V compound semiconductor. (b) Transition matrix elements between the valence and the conduction bands. The angular momentum is quantized along the wavevector ($\mathbf{p}$) of the valence band state.

The Hamiltonian for the optical interband transition[23] between the conduction band and the heavy-hole and light-hole bands is expressed as,

$$\hat{H}_1 = d_{\rm cv}^* \sum_{\mu,\nu,\mathbf{p}} F_\nu^* \mathbf{n}_\nu^* \cdot \left(\mathbf{n}_{\mu,\mathbf{p}} \hat{h}_{\mu-\mathbf{p}} \hat{e}_{\bar{\mu}\mathbf{q}+\mathbf{p}} + \frac{1}{\sqrt{3}} \mathbf{n}_{\mu,\mathbf{p}} \hat{l}_{\mu-\mathbf{p}} \hat{e}_{\mu\mathbf{q}+\mathbf{p}} \right.$$
$$\left. - \sqrt{\frac{2}{3}} \mathbf{z}_\mathbf{p} \hat{l}_{\mu-\mathbf{p}} \hat{e}_{\bar{\mu}\mathbf{q}+\mathbf{p}} \right) + \text{h.c.}, \tag{6}$$

with $\mathbf{n}_{+,\mathbf{p}}$ ($\mathbf{n}_{-,\mathbf{p}}$) denoting the right (left) circular polarization with respect to $\mathbf{p}$ which are defined as $\mathbf{n}_{\pm,\mathbf{p}} \equiv (\mp \mathbf{x}_\mathbf{p} - i\mathbf{y}_\mathbf{p}) / \sqrt{2}$, $\mathbf{z}_\mathbf{p} \equiv \mathbf{p}/p$, and $\bar{\mu} \equiv -\mu$. We have neglected the small wavevector dependence of the dipole matrix element $d_{\rm cv}$.[22] Since the total angular momentum is conserved in the optical transition processes, we have the selection rules and the relative transition strengths[23] as illustrated in Fig. 2 (b).

Under the condition that the optical transition strength is much weaker than the detuning of the light from the band edge, the effective Hamiltonian is obtained by the second-order perturbation as

$$\mathcal{H}_{\rm eff} = \text{Tr}\left[\hat{\rho} \hat{H}_1 \left(\hat{H}_0 - \hbar\omega_q \right)^{-1} \hat{H}_1 \right]. \tag{7}$$

Without loss of generality, we assume the system has translational symmetry so that

$$\mathrm{Tr}\left[\hat{\rho}\hat{e}^{\dagger}_{\mu\mathbf{k}}\hat{e}_{\nu\mathbf{k}'}\right] = \delta_{\mathbf{k},\mathbf{k}'} f_{\mu\nu,\mathbf{k}}.$$

Then the effective Hamiltonian is explicitly worked out as

$$\begin{aligned}
\mathcal{H}_{\text{eff}} &= |d_{\text{cv}}|^2 \sum_{\sigma,\sigma'} F^*_{\sigma} F_{\sigma'} \mathbf{n}_{\sigma'} \mathbf{n}^*_{\sigma} : \\
&\sum_{\mu,\mathbf{p}} \Bigg[\left(1 - f_{\bar{\mu}_{\mathbf{p}}\bar{\mu}_{\mathbf{p}},\mathbf{q}+\mathbf{p}}\right) \frac{\mathbf{n}_{\mu,\mathbf{p}} \mathbf{n}^*_{\mu,\mathbf{p}}}{E_{e\mathbf{q}+\mathbf{p}} + E_{h-\mathbf{p}} - \hbar\omega_q} \\
&+ \frac{1}{3} \left(1 - f_{\mu_{\mathbf{p}}\mu_{\mathbf{p}},\mathbf{q}+\mathbf{p}}\right) \frac{\mathbf{n}_{\mu,\mathbf{p}} \mathbf{n}^*_{\mu,\mathbf{p}}}{E_{e\mathbf{q}+\mathbf{p}} + E_{l-\mathbf{p}} - \hbar\omega_q} \\
&+ \frac{\sqrt{2}}{3} f_{\bar{\mu}_{\mathbf{p}}\mu_{\mathbf{p}},\mathbf{q}+\mathbf{p}} \frac{\mathbf{n}_{\mu,\mathbf{p}} \mathbf{z}^*_{\mathbf{p}}}{E_{e\mathbf{q}+\mathbf{p}} + E_{l-\mathbf{p}} - \hbar\omega_q} \\
&+ \frac{\sqrt{2}}{3} f_{\mu_{\mathbf{p}}\bar{\mu}_{\mathbf{p}},\mathbf{q}+\mathbf{p}} \frac{\mathbf{z}_{\mathbf{p}} \mathbf{n}^*_{\mu,\mathbf{p}}}{E_{e\mathbf{q}+\mathbf{p}} + E_{l-\mathbf{p}} - \hbar\omega_q} \\
&+ \frac{2}{3} \left(1 - f_{\bar{\mu}_{\mathbf{p}}\bar{\mu}_{\mathbf{p}},\mathbf{q}+\mathbf{p}}\right) \frac{\mathbf{z}_{\mathbf{p}} \mathbf{z}^*_{\mathbf{p}}}{E_{e\mathbf{q}+\mathbf{p}} + E_{l-\mathbf{p}} - \hbar\omega_q} \Bigg],
\end{aligned} \tag{8}$$

where the subscript $\mathbf{p}$ of the spin index μ indicates that the spin is quantized along $\mathbf{p}$. The physical process for each term can readily be recognized. For example, the term with $\mathbf{n}_{\mu,\mathbf{p}}\mathbf{z}^*_{\mathbf{p}}$ involves in sequence such virtual processes: the excitation of an electron-light-hole pair by a photon with linear polarization along $\mathbf{p}$, the propagation of an electron with spin coherence, and the electron-hole recombination with a circularly polarized photon emitted. To proceed the derivation, we omit the trivial background constant, and consider only the electron population deviating from the equilibrium, i.e.,

$$f'_{\mu\nu,\mathbf{p}} \equiv f_{\mu\nu,\mathbf{p}} - f^{(0)}_{\mu\nu,\mathbf{p}}.$$

Furthermore, as we are not interested in charge effects such as a charge current, we ignore the terms with $f'_{++,\mathbf{p}} + f'_{--,\mathbf{p}}$, but keep those associated with the spin population

$$f'_{\mathbf{N}}(\mathbf{p}) \equiv \sum_{\mu,\nu} \mathbf{N} \cdot \boldsymbol{\sigma}_{\mu,\nu} f'_{\mu\nu,\mathbf{p}},$$

for example,

$$f'_{\mathbf{X}}(\mathbf{p}) = f'_{+-,\mathbf{p}} + f'_{-+,\mathbf{p}}.$$

For the moment, we neglect the small light wavevector $\mathbf{q}$ in Eq. (8). The first term in the square bracket gives rise to the coupling to a net spin polarization. To show this, we sum the terms with $\pm\mathbf{p}$. The result relevant to the spin effect is

$$\propto \left[f'_{\mathbf{z}_p}(\mathbf{p}) + f'_{\mathbf{z}_p}(-\mathbf{p})\right]\left(\mathbf{n}_{+,\mathbf{p}}\mathbf{n}^*_{+,\mathbf{p}} - \mathbf{n}_{-,\mathbf{p}}\mathbf{n}^*_{-,\mathbf{p}}\right),$$

which contributes to the spin polarization but not to the spin current. Similarly, note

$$\mathbf{z}_\mathbf{p}\mathbf{z}^*_\mathbf{p} = 1 - \mathbf{n}_{+,\mathbf{p}}\mathbf{n}^*_{+,\mathbf{p}} - \mathbf{n}_{-,\mathbf{p}}\mathbf{n}^*_{-,\mathbf{p}},$$

the second and the fifth terms also contribute to the spin polarization. The physical mechanisms for theses terms are clear. For the HH-CB transitions, where a (virtually) absorbed photon has to be emitted with the same circular polarization, the electron spin is conserved. An LH state contains both circular and linear orbital components, so an LH-CB transition can be either circularly or linearly polarized. For the LH-CB transitions in the second and the fifth terms the photons involved in the virtual absorption and emission have the same polarization, so the electron spin is conserved. In this paper, we are interested in the effects of a pure spin current only, as the net spin polarization is set to be zero.

As the third and the fourth terms in Eq. (8), one can verify this by noticing that the summation over $\pm\mathbf{p}$ leads to terms containing

$$f'_{\mathbf{X}_\mathbf{p}}(\mathbf{p}) + f'_{\mathbf{X}_\mathbf{p}}(-\mathbf{p})$$

and

$$f'_{\mathbf{Y}_\mathbf{p}}(\mathbf{p}) + f'_{\mathbf{Y}_\mathbf{p}}(-\mathbf{p}),$$

which is connected with the spin polarization. This process excites the same electron spins for opposite momenta $\pm\mathbf{p}$, and the spin current is not coupled to such processes.

Next we consider the effect of the small light wavevector $\mathbf{q}$ up to the first order by the expansion

$$E_{e\ \mathbf{q}+\mathbf{p}} \approx E_{e\ \mathbf{p}} + \mathbf{q}\cdot\nabla_\mathbf{p}E_{e\ \mathbf{p}},$$

and

$$f'_\mathbf{N}(\mathbf{q}+\mathbf{p}) \approx f'_\mathbf{N}(\mathbf{p}) + \mathbf{q}\cdot\nabla_\mathbf{p}f'_\mathbf{N}(\mathbf{p}).$$

The gradient in the momentum space $\nabla_\mathbf{p}$ contributes the electron velocity which is opposite for opposite momenta. So the light couples to opposite electron spin for opposite momenta, and in turn to a pure spin current.

The effective coupling between the light and the spin current in an n-doped direct-gap semiconductors can also be reformulated as

$$\mathcal{H}_{\text{eff}} = \zeta_1 q I_z \mathbf{z} \cdot \mathbb{J} \cdot \mathbf{z} + \zeta_2 q I_z J_Z, \tag{9}$$

and the coefficients ζ_i ($i = 1, 2$), as given later in Eq. (11), are determined by material parameters and the light frequency. It is easy to check the internal rotational, spatial inversion, and time inversion symmetry of the Hamiltonian as a scalar.

All terms in the effective Hamiltonian shown by Eq. (9) depend on the small light wave vector q, which is characteristic of coupling to the magnetic dipole moment of a chiral quantity.

To determine the coupling constants ζ_j in the effective Hamiltonian, we assume that the electrons are driven only slightly away from the equilibrium with Fermi wavevector k_F (or Fermi energy E_F), and the spin distribution is expressed as

$$f'_{\mathbf{N}}(\mathbf{p}) = f'_{\mathbf{N}}(p) \cos\theta_{\mathbf{p}}, \tag{10}$$

($\theta_{\mathbf{p}}$ denoting the angle between $\mathbf{p}$ and $\mathbf{Z}$), which is usually the case for weak currents. The light frequency is lower by Δ_h and Δ_l than the transition energy from the HH bands and the LH bands to the Fermi level of the electron gas, respectively. A straightforward calculation yields

$$\zeta_1 = \frac{\hbar}{e}|d_{\text{cv}}|^2 \left(\frac{4m_e}{5\Delta_h^2 m_h} - \frac{1}{5\Delta_h E_F} + \frac{8m_e}{15\Delta_l^2 m_l} + \frac{1}{5\Delta_l E_F} \right), \tag{11a}$$

$$\zeta_2 = \frac{\hbar}{e}|d_{\text{cv}}|^2 \left(\frac{2m_e}{5\Delta_h^2 m_h} - \frac{3}{5\Delta_h E_F} - \frac{2m_e}{5\Delta_l^2 m_l} + \frac{3}{5\Delta_l E_F} \right), \tag{11b}$$

where m_e, m_h, and m_l are in turn the electron, HH, and LH effective mass. For a spin distribution different from Eq. (10), the coupling constants shown above will only be quantitatively changed in some form factors.

3. Spin Current Faraday Rotation

With the effective coupling between the light beam and the spin current derived above, the linear optical susceptibility is obtained as

$$\chi_{\mu,\nu} + \chi^*_{\nu,\mu} = \frac{1}{\epsilon_0} \frac{\partial^2 \mathcal{H}_{\text{eff}}}{\partial F^*_\mu \partial F_\nu}, \tag{12}$$

where ϵ_0 is the vacuum permittivity. As we show below, this will produce the circular birefringence effect due to the spin current $\mathbb{J}$.

Both terms in Eq. (9) contain the light "spin" component $I_z \equiv (F^*_+ F_+ - F^*_- F_-)/2$. Therefore the susceptibility calculated with Eq. (12)

has opposite signs for the clockwise and anti-clockwise circular polarizations,

$$\chi_{++} = -\chi_{--} = \frac{1}{4\epsilon_0} q \left(\zeta_1 \mathbf{z} \cdot \mathbf{JZ} \cdot \mathbf{z} + \zeta_2 J_z \right). \tag{13}$$

This is quite similar to the Faraday rotation,[24] and is responsible for circular birefringence. In contrast to the usual Faraday rotation in magnetooptics,[24] the circular birefringence effect due to the pure spin current involves no net magnetization.

We can ecaluate the Faraday rotation angle as

$$\delta_{\mathrm{F}} = \omega_q L \left(\chi_{++} - \chi_{--} \right) / \left(4nc \right), \tag{14}$$

where L is the light propagation distance, n is the material refractive index, and c is the light velocity in vacuum.

In general, by measuring the Faraday rotation through choosing appropriate propagation and polarization directions of the light beam, one can determine the direction, polarization, and amplitude of a spin current, and distinguish it from a net spin polarization. To demonstrate the feasibility of such methods, we consider a realistic case that was studied in Ref. 12. As shown in Fig. 1 (b), a transverse spin current is caused by the spin Hall effect. In Ref. 12, the Faraday rotation of a light normal to the sample surface (and parallel to the spin polarization) is measured, with non-vanishing results only near the edges where spins are accumulated. The absence of Faraday rotation in the middle region where the spin current flows with no net spin polarization can be readily explained by Eq. (13): In this experimental setup, $\mathbf{Z} \cdot \mathbf{z} = 0$ and $J_Z = 0$. To directly detect the spin current where it flows, we propose to tilt the light beam with a zenith angle β from the normal direction and an azimuth angle γ, and detect the Faraday rotation δ_{F} (see Fig. 1 (b) for the configuration). The results is predicted to vary with the angles as

$$\delta_{\mathrm{F}}(\beta, \gamma) = \delta_{\mathrm{F},0} \sin\beta \cos\gamma. \tag{15}$$

To estimate the amplitudes of the effect, we take parameters for bulk GaAs sample as in Ref. 12, i.e., $m_e = 0.067 m_0$ (m_0 being the free electron mass), $m_h = 0.45 m_0$, $m_l = 0.082 m_0$, $L = 2.0\ \mu\text{m}$, $n = 3.0$, $d_{\text{cv}} = 6.7\ e\text{Å}$, and $E_F = 5.3$ meV ($k_F = 0.96 \times 10^6\ \text{cm}^{-1}$) for doping density $3 \times 10^{16}\ \text{cm}^{-3}$. We take the light wavelength to be around 800 nm, and the detuning $\Delta_h = 1.0$ meV and $\Delta_l = 4.5$ meV. For a spin current with amplitude $J_s = 20\ \text{nA}\mu\text{m}^{-2}$,[12] the maximum Faraday rotation, reached when $\beta \to \pi/2$ and $\gamma \to 0$, is $\delta_{F,0} = 0.38 \mu\text{rad}$. This phase shift is detectable experimentally.[12]

4. Summary

In conclusion, the intrinsic interaction between a polarized light and a spin current may induce measurable circular birefringence as a direct non-demolition measurement of a pure spin current. Unlike the optical injection of spin currents,[25–27] the optical measurement scheme proposed here does not rely on the inversion asymmetry of the sample.

Acknowledgments

This work was supported by the NSFC Grant No.10774086,10574076, and the Basic Research Program of China Grant 2006CB921500 and the Hong Kong RGC Direct Grant 2060284.

References

1. S. A. Wolf, D. D. Awschalom, R. A. Buhrman, J. M. Daughton, S. von Molnár, M. L. Roukes, A. Y. Chtchelkanova, and D. M. Treger, *Science* **294**, 1488 (2001).
2. J. M. Kikkawa and D. D. Awschalom, *Nature* **397**, 139 (1999).
3. J. Stephens, J. Berezovsky, J. P. McGuire, L. J. Sham, A. C. Gossard, and D. D. Awschalom, *Phys. Rev. Lett.* **93**, 097602 (2004).
4. S. A. Crooker, M. Furis, X. Lou, C. Adelmann, D. L. Smith, C. J. Palmstrøm, and P. A. Crowell, *Science* **309**, 2191 (2005).
5. X. H. Lou, C. Adelmann, S. A. Crooker, E. S. Garlid, J. Zhang, K. S. M. Reddy, S. D. Flexner, C. J. Palmstrøm, and P. A. Crowell, *Nature Phys.* **3**, 197 (2007).
6. I. Appelbaum, B. Q. Huang, and D. J. Monsma, *Nature* **447**, 295 (2007).
7. S. O. Valenzuela and M. Tinkham, *Nature* **442**, 176 (2006).
8. M. I. Dyakonov and V. I. Perel, *Phys. Lett. A* **35**, 459 (1971).
9. J. E. Hirsch, *Phys. Rev. Lett.* **83**, 1834 (1999).
10. S. Murakami, N. Nagaosa, and S. C. Zhang, *Science* **301**, 1348 (2003).
11. J. Sinova, D. Culcer, Q. Niu, N. A. Sinitsyn, T. Jungwirth, and A. H. Macdonald, *Phys. Rev. Lett.* **92**, 126603 (2004).
12. Y. K. Kato, R. C. Myers, A. C. Gossard, and D. D. Awschalom, *Science* **306**, 1910 (2004).
13. J. Wunderlich, B. Kaestner, J. Sinova, and T. Jungwirth, *Phys. Rev. Lett.* **94**, 047204 (2005).
14. M. J. Stevens, A. L. Smirl, R. D. R. Bhat, A. Najmaie, J. E. Sipe, and H. M. van Driel, *Phys. Rev. Lett.* **90**, 136603 (2003).
15. H. Zhao, E. J. Loren, H. M. van Driel, and A. L. Smirl, *Phys. Rev. Lett.* **96**, 246601 (2006).
16. Y. A. Bychkov and E. I. Rashba, Pis'ma Zh. *Eksp. Teor. Fiz.* **39**, 66 (1984) [*Sov. Phys. JETP Lett.* **39**, 78 (1984)].
17. G. Dresselhaus, *Phys. Rev.* **100**, 580 (1955).

18. Strictly speaking, a light beam does not conserve the timeinversion symmetry. But its angular-momentum flux apart from the energy flux, being a pure spin current of interest here, does.
19. Jing Wang, Bang-fen Zhu and Ren-Bao Liu, *Phys. Rev. Lett.* **100**, 086603 (2008).
20. R. C. Jones, *J. Opt. Soc. Am.* **31**, 488 (1941).
21. G. E. Pikus, V. A. Marushchak, and A. N. Titkov, *Sov. Phys. Semicond.* **22**, 115 (1988).
22. W. Paul and T. S. Moss, eds., in *Handbook on Semicondutors: Band Theory and Transport Properties, vol. 1* (North-Holland, Amsterdam, 1982).
23. F. Meier and B. P. Zakharchenya, eds., in *Optical Orientation* (Elsevier, Amsterdam, 1984).
24. A. K. Zvezdin and V. A. Kotov, in *Modern Magnetooptics and Magnetooptical Materials* (Taylor and Francis Group, New York, 1997).
25. X. D. Cui, S. Q. Shen, J. Li, Y. Ji, W. Ge, and F. C. Zhang, *Appl. Phys. Lett.* **90**, 242115 (2007).
26. R. D. R. Bhat and J. E. Sipe, *Phys. Rev. Lett.* **85**, 5432 (2000).
27. R. D. R. Bhat, F. Nastos, A. Najmaie, and J. E. Sipe, *Phys. Rev. Lett.* **94**, 096603 (2005).

FROM BCS TO HTS AND RTS

C. W. CHU

Hong Kong University of Science and Technology,
University of Houston, and
Lawrence Berkeley National Laboratory

Great progress has been made in high temperature superconductivity (HTS) science, material and technology in the 20 years since its discovery. The next grand challenge will be room temperature superconductivity (RTS). Room temperature superconductivity, if achieved, can change the world both scientifically and technologically. Unfortunately, it has long been considered by some to belong to the domain of science fiction and to occur only "at an astronomical temperature and at an astronomical distance". With the advent of HTS in 1987, the outlook for RTS has become much brighter. Currently, there appears to be no reason, either theoretical or experimental, why room temperature superconductivity should be impossible. BCS theory has provided the basic framework for the occurrence and understanding of superconductivity, but, since its inception, it has failed to show where and how to find superconductivity at higher temperatures. To date, empiricism remains the most effective way to discover superconductors with high transition temperatures. In this paper based on the talk given at the Professor Yang's 85th birthday celebration on October 31, 2007 in Singapore, I shall summarize the search for superconductors of higher T_c prior to and after the discovery of HTS, list the common features of HTS and describe some approaches toward RTS that we are currently pursuing.

1. Introduction

It is rather fitting for us to discuss the possibility of achieving room temperature superconductivity (RTS) in this auspicious year (2007). During 2007, we are celebrating the 50th anniversary of the BCS theory of superconductivity [1], the 50th anniversary of the proposition of parity nonconservation [2] (this is particularly so as we are celebrating Professor Yang's 85th birthday here in Singapore) and the 20th anniversary of the discovery of the first liquid nitrogen superconductor $YBa_2Cu_3O_7$ [3] (Fig. 1). It is also interesting to note that the discovery of the first cuprate high temperature superconductor (HTS) Ba-doped La_2CuO_4 in 1986 [4] was exactly 300 years after

Microscopic Theory of Superconductivity*

J. Bardeen, L. N. Cooper, and J. R. Schrieffer
Department of Physics, University of Illinois, Urbana, Illinois
(Received February 18, 1957)

Phys. Rev. 106, 162 (1957)

Parity Nonconservation and a Two-Component Theory of the Neutrino

T. D. Lee, *Columbia University, New York, New York*
and
C. N. Yang, *Institute for Advanced Study, Princeton, New Jersey*
(Received January 10, 1957; revised manuscript received January 17, 1957)

Phys. Rev. 105, 1671 (1957)

Superconductivity at 93 K in a New Mixed-Phase Y-Ba-Cu-O Compound System at Ambient Pressure

M. K. Wu, J. R. Ashburn, and C. J. Torng
Department of Physics, University of Alabama, Huntsville, Alabama 35899
and
P. H. Hor, R. L. Meng, L. Gao, Z. J. Huang, Y. Q. Wang, and C. W. Chu [(a)]
Department of Physics and Space Vacuum Epitaxy Center, University of Houston, Houston, Texas 77004
(Received 6 February 1987; Revised manuscript received 18 February 1987)

Phys. Rev. Lett. 58, 908 (1987)

Fig. 1. 2007 is a special year: 50 years after the BCS theory, 50 years after the noconservation of parity; and 20 years after YBCO.

Newton published his Principia Mathematica [5] that had formally ushered in the era of modern science (Fig. 2).

Ever since the discovery of superconductivity in 1911 [6], the search for superconductors with higher superconducting transition temperatures (T_c) has been the major driving forces in the long sustained research effort on superconductivity. The rise of T_c with time is summarized in Fig. 3. Before 1986, through mainly Matthias's effort, a generation of superconducting inter-metallic alloys and compounds was born, giving rise to the then-record T_c of 23.2 K in Nb_3Ge in 1973 [7] attainable in a liquid hydrogen environment. Mueller and Bednorz inaugurated the new era of high temperature superconductivity (HTS) by discovering the new generation of superconducting perovskite-like cuprates with a T_c up to a new record of 35 K in Ba-doped La_2CuO_4 in 1986 [4]. Twenty-one years later, people can easily accept the report as matter-of-factly. However, when their

Possible High T_c Superconductivity in the Ba – La – Cu – O System

J.G. Bednorz and K.A. Müller
IBM Zürich Research Laboratory, Rüschlikon, Switzerland

Received April 17, 1986

Z. Physik B 64, 189 (1986)

I. S. Newton
July 5,1686

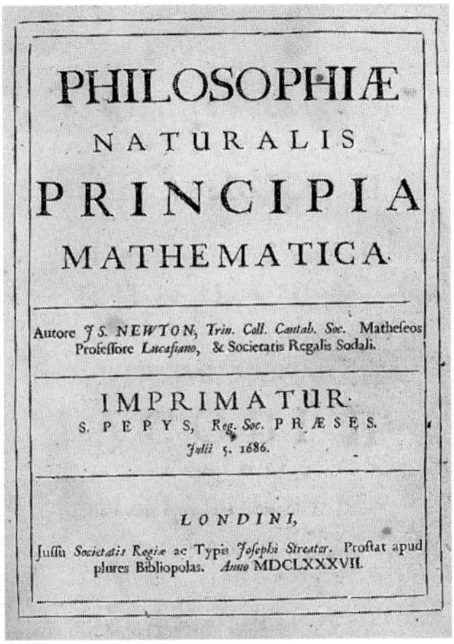

PHILOSOPHIÆ
NATURALIS
PRINCIPIA
MATHEMATICA.

Autore J S. NEWTON, Trin. Coll. Cantab. Soc. Matheſeos Profeſſore Lucaſiano, & Societatis Regalis Sodali.

IMPRIMATUR.
S. PEPYS, Reg. Soc. PRÆSES.
Julii 5. 1686.

LONDINI,
Juſſu Societatis Regiæ ac Typis Joſephi Streater. Proſtat apud plures Bibliopolas. Anno MDCLXXXVII.

Fig. 2. HTS was discovered 300 years after Newton's Philosophia Naturalis Principia Mathematica.

seminal discovery first appeared in September 1986, it was met with skepticism except by a very few groups, since oxides are mostly insulators, and are not even metallic, let alone superconducting at high temperature. Our group in Houston was among these very lucky few non-skeptics [8]. This is because we had been actively investigating the unstable perovskite and related oxide systems, such as $BaPb_xBi_{1-x}O_3$ and $Li_{1+x}Ti_{2-x}O_4$, since the mid-70s, from which not only did we learn that superconductivity is possible in oxides, but also mastered the oxide synthesis skill crucial for later HTS studies. Our previous extensive studies on the correlation of lattice instabilities with T_c using the high pressure technique had convinced us that lattice instabilities should not be the absolute deterrence to higher T_c, in contrast to the then-prevailing theoretical prediction, and that higher

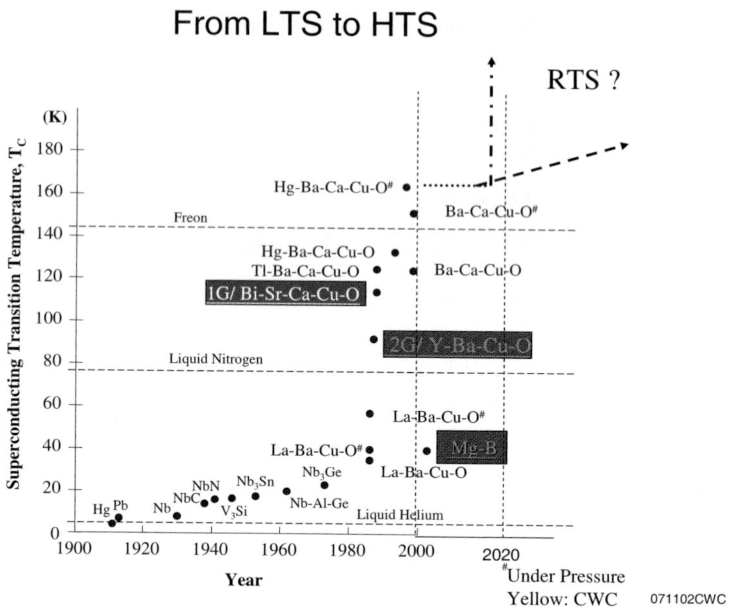

Fig. 3. T_c-evolution with time.

T_c was possible [9]. Therefore, we took the results of Mueller and Bednorz seriously and reproduced them soon afterward. We quickly raised the T_c to 40 K [10] and then 52 K [11] by the application of pressures at a rate more than ten times that on the inter-metallic superconductors. Our observation of a T_c higher than 40 K shattered the then-theoretical T_c-limit of 30's K [12] and raised serious doubts about the validity of the assumptions on which the prediction was made. The unusually large positive pressure effect on T_c suggested to me right away that higher T_c may be achievable through chemical pressures by replacing elements in the compound with smaller ones of the same valence, such as Ba by Sr or Ca; or La by Y or Lu. The Ba – Sr replacement to raise T_c was quickly confirmed by us and others, but the Ba-Ca substitution was unfortunately found to suppress T_c. The effect of other suggested replacements to enhance T_c was not carried out until a month later. The unusually large positive pressure effect on T_c observed also suggested that these oxide superconductors may belong to a new class of materials that warranted further studies, in contrast to some who thought that the Ba – doped La_2CuO_4 was not unusual since its 35 K at ambient pressure which fell within the range predicted by the theory [13].

At the 1986 Fall MRS Meeting in Boston on December 4, I made an oral presentation on our work on $BaPb_xBi_{1-x}O_3$ but, before concluding the talk, I disclosed our duplication of the results of Mueller and Bednorz. During question-and-answer at the end of my talk, Kitazawa from Tokyo announced that the phase responsible for the 35 K superconductivity in the mixed-phase La-Ba-Cu-O samples of Mueller and Bednorz had been identified as the Ba-doped La_2CuO_4 or $La_{2-x}Ba_xCuO_4$ known as the 214 phase. With this piece of information on hand, it was quite natural for people, including ourselves, to focus on making the pure 214-phase samples and to examine the origin of the unusually high T_c in this unusual compound. We determined that to get 214 single crystals would be the best approach. Unfortunately, we failed to grow 214 single crystals. So, following the destruction of two of our three crystal-growing Pt-crucibles, I decided to focus instead on stabilizing the high temperature resistivivity drops, indicative but not a proof of superconductivity, that were detected sporadically in the multiphase samples based on the nominal compositions of Bednorz and Mueller, by replacing La with Y and Lu for the reason mentioned earlier. It should be noted that the first sign of superconductivity as evidenced by a resistivity drop at a temperature above 70 K was detected in mid-November 1986, although it was too fleeting to make a definitive characterization due to the unstable nature of the samples. Nonetheless, I showed the preliminary data to M. K. Wu of Alabama at the Boston meeting and successfully convinced him to join the search. In mid-January 1987, we observed a large diamagnetic shift or Meissner signal in one of our mixed-phase La-Ba-Cu-O samples up to $\sim$ 96 K, representing the first definitive superconductivity signal detected above the liquid nitrogen temperature of 77 K (Fig. 4). Unfortunately, the sample degraded and the signal was lost the following day. Nonetheless, the X-ray pattern of the fresh sample was taken and later identified to possess the 123 structure (see below). In late January 1987, the 93 K superconductivity was stabilized [3] in mixed-phase Y-Ba-Cu-O samples, almost tripling the T_c of the 214 compound. The discovery broke the liquid nitrogen temperature barrier of 77 K and posed serious challenges to physicists concerning the cause for the observation. It also brought superconductivity technology a giant step closer to applications that could use the practical liquid nitrogen as their coolant. In less than a month, the superconducting phase $YBa_2Cu_3O_7$ (known as 123 or YBCO) was identified and its structure resolved with Bob Hazen et al. from the Carnegie Geophysical Laboratory in Washington [14]. With the structure information on hand, we quickly found that Y in YBCO is electronically isolated from

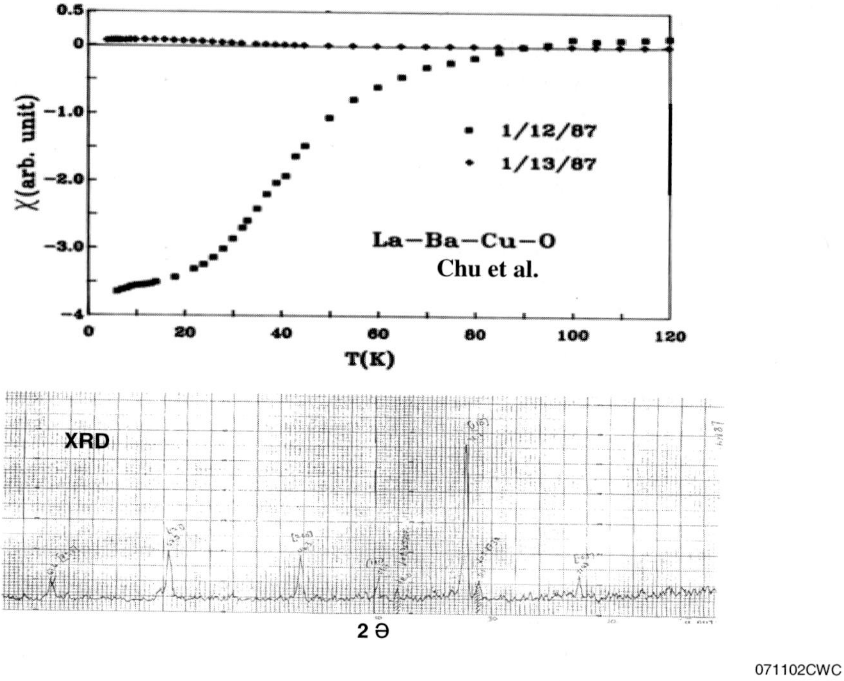

Fig. 4. The first definitive sign of superconductivity above 77 K.

the superconducting component of the compound and discovered the whole cuprate series of $RBa_2Cu_3O_7$ (RBCO or R123) with a $T_c \sim 90$s K where R = Y and rare-earth elements [15]. The T_c has since been advanced first in 1988 to 115 K in $Bi_2Sr_2Ca_2Cu_3O_{10}$ and to 125 K in $Tl_2Ba_2Ca_2Cu_3O_{10}$ by Maeda et al. [16] and Herman and Sheng [17], respectively, and then in 1993 to 134 K in $HgBa_2Ca_2Cu_3O_9$ by Schilling et al. [18]. The T_c of $HgBa_2Ca_2Cu_3O_9$ was further advanced to 164 K by us by the application of pressures up to 30 GPa [19]. This is the record T_c to date, albeit under pressure, and can be attained through household air-conditioning technology using Freon (CF_4) with a boiling point of 148 K.

In the last 20 years, many theoretical models have been advanced to account for numerous unusual observations in high temperature superconductors (HTS) [20] and many superconducting prototype devices have been constructed and demonstrated successfully with superior performance to their non-superconducting counterparts. To take advantage of the full prowess of superconductivity for our daily lives, room temperature superconductivity (RTS) will be a natural target in order to avoid completely the inconvenience of cooling. Even before the discovery of superconductivity, people had already been fascinated by the concept of perpetual motion machines and RTS, because the flow of a persistent electric current in a superconducting ring is the closest thing to a perpetual machine that we have. Therefore, room temperature superconductors (RTS) have long found their way into popular culture through science-fiction and cinema before entering into serious science. RTS, if achieved, could profoundly change the world scientifically and technologically as well.

2. A Practical Room Temperature Superconductor

I remember that when I was a graduate student in the late 1960s, I asked my thesis advisor, the late Professor Bernd T. Matthias, whether there existed a RTS and if yes, where. His answer was brief and direct "Yes, just go to the edge of the universe." At the time, the highest superconducting transition temperature (T_c) was 21 K found by him in the pseudoternary inter-metallic compound, $Nb_3(Al, Ge)$ [21] and the ambient temperature at the edge of the universe is 3 K, due to the cosmic microwave background radiation resulting from the residue of Big Bang. A 21 K superconductor is thus a room temperature superconductor at the edge of the universe. However the edge of the universe is more than 13 billion light years away from us, indeed an astronomical distance that can be reached by us only in our dream.

Therefore, strictly speaking, RTS is a relative term, depending on the environment of the superconductor. It denotes a superconductor with a T_c equal to or above the ambient temperature of the environment in which the superconductor is located and used. In principle, one can thus achieve RTS either by raising the T_c of a superconductor or by lowering the ambient temperature of the environment so that the two temperatures can meet. In this presentation, I shall focus on raising the T_c.

Over the years, various target T_cs have been set at different time, e.g. 77 K (liquid nitrogen boiling point), 100 K (inside the cargo bay of the space shuttle or on the moon's surface, opposite to the sun), 120 K (liquid natural gas boiling point), 148 K (liquid Freon boiling point), 198 K (dry ice temperature) and 300 K (temperature of our living environment). With the superconductors we have, many of these target temperatures have been reached. Unfortunately, they are still not readily practical for the ubiquitous applications of superconducting devices envisioned. For instant, $HgBa_2Ca_2Cu_3O_9$ with the current record T_c of 164 K [19] under high pressure can be considered a RTS in a liquid Freon environment, achievable by an air conditioner. However, the high pressure required renders it impractical not to mention the undesirable effect of Freon on the protective ozone layer in our upper atmosphere. If one could enhance the T_c to 198 K, the simple dry ice cooling would suffice. Unfortunately, for the environmentally conscious generation, such a superconductor is not acceptable due to the greenhouse carbon dioxide gas released from the dry ice.

As a result, the practical and desirable RTS that we want today is one that has a T_c of 300 K, high enough so that its superconducting state can be achieved in our living environment without the burden of using any cryogenic cooling. On the other hand, in order to take advantage of 90% of the maximum current-carrying capacity of a superconductor, the operating temperature usually should be kept at $\sim$ 70% of its T_c or lower. For an operating temperature of 300 K of our living environment, the T_c required therefore will be $\sim$ 430 K.

The discovery of such a superconductor will have an all-encompassing impact on our lives whenever we use electricity and a new industrial revolution will follow. According to our current theoretical understanding and experimental data, there exists no reason why RTS should be impossible. It is therefore not surprising to find that the 2006 DoE Workshop on Basic Science Needs for Superconductivity Report has identified the discovery of RTS together with the unraveling of the mechanism of HTS as two grand challenges in our future superconductivity research. Last summer, the US Air Force Office of Scientific Research, NATO and the Texas Center for Superconductivity at the University of Houston had jointly held in Norway a very successful "Workshop on the Road to RTS." For simplicity in later discussions, I shall take RTS to be possession of a T_c at or above 300 K.

3. Some Interesting Claims

The long and tortuous path in the search for superconductors with higher T_c has been dotted with triumphs of success and agonies of failure, including extravagant claims. Successes have been reported in various articles and books; and failures are too many to document and are often unreported. Being an optimist and a strong believer that whatever is not prohibited by the laws of physics will happen, I often give the benefit of the doubt to many claims of RTS, even the extravagant ones. I would not dismiss them outright until they are proven false by reasoning or by testing them experimentally to the best I can. Consequently, I have been contacted by many such claimers. Unfortunately, so far, to find them not superconducting is the norm let alone superconducting at room temperature. Examples are abundant, but let me just cite just a few for amusement. A few years ago, a California based company (which may still exist) that was raising capital to commercialize its alleged RTS based on a modified polymer material supposedly developed in the former Soviet Union with published references. The head of the company contacted me and asked me to test the materials. I did but found them to be insulating. In another example, the anchorwoman of a reputable TV program asked me to test a piece of material which was allegedly to have been left by an extraterrestrial vehicle in an Arizona desert and determined by a few reputable labs to be superconducting at room temperature in the presence of a strong microwave. I found it to be a rather ordinary metal containing elements such as Fe, Mn, In etc. and not superconducting. Yet another example was a call several years ago from a Croatian physicist who asked me to sign a disclosure agreement before faxing me information about his RTS claim. After seeing his faxed message, I had some doubts but still repeated what was described in the message. When told of our negative test results, he attributed them to our alleged poor sample quality. Still giving him the benefit of doubt, I asked for a piece of his sample, but he wanted me to purchase it when he started to mass produce it and put it on the market before Easter the following year. Unfortunately I have yet to hear from him since that Easter.

While many of the claims can be ignored outright due to experimental artifacts or dismissed after cursory examinations, others are interesting and may be worth further investigation. Revisiting these may lead to new understanding of HTS and new mechanisms for HTS, in addition to higher T_c. I shall briefly describe four of these other reports of very high T_c, occasionally exceeding 300 K below:

(1) In 1946, Ogg reported [22] a large drop in resistance and the presence of persistent current in their sodium-ammonia solution at temperatures as high as 180 K upon rapid cooling but not on slow cooling. He attributed the observations to superconductivity. According to him the superconductivity detected was due to Bose-Einstein condensation of Bosons of electron-pairs in vacancies that resulted from dissolution of Na in NH_3 [23]. It should be noted that Ogg was the first to suggest electron pairing for superconductivity. Later in the same year, conflicting reports appeared [24]: some disputed Ogg's results, some saw resistance drop but not to zero, yet another found corroboration for Ogg's observation. Interest in this material system resurged [25] in the earlier 1970's mostly in the former Soviet Union but waned in the late

1970's. However, the topics has been picked up by some recently in the West [26]. Unfortunately, the issue concerning the suggested existence of superconductivity in the solution remains unresolved. On the other hand, the phase separation takes place in the Na-NH_3 solution separating the metallic from the insulating ones, a reflection of the presence of some kind of instabilities in the system that are reminiscent of all high temperature superconductors known today. If electron pairs do form in the vacancies in the insulating NH_3-background prior to the onset of phase coherence as initially proposed by Ogg, this may be another similarity to what takes place in the HTS cuprates below the pseudo-gap temperature, not to mention the possible pairing in real space proposed in cuprates. Given the tremendous improvement in experimental techniques today compared with those used by Ogg and his contemporaries and the possible resemblance between the known HTSs and the Na-NH_3 solution, a revisit may be warranted.

(2) A large drop of resistivity associated with a small ac diamagnetic susceptibility shift in CuCl at $\sim$ 40 kbar and 300 K was reported in 1975 [27]. A series of pressure-induced phases was also proposed, although without any X-ray diffraction structural change under pressure to 90 kbar [28]. Later in 1978, Brandt et al. claimed the observation of superconductivity up to 170 K in CuCl under pressures based on the large resistive and magnetic anomalies detected and attributed the observation to electron pairing via the exchange of excitons [29]. This was in contrast with the earlier failed search for the metallic phase in CuCl under pressure in a thermal equilibrium condition below room temperature [28]. However, for samples in their thermally non-equilibrium state during rapid warming, an ac susceptibility anomaly above 90 K over a temperature range of 10-20 K was detected [30], corresponding to a paramagnetic-diamagnetic-paramagnetic transition. The magnitude of the diamagnetic shift was estimated to be about 7% that of a bulk superconductor. In the temperature region where the diamagnetic shift occurred, the resistivity was also observed to undergo a sharp decrease. The simultaneous transient appearance of the resistive and magnetic anomalies upon rapid warming is rather intriguing and consistent with, although not proof of, a possible superconducting transition in a minute part of the sample. Unfortunately, no definitively confirmation or refutation has been reported. Ginzburg proposed that a possible new state of superdiagmanetism might have existed in CuCl [31]. The transient nature of the anomalies reported [30] may re-

sult from a temperature induced strain/stress relaxation which in turn gives rise to a disproportionation of the CuCl compound, creating a rather complex metastable metal/insulator or insulator/insulator composite material system rather different from its simple parent CuCl that was originally examined. The proposed role of interfaces in these composite material systems for superconductivity is encouraged by the recent observation of superconductivity at the interface of two insulating compounds [32]. Disproportionation also represents the presence of instabilities that commonly occur in high T_c and related oxide materials. Furthermore, given the important role of Cu in cuprate HTS, reexamination of CuCl may be warranted. (I still remember that Phil Anderson dropped me a short note "Cu rises again!" immediately after the discovery of the 93 K superconductivity in YBCO was announced in 1987.) To develop a proper experimental environment to simulate the transient effects induced by rapid warming of CuCl under pressure, one may be able to turn the transient anomalies into steady ones for definitive characterization. For example, one may be able to simulate the pressure gradient induced by the rapid warming by steady physical means.

(3) Before 1986-87, physicists in the field of superconductivity research were very pessimistic about high T_c due to the long stagnation of T_c at 23.2 K coupled with the prevailing belief of the existence of the then-theoretical T_c-ceiling of 30's K [20]. I still remember the extremely emotional burden in writing the YBCO paper to make claim of the first 93 K superconductivity [3]. Even after the acceptance of the paper, I continued to repeatedly ask myself and my students "Could there be phenomena other than superconductivity that are able to account for our observations? Please think and think hard," knowing that a mistake of this magnitude could send me into life time exile from my superconductivity career. However, the discovery and the immediate confirmation of YBCO with a T_c at 93 K in the early 1987 drastically changed the mental state of people in the field. Physicists were suddenly in a state of euphoria and became extremely bullish about T_c, thinking that only the sky was the limit. For example, J. T. Chen et al. reported in May 1987 the observation of a two-step resistive transition in the Y-Ba-Cu-O compounds with one transition beginning at 240 K [33]. They attributed the transition to superconductivity in the yet-to-be identified unstable Y-Ba-Cu-O compound in its granular form based on the ac Josephson-effect they claimed to have detected. Many other

similar reports of anomalies indicating superconductivity at very high temperatures in cuprates appeared in the ensuing years. They include the observation of superconductivity above 200 K in BSCCO [34] and HBCCO [35]. All reports share some but not all of the following features: a sharp resistive drop (but not to zero), a diamagnetic shift (but small and superimposed on a large paramagnetic background), and a poor reproducibility (not just from lab to lab and from sample to sample, but also from run to run of the same sample). They definitely cannot satisfy simultaneously the four criteria I set in 1987 for a serious claim of superconductivity, i.e. zero resistivity, large diamagnetic shift decisively showing the Meissner effect, stability high enough for a definitive diagnosis and good reproducibility from sample to sample and from lab to lab. Therefore in 1987, I dubbed these fleeting anomalies at best Unidentified Superconducting Objects (USOs) to be parallel to the Unidentified Flying Objects (UFOs) or Unidentified Superconducting Anomalies (USAs) to be patriotic. Later Koichi Kitazawa told me that he had already coined the name USO for anomalies of this kind and that "uso" sounds like "a big lie" in Japanese. Some of the USOs have been shown to arise from experimental artifacts or from misinterpretation of the data, while the cause for others remains unclear. The high frequency of the reports, the similar high temperature range (between $\sim 240 - 300$ K) for the sightings of USOs and the many reputable labs from which the reports originated make these USOs too tantalizing to ignore. This is especially true for the low reproducibility of the USOs in view of the extremely unstable chemical nature of the cuprates, and for the possibly missing Meissner effect in view of the ever-decreasing of coherence length as the T_c increases.

(4) The great majority of the USOs reported have been in cuprates as described above. One exception is the report by Reich et al. in Jerusalem of the detection of a superconducting transition with a T_c of 91 K in the Na-doped surface of WO_3, as evidenced by the diamagnetic signal and the drop of resistance below the transition temperature [36]. Careful characterization of the samples and examination of the transition were performed later by groups in Jerusalem and Zurich, using the XDR, XPS, EPR, STM, MOI and STS techniques in the ensuing two years [37]. All observations appear to be consistent with the existence of superconductivity in the alleged Na-doped surface of WO_3. Unfortunately, all reports were done on samples from the same source, i.e. Reich's lab, and no other confirmations have appeared. Recognizing

the chemically sensitive nature of the Na/WO_3 interface, the difficulty in reproducing the observation is not unexpected. The implication of a high T_c in this non-cuprate material system would be profound, if its existence is proven and the possibility of contamination by cuprate HTS is excluded.

4. Some Visionary Predictions

The discovery of YBCO in 1987 changed the outlook for higher T_c. Unfortunately T_c has stopped rising since 1994 and the feeling of gloom and doom has started to return in recent years. This seems to have been epitomized by the recent appearance of the article by Barth and Max titled "Mapping High Temperature Superconductors: A Scientometric Approach" [38]. According to their scientometric analysis of the time-dependence of the overall number of articles and patents and the time-variation of publications related to specific compound subsets and subject categories, they showed beautiful figures that predicted the death of the field of HTS between 2010 and 2015 by linear extrapolation. While people in database collection and research consider the analysis accurate and sound, many scientists point out the blindness of "bean counters" in scientific research. In fact by applying the same approach, many fields of science should have been dead long ago. For example, atomic physics and superconductivity would have been over long before Bose-Einstein condensation and HTS entered the stage of exciting research, scientific breakthrough goes beyond statistics. I often resonate with the statement by Mark Twain about figures– "there are three kinds of lies: lies, damned lies and statistics." Fortunately, Barth and Marx also pointed out that the situation can drastically change if a breakthrough were to take place, such as when superconductors are found to work at a higher temperature or in a new class of materials, or a theory is found to explain the HTS. To discover a RTS would definitely be such a breakthrough and more. Since the 1960s, some visionary theorists have already proposed that RTS may be achievable through a variety of schemes. A few examples are given below:

(1) In 1964, Little examined the question originally posed by London, whether superconductivity occurs in organic macromolecules within the framework of the BCS theory and concluded that superconductivity above room temperature is not only possible but also expected in organic macromolecules of a special design [39]. The proposed model macromolecule consists of a long spine and a series of side chains. The

long spine may or may not be a conducting system. By choosing the molecules in the side chains with proper oscillation of charges, the electrons in the spine can be polarized to form pairs via the exchange of excitons with the side chain molecules. According to Little's estimation of the matrix elements and density of states in his model polymer, a superconducting transition should occur at temperature well above room temperature from an insulating, a semiconducting or a metallic state of the spine. He also pointed out the challenges in establishing electric contacts for resistive measurements and in determining the Meissner effect. While superconductivity has been discovered [40] in organic metals often under pressure, the T_c remains low but in materials with structures different from that suggested by Little and due to different mechanisms. One of the challenges in realizing Little's vision is in synthesizing the macromolecules with the proposed design. With the great advancement in material synthesis and diagnostic techniques for nanomaterials made in recent years, an attempt to produce macromolecules including the model macromolecule proposed by Little may be worthwhile and timely. The suggested transition directly from a localized state to the superconducting state reminiscent of what takes place in cuprate HTS in the presence of a high field at low temperature [41] and of field-induced superconductivity in λ-$(BETS)_2FeCl_2$ is intriguing.

(2) In 1964, Ginzburg noticed the possible drawbacks of the one-dimensional organic macromolecule superconductor proposed by Little, such as fluctuations, instabilities and lack of Coulomb screening, and therefore focused instead on two-dimensional materials that might alleviate the impasse in part. He proposed that surface superconductivity could occur in the surface of a metal, especially when covered by a dielectric material and suggested a possible T_c of $10^2 - 10^3$ K after making a few estimations [42]. This is different from the surface superconductivity in the surface of a bulk homogeneous superconductor existing slightly above H_{c2} proposed by Saint-James and de Gennes a year earlier. Ginzburg wrote that $T_c \sim \Theta \exp(-1/g)$ with Θ being the characteristic temperature of the excitation energy spectrum responsible for electron pairing and g the effective attraction between electrons. He argued that for electron-phonon interaction the characteristic temperature is the Debye temperature Θ_D, which is $\sim 10^2$ K. In the weak coupling approximation of BCS, the maximum T_c was estimated to be below 40 K. He proposed a novel mechanism to enhance T_c within the weak limit of BCS, i.e. to increase T_c by taking advantage of the

high characteristic temperature Θ_e of the electron-electron interaction via the exchange of excitons, which can be as high as 10^5 K. By taking a realistic value of g $\sim 1/4$, Ginzburg obtained a $T_c \sim 1.8 \times 10^3$ K for a $\Theta_e \sim 10^5$ [43]. In order to facilitate the so-called exciton mechanism, he conjectured that one should consider the following material systems: metallic thin films, metal surfaces covered by dielectrics, metal sandwiches with dielectrics in between and two-dimensional layer compounds. Many experiments were done following the suggestion. For example, in the early 1970's, the layered compounds of transition-metal dichalcogenides were considered to be the natural candidate and extensive research was conducted. Unfortunately, the T_c remained low and the superconductivity observed could be explained by the conventional electron-phonon mechanism. In spite of the disappointing outcome, the study led to the discovery of charge density waves in 1973, which became one of the most studied topics in the ensuing decades [44]. It may not be surprising if there exists a connection between the layered materials Ginzburg proposed and the cuprates in which HTS was discovered a little more than three decades later in view of the common metal-semiconductor layer sequence present. For the two-dimensional materials proposed by Ginzburg to work, close coupling between the layers is needed and the metallic and semiconducting layers in cuparate HTSs happen to assemble themselves naturally. The new development in thin-film deposition may provide another avenue to synthesize these layers with better coupling to test the proposal.

(3) In 1968, Ashcroft proposed that metallic hydrogen could have a T_c of a few hundred degrees Kelvin based on the standard BCS formula [45]. Assuming that metallic monatomic hydrogen could be achieved by compressing hydrogen under high pressure, he pointed out that metallic hydrogen could be considered as an element of the alkaline metal group but rather different from other members of the group. While all alkali metal elements (except Li with a $T_c = 0.0004$ K) are not superconducting at ambient pressure due to the complete cancellation of the electron - phonon attraction by the electron - electron repulsion, metallic hydrogen has instead: a high Debye temperature Θ_D as the prefactor in the BCS formula due to its low ionic mass; a strong electron-phonon coupling λ due to the absence of the inner core structure; and a relatively large density of states N(0) at the Fermi energy due to its high electron density, all contributing to a high T_c proposed. By using reasonable estimated values of Θ_D, λ, N(0) and the Coulomb pseudo-

potential, Ashcroft obtained a lower bounded T_c on the order of 10^2 K. In the ensuing years, numerous analyses of metallic hydrogen with different sophistications point to the same temperature range for T_c. In 1997, Ashcroft and Richardson showed that the correlated fluctuations between electrons and holes in the metallic modification of diatomic hydrogen through band-overlap under high pressure will reduce the Coulomb pseudopotential and result in an enhanced T_c higher than its monatomic counterpart [46]. Recently, Ashcroft further suggested that metallic group IVa hydrides with hydrogen as the major constituent would be high T_c superconductors under pressure, following the similar approach that predicts metallic hydrogen to be a high temperature superconductor [47]. He also suggested that the group IVa hydrides have the possible added advantage of requiring a lower critical pressure to achieve their metallic state. It should be pointed out that, in spite of the extensive studies of hydrogen under pressures in the 100s GPa-range over the past few decades, the existence of the metallic state of hydrogen remains elusive, and superconductivity has yet to be detected. However, in view of the great advancements made in computational science and high pressure techniques in recent years, it is time to tackle this exciting problem head-on.

(4) Although several complex material systems were suggested for raising the T_c by taking advantage of the possibly high characteristic temperatures of their exciton energy spectra after Ginzburg's 1964 proposition, detailed analyses of the feasibility of superconductivity due to the exciton mechanism were lacking. In 1973, Allender, Bray and Bardeen had carried out a more rigorous analysis on a simple model system to explore the existence of the exciton mechanism and its impact on T_c, if it exists [48]. Their model material consists of a metal thin film of Fermi energy E_F on a semiconductor of a narrow gap of E_g. When the metal thin film is in perfect contact with the semiconductor to form a chemical bond at the interface and the Fermi level of the metal thin film lies in the middle of the semiconductor gap, the electron wave function will have maximum penetration into the semiconductor without localization or the detrimental effect due to band bending in the semiconductor. These electrons that tunnel into and spend enough time inside the semiconductor gap will form pairs via the exchange of excitons. They concluded that T_c can indeed be enhanced by adding the exciton mechanism to the phonon mechanism when the material conditions are optimized and that the exciton mechanism is a promising vehicle for higher T_c. While

the exciton mechanism may be as promising as they have shown, they also point out the difficulty in creating the model material system with the proper metal/semiconductor interfaces that satisfy the stringent requirements so that the full effectiveness of the exciton mechanism can be realized. Several experimental attempts had been made until the late 80s but with no success reported. With the recent advancement in multi-thin-film synthesis, one should be able to artificially prepare samples satisfying the stringent criteria specified for the exciton mechanism, Naturally occurring metal/semiconductor layered compounds may be a good alternative if the E_F is adjusted to be located in the middle of E_g.

5. Common Features of Superconductivity with High T_c

Many superconductors of various material families have been found: some have higher T_c than others and some are easier to be shaped into practical forms for applications than others. However, superconductors with a relatively high T_c appear to share some common features, independent of the compound families to which they belong.

At the moment, we do not yet know enough about a RTS other than that it has a $T_c > 300$ K, as we have chosen, to ask intelligent questions with answers to which will lead us to the promised land of RTS. It is also not unlikely that a RTS to be discovered may turn out to be a very different material from the HTSs to which we are accustomed to. However, being a superconductor, RTS has to have the basic superconducting characteristics and very likely has to share at least some features with the HTSs we have. A review of these features is considered most helpful and can serve as a launching pad for our attempt to look for RTS:

(1) Electron-pairing and phase-coherence

According to the BCS theory, electron-pairing and phase-coherence are the very basic features of a superconducting state that exhibits zero resistivity ($\rho = 0$) and perfect diamagnetism (magnetic induction $B = 0$). Electrons in the presence of an attraction, no matter how small, will form pairs, lower the energy and result in an energy gap (Δ) immediately below the Fermi energy [40], although nodes or lines of nodes occur in superconductors with non-s symmetric pairing orders. Electron pairing takes place in the k-space via the virtual exchange of bosons, although in cuprate HTS, pairing of electrons in the real-space has also been proposed [50]. When the wave functions of the electron pairs overlap, phase coherence is established and the compound undergoes a bulk phase transition to the macroscopic superconducting state at T_c. It has been long accepted that electron-pairing and phase-coherence occur at the same temperature, T_c for the conventional low temperature superconductors. However, suggestions have been made that electron-pairing may proceed phase-coherence at a temperature higher than T_c in the cuprate high temperature superconductors [51]. If true, the feature provides a new degree of freedom for the search for superconductors of higher T_c.

(2) Strongly correlated electron systems

The current HTSs with a T_cs above 77 K are cuprate oxides with a strong interaction between electrons due to the incompletely filled 3d-shell. They can therefore be considered to belong to the class of materials of transition metal oxides where strongly correlated electrons exist. This is further evidenced by the failure of band calculations, which neglect the correlations between electrons, to predict the physical properties of the undoped parent compound of the 214 HTS, La_2CuO_4. It is an antiferromagnetic insulator with a sensitive dependence on doping, in contrast to predictions by band calculations. Strongly correlated electron systems are rich in electronically induced phase transitions with a wide range of transition temperatures and therefore may provide various avenues to HTS. Figure 5 shows schematically the linkages between magnetism and superconductivity and between magnetism and ferroelectricity in strongly correlated electron systems. In contrast to the original thinking that magnetism is detrimental to superconductivity, we now find that magnetism can help superconductivity [52] and that magnetic field can induced superconductivity [53]. Similarly, magnetism and ferroelectricity had been considered to be antithetic to each other until recent experiments showed that they can help each other under the proper conditions [54]. I believe that if the missing link between ferroelectricity and superconductivity is found, higher T_c can be achieved.

(3) Instabilities

In the strongly correlated electron systems, there exist various electronically or phonoically induced transitions of different kinds, e.g. superconducting, structural, antiferromagnetic, ferromagnetic, ferroelectric, antiferroelectric, metal-insulator, charge-density-waves, spin-density-waves, charge-order, orbital-order, etc. [55] They arise from

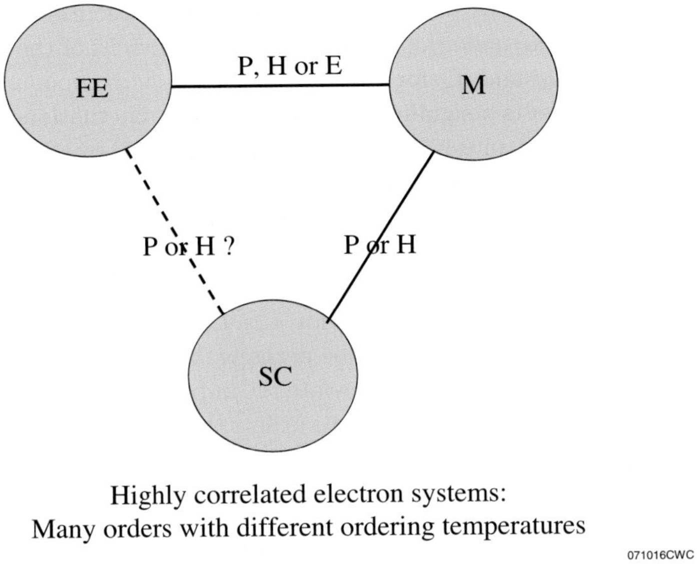

Fig. 5. The relationship between superconductivity, magnetism and ferroelectricity.

instabilities due to the proximity of their Fermi levels to the singularities in their respective energy spectra, such as phonon, electron, spin, exciton and charge. The dominant instability will win over the weaker ones and determines the nature of the transition. The transition can be induced by variation in temperature, pressure, chemical doping, magnetic field and/or electric field. In other words, there can be more than one type of interactions in a solid and the strengths (absolute or relative) of these interaction can be adjusted by changing the above physical or chemical parameters (see e.g. Fig. 5).

Apparently, superconductors with high T_c are intrinsically unstable, physically or chemically due to the strong attractive interaction involved whether they are intermetallic or non-intermetallic. A case in point is the A15 system of inter-metallic compounds with relatively high T_cs prior to 1986, such as V_3Si (17 K) and Nb_3Sn (18 K), which undergo a structural transformation at T_m due to the softening of the phonons prior to entering the superconducting state on cooling. When pressure brings T_m closer to T_c, T_c increases as for the case of V_3Si; whereas when T_m is moved away from T_c by pressure, T_c decreases as in the case of Nb_3Sn, demonstrating a significant positive role of phonons in superconductivity of these compounds [56]. HTS cuprates exhibit a magnetically driven pseudogap opening at T_p that decreases while the T_c increases in the underdoped region [57], reminiscent of the relationship between T_m and T_c for the A15 intermetallic compounds under pressure. It suggests a significant role of magnetic fluctuations in HTS. In addition to the physical instabilities discussed above, instabilities can be chemical in nature. For example Nb_3Ge cannot achieve its optimal Nb:Ge = 3:1 stoichiometry in bulk synthesized under the ambient condition. A similar situation occurs for the HTS cuprates, i.e. YBCO can loose its oxygen easily, and doping in the $HgBa_2Ca_{n\text{-}1}Cu_nO_{2n+3\text{-}\delta}$ tends to reduce n for $n > 3$, while for $n > 4$ high pressure synthesis is required usually. Instabilities will be perhaps the most serious challenge for achieving RTS. One way to alleviate the impasse in part is to develop a complex material structure and to utilize the extreme conditions.

(4) Fluctuations

As has been pointed out above, HTSs are strongly correlated electron systems that can possess interactions of different nature. Instabilities in the energy spectra of these interactions result in transitions of different types. Near the phase transition, fluctuations set in. In many compounds, instabilities associated with a transition result in fluctuations prior to its formation of a long-rage order, such as magnetic-order, charge-density-waves, spin-density-waves, or charge-order. These fluctuations may become a source of electron pairing or superconductivity, especially when the superconducting transition gets close to the transitions for the above orders to take advantage of the associated fluctuations. Various experiments have shown that many of these interactions or transitions are tunable by both physical and chemical means. They can be coupled to one another with one being able to enhance or suppress the other. Such a coupling becomes most effective through the fluctuations near the phase transition. For instance, the soft phonon modes associated with a structural transition have been observed to enhance superconductivity as evident, for example, from the detection of higher T_c in the A15 intermetallic compounds [56] when the structural and superconducting transitions that occur simultaneously in these compounds are brought closer to one another by pressures, and from the observation of the highest T_c near the alkaline-doping-induced metal-insulator transition in WO_3 [58]. The ever- presence of magnetic fluctuations associated with the antiferromagnetic transition that takes place in the undoped parent compounds of cuprate HTSs

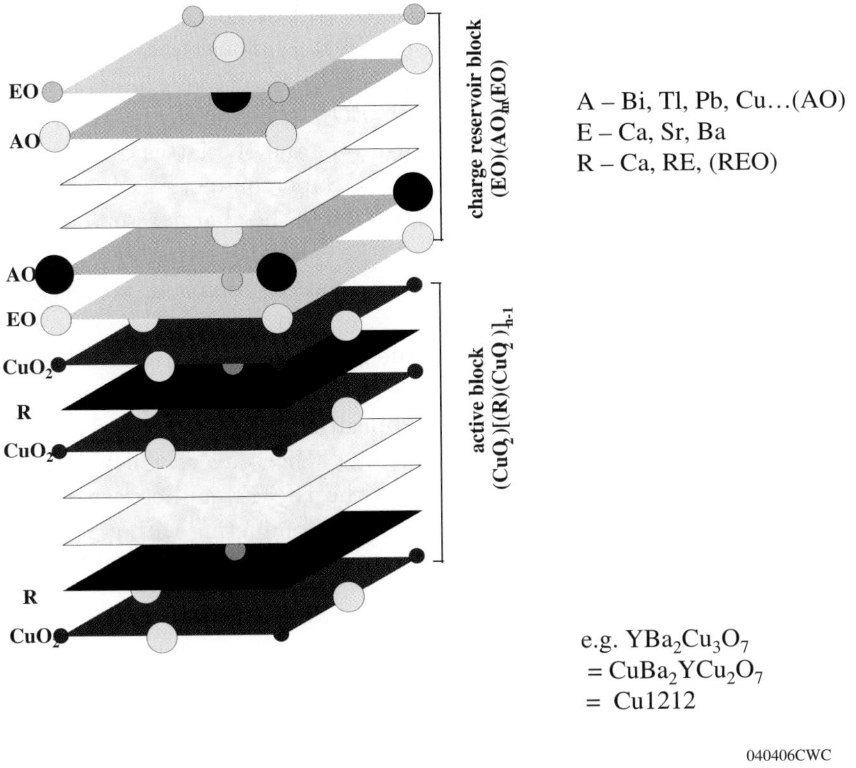

Fig. 6. The layered structure of HTS cuprates.

or associated with the spin-density-waves transition that occurs in the insulator parent of the organic superconductors have strongly demonstrated the significant positive role of magnetic fluctuations in superconductivity.[52]

(5) Layered structure with two different blocks

Reduced dimensionality has been shown to facilitate the enhancement of T_c, generally due to the relatively large density of states and the stronger electron-electron interaction associated with a 2D system. For instance, all HTS cuprates exhibit a layered structure as shown in Fig. 6 [59]. They can be represented by a generic layered formula $A_mE_2R_{n-1}Cu_nO_{2n+m+2}$, where A = Bi, Tl, Pb or Cu; E = Ca, Sr or Ba, and R = Ca or a rare-earth element.

The generic formula can be rewritten as $A_mE_2R_{n-1}Cu_nO_{2n+m+2}$ = $[(EO)(AO)_m(EO)]+[(CuO_2)R_{n-1}(CuO_2)_{n-1}]$, consisting of two main blocks, namely the active block $[(CuO_2)R_{n-1}(CuO_2)_{n-1}]$ and the charge reservoir block $[(EO)(AO)_m(EO)]$. The former consists of n (CuO_2)-layers interleaved by n-1 R-layers and the latter comprises m (AO)-layers bracketed by 2 (EO)-layers. Superconducting current flows in the CuO_2-layers in the active block while the charge reservoir block enables the introduction of carriers to the CuO_2-layers without affecting the layer-integrity (also known as modulation doping in multilayered structures of semiconductors).

The above appears also to be true for transition metal nitrides. For example, ß-HfNCl has a structure related to the $CdCl_2$-type with the [Hf-N-N-Hf] layers occupying the Cd-positions and sandwiched between the loosely packed Cl-layers. When $Li_x(THF)_y$-layers which act as a charge-reservoir, are intercalated into the HfN-layers between the Cl-layers, ß-HfNCl/$Li_x(THF)_y$ becomes superconducting with a T_c reaching 25.5 K.[60] This is in strong contrast to the T_c = 8.8 K of the cubic HfN.

(6) Near the Metal-Insulator phase boundary, low carrier density and high degree of covalence

The metal-insulator transition represents one of the electronically or phononically induced extreme instabilities associated with HTS as discussed above. For instance, superconductivity evolves from an antiferromagnetic Mott insulator for the cuprates with a T_c up to 134 K, from a charge-density-wave insulator for $Ba_{1-x}K_xBiO_3$ and $BaPb_{1-x}Bi_xO_3$ with a T_c up to 30 K and 13 K,[61] respectively, from a yet-unknown insulating phase near $x \sim 0.1$ and for $Li_{1+x}Ti_{2-x}O_4$ with a T_c up to 14 K [62] and from a ferroelectric insulator for A_xWO_3 with a T_c up to 7 K, where A = alkaline elements [58]. A similar situation is found in the organic salts when the spin-density-wave (SDW) gap is quenched by pressure they become superconducting. [59] Although the T_cs of the non-cuprates is not high compared with those of the cuprate HTSs, they are high within their own material groups. As a result, the carrier concentrations of the HTS and related compounds are low, and the degree of covalency is high for carriers in the conduction band. As a precursor with doping, these superconducting compounds usually exhibit a large temperature-independent Pauli susceptibility at room temperature indicative of a large density of states near the Fermi surface.

(7) Mixed valence

Mixed valence states are often found in superconductors with a relatively high T_c. The appearance of mixed valence states in a compound also represents an example of the electron-induced instabilities discussed earlier. This is evident, e.g. in (Cu^{+2} & Cu^{+3}) in cuprate HTSs, (Bi^{+3} & Bi^{+4}) in $Ba_{1-x}K_xBiO_3$ and $BaPb_{1-x}Bi_xO_3$, (Ti^{+3} & Ti^{+4}) in $Li_{1+x}Ti_{2-x}O_4$ and (W^{+4} & W^{+6}) in A_xWO_3. It is interesting to note that doping into the Ba-site of $BaBiO_3$ by K gives a maximum T_c of ~ 34 K while doping into the Bi-site by Pb leads to a maximum T_c of only 13 K. This observation suggests that in addition to changing the carrier concentration, doping in $BaPbBiO_3$ must affect other factors that are important to T_c. Determining the cause of this observation will also provide insights into the occurrence of high T_c in oxide superconductors.

(8) High polarizability

The crucial role of the active block of HTS cuprates has long been recognized. However, different cuprate systems with identical active blocks but different charge reservoir blocks display different T_cs. For example, the maximum T_c's of $HgBa_2Ca_2Cu_3O_{9-\delta}$ and $Bi_2Sr_2Ca_2Cu_3O_{10}$ are 134 K and 115 K, respectively, although both have the same active block of $[(CuO_2)Ca(CuO_2)Ca(CuO_2)]$. The charge reservoir blocks must play a role in their T_cs. It is interesting to note that HTS cuprates with Ba in their charge reservoir blocks, i.e. $[(BaO)(AO)_m(BaO)]$, usually exhibits a higher T_c than their iso-electronic and iso-structural counterpart with Sr containing reservoir blocks, i.e. $[(SrO)(AO)_m(SrO)]$. For instance, $YBa_2Cu_3O_7 = [(BaO)(CuO)(BaO)]+[(CuO_2)Y(CuO_2)]$ possesses a $T_c \sim 93$ K whereas $YSr_2Cu_3O_7$ displays a $T_c \leq 60$ K. A similar observation is found in ferroelectrics, e.g. $BaTiO_3$ becomes ferroelectric at a Curie temperature $T_C = 393$ K while $SrTiO_3$ remains non-ferroelectric down to 1 K in its unstrained state. This has been attributed to the larger polarizability of the Ba-ion. It may not be surprising to see a similar reason for the case of superconductivity although it has not yet been included in any theoretical model. Phenomenologically, a general correlation between superconductivity and ferroelectricity has been previously suggested by Jim Phillips.

(9) Magnetic Interactions

Many have demonstrated that the conventional electron-phonon interaction derived from the normal state properties of the cuprates is not sufficient to produce a T_c as high as those observed in these superconductors. Magnetic interactions have therefore been proposed. This is certainly consistent with the fact that all superconductors known to date with a T_c above 77 K, liquid nitrogen temperature, are cuprates whose Cu^{2+}-ions with a spin $S = {}^1/_2$ may provide the suggested antiferromagnetic fluctuations needed for the attractive interaction for the high T_c. The question whether the Cu^{2+}-ion with $S = {}^1/_2$ is synonymous to a high temperature superconductivity remains unanswered. Given the ubiquitous nature of superconductivity and magnetism, it is not unthinkable that high temperature superconductivity can be found in materials containing other types of magnetic ions with possibly different S that can generate the needed ferromagnetic fluctuations.

6. The Enlightened Empirical Approach

In philosophy, "Empiricism or Rationalism" has been debated among scholars for a long time. The difference between the two lies mainly in the different relationships between reason and knowledge and between experience and knowledge: whether reason or experience is the source of knowledge and whether knowledge can be gained independent of experience. Empiricists consider experience to be the source of knowledge and to be independent of knowledge, while rationalists claim that reason is the source of knowledge and is independent of experience. In the long search for superconductors of higher T_c, Matthias, Geballe and Mueller reminded us at different times of the effectiveness of the empirical approach.

I think the most effective empirical approach toward RTS is better conceptualized as an enlightened empirical approach that embodies both experience and reason. Looking back at the history of superconductivity or of science in general, the line has always been blurred between empiricists and rationalists, or between Edisonians and Einsteinians, i.e. between experimentalists and theorists. Synergistic effect between the two has long been recognized as was well illustrated, for example, by Bardeen in the development of the BCS theory [J. Bardeen, *Impact of Basic Research on Technology, ed. B. Kursunoglu and A. Perlmutt (Plenum Press, 1973) p.15]*. Even when Matthias said "don't listen to theorists," he reminded people that "I do have good theorist friends."

In Sections 3, 4 and 5, I have summarized what we have learned from past experiments and theories. They can serve as a guide to project enlightened-empirically into the future paths that may be most fruitful to achieve HTS with higher T_c or even RTS. The summaries include some of the interesting claims that may warrant further tests, some visionary predictions that may be pursued and the common features of superconductors with high T_c that may provide directions for the search for novel superconductors. The worldwide extensive studies on HTS cuprates and related materials over the last two decades have given us unusual physical insight to materials, powerful computational capabilities for materials, sensitive material characterization techniques and new material synthesis tools. Time is ripe to bring all these skills to bear in the search for novel superconductors with higher T_c, preferably at room temperature. If history is any guide, I strongly believe that, in the process, in addition to higher T_cs, new physics will be discovered and novel materials found with both scientific and technological significance.

While the exact path to be taken depends on the style and taste of the practitioner, to paraphrase of Professor Yang's statement on doing physics, I would like to list some of my thoughts below:

(1) Some interesting claims that may be worth revisiting

- The Na-NH_3 system reported to display persistent current suggestive of superconductivity at temperatures up to 180 K may bear a certain possible resemblance to the cuprate HTS system, namely, phase separation, pseudo-gap and real-space pairing. Given the tremendous improvement in diagnostic tools and experiment-controls today, a revisit may be worthwhile.
- The large resistive and diamagnetic anomalies suggestive of superconductivity at temperatures up to 160 K reported in CuCl during rapid warming reported may be associated with some of the common features of HTS, such as instability, disproportionation, and interface effect. The role of Cu in the cuprate HTS era is particularly intriguing. It will make easier a definitive diagnosis of the nature of the anomalies by developing a technique to transform the transient nature of the anomalies to a steady one.
- The frequent report of resistive anomalies in cuprates suggestive of superconductivity in the similar temperature range of 230-350 K by different reputable labs in different countries is too tantalizing and does not seem to be pure coincident. With the recent development of ultra-sensitive diagnostic tools such as microwave spectrometry and of powerful synthesis technique, a systematic revisit is worthwhile.
- The resistive, magnetic, and microwave anomalies and magneto-optical imaging data suggestive of superconductivity at temperatures

up to 91 K reported in WO_3 surface-doped by Na remain a puzzle. In spite of the relatively low temperature of the anomalies, the system appears to be simple and represents the first non-cuprate system to display such a high T_c, if proven to be superconducting and not due to contamination by cuprate HTS. Controlled surface doping can be deployed to resolve the puzzle easily.

(2) Some visionary predictions that may be worth testing

- The organic macromolecules that consist of a long spine and a series of side chains were predicted to superconduct above room temperature. They may be designed using powerful computational skills, and synthesized and characterized by advanced techniques acquired over the last two decades. At the same time we should keep in mind how to overcome several serious challenges for a one-dimensional electron system: i. the inherent fluctuations associated with a one-dimensional system that prevents long-range order and a coherent phase transition; ii. the inherent Peierls instability that will result in the opening of a gap and a transition to an insulating state; and iii. the lack of screening of the Coulomb repulsion that will overwhelm the attractive interaction for electron pairing.
- The surface superconductivity at high T_cs above room temperature was predicted to take place in the surface of a metal, especially when covered by a dielectric material. This may be realized today by the powerful molecular beam epitaxial technique now available to grow samples with desired interfaces in a controlled atmosphere available.
- Superconductivity with a T_c above room temperature was predicted to take place in light ionic mass materials such as H, Li and group IVa hydrides with H as a major constituent under very high pressures. With the advancement in ultrahigh pressure and characterization techniques, the theoretically predicted conditions are now within reach of experiments.
- The interfacial superconductivity with enhanced T_c through the exchange of excitons was predicted to take place in a materials system that consists of a metallic thin film on a semiconductor with Fermi energy E_F and energy gap E_g, respectively. The stringent requirements on the coupling between the metal film and the semiconductor and on the E_F-E_g relation can now be achieved with advanced experimental and computational techniques.

(3) Common features of superconductivity with high T_c as a guide

Besides what has been pointed out in the above two sub-sections, I shall list the general directions that are most likely to yield results in our search for novel superconductors with higher T_c or even RTS.

- Pay attention to layered strongly correlated two-dimensional electron systems displaying multiple interactions and instabilities, and seek ways, physical and chemical, to optimize the multiple interactions and to control the instabilities.
- Pay particular attention to layered strongly correlated material systems that exhibit antiferromagnetic or ferromagnetic fluctuations and strong covalent bonding.
- Pay attention to multi-component layered strongly correlated material systems.
- Pay attention to multi-scale material systems.
- Examine materials that may show the appearance of psudo-gap at high temperature signaling the onset of electron pairing.
- Try to pay more attention to noncuprate layered materials systems
- Try to design and synthesize layered inorganic/organic hybrid systems.
- Develop conventional or nonconventional doping techniques for the above systems.
- Try to take advantage of pressure, fields, and chemical and physical means in the search.
- Improve the cuprate HTS.

7. Conclusion

To date, there exists neither theoretical nor experimental evidence to exclude the existence of novel superconductors with a T_c exceeding room temperature. Due to the past two decades of extensive work on HTS worldwide, we have acquired better insight into materials, achieved powerful material computational methods and developed powerful materials characterization and diagnostic tools, and novel material synthesis techniques. What we have learned in the study of HTS over the last 20 years may now bear fruits in our search for novel superconductors of higher T_c, especially in layered strongly correlated electron systems with multi-subcomponents and magnetic fluctuations. By using different pressure, fields, and chemical and physical means of tuning, I believe that whatever the law of physics do not say will not happen, e.g. RTS, will happen.

Acknowledgments

I would like to express my deep appreciation to all my hard working colleagues in the Texas Center for Superconductivity at the University of Houston and those in other laboratories in different continents for discussions and inspirations over the last decades. The work in Houston is supported in part by the T. L. L. Temple Foundation, the John J. and Rebecca Moores Endowment, the United States Air Force Office of Scientific Research, and the State of Texas through the Texas Center for Superconductivity at the University of Houston; at Lawrence Berkeley National Laboratory through the US Department of Energy.

Note added: After the delivery of the above presentation in Singapore in October 2007, the exciting news of the discovery by Hosono et al. in Japan of a new class of superconductor at a relatively high temperature reached me late February 2008. Guided by the rule that high temperature superconductivity usually occurs in the strongly correlated electron layered systems with magnetic fluctuations as the case of cuprates, they examined the layered rare-earth transition-metal oxypnictides (ROTPn, where R = rare-earth element, T = transition-metal element, and Pn = pnictogen) and discovered superconductivity in several of these compounds with a T_c as high as 26 K in the doped La(O,F)FeAs. The presence of the large concentration of magnetic Fe appears to be consistent with the suggested important role of magnetism in superconductivity with high T_c. In the few weeks following the news, the T_c was raised to 52 K by replacing La by rare-earth elements of smaller radii. Pressure was also found to raise the T_c of La(O,F)FeAs. Euphoria permeated in the community such that, as far as T_c is concerned, only the sky would be the limit. By carrying out a systematic pressure study and analyzing the data, we concluded that the maximum T_c of the doped ROFeAs (R1111) can only be in the 50s K and almost independent of R, similar to the cuprates (R123). Later a similar layered A'Fe_2As_2 (A'122) where A' = alkaline-earth element Ba or Sr was found superconducting with a T_c up to 38 K when it is hole-doped by partial replacement of A' by alkaline metal A = K or Cs. Subsequently, we found that K122 and Cs122 were superconducting at 2.8 and 3.7 K, respectively and form complete phase diagrams of (A,Sr)Fe_2As_2 with A = K and Cs, showing that a continued evolution from a superconducting state at A122 to a spin-density-wave (SDW) state at A'122 with an intermediate region where the SDW state coexists with the superconducting state through continuous electron doping by partial replacement of A with A'.

Later, yet another FeAs-layered compound system AFeAs (or A111) with A = Li and Na was discovered by us to be superconducting at 20 and 17 K, respectively. The above three new compounds, R1111, A'122 and A111, can be considered to form a homologous compound series with FeAs-layers, similar to the cuprates with the CuO_2-layers. A111 may then be the equivalent of the infinite-layered cuprate, $SrCuO_2$. We have therefore recently proposed that higher T_c may be achievable by designing and synthesizing more complex compounds that consist of TPn-layers.

References

1. J. Bardeen, L.N. Cooper and J.R. Schrieffer, Phys. Rev. 106, 162 (1957)
2. T.D. Lee and C.N. Yang, PR105, 1671 (1957
3. M.K. Wu *et al.*, Phys. Rev. Lett. 58, 908 (1987)
4. J.G. Bednorz and K.A. Müller, Z. Phys. B 64, 189 (1986)
5. Sir Isaac Newton, Philosophiæ Naturalis Principia Mathematica, 1686
6. H.K. Onnes, Leiden Comm. 120b, 122b, 124c (1911)
7. J.R. Gavaler *et al.*, J. Appl. Phys. 45, 3009 (1974); L.R. Testardi *et al.*, Solid State Comm. 15, 1 (1974)
8. Tokyo, Houston and Beijing groups
9. C.W. Chu and L.R. Testardi, Phys. Rev. Lett. 32, 766 (1974); C.W. Chu, Phys. Rev. Lett. 33, 1283 (1974); C.W. Chu and V. Diatschenko, Phys. Rev. Lett. 41, 572 (1978)
10. C.W. Chu *et al.*, Phys. Rev. Lett. 58, 405 (1987)
11. C.W. Chu *et al.*, Science 235, 567 (1987)
12. W.L. McMillan, Phys. Rev. 167, 331 (1968)
13. R.J. Cava *et al.*, Phys. Rev. Lett. 58, 408 (1987)
14. R.M. Hazen *et al.*, Phys. Rev. B (RC) 35, 7238 (1987)
15. P. H. Hor *et al.*, Phys. Rev. Lett. 58, 1891 (1987)
16. H. Maeda *et al.*, Jpn. J. Appl. Phys. 27, L209 (1988)
17. Z.Z. Sheng and A.M. Hermann, Nature 332, 138 (1988)
18. A. Schilling *et al.*, Nature 363, 56 (1993)
19. L. Gao *et al.*, Phys. Rev. B (RC) 50, 4260 (1994)
20. For a review, see for example Superconductivity, by C.P. Poole Jr., H.A. Farach, R.J. Creswick and R. Prozorov, Academic Press (2008); - Appl.
21. B.T. Matthias, T.H. Geballe, L.D. Longinotti, E. Corenzwit, G.W. Hull (1966)
22. R.A. Ogg, Phys. Rev. 69, 243 (1946)
23. R.A. Ogg, J. Am. Chem. Soc. 68, 155 (1946)
24. H.W. Boore *et al.*, Phys. Rev. 70, 72 (1946); J.G. Daunt *et al.* Phys. Rev. 70, 219 (1946); and J.W. Hodgins, Phys. Rev. 70, 568 (1946)
25. D.A. Kirzhnitz and Yu. V. Kopaev, JETP Lett. 17, 270 (1973); I.M. Dmitrenko and I.S. Shchetkin, JETP Lett. 18, 292 (1974)
26. For example, P. Edwards *et al.*, Chem. Phys. Chem. 7,2015 (2006)
27. A.P. Rusakov *et al.*, Phys. Status Solid B75, K191 (1975)

28. C.W. Chu *et al.*, J. Phys. C: Solid State Phys. 8, L241 (1975); A.P. Rusakov *et al.* Sov. Phys.: Solid State 19, 680 (1977)
29. N.B. Brandt *et al.*, JETP Lett. 27,37 (1978)
30. C.W. Chu *et al.*, Phys. Rev. B18, 3116 (1978)
31. V.L. Ginzburg, S. S. Comm. 50, 339 (1984)
32. N. Reyren *et al.*, Science 317, 1196 (2007)
33. J.T. Chen *et al.*, Phys. Rev. Lett. 58, 1972 (1987)
34. M. Lagues *et al.*, Science 262, 1850 (1993)
35. J.L. Tholence *et al.*, Phys. Lett. A184, 215 (1994)
36. S. Reich and Y. Tsabba, Eur. Phys. J. B9, 1 (1999)
37. See for example, S. Reich *et al.*, J. Superconductivity: Incorporating Novel Magnetism 13, 855 (2002)
38. H. Barth and W. Marx, Cond.-Mat/0609114 (2006)
39. W.A. Little, Phys. Rev. 164, A1416 (1964)
40. D. Jerome *et al.* J. Physique Lett, 41, 95 (1980); M. Lang and J. Müller, Cond.-Mat/0302157 (2003)
41. Y. Ando *et al.*, Phys. Rev. Lett. 77, 2065 (1996)
42. V.L. Ginzburg, JETP 47, 2318 (1964)
43. V.L. Ginzburg, Soviet Phys., Uspekhi 13, 335 (1970)
44. J.A. Wilson *et al.*, Phys. Rev. Lett. 32, 882 (1974)
45. N.W. Ashcroft, Phys. Rev. Lett. 21, 1748 (1968)
46. C.F. Richardson and N.W. Ashcroft, Phys. Rev. Lett. 78, 119 (1997)
47. N.W. Ashcroft, Phys. Rev. Lett. 92, 187002 (2004)
48. D. Allender, J. Bray and J. Bardeen, Phys. Rev. B7, 1020 (1973)
49. L.N. Cooper, Phys. Rev. 104, 1189 (1956)
50. M. Rice and L. Sneddon, Phys. Rev. Lett. 47, 689 (1981)
51. Z.A. Xu *et al.*, Nature 406, 486 (2000)
52. K.S. Bedell and D. Pines, Phys. Rev. B 37, 3730 (1988)
53. H.W. Meul *et al.*, Phys. Rev. Lett. 53, 497 (1984)
54. R.P. Chaudhury *et al.*, Phys. Rev. B, in press (2008)
55. M. Rice, Physica C 282-287, xix (1997)
56. C.W. Chu and V. Diatschenko, Phys. Rev. Lett. 41, 572 (1978)
57. C.Y. Huang *et al.*, J. Korean Phys. Soc. 31, 16 (1997)
58. R.G. Sweedler *et al.*, Phys. Lett. 15, 108 (1965)
59. C.W. Chu, History of Original Ideas and Basic Discoveries in Particle Physics, ed. H.B. Newman and T. Ypsilantis, Plenum Press, New York (1996), p. 793
60. S. Yamada *et al.*, Nature 239, 530 (1998)
61. L. F. Mattheiss *et al.*, Phys. Rev. B 37, 3745 (1988); A.W. Sleight *et al.*, Solid State Comm. 17, 27 (1975)
62. D.C. Johnston *et al.*, Mater. Res. Bull. 8, 777 (1973)
63. D. Jérome *et al.*, J. Physique Lett. 41, 95 (1980)

THE FIBONACCI MODEL AND THE TEMPERLEY-LIEB ALGEBRA

LOUIS H. KAUFFMAN

Department of Mathematics, Statistics and Computer Science (m/c 249), 851 South Morgan Street, University of Illinois at Chicago, Chicago, Illinois 60607-7045, USA
E-mail: kauffman@uic.edu

SAMUEL J. LOMONACO, Jr.

Department of Computer Science and Electrical Engineering, University of Maryland Baltimore County, 1000 Hilltop Circle, Baltimore, MD 21250, USA
E-mail: lomonaco@comcast.net

We give an elementary construction of the Fibonacci model, a unitary braid group representation that is universal for quantum computation. This paper is dedicated to Professor C. N. Yang, on his 85-th birthday.

Keyword: Knots; links; braids; quantum computing; unitary transformation; Jones polynomial; Temperley-Lieb algebra.

1. Introduction

This paper gives an elementary construction for the unitary representation of the Artin braid group that constructs the well-known Fibonacci model. This model gives a topological basis for quantum computation. The present paper is an outgrowth of our papers[9–11] and is related to the analysis of the quantum algorithms for the Jones polynomial in the paper by Shor and Jordan.[18] We show that unitary representations of the braid group arise naturally in the context of the Temperely-Lieb algebra. The Fibonacci model is usually constructed by using recoupling theory. Here we show how the model emerges naturally from braid group representations to the Temperley-Lieb algebra. We use the idea of recoupling and change of basis in process spaces to motivate the construction, but we do not have to rely on any machinery of recoupling theory. This makes the present paper self-contained.

For the reader interested in the relevant background in topological quantum computing we recommend the following Refs. 1–6, 14–17, 19, 20.

Here is a very condensed presentation of how unitary representations of the braid group are constructed via topological quantum field theoretic methods. For simplicity assmue that one has a single (mathematical) particle with label P that can interact with itself to produce either itself labeled P, or itself with the null label $*$. When $*$ interacts with P the result is always P. When $*$ interacts with $*$ the result is always $*$. One considers process spaces where a row of particles labeled P can successively interact, subject to the restriction that the end result is P. For example the space $V[(ab)c]$ denotes the space of interactions of three particles labeled P. The particles are placed in the positions a, b, c. Thus we begin with $(PP)P$. In a typical sequence of interactions, the first two P's interact to produce a $*$, and the $*$ interacts with P to produce P.

$$(PP)P \longrightarrow (*)P \longrightarrow P.$$

In another possibility, the first two P's interact to produce a P, and the P interacts with P to produce P.

$$(PP)P \longrightarrow (P)P \longrightarrow P.$$

It follows from this analysis that the space of linear combinations of processes $V[(ab)c]$ is two dimensional. The two processes we have just described can be taken to be the the qubit basis for this space. One obtains a representation of the three strand Artin braid group on $V[(ab)c]$ by assigning appropriate phase changes to each of the generating processes. One can think of these phases as corresponding to the interchange of the particles labeled a and b in the association $(ab)c$. The other operator for this representation corresponds to the interchange of b and c. This interchange is accomplished by a *unitary change of basis mapping*

$$F : V[(ab)c] \longrightarrow V[a(bc)].$$

If

$$A : V[(ab)c] \longrightarrow V[(ba)c : d]$$

is the first braiding operator (corresponding to an interchange of the first two particles in the association) then the second operator

$$B : V[(ab)c] \longrightarrow V[(ac)b]$$

is accomplished via the formula $B = F^{-1}AF$ where the A in this formula acts in the second vector space $V[a(bc)]$ to apply the phases for the interchange of b and c. In this scheme, vector spaces corresponding to associated strings of particle interactions are interrelated by *recoupling transformations* that generalize the mapping F indicated above. A full representation of the Artin braid group on each space is defined in terms of the local interchange phase gates and the recoupling transfomations. These gates and transformations have to satisfy a number of identities in order to produce a well-defined representation of the braid group. These identities were discovered originally in relation to topological quantum field theory. In our approach the structure of phase gates and recoupling transformations arise naturally from the structure of the bracket model for the Jones polynomial[7] and a corresponding representation of the Temperley-Lieb algebra. Thus we obtain an entry into topological quantum computing that is directly related to the original construction of the Jones polynomial.

The present paper has two sections. The first section discusses the structure of the Temperley-Lieb algebra in relation to the Jones polynomial and the bracket polynomial model for the Jones polynomial. The second section constructs the Fibonacci model via a representation of the Temperley-Lieb algebra.

2. The Bracket Polynomial and the Jones Polynomial

We now discuss the Jones polynomial. We shall construct the Jones polynomial by using the bracket state summation model.[8] The bracket polynomial, invariant under Reidmeister moves II and III, can be normalized to give an invariant of all three Reidemeister moves. This normalized invariant, with a change of variable, is the Jones polynomial.[7] The Jones polynomial was originally discovered by a different method than the one given here.

The *bracket polynomial*,[8] $< K > = < K > (A)$, assigns to each unoriented link diagram K a Laurent polynomial in the variable A, such that

(1) If K and K' are regularly isotopic diagrams, then $< K > = < K' >$.
(2) If $K \sqcup O$ denotes the disjoint union of K with an extra unknotted and unlinked component O (also called 'loop' or 'simple closed curve' or 'Jordan curve'), then

$$< K \sqcup O > = \delta < K >,$$

where

$$\delta = -A^2 - A^{-2}.$$

(3) $< K >$ satisfies the following formulas

$$< \chi > = A < \asymp > + A^{-1} <)(>$$

$$< \overline{\chi} > = A^{-1} < \asymp > + A <)(>,$$

where the small diagrams represent parts of larger diagrams that are identical except at the site indicated in the bracket. We take the convention that the letter chi, χ, denotes a crossing where *the curved line is crossing over the straight segment.* The barred letter denotes the switch of this crossing, where *the curved line is undercrossing the straight segment.* See Figure 1 for a graphic illustration of this relation, and an indication of the convention for choosing the labels A and A^{-1} at a given crossing.

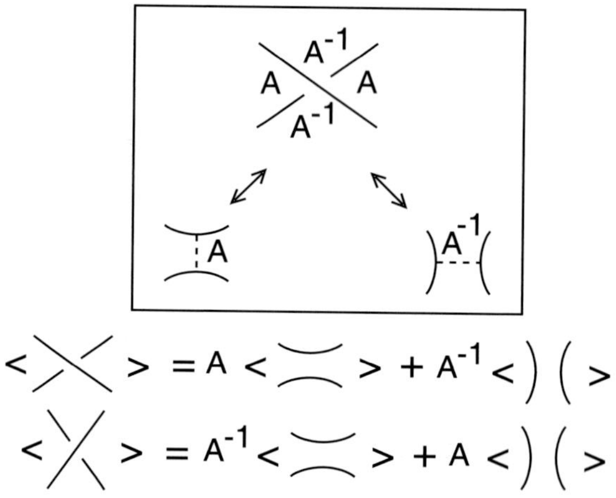

Fig. 1. Bracket smoothings.

It is easy to see that Properties 2 and 3 define the calculation of the bracket on arbitrary link diagrams. The choices of coefficients (A and A^{-1}) and the value of δ make the bracket invariant under the Reidemeister moves II and III. Thus Property 1 is a consequence of the other two properties.

In computing the bracket, one finds the following behaviour under Reidemeister move I:

$$< \gamma > = -A^3 < \smile >$$

and

$$< \overline{\gamma} >= -A^{-3} < \smile >$$

where γ denotes a curl of positive type as indicated in Figure 2, and $\overline{\gamma}$ indicates a curl of negative type, as also seen in this figure. The type of a curl is the sign of the crossing when we orient it locally. Our convention of signs is also given in Figure 2. Note that the type of a curl does not depend on the orientation we choose. The small arcs on the right hand side of these formulas indicate the removal of the curl from the corresponding diagram. The bracket is invariant under regular isotopy and can be normalized to an invariant of ambient isotopy by the definition

$$f_K(A) = (-A^3)^{-w(K)} < K > (A),$$

where we chose an orientation for K, and where $w(K)$ is the sum of the crossing signs of the oriented link K. $w(K)$ is called the *writhe* of K. The convention for crossing signs is shown in Figure 2.

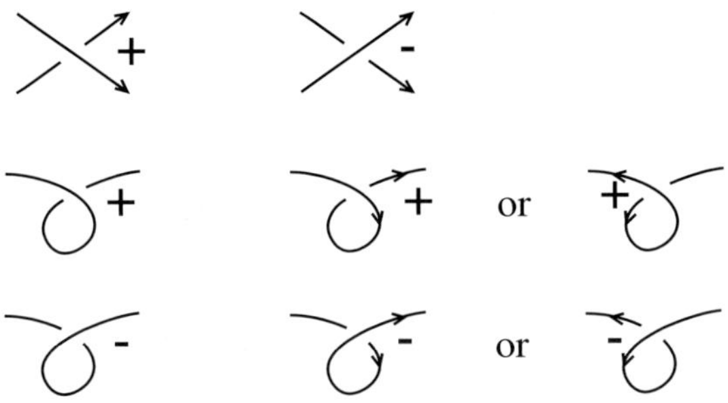

Fig. 2. Crossings signs and curls.

Remark. By a change of variables one obtains the original Jones polynomial, $V_K(t)$, for oriented knots and links from the normalized bracket:

$$V_K(t) = f_K(t^{-\frac{1}{4}}).$$

The bracket model for the Jones polynomial is quite useful both theoretically and in terms of practical computations. One of the neatest applications is to simply compute $f_K(A)$ for the trefoil knot K and determine that $f_K(A)$ is not equal to $f_K(A^{-1}) = f_{-K}(A)$. That computation shows that the trefoil is not ambient isotopic to its mirror image, a fact that is much harder to prove by classical methods.

The State Summation. In order to obtain a closed formula for the bracket, we now describe it as a state summation. Let K be any unoriented link diagram. Define a *state*, S, of K to be a choice of smoothing for each crossing of K. There are two choices for smoothing a given crossing, and thus there are 2^N states of a diagram with N crossings. In a state we label each smoothing with A or A^{-1} according to the left-right convention discussed in Property 3 (see Figure 1). The label is called a *vertex weight* of the state. There are two evaluations related to a state. The first one is the product of the vertex weights, denoted

$$< K|S > .$$

The second evaluation is the number of loops in the state S, denoted

$$||S||.$$

Define the *state summation*, $< K >$, by the formula

$$< K > = \sum_S < K|S > \delta^{||S||-1}.$$

It follows from this definition that $< K >$ satisfies the equations

$$< \chi > = A < \asymp > + A^{-1} <)(>,$$

$$< K \sqcup O > = \delta < K >,$$

$$< O > = 1.$$

The first equation expresses the fact that the entire set of states of a given diagram is the union, with respect to a given crossing, of those states with an A-type smoothing and those with an A^{-1}-type smoothing at that crossing. The second and the third equation are clear from the formula defining the state summation. Hence this state summation produces the bracket polynomial as we have described it at the beginning of the section.

The Temperley Lieb Algebra. The Temperely Lieb algebra TL_n is an algebra over the ring $Z[A, A^{-1}]$ and $\delta = -A^2 - A^{-2}$ with multiplicative generators $\{I, U_1, U_2, \cdots, U_{n-1}\}$ where I is an identity element and the other generators satisfy the relations

$$U_i^2 = \delta U_i,$$

$$U_i U_{i\pm 1} U_i = U_i,$$

$$U_i U_j = U_j U_i,$$

where $|i - j| > 1$ in the last equation, and i runs through all values from 1 to $n-1$ for the first two equations whenever they are defined. The additive structure of the Temperely-Lieb algebra makes it a free module over the base ring.

The Temperley-Lieb algebra TL_n can be interpreted as shown in Figure 3 in terms of planar connection patterns between two rows of n points. In this Figure we illustrate the multiplicative relations and we show how a closure of a multiplicative element of the algebra leads to a collection of loops in the plane. There is a trace function $tr : TL_n \longrightarrow Z[A, A^{-1}]$ defined in the diagrammatic interpretation by the formula

$$tr(P) = \delta^{||P||-1}$$

where P is a product of gererators of the Temperley-Lieb algebra, and $||P||$ denotes the number of loops in the closure of the diagrammatic version of P, as illustrated in Figure 3. This trace function is extended linearly to the whole algebra, and it has the property that $tr(ab) = tr(ba)$ for any elements a and b of the algebra.

Another way to think about the bracket polynomial is to first make a representation of the n-strand Artin braid group to the Temperley-Lieb algebra TL_n on n-strands and then take the trace (tr as defined above) of this representation. The representation

$$rep : B_n \longrightarrow TL_n$$

is given by the formula

$$rep(\sigma_i) = AI + A^{-1}U_i$$

where σ_i denotes the i-th braid generator and I denotes the identity element in the Temperley-Lieb algebra. In Figure 3 we have illustrated this formula in the case $i = 1$. This illustration should suffice for the reader to see what is our orientation convention for braid generators, and our convention for the standard closure of a diagrammatic element of the Temprely-Lieb algebra. This same notion of closure applies to braids. The reader will note that this braid representation parallels the definition of the bracket polynomial, so that the expansion formula for the bracket leads directly to the braid representation when applied to a diagram for the braid. The following Theorem is a consequence of this correspondence.

Theorem 1. Let b be a braid in B_n and let $\bar{b}$ denote its standard closuure. Let *rep* denote the representation of the braid group discussed above, and let tr denote the trace on the Temperley-Lieb algebra discussed above. Then the bracket polynomial for the braid closure is given by the formula

$$<\bar{b}>= tr(rep(b)).$$

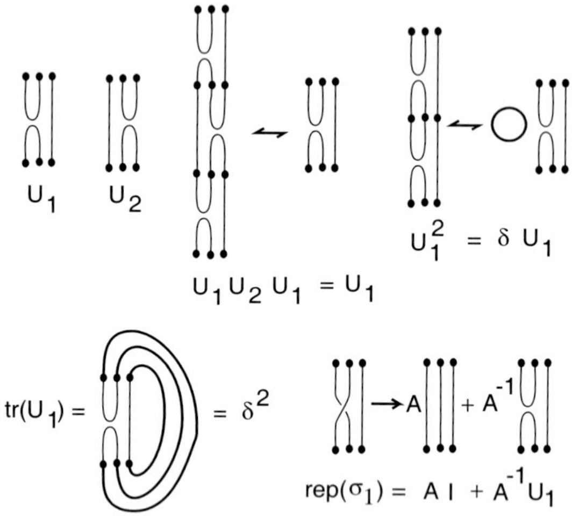

Fig. 3. Diagrammatic Temperley-Lieb algeba.

Remark. We can now explain how to produce unitary representations of the braid group in relation to the Temperley-Lieb algebra and the Jones polynomial. In order to make a unitary representation of the braid group, it is sufficient to find a representation $\Gamma : TL_n \longrightarrow Aut(V)$ of the Temperley-Lieb algebra to a complex vector space V such that $\Gamma(U_i)$ is a real and symmetric. Then, for any element $A = e^{i\theta}$ on the unit circle in the complex plane, we see that $\rho = \Gamma \circ rep$ is a unitary representaton of the Artin braid group. This statement follows from the fact, that under the above conditions $AI + A^{-1}\Gamma(U_i)$ is unitary. In the next section we give an elementary construction for a large class of unitary representations of the three-strand braid group, and then extend this class to include the Fibonacci model discussed in the Introduction.

3. Unitary Representations of the Braid Group and the Fibonacci Model

The constructions in this section are based on the combinatorics of the Fibonacci model. In this model we have a (mathematical) particle P that interacts with itself either to produce P or to produce a neutral particle $*$. If X is any particle then $*$ iteracts with X to produce X. Thus $*$ acts as an identity trasformation. These rules of interaction are illustrated in Figure 4.

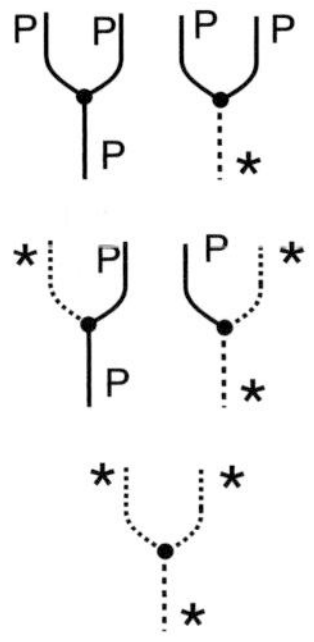

Fig. 4. The Fibonacci particle P.

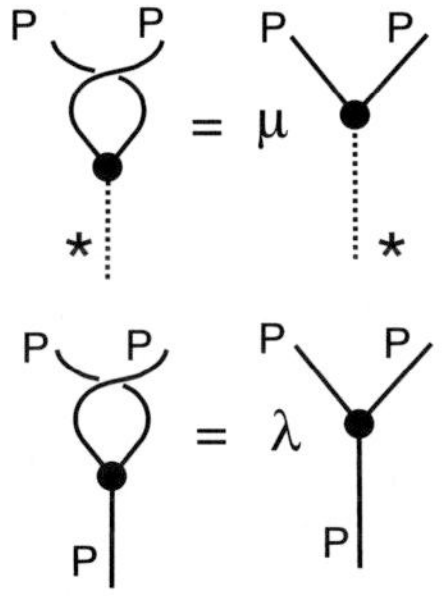

Fig. 5. Local braiding.

The braiding of two particles is measured in relation to their interaction. In Figure 5 we illustrate braiding of P with itself in relation to the two possible interactions of P with itself. If P interacts to produce $*$, then the braiding gives a phase factor of μ. If P interacts to produce P, then the braiding gives a phase factor of λ. We assume at the outset that μ and λ are unit complex numbers. One should visualize these particles as moving in a plane and the diagrams of interaction are either creations of two particles from one particle, or fusions of two particles to a single particle (depending on the choice of temporal direction). Thus we have a braiding matrix for these "local" particle interactions:

$$R = \begin{pmatrix} \mu & 0 \\ 0 & \lambda \end{pmatrix}$$

written with respect to the basis $\{|*\rangle, |P\rangle\}$ for this space of particle interactions.

We want to make this braiding matrix part of a larger representation of the braid group. In particular, we want a representation of the three-strand braid group on the process space V_3 illustrated in Figure 6. This space starts with three P particles and considers processes associated in the patttern $(PP)P$ with the stipulation that the end product is P. The possible pathways are illustrated in Figure 6. They correspond to $(PP)P \longrightarrow (*)P \longrightarrow P$ and $(PP)P \longrightarrow (P)P \longrightarrow P$. This process space has dimension two and can support a second braiding generator for the second two strands on the top of the tree. In order to articulate the second braiding we change basis to the process space corresponding to $P(PP)$ as shown in Figures 7 and 8. The change of basis is shown in Figure 7 and has matrix F as shown below. We want a unitary representation ρ of three-strand braids so that $\rho(\sigma_1) = R$ and $\rho(\sigma_2) = S = F^{-1}RF$. See Figure 8. We take the form of the matrix F as follows.

$$F = \begin{pmatrix} a & b \\ b & -a \end{pmatrix}$$

where $a^2 + b^2 = 1$ with a and b real. This form of the matrix for the basis change is determined by the requirement that F is symmetric with $F^2 = I$. The symmetry of the change of basis formula essentially demands that $F^2 = I$. If F is real, symmetric and $F^2 = I$, then F is unitary. Since R is unitary we see that $S = FRF$ is also unitary. Thus, if F is constructed in this way then we obtain a unitary representation of B_3.

Now we try to simultaneously construct an F *and* construct a representation of the Temperley-Lieb algebra as described in section 2. We begin by noting that

$$R = \begin{pmatrix} \mu & 0 \\ 0 & \lambda \end{pmatrix} = \begin{pmatrix} \lambda & 0 \\ 0 & \lambda \end{pmatrix} + \begin{pmatrix} \mu - \lambda & 0 \\ 0 & 0 \end{pmatrix} = \begin{pmatrix} \lambda & 0 \\ 0 & \lambda \end{pmatrix} + \lambda^{-1} \begin{pmatrix} \delta & 0 \\ 0 & 0 \end{pmatrix}$$

where $\delta = \lambda(\mu - \lambda)$. Thus $R = \lambda I + \lambda^{-1} U$ where $U = \begin{pmatrix} \delta & 0 \\ 0 & 0 \end{pmatrix}$ so that $U^2 = \delta U$. For the Temperley-Lieb representation, we want $\delta = -\lambda^2 - \lambda^{-2}$ as explained in section 2. Hence we need $-\lambda^2 - \lambda^{-2} = \lambda(\mu - \lambda)$, which implies that $\mu = -\lambda^{-3}$. With this restriction on μ, we have the Temperley-Lieb representation and the corresponding unitary braid group representation for 2-strand braids and the 2-strand Temperley-Lieb algebra.

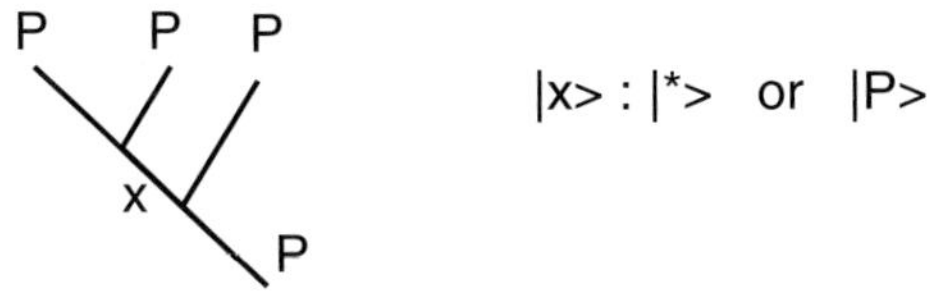

Fig. 6. Three strands at dimension two.

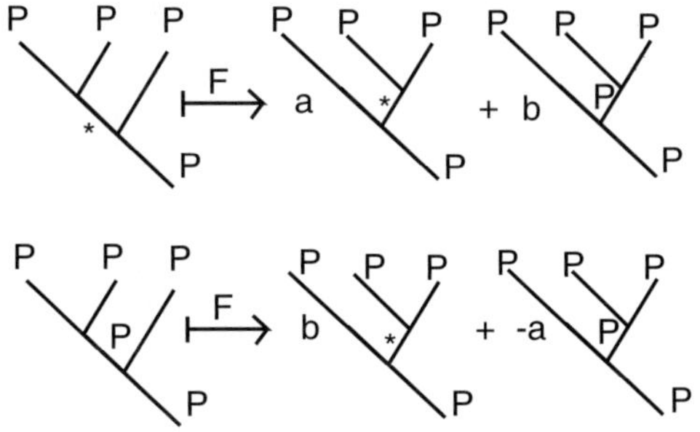

Fig. 7. Recoupling formula.

Now we can go on to B_3 and TL_3 via $S = FRF = \lambda I + \lambda^{-1}V$ with $V = FUF$. We must examine V^2, UVU and VUV. We find that

$$V^2 = FUFFUF = FU^2F = \delta FUF = \delta V,$$

as desired and

$$V = FUF = \begin{pmatrix} a & b \\ b & -a \end{pmatrix} \begin{pmatrix} \delta & 0 \\ 0 & 0 \end{pmatrix} \begin{pmatrix} a & b \\ b & -a \end{pmatrix} = \delta \begin{pmatrix} a^2 & ab \\ ab & b^2 \end{pmatrix}.$$

Thus $V^2 = V$ and since $V = \delta|v\rangle\langle v|$ and $U = \delta|w\rangle\langle w|$ with $w = (1,0)^T$ and $v = Fw = (a,b)^T$ (T denotes transpose), we see that

$$VUV = \delta^3|v\rangle\langle v|w\rangle\langle w|v\rangle\langle v| = \delta^3 a^2|v\rangle\langle v| = \delta^2 a^2 V.$$

Similarly $UVU = \delta^2 a^2 U$. Thus, we need $\delta^2 a^2 = 1$ and so we shall take $a = \delta^{-1}$. With this choice, we have a representation of the Temperley-Lieb algebra TL_3 so that $\sigma_1 = AI + A^{-1}U$ and $\sigma_2 = AI + A^{-1}V$ gives a unitary

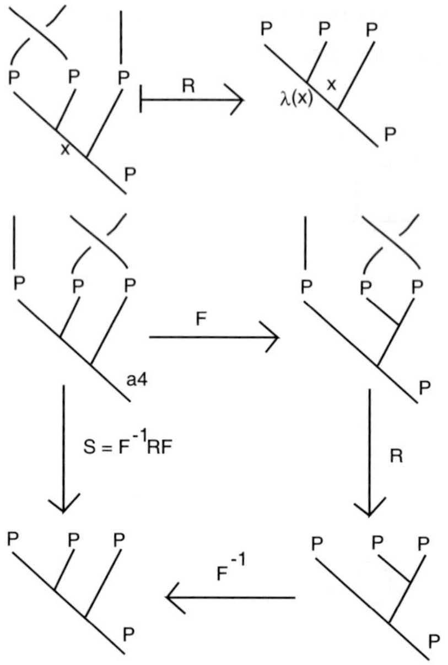

Fig. 8. Change of basis.

representation of the braid group when $A = \lambda = e^{i\theta}$ and $b = \sqrt{1-\delta^{-2}}$ is real. This last reality condition is equivalent to the inequality

$$cos^2(2\theta) \geq \frac{1}{4},$$

which is satisfied for infinitely many values of θ in the ranges

$$[0, \pi/6] \cup [\pi/3, 2\pi/3] \cup [5\pi/6, 7\pi/6] \cup [4\pi/3, 5\pi/3].$$

With these choices we have

$$F = \begin{pmatrix} a & b \\ b & -a \end{pmatrix} = \begin{pmatrix} 1/\delta & \sqrt{1-\delta^{-2}} \\ \sqrt{1-\delta^{-2}} & -1/\delta \end{pmatrix}$$

real and unitary, and for the Temperley-Lieb algebra,

$$U = \begin{pmatrix} \delta & 0 \\ 0 & 0 \end{pmatrix}, V = \delta \begin{pmatrix} a^2 & ab \\ ab & b^2 \end{pmatrix} = \begin{pmatrix} a & b \\ b & \delta b^2 \end{pmatrix}.$$

Now examine Figure 9. Here we illustrate the action of the braiding and the Temperley-Lieb Algebra on the first Fibonacci process space with basis $\{|*\rangle, |P\rangle\}$. Here we have $\sigma_1 = R, \sigma_2 = FRF$ and $U_1 = U, U_2 = V$ as described above. Thus we have a representation of the braid group on three strands and a representation of the Temperley-Lieb algebra on three strands with no further restrictions on δ.

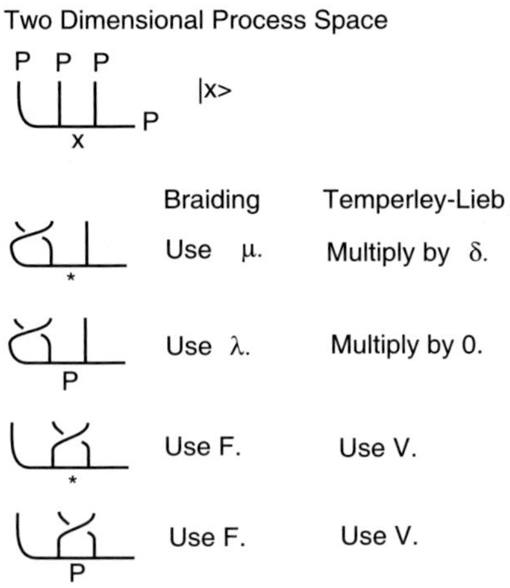

Fig. 9. Algebra for a two dimensional process space.

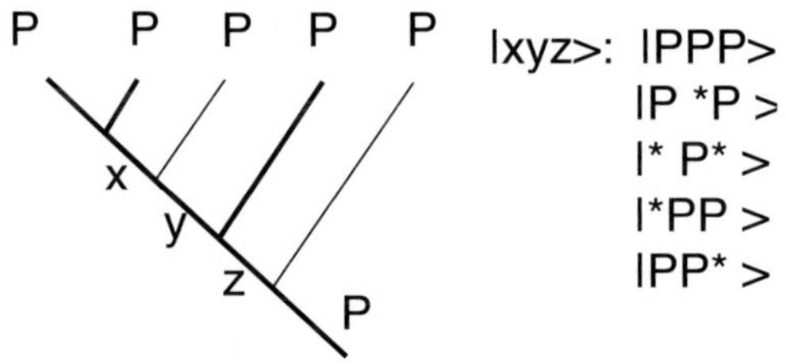

Fig. 10. A five dimensional process space.

So far, we have arrived at exactly the 3-strand braid representations that we used in our papers[12,13] giving a quantum algorithm for the Jones polynomial for three-strand braids. In this paper we are working in the context of the Fibonacci process spaces and so we wish to see how to make a representation of the Temperley-Lieb algebra to this model as a whole, not restricting ourselves to only three strands. The generic case to consider is the action of the Temperley-Lieb algebra on process spaces of higher dimension as shown in Figures 10 and 11. In the Figure 11 we have illustrated the triplets from the previous figure as part of a possibly larger tree and

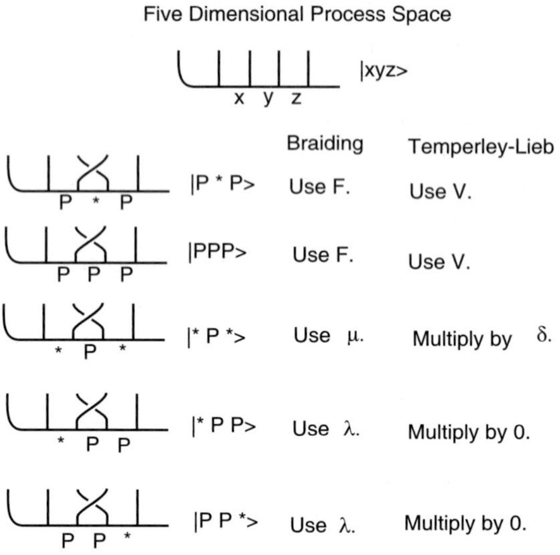

Fig. 11. Algebra for a five dimensional process space.

have drawn the strings horizontally rather than diagonally. In this figure we have listed the effects of braiding the vertical strands 3 and 4. We see from this figure that the action of the Temperley-Lieb algebra must be as follows:

$$U_3|P*P\rangle = a|P*P\rangle + b|PPP\rangle,$$

$$U_3|PPP\rangle = b|P*P\rangle + \delta b^2|PPP\rangle,$$

$$U_3|*P*\rangle = \delta|*P*\rangle,$$

$$U_3|*PP\rangle = 0,$$

$$U_3|PP*\rangle = 0.$$

Here we have denoted this action as U_3 because it connotes the action on the third and fourth vertical strands in the sequences shown in Figure 11. Note that in a larger sequence we can recognize U_j by examining the triplet surrounding the $j-1$-th element in the sequence, just as the pattern above is governed by the elements surrounding the second element in the sequence. For simplicity, we have only indicated three elements in the sequences above. Note that in a sequence for the Fibonacci process there are never two consecutive appearances of the neutral element $*$.

We shall refer to a sequence of $*$ and P as a *Fibonacci sequence* if it contains no consecutive appearances of $*$. Thus $|PP * P * P * P\rangle$ is a Fibonacci sequence. In working with this representation of the braid group and Temperley-Lieb algebra, it is convenient to assume that the ends of the sequence are flanked by P as in Figures 10 and 11 for sequences of length 3. It is convenient to leave out the flanking P's when notating the sequence.

Using these formulas we can determine conditions on δ such that this is a representation of the Temperley-Lieb algebra for all Fibonacci sequences. Consider the following calculation:

$$U_4U_3U_4|PPPP\rangle = U_3U_2(b|PP * P\rangle + \delta b^2|PPPP\rangle)$$

$$= U_4(bU_3|PP * P\rangle + \delta b^2 U_3|PPPP\rangle)$$

$$= U_4(0 + \delta b^2(b|P * PP\rangle + \delta b^2|PPPP\rangle))$$

$$= \delta b^2(bU_4|P * PP\rangle + \delta b^2 U_4|PPPP\rangle)$$

$$= \delta^2 b^4 U_4|PPPP\rangle.$$

Thus we see that in order for $U_4U_3U_4 = U_4$, we need that $\delta^2 b^4 = 1$.

It is easy to see that $\delta^2 b^4 = 1$ is the only remaining condition needed to make sure that the action of the Temperley-Lieb algebra extends to all Fibonacci Model sequences.

Note that $\delta^2 b^4 = \delta^2(1 - \delta^{-2})^2 = (\delta - 1/\delta)^2$. Thus we require that

$$\delta - 1/\delta = \pm 1.$$

When $\delta - 1/\delta = 1$, we have the solutions $\delta = \frac{1 \pm \sqrt{5}}{2}$. However, for the reality of F we require that $1 - \delta^{-2} \geq 0$, ruling out the choice $\delta = \frac{1-\sqrt{5}}{2}$. When $\delta - 1/\delta = -1$, we have the solutions $\delta = \frac{-1 \pm \sqrt{5}}{2}$. This leaves only $\delta = \pm\phi$ where $\phi = \frac{1+\sqrt{5}}{2}$ (the Golden Ratio) as possible values for δ that satisfy the reality condition for F. Thus, up to a sign we have arrived at the well-known value of $\delta = \phi$ (the Fibonacci model) as essentially the only way to have an extension of this form of the representation of the Temperley-Lieb algebra for n strands. Let's state this positively as a Theorem.

Theorem 2. Let V_{n+2} be the complex vector space with basis $\{|x_1 x_2 \cdots x_n\rangle\}$ where each x_i equals either P or $*$ and there do *not* occur two consecutive appearances of $*$ in the sequence $\{x_1, \cdots x_n\}$. We refer to this basis for V_n as the set of *Fibonacci sequences* of length n. Then the dimension of V_n is equal to f_{n+1} where f_n is the n-th Fibonacci number: $f_0 = f_1 = 1$ and $f_{n+1} = f_n + f_{n-1}$. Let $\delta = \pm\phi$ where $\phi = \frac{1+\sqrt{5}}{2}$. Let $a = 1/\delta$ and $b = \sqrt{1-a^2}$. Then the Temperley-Lieb algebra on $n+2$ strands with loop value δ acts on V_n via the formulas given below. First we give the left-end actions.

$$U_1|*x_2x_3\cdots x_n\rangle = \delta|*x_2x_3\cdots x_n\rangle,$$

$$U_1|Px_2x_3\cdots x_n\rangle = 0,$$

$$U_2|*Px_3\cdots x_n\rangle = a|*Px_3\cdots x_n\rangle + b|PPx_3\cdots x_n\rangle,$$

$$U_2|P*x_3\cdots x_n\rangle = 0,$$

$$U_2|PPx_3\cdots x_n\rangle = b|*Px_3\cdots x_n\rangle + \delta b^2|PPx_3\cdots x_n\rangle.$$

Then we give the general action for the middle strands.

$$U_i|x_1\cdots x_{i-3}P*Px_{i+1}\cdots x_n\rangle = a|x_1\cdots x_{i-3}P*Px_{i+1}\cdots x_n\rangle$$

$$+b|x_1\cdots x_{i-3}PPPx_{i+1}\cdots x_n\rangle,$$

$$U_i|x_1\cdots x_{i-3}PPPx_{i+1}\cdots x_n\rangle = b|x_1\cdots x_{i-3}P*Px_{i+1}\cdots x_n\rangle$$

$$+\delta b^2|x_1\cdots x_{i-3}PPPx_{i+1}\cdots x_n\rangle,$$

$$U_i|x_1\cdots x_{i-3}*P*x_{i+1}\cdots x_n\rangle = \delta|x_1\cdots x_{i-3}*P*x_{i+1}\cdots x_n\rangle,$$

$$U_i|x_1\cdots x_{i-3}*PPx_{i+1}\cdots x_n\rangle = 0,$$

$$U_i|x_1\cdots x_{i-3}PP*x_{i+1}\cdots x_n\rangle = 0.$$

Finally, we give the right-end action.

$$U_{n+1}|x_1\cdots x_{n-2}*P\rangle = 0,$$

$$U_{n+1}|x_1\cdots x_{n-2}P*\rangle = 0,$$

$$U_{n+1}|x_1\cdots x_{n-2}PP\rangle = b|x_1\cdots x_{n-2}P*\rangle + \delta b^2|x_1\cdots x_{n-2}PP\rangle.$$

Remark. Note that the left and right end Temperley-Lieb actions depend on the same basic pattern as the middle action. The Fibonacci sequences

$|x_1x_2\cdots x_n\rangle$ should be regarded as flanked left and right by P's just as in the special cases discussed prior to the proof of Theorem 2.

Corollary. With the hypotheses of Theorem 2, we have a unitary representation of the Artin Braid group B_{n+2} to TL_{n+2}, $\rho : B_{n+2} \longrightarrow TL_{n+2}$ given by the formulas

$$\rho(\sigma_i) = AI + A^{-1}U_i,$$

$$\rho(\sigma_i^{-1}) = A^{-1}I + AU_i,$$

where $A = e^{3\pi i/5}$ where the U_i connote the representation of the Temperley-Lieb algebra on the space V_{n+2} of Fibonacci sequences as described in the Theorem above.

Remark. The Theorem and Corollary give the original parameters of the Fibonacci model and shows that this model admits a unitary representation of the braid group via a Jones representation of the Temperley-Lieb algebra.

In the original Fibonacci model,[11] there is a basic non-trivial recoupling matrix F.

$$F = \begin{pmatrix} 1/\delta & 1/\sqrt{\delta} \\ 1/\sqrt{\delta} & -1/\delta \end{pmatrix} = \begin{pmatrix} \tau & \sqrt{\tau} \\ \sqrt{\tau} & -\tau \end{pmatrix}$$

where $\delta = \frac{1+\sqrt{5}}{2}$ is the golden ratio and $\tau = 1/\delta$. The local braiding matrix is given by the formula below with $A = e^{3\pi i/5}$.

$$R = \begin{pmatrix} A^8 & 0 \\ 0 & -A^4 \end{pmatrix} = \begin{pmatrix} e^{4\pi i/5} & 0 \\ 0 & -e^{2\pi i/5} \end{pmatrix}.$$

This is exactly what we get from our method by using $\delta = \frac{1+\sqrt{5}}{2}$ and $A = e^{3\pi i/5}$. Just as we have explained earlier in this paper, the simplest example of a braid group representation arising from this theory is the representation of the three strand braid group generated by $\sigma_1 = R$ and $\sigma_2 = FRF$ (Remember that $F = F^T = F^{-1}$.). The matrices σ_1 and σ_2 are both unitary, and they generate a dense subset of $U(2)$, supplying the local unitary transformations needed for quantum computing. The full braid group representation on the Fibonacci sequences is computationally universal for quantum computation. In our earlier paper[11] we gave a construction for the Fibonacci model based on Temperely-Lieb recoupling theory. In this paper, we have reconstructed the Fibonacci model on the more elementary grounds of the representation of the Temperley-Lieb algebra summarized in the statement of the Theorem 2 and its Corollary.

References

1. D. Aharonov, V. Jones, Z. Landau, A polynomial quantum algorithm for approximating the Jones polynomial, quant-ph/0511096.
2. M. Freedman, A magnetic model with a possible Chern-Simons phase, quant-ph/0110060v1 9 Oct 2001, (2001), preprint
3. M. Freedman, *Topological Views on Computational Complexity*, Documenta Mathematica - Extra Volume ICM, 1998, pp. 453–464.
4. M. Freedman, M. Larsen, and Z. Wang, A modular functor which is universal for quantum computation, quant-ph/0001108v2, 1 Feb 2000.
5. M. H. Freedman, A. Kitaev, Z. Wang, Simulation of topological field theories by quantum computers, *Commun. Math. Phys.*, **227**, 587-603 (2002), quant-ph/0001071.
6. M. Freedman, Quantum computation and the localization of modular functors, quant-ph/0003128.
7. V.F.R. Jones, A polynomial invariant for links via von Neumann algebras, Bull. Amer. Math. Soc. **129** (1985), 103–112.
8. L.H. Kauffman, State models and the Jones polynomial, Topology **26** (1987), 395–407.
9. L. H. Kauffman and S. J. Lomonaco Jr., Spin Networks and anyonic topological computing, In "Quantum Information and Quantum Computation IV", (Proceedings of Spie, April 17-19,2006) edited by E.J. Donkor, A.R. Pirich and H.E. Brandt, Volume 6244, Intl Soc. Opt. Eng., pp. 62440Y-1 to 62440Y-12.
10. L. H. Kauffman and S. J. Lomonaco Jr., Spin Networks and anyonic topological computing II, In "Quantum Information and Quantum Computation V", (Proceedings of Spie, April 10-12,2007) edited by E.J. Donkor, A.R. Pirich and H.E. Brandt, Volume 6573, Intl Soc. Opt. Eng., pp. 65730U-1 to 65730u-13.
11. L. H. Kauffman and S. J. Lomonaco Jr., q - Deformed Spin Networks, Knot Polynomials and Anyonic Topological Quantum Computation, *JKTR* Vol. 16, No. 3 (March 2007), pp. 267-332.
12. L. H. Kauffman, Quantum computing and the Jones polynomial, math.QA/0105255, in *Quantum Computation and Information*, S. Lomonaco, Jr. (ed.), AMS CONM/305, 2002, pp. 101–137.
13. L.H. Kauffman and Samuel J. Lomonaco Jr. A Three-stranded quantum algorithm for the Jones polynonmial, in "Quantum Information and Quantum Computation V", (Proceedings of Spie, April 2007) edited by E.J. Donkor, A.R. Pirich and H.E. Brandt, Intl Soc. Opt. Eng. , 65730T-1-16.
14. A. Kitaev, Anyons in an exactly solved model and beyond, *arXiv.cond-mat/0506438 v1 17 June 2005.*
15. G. Moore and N. Read, Non-abelions in the fractional quantum Hall effect, *Nuclear PhysicsB* 360 (1991), 362-396.
16. A. Marzuoli and M. Rasetti, Spin network quantum simulator, *Physics Letters A* **306** (2002) 79–87.

17. J. Preskill, Topological computing for beginners, (slide presentation), Lecture Notes for Chapter 9 - Physics 219 - Quantum Computation. *http://www.iqi.caltech.edu/ preskill/ph219*
18. P. W. Shor and S. P. Jordan, Estimating Jones polynomials is a complete problem for one clean qubit. *arxiv:0707.2831v1 [quqnt-ph]* 19 Jul 2007.
19. F. Wilczek, *Fractional Statistics and Anyon Superconductivity,* World Scientific Publishing Company (1990).
20. E. Witten, Quantum field Theory and the Jones Polynomial, *Commun. Math. Phys.*, vol. 121, 1989, pp. 351-399.

YANG-BAXTER EQUATION AND QUANTUM PERIODIC TODA LATTICE

LEON TAKHTAJAN

Stony Brook State University of New York
E-mail: leontak@math.sunysb.edu

We review old and recent results on periodic quantum Toda lattice: a completely integrable system of N particles on the line with the nearest neighbors interaction with exponential potential.

ATOMIC-SCALE STRUCTURE: FROM SURFACES TO NANOMATERIALS

M.A. VAN HOVE

Department of Physics and Materials Science, City University of Hong Kong
E-mail: vanhove@cityu.edu.hk

This presentation will sketch past progress and explore possible future directions in the atomic-scale structure determination of surfaces, interfaces and nanostructures. Atomic-scale structure is the basis of understanding for a wide range of phenomena in physics, chemistry, materials science and other fields, and needs to be determined from experiment, by efficient theoretical and computational interpretation.

Comparisons will be made between different available techniques for surface structure determination, in particular regarding their relative capabilities. Highlighted will be Scanning Tunneling Microscopy and Low-Energy Electron Diffraction. Both of these techniques are capable of detailed atomic-scale structure determination of nanostructures, by comparison between experiment and theoretical simulation.

Examples will be given to illustrate recent progress in structure determination of several nanostructures, including buckminsterfullerenes (C60), carbon nanotubes (CNTs) and silicon nanowires (SiNWs).

TOPOLOGICAL QUANTUM NUMBERS AND PHASE TRANSITIONS IN MATTER

DAVID THOULESS

University of Washington
E-mail: thouless@u.washington.edu

In the early years of the last century the discreteness of matter, of electric charge, and of mechanical action became firmly established, and slowly some of the more subtle implications of the interplay of these three were worked out. Dirac showed that magnetic monopoles also had to be quantized, the importance of dislocations in solids was shown, and the quantization of circulation in neutral superfluids and of magnetic flux in superconductors was predicted and demonstrated. Such topological defects can be a sign of a symmetry broken by a phase transition, or, as Onsager suggested in his first exposition of quantized circulation, can themselves drive a phase transition. I discuss circulation in superfluids, flux in superconductors and Hall conductance in inversion layers as examples of such quantum numbers. I show why there is a topological quantum number, and ask how the mathematical quantum number is related to measurable quantities. Recently there has been interest in whether the robustness of such topological defects make them suitable for quantum manipulation.

PROFESSOR C. N. YANG AND STATISTICAL MECHANICS

F. Y. WU

Department of Physics, Northeastern University,
Boston, Massachusetts 02115, U.S.A.
E-mail: fywu@neu.edu

Professor Chen Ning Yang has made seminal and influential contributions in many different areas in theoretical physics. This talk focuses on his contributions in statistical mechanics, a field in which Professor Yang has held a continual interest for over sixty years. His Master's thesis was on a theory of binary alloys with multi-site interactions, some 30 years before others studied the problem. Likewise, his other works opened the door and led to subsequent developments in many areas of modern day statistical mechanics and mathematical physics. He made seminal contributions in a wide array of topics, ranging from the fundamental theory of phase transitions, the Ising model, Heisenberg spin chains, lattice models, and the Yang-Baxter equation, to the emergence of Yangian in quantum groups. These topics and their ramifications will be discussed in this talk.

Keyword: Phase transition; Ising and lattice models; Yang-Baxter equation.

1. Introduction

Statistical mechanics is the subfield of physics that deals with systems consisting of large numbers of particles. It provides a framework for relating macroscopic properties of a system, such as the occurrence of phase transitions, to microscopic properties of individual atoms and molecules.

The theory of statistical mechanics was founded by Gibbs (1834-1903) who based his considerations on earlier works of Boltzmann (1844-1906) and Maxwell (1831-1879). By the end of the 19th century, classical mechanics was fully developed and applied successfully to rigid body motions. However, after it was recognized that ordinary materials consist of 10^{23} molecules, it soon became apparent that the application of traditional classical mechanics is fruitless in explaining physical phenomena on the basis of molecular considerations. To overcome this difficulty, Gibbs proposed a statistical theory for computing bulk properties of real materials.

Statistical mechanics as proposed by Gibbs applies to all physical systems regardless of their macroscopic states. But in early years there had been doubts about whether it could fully explain physical phenomena such as phase transitions. In 1937, Mayer[1] developed the method of cluster expansions for analyzing the statistical mechanics of a many-particle system which worked well for systems in the gas phase. This offered some hope of explaining phase transitions, and the Mayer theory subsequently became the main frontier of statistical mechanical research. This was unfortunate in hindsight since, as Yang and Lee would later show (see Sec. 4), the grand partition function used in the Mayer theory cannot be continued into the condensed phase, and hence it does not settle the question it set out to answer.

This was the stage and status of statistical mechanics in the late 1930's when Professor C. N. Yang entered college.

2. A Quasi-chemical Mean-field Model of Phase Transition

In 1938, Yang entered the National Southwest Associate University, a university formed jointly by National Tsing Hua University, National Peking University and Nankai University during the Japanese invasion, in Kunming, China. As an undergraduate student Yang attended seminars given by J. S. (Zhuxi) Wang, who had recently returned from Cambridge, England, where he had studied theory of phase transitions under R. H. Fowler. These lectures brought C. N. Yang in contact with the Mayer theory and other latest developments in statistical mechanics.[2–4]

After obtaining his B.S. degree in 1942, Yang continued to work on an M.S. degree in 1942-1944, and he chose to work in statistical mechanics under the direction of J. S. Wang. His Master's thesis included a study of phase transitions using a quasi-chemical method of analysis, and led to the publication of his first paper.[5]

In this paper, Yang generalized the quasi-chemical theory of Fowler and Guggenheim[6] of phase transitions in a binary alloy to encompass 4-site interactions. The idea of introducing multi-site interactions to a statistical mechanical model was novel and new. In contrast, the first mentioning of a lattice model with multi-site interactions was by myself[7] and by Kadanoff and Wegner[8] in 1972 - that the 8-vertex model solved by Baxter[9] is also an Ising model with 4-site interactions. Thus, Yang's quasi-chemical analysis of a binary alloy, an Ising model in disguise, predated the important study of a similar nature by Baxter in modern-day statistical mechanics by three decades!

3. Spontaneous Magnetization of the Ising Model

The two-dimensional Ising model was solved by Onsager in 1944.[10] In a legendary footnote of a conference discussion, Onsager[11] announced without proof a formula of the spontaneous magnetization of the two-dimensional Ising model with nearest-neighbor interactions K,

$$I = \left(1 - \sinh^{-4} 2K\right)^{1/8}. \tag{1}$$

Onsager never published his derivation since, as related by him later, he had made use of some unproven results on Toeplitz determinants which he did not feel comfortable to put in print. Since the subject matter was close to his Master's thesis, Yang had studied the Onsager paper extensively and attempted to derive (1). But the Onsager paper was full of twists and turns offering very few clues to the computation of the spontaneous magnetization.[12]

A simplified version of the Onsager solution by Kauffman[13] appeared in 1949. With the new insight to Onsager's solution, Yang immediately realized that the spontaneous magnetization I can be computed as an off-diagonal matrix element of Onsager's transfer matrix. This started Yang on the most difficult and the longest calculation of his career.[12]

After almost 6 months of hard work off and on, Yang eventually succeeded in deriving the expression (1) and published the details in 1952.[14] Several times during the course of the work, the calculation stalled and Yang almost gave up, only to have it picked up again days later with the discovery of new tricks or twists.[12] It was a most formidable *tour de force* algebraic calculation in the history of statistical mechanics.

3.1. *Universality of the critical exponent β*

At Yang's suggestion, C. H. Chang[15] extended Yang's analysis of the spontaneous magnetization to the Ising model with anisotropic interactions K_1 and K_2, obtaining the expression

$$I = \left(1 - \sinh^{-2} 2K_1 \sinh^{-2} 2K_2\right)^{1/8}. \tag{2}$$

This expression exhibits the same critical exponent $\beta = 1/8$ as in the isotropic case, and marked the first ever recognition of universality of critical exponents, a fundamental principle of critical phenomena proposed by Griffiths 20 years later.[16]

3.2. *An integral equation*

A key step in Yang's evaluation of the spontaneous magnetization is the solution of an integral equation (Eq. (84) in Ref. [13]) whose kernel is a product of 4 factors I, II, III, and IV. Yang pioneered the use of Fredholm integral equations in the theory of exactly solved models (see also Sec. 7.1 below). This particular kernel and similar ones have been used later by others, as they also occurred in various forms in studies of the susceptibility[17] and the n-spin correlation function of the Ising model.[18]

4. Fundamental Theory of Phase Transitions

As described in the above, the frontier of statistical mechanics in the 1930's focused on the Mayer theory and the question whether the theory was applicable to all phases of a matter. Being thoroughly versed in the Mayer theory as well as the Ising lattice gas, Yang investigated this question in collaboration with T. D. Lee. Their investigation resulted in two fundamental papers on the theory of phase transitions.[19,20]

In the first paper,[19] Yang and Lee examined the question whether the cluster expansion in the Mayer theory can apply to both the gas and condensed phases. This led them to examine the convergence of the grand partition function series in the thermodynamic limit, a question that had not previously been closely investigated. To see whether a single equation of state can describe different phases, they looked at zeroes of the grand partition function in the complex fugacity plane, again a consideration that revolutionized the study of phase transitions. Since an analytic function is defined by its zeroes, under this picture the onset of phase transitions is signified by the pinching of zeroes on the real axis. This shows that the Mayer cluster expansion, while working well in the gas phase, cannot be analytically continued, and hence does not apply, in the condensed phase. It also rules out any possibility in describing different phases of a matter by a single equation of state.

In the second paper,[20] Lee and Yang applied the principles formulated in the first paper to the example of an Ising lattice gas. By using the spontaneous magnetization result (1), they deduced the exact two-phase region of the liquid-gas transition. This established without question that the Gibbs statistical mechanics holds in all phases of a matter. The analysis also led to the discovery of the remarkable Yang-Lee circle theorem, which states that zeroes of the grand partition function of a ferromagnetic Ising lattice gas always lie on a unit circle.

These two papers have profoundly influenced modern-day statistical mechanics as described in the following:

4.1. *The existence of the thermodynamic limit*

Real physical systems typically consist of $N \sim 10^{23}$ particles confined in a volume V. In applying Gibbs statistical mechanics to real systems one takes the thermodynamic (bulk) limit $N, V \to \infty$ with N/V held constant, and implicitly assumes that such a limit exists. But in their study of phase transitions,[19] Yang and Lee demonstrated the necessity of a closer examination of this assumption. This insight initiated a host of rigorous studies of a similar nature.

The first comprehensive study was by Fisher[21] who, on the basis of earlier works of van Hove[22] and Groeneveld,[23] established in 1964 the existence of the bulk free energy for systems with short-range interactions. For Coulomb systems with long-range interactions the situation is more subtle, and Lebowitz and Lieb established the bulk limit by making use of the Gauss law unique to Coulomb systems.[24] The existence of the bulk free energy for dipole-dipole interactions was subsequently established by Griffiths.[25] These rigorous studies led to a series of later studies on the fundamental question of the stability of matter.[26]

4.2. *The Yang-Lee circle theorem and beyond*

The consideration of Yang-Lee zeroes of the Ising model opened a new window in statistical mechanics and mathematical physics. The study of Yang-Lee zero loci has been extended to Ising models of arbitrary spins,[27] to vertex models,[28] and to numerous other spin systems.

While the Yang-Lee circle theorem concerns zeroes of the grand partition function, in 1964 Fisher[29] proposed to consider zeroes of the partition function, and demonstrated that they also lie on circles. The Fisher argument has since been made rigorous with the density of zeroes explicitly computed by Lu and myself.[30,31] The partition function zero consideration has also been extended to the Potts model by numerous authors.[32]

The concept of considering zeroes has also proven to be useful in mathematical physics. A well-known intractable problem in combinatorics is the problem of solid partitions of an integer.[33] But a study of the zeroes of its generating function by Huang and myself[34] shows they tend towards a unit circle as the integer becomes larger. Zeroes of the Jones polynomial in knot theory have also been computed, and found to tend towards the unit circle as the number of nodes increases.[35] These findings appear to pointing to some unifying truth lurking beneath the surface of many unsolved problems in mathematics and mathematical physics.

5. The Quantization of Magnetic Flux

During a visit to Stanford University in 1961, Yang was asked by W. M. Fairbank whether or not the quantization of magnetic flux, if found, would be a new physical principle. The question arose at a time when Fairbank and B. S. Deaver were in the middle of an experiment investigating the possibility of magnetic flux quantization in superconducting rings. Yang, in collaboration with N. Byers, began to ponder over the question.[36]

By the time Deaver and Fairbank[37] successfully concluded from their experiment that the magnetic flux is indeed quantized, Byers and Yang[38] have also reached the conclusion that the quantization result did not indicate a new property. Rather, it can be deduced from usual quantum statistical mechanics. This was the "first true understanding of flux quantization".[39]

6. The Off-Diagonal Long-Range Order

The physical phenomena of superfluidity and superconductivity have been among the least-understood macroscopic quantum phenomena occurring in nature. The practical and standard explanation has been based on bosonic considerations: the Bose condensation in superfluidity and Cooper pairs in the BCS theory of superconductivity. But there had been no understanding of a fundamental nature in substance. That was the question Yang pondered in the early 1960's.[40]

In 1962, Yang published a paper[41] with the title *Concept of off-diagonal long-range order and the quantum phases of liquid helium and of superconductors*, which crystallized his thoughts on the essence of superfluidity and superconductivity. While the long-range order in the condensed phase in a real system can be understood, and computed, as the diagonal element of the two-particle density matrix, Yang proposed in this paper that the quantum phases of superfluidity and superconductivity are manifestations of a long-range order in off-diagonal elements of the density matrix. Again, this line of thinking and interpretation was totally new, and the paper has remained to be one that Yang has "always been fond of".[40]

7. The Heisenberg Spin Chain and the 6-vertex Model

After the publication of the paper on the long-range off-diagonal order, Yang experimented using the Bethe ansatz in constructing a Hamiltonian which can actually produce the off-diagonal long-range order.[42] Instead, this effort led to ground-breaking works on the Heisenberg spin chain, the 6-vertex model, and the one-dimensional delta function gas described below.

7.1. *The Heisenberg spin chain*

In a series of definitive papers in collaboration with C. P. Yang, Yang studied the one-dimensional Heisenberg spin chain with the Hamiltonian

$$H = -\frac{1}{2} \sum \left(\sigma_x \sigma_x' + \sigma_y \sigma_y' + \Delta \sigma_z \sigma_z' \right) . \tag{3}$$

Special cases of the Hamiltonian had been considered before by others. But Yang and Yang analyzed the Bethe ansatz solution of the eigenvalue equation of (3) with complete mathematical rigor, including a rigorous analysis of a Fredholm integral equation arising in the theory in the full range of Δ. The ground state energy is found to be singular at $\Delta = \pm 1$. Furthermore, this series of papers has become important as it formed the basis of ensuing studies of the 6-vertex model, the one-dimensional delta function gas and numerous other related problems.

7.2. *The 6-vertex model*

In 1967, Lieb[44] solved the residual entropy problem of square ice, a prototype of the two-dimensional 6-vertex model, using the method of Bethe ansatz. Subsequently, the solution was extended to 6-vertex models in the absence of an external field.[45] These solutions share the characteristics that they are all based on Bethe ansatz analyses involving real momentum k.

In the same year 1967, Yang, Sutherland and C. P. Yang[46] published a solution of the general 6-vertex model in the presence of external fields, in which they used the Bethe ansatz with complex momentum k. But the Sutherland-Yang-Yang paper did not provide details of the solution. This led others to fill in the gap in ensuing years, often with analyses starting from scratch, to understand the thermodynamics. Thus, the $\Delta < 1$ case was studied by Nolden,[47] the $\Delta \geq 1$ case by Shore and Bukman,[48] and the case $|\Delta| = \infty$ by myself in collaboration with Huang *et al.*[49] The case of $|\Delta| = \infty$ is of particular interest, since it is also a 5-vertex model as well as an honeycomb lattice dimer model with a nonzero dimer-dimer interaction. It is the only known soluble interacting close-packed dimer model.

8. One-dimensional Delta Function Gases

8.1. *The Bose gas*

The first successful application of the Bethe ansatz to a many-body problem was the one-dimensional delta function Bose gas solved by Lieb and Liniger.[50] Subsequently, by extending considerations to include all excitations, Yang and C. P. Yang deduced the thermodynamics of the Bose gas.[51] Their theoretical prediction has recently been found to agree very well with experiments on a one-dimensional Bose gas trapped on an atom chip.[52]

8.2. *The Fermi gas*

The study of the delta function Fermi gas was more subtle. In a seminal work having profound and influential impacts in many-body theory, statistical mechanics and mathematical physics, Yang in 1967 produced the full solution of the delta function Fermi gas.[53] The solution was obtained as a result of the combined use of group theory and the nested Bethe ansatz, a repeated use of the Bethe ansatz devised by Yang.

One very important ramification of the Fermi gas work is the exact solution of the ground state of the one-dimensional Hubbard model obtained by Lieb and myself.[54] The solution of the Hubbard model is similar to that of the delta function gas except with the replacement of the momentum k by $\sin k$ in the Bethe ansatz solution. Due to its relevance in high T_c superconductivity, the Lieb-Wu solution has since led to a torrent of further works on the one-dimensional Hubbard model.[55]

9. The Yang-Baxter Equation

The two most important integrable models in statistical mechanics are the delta function Fermi gas solved by Yang[53] and the 8-vertex model solved by Baxter.[9,56] The key to the solubility of the delta function gas is an operator relation[57] of the S-matrix,

$$Y_{jk}{}^{ab}\, Y_{ik}{}^{bc}\, Y_{ij}{}^{ab} = Y_{ij}{}^{bc}\, Y_{ik}{}^{ab}\, Y_{jk}{}^{bc}, \tag{4}$$

and for the 8-vertex model the key is a relation[58] of the 8-vertex operator,

$$U_{i+1}(u)U_i(u+v)U_{i+1}(v) = U_i(v)U_{i+1}(u+v)U_i(u). \tag{5}$$

Noting the similarity of the two relations and realizing they are fundamentally the same, in a paper on the 8-vertex model Takhtadzhan and Faddeev[59] called it the Baxter-Yang relation. Similar relations also arise in other quantum and lattice models. These relations have since been referred to as the *Yang-Baxter equation.*[60]

The Yang-Baxter equation is an internal consistency condition among parameters in a quantum or lattice model, and can usually be written down by considering a star-triangle relation.[60] The solution of the Yang-Baxter equation, if found, often aids in solving the model itself. The Yang-Baxter equation has been shown to play a central role in connecting many subfields in mathematics, statistical mechanics and mathematical physics.[61]

9.1. *Knot invariants*

One example of the role played by the Yang-Baxter equation in mathematics is the construction of knot (link) invariants. Knot invariants are algebraic quantities, often in polynomial forms, which preserve topological properties of three-dimensional knots. In the absence of definite prescriptions, very few knot invariants were known for decades. The situation changed dramatically after the discovery of the Jones polynomial by Jones in 1985,[62] and the subsequent revelation that knot invariants can be constructed from lattice models in statistical mechanics.[63]

The key to constructing knot invariant from statistical mechanics is the Yang-Baxter equation. Essentially, from each lattice model whose Yang-Baxter equation possesses a solution, one constructs a knot invariant. One example is the Jones polynomial which can be constructed from a solution of the Yang-Baxter equation of the Potts model, even though the solution is in an unphysical regime.[64] Other examples are described in a 1992 review on knot theory and statistical mechanics by myself.[65]

9.2. *The Yangian*

In 1985, Drinfeld[66] showed that there exists a Hopf algebra (quantum group) over $SL(n)$ associated with the Yang-Baxter equation (4) after the operator Y is expanded into a series. Since Yang found the first rational solution of the expanded equation, he named the Hopf algebra the *Yangian* in honor of Yang.[66]

Hamiltonians with the Yangian symmetry include, among others, the one-dimensional Hubbard model, the delta function Fermi gas, the Haldane-Shastry model,[67] and the Lipatov model.[68] The Yangian algebra is of increasing importance in quantum groups, and has been used very recently in a formulation of quantum entangled states.[69]

10. Conclusion

In this talk I have summarized the contributions made by Professor Chen Ning Yang in statistical mechanics. It goes without saying that it is not possible to cover all aspects of Professor Yang's work in this field, and undoubtedly, there are omissions. But it is clear from what is presented, however limited, that Professor C. N. Yang has made immense contributions to this relatively young field of theoretical physics.

A well-known treatise in statistical mechanics is the 20-volume *Phase Transitions and Critical Phenomena* published in 1972 - 2002.[70] The series covers almost every subject matter of traditional statistical mechanics. The first chapter of Volume 1 is an introductory note by Professor Yang, in which he assessed the status of the field and remarked about possible future directions of statistical mechanics.

At the conclusion he wrote:

One of the great intellectual challenges for the next few decades is the question of brain organization.

As research in biophysics and brain memory functioning has mushroomed into a major field in recent years, this is an extraordinary prophecy and a testament to the insight and foresight of Professor Chen Ning Yang.

Acknowledgments

I would like to thank Dr. K. K. Phua for inviting me to the Symposium. I am grateful to M.-L. Ge and J. H. H. Perk for inputs in the preparation of the talk, and to J. H. H. Perk for a critical reading of the manuscript.

References

1. J. E. Mayer, *J. Chem. Phys.* **5**, 67 (1937).
2. C. N. Yang, in *Selected Papers (1945-1980) with Commentary* (World Scientific, Singapore, 2005).
3. C. N. Yang, *Int. J. Mod. Phys. B* **2**, 1325 (1988).
4. T. C. Chiang, *Biography of Chen-Ning Yang: The Beauty of Gauge and Symmetry* (in Chinese), (Tian Hsia Yuan Jian Publishing Co., Taipei, 2002).
5. C. N. Yang, *J. Chem. Phys.* **13**, 66 (1943).
6. R. H. Fowler and E. A. Guggenheim, *Proc. Roy. Soc.* **A174**, 187 (1940).
7. F. Y. Wu, *Phys. Rev. B* **4**, 2312 (1971).
8. L. P. Kadanoff and F. Wegner, *Phys. Rev. B* **4**, 3989 (1972).
9. R. J. Baxter, *Phys. Rev. Lett.* **26**, 832 (1971).
10. L. Onsager, *Phys. Rev.* **65**, 117 (1944),
11. L. Onsager, *Nuovo Cimento* **6**, Suppl. p. 261 (1949).
12. Ref. [2], p. 12.

13. B. Kauffman, *Phys. Rev.* **76**, 1232 (1949).
14. C. N. Yang, *Phys. Rev.* **85**, 808 (1952).
15. C. H. Chang, *Phys. Rev.* **88**, 1422 (1952).
16. R. B. Griffiths, *Phys. Rev. Lett* **24**, 1479 (1970).
17. E. Barouch, B. M. McCoy, and T. T. Wu, *Phys. Rev. Lett.* **31**, 1409 (1973).
18. B. M. McCoy, C. A. Tracy, and T. T. Wu, *Phys. Rev. Lett.* **38**, 793 (1973); D. B. Abraham *Commun. Math. Phys.* **59**, 17 (1978); *ibid.* **60**, 205 (1978).
19. C. N. Yang and T. D. Lee, *Phys. Rev.* **87**, 404 (1952).
20. T. D. Lee and C. N. Yang, *Phys. Rev.* **87**, 410 (1952).
21. M. E. Fisher, *Arch. Rat. Mech. Anal.* **17**, 377 (1964).
22. L. van Hove, *Physica* **15**, 951 (1949).
23. J. Groeneveld, *Phys. Lett.* **3**, 50 (1962).
24. J. L. Lebowitz and E. H. Lieb, *Phys. Rev. Lett.* **22**, 631 (1969).
25. R. B. Griffiths, *Phys. Rev.* **176**, 655 (1968).
26. See E. H. Lieb, *Rev. Mod. Phys.* **48**, 553 (1976).
27. R. B. Griffiths, *J. Math. Phys.* **10**, 1559 (1969).
28. M. Suzuki and M. E. Fisher, *J. Math. Phys.* **12**, 235 (1971).
29. M. E. Fisher, in *Lecture Notes in Theoretical Physics*, Vol. 7c, ed. W. E. Brittin (University of Colorado Press, Boulder, 1965).
30. W. T. Lu and F. Y. Wu, *Physica A* **258**, 157 (1998).
31. W. T. Lu and F. Y. Wu, *J. Stat. Phys.* **102**, 953 (2001).
32. See, for example, C. N. Chen, C. K. Hu and F. Y. Wu, *Phys. Rev. Lett.* **76**, 169 (1996).
33. P. A. MacMahon, *Combinatory Analysis*, Vol. 2 (Cambridge University Press, United Kingdom, 1916).
34. H. Y. Huang and F. Y. Wu, *Int. J. Mod. Phys. B* **11**, 121 (1997).
35. F. Y. Wu and J. Wang, *Physica A*, **296**, 483 (2001).
36. Ref. [2], pp. 49-50; Ref. [3], p. 1328.
37. B. S. Deaver and W. M. Fairbank, *Phys. Rev. Lett.* **7**, 43 (1961).
38. N. Byers and C. N. Yang, *Phys. Rev. Lett.* **7**, 46 (1961).
39. Ref. [3]. p.1328.
40. Ref. [2], p. 54; Ref. [3], p. 1328.
41. C. N. Yang, *Rev. Mod. Phys.* **34**, 694 (1962),
42. Ref. [2], p. 63.
43. C. N. Yang and C. P. Yang, *Phys. Rev.* **150**, 321, 327 (1966); *ibid.* **151**, 258 (1966).
44. E. H. Lieb, *Phys. Rev. Lett.* **18**, 692 (1967).
45. E. H. Lieb, *Phys. Rev. Lett.* **18**, 1046, (1967); *ibid.* **19**, 588 (1967).
46. B. Sutherland, C. N. Yang and C. P. Yang, *Phys. Rev. Lett.* **19**, 588 (1967).
47. I. Nolden, *J. Stat. Phys.* **67**, 155 (1992).
48. J. D. Shore and D. J. Bukman, *Phys. Rev. Lett.* **72**, 604 (1994); D. J. Bukman and J. D. Shore, *J. Stat. Phys.* **78**, 1227 (1995).
49. H. Y. Huang, F. Y. Wu, H. Kunz and D. Kim, *Physica A* **228**, 1 (1996).
50. E. H. Lieb and W. Liniger, *Phys. Rev.* **130**, 1605 (1963); E. H. Lieb, *Phys. Rev.* **130**, 1616 (1963).
51. C. N. Yang and C. P. Yang, *J. Math, Phys.* **10**, 1315 (1969).

52. A. H. van Amerongen, J. J. P. van Es, P. Wicke, K. V. Kheruntsyan, and N. J. van Druten, *Phys. Rev. Lett.* **100**, 090402 (2008).
53. C. N. Yang, *Phys. Rev. Lett.* **19**, 1312 (1967).
54. E. H. Lieb and F. Y. Wu, *Phys. Rev. Lett.* **20**, 1445 (1968); *ibid.* **21**, 192 (1968); *Physica A* **321**, 1 (2003).
55. See, for example, F. H. L. Essler, H. Frahm, F. Göhmann, A. Klümper, and V. E. Korepin, *The One-dimensional Hubbard Model*, (Cambridge University Press, United Kingdom, 2005).
56. R. J. Baxter, *Exactly Solved Models*, (Academic Press, London, 1980).
57. Equation (8) in Ref. [53].
58. Equation (10.4.31) in Ref. [56]
59. L. A. Takhtadzhan and L. D. Faddeev, *Russian Math. Surveys* **34**:5, 11 (1979).
60. J. H. H. Perk and H. Au-Yang, *Yang-Baxter Equations* in *Encyclopedia of Mathematical Physics*, eds. J.-P. Francoise, G. L. Naber and S. T. Tsou (Oxford:Elsevier, 2006); arXiv: math-ph/0606053.
61. C. N. Yang and M.-L. Ge, *Int. J. Mod. Phys.* **20**, 2223 (2006).
62. V. F. R. Jones, *Bull.Am. Math. Soc.* **12**, 103 (1985).
63. L. H. Kauffman, *Topology* **26**, 395 (1987).
64. L. H. Kauffman, *Contemp. Math.* **78**, 263 (1988).
65. F. Y. Wu, *Rev. Mod. Phys.* **64** 1099 (1992).
66. V. G. Drinfeld, *Soviet Math. Dokl.* **32**:1, 254 (1985).
67. See F. D. M. Haldane, in *Proceedings of 16th Taniguchi Symposium on Condensed Matter Physics*, eds. O. Okiji and N. Kawakami (Springer, Berlin, 1994).
68. See L. Dolan, C. R. Nappi, and E. Witten, *J. High Energy Phys.* **10**-017 (2003).
69. C. M. Bai, M.-L. Ge and X. Kang, this volume.
70. *Phase Transitions and Critical Phenomena*, eds. C. Domb and M. S. Green, Vols. 1 - 6; eds. C. Domb and J. L. Lebowitz, Vols. 7 - 20 (Academic Press, New York, 1972 - 2002).

A FEW PIECES OF MATHEMATICS INSPIRED BY REAL BIOLOGICAL DATA

BAILIN HAO
T-Life Research Center, Fudan University, Shanghai 200433, China
Institute of Theoretical Physics, Beijing 100080, China
Santa Fe Institute, Santa Fe, NM 87501, USA
E-mail: hao@mail.itp.ac.cn
http://www.itp.ac.cn/~hao/

Biology now produces huge amount of data mainly in the form of symbolic sequences. These data are noisy and incomplete so that statistics is inevitable as a first step in any analysis. However, in order to reveal the biological regularities masked by billion years of random mutations and natural selection one must invoke more or less deterministic approaches. We will draw a few simple examples from our recent bioinformatics work that make use of Poisson distribution and Markov prediction in statistics, Goulden-Jackson cluster method in enumerative combinatorics, number of Eulerian loops in graph theory, and factorizable language in formal language theory.

Keyword: DNA; proteins; Poisson distribution; Markov prediction; Eulerian loop; graph theory; Goulden-Jackson cluster method; factorizable language.

1. Symbolic Sequences from Biology

It is an amazing fact that a substantial part of fundamental biological knowledge is embodied in symbolic sequences such as DNAs and proteins. These sequences are obtained as results of far-reaching coarse-graining by ignoring many details at atomic level. For a theoretical physicist DNAs are one-dimensional, directed, unbranching heteropolymers made of four kinds of monomers (the nucleotides a, c, g and t) with typical length ranging from 10^4 to 10^8. Proteins are one-dimensional, directed, unbranching heteropolymers made of 20 kinds of monomers (the amino acids A, C, $\cdots$ to Y) with typical length from 10^2 to 10^4.

By the end of December 2007 there have been more than 700 finished and 2700 on-going genome sequencing projects worldwide (see the GOLD database[1]). In addition to genomic projects there are many proteomic, transcriptomic, metabolomic, and other "-omic" projects that produce comparably huge amount of data. All these projects generate symbols and symbolic sequences counted in Giga (10^8) everyday. Without the use of computers, supported by physical consideration and mathematical analysis it would be progressively impossible to comprehend the biological knowledge buried in these data. These data are the main driving force that is transforming biology from an experimental endeavor to a science standing on the three mainstays of experiment, theory, and computation.

2. Necessity to Study Real Biological Data

Since we have come to the notion of symbolic sequences it is appropriate to recollect a basic fact on big sets of long symbolic sequences. In Claude Shannon's seminal 1948 paper[2] that laid the foundation of modern information theory, in addition to the famous definition of information, Shannon stated a few other theorems. His Theorem 3 can be roughly interpreted as follows. Consider all possible sequences of length N made of letters from a finite alphabet Σ. There are $|\Sigma|^N$ such sequences, where $|\Sigma|$ is the cardinality of the alphabet ($|\Sigma| = 4$ for DNAs and $|\Sigma| = 20$ for proteins). Generally speaking, when N gets very large, these $|\Sigma|^N$ sequences can be divided into two subsets: a huge subset of typical sequences and a small subset of "atypical" sequences. The statistical characteristics of a typical sequence resembles that of any other typical sequence or the bulk of the huge subset, while the property of any atypical sequence is very specific and has to be scrutinized almost individually. The simplest members of the atypical subset are sequences made of repetitions of one and the same letter as well as various kinds of periodic and quasi-periodic sequences. However, the most significant ones from the atypical subset are those with hidden regularities mixed with seemingly random background. These sequences are the really complex ones.

Biological sequences are the result of several billions years of evolution and natural selection. They must belong to the subset of atypical sequences in the space of all possible sequences of similar lengths. Due to the huge volume of data, inevitable noisy background and experimental errors, statistical tools must be invoked in the beginning of any analysis of biological sequences. However, one needs more "deterministic" approaches to reveal hidden regularities in the real data. In fact, by looking at real sequence data one may encounter surprises and discover peculiar features that cannot be seen in statistical studies alone. In this brief review we show a few examples unearthed from real bacterial genomes in our recent genomic study. Some neat mathematical problems may be posed and solved exactly. It is remarkable that not much biological knowledge is required in order to appreciate this kind of biology-inspired mathematics.

3. Fine Structure in One-Dimensional K-String Histograms of Randomized Bacterial Genomes

We commence with simple-minded counting of short K-strings in a given genome, where K is a fixed integer, say, $K = 7$ or 8. Take for example, the genome of the harmless K12 strain of *E. coli*. It is a DNA loop made of $N = 4639675$ letters. We collect short strings of length $K = 8$ along the loop, shifting one letter at a time, one would get a total of N strings belonging to $4^8 = 65\,536$ different types. If the *E. coli* genome is a random sequence, each string type would on average appear $N/4^8 \approx 71$ times. We expect a bell-shaped distribution centered around 71. To check whether this is true we draw a histogram by plotting the number of appearance of a given string type against its real count in the genome. This histogram is given in the left part of Fig. 1. We compare this histogram with another one obtained after randomization of the original genome with the same number of individual nucleotides, see right part of Fig. 1. In the left histogram the count goes along the x-axis from zero (missing string types) to a certain maximum, namely, 777 in this case. The distribution is biased towards small counts and the maximum appears around 35. The histogram of the randomized sequence shows a sharp contrast to the original one. Indeed, its peak locates around 71 and the count ranges from 37 to 111, i.e., there are no missing string types at $K = 8$. Everything seems to be normal and as expected.

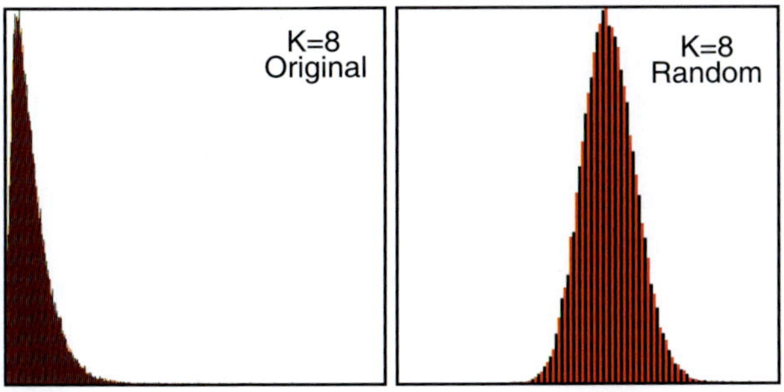

Fig. 1. One-dimensional histogram of 8−strings in the *E. coli* genome. Left: original; Right: randomized. For coordinates see text.

Now let us take another bacterium, say, *Mycobacterium tuberculosis*. The one-dimensional histogram resembles that of *E. coli*, i.e., a continuous distribution biased towards small counts as shown in the left part of Fig. 2. However, if we randomize this genome, keeping the number of a, c, g and t unchanged, strange thing happens: there appears fine structure in the right part of Fig. 2. This seems to be anti-intuitive at first glance as we have got used to the fact that usually fine structures in any "spectrum" go away after randomization.

However, a little more reflection would tell that the fine structure is caused by the uneven distribution of nucleotides in this genome. The letters g and c make up almost 2/3 of the genome. When forming various 8-strings, one simply does not have enough letters for some string types. This understanding may be made more precise by adopting a certain model to describe the randomized sequence.[3]

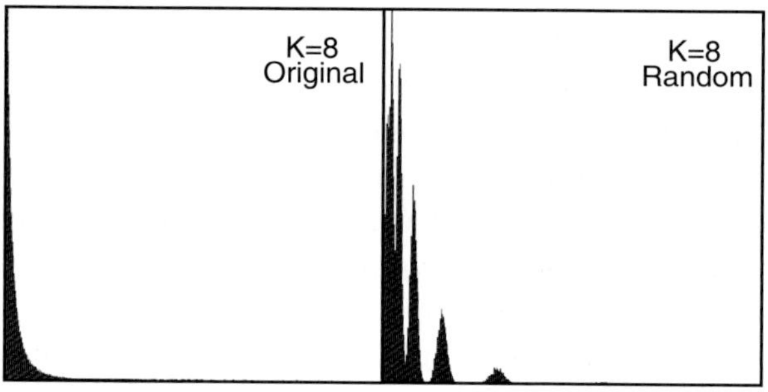

Fig. 2. One-dimensional histogram of 8−strings in the *M. tuberculosis* genome. Left: original; Right: randomized. For coordinates see text.

A simple stochastic model would have three parameters as there are four probabilities normalized to 1: $p_a + p_c + p_g + p_t = 1$. It follows from the Watson-Crick paring rule that in a double-helix DNA the number of g equals that of c and the number of a equals that of t. However, these equalities still hold approximately for single-strand DNAs (the empirical "Chargaff Parity Rule II"). Taking this rule as exact equalities, we arrive at a model that contains two types of letters a and c with one parameter p_c as $p_c + p_a = 0.5$. The probability p_i for a string of length K to contain i letters c is $p_i = p_c^i(0.5 - p_c)^{K-i}$. When the sequence length L is long enough, the one-dimensional histogram will consist of $(K+1)$ peaks, each described by a Poisson distribution with parameter $\lambda_i = (L - K + 1)p_i$ which determines the location as well as the width of the corresponding peak. For the *M. tuberculosis* genome $p_c = 0.328$ and the peaks seen in the right part of Fig. 2 are indeed described by 9 peaks. In the *E. coli* genome the frequencies of the four letters appear almost the same, therefore all λ_i happen to be very close to each other. The "single peak" seen in the right part of Fig. 1 is actually formed by 9 overlapping Poisson distributions with slightly different parameters. We can further improve the stochastic model to get even better description of the one-dimensional histogram.[4] However, this must involve more combinatorial consideration such as the Goulden-Jackson cluster method to be introduced in the next sections.

4. Number of True and Redundant Missing Strings in Bacterial Genomes

By looking at the actual counts in an one-dimensional histogram we get some more problems to think about. For example, in the *E. coli* genome at string length $K = 7$ there is only 1 missing string, namely, *gcctagg*. At the next length $K = 8$ there are 176 missing string types. These two integers, 1 and 176, raise a question: among the 176 missing 8-strings there are 8 redundant ones whose absence is simply the consequence of *gcctagg* being missing. It is natural to ask how many missing strings at a given length K are redundant, i.e., caused by the missing of shorter strings. Put in another way, suppose that one K-string is missing at length $K + 0$, how many strings it would take away at string length $K + i$? By mathematical induction one derives a simple formula $4^i(i+1)$ for the number of redundant missing strings at length $K + i$. However, in some cases it provides only an approximate solution as the derivation does not take into account possible overlaps of prefixes and suffixes of the string as it happens with *gcctagg*. When there are more than one missing strings at length K, the problem may become even more prominent. For example, in the *Aquifex aeolicus* genome four strings are missing at $K = 7$: *gcgcgcg*, *gcgcgca*, *agcgcgc*, and *tgcgcgc*. The overlapping of their pre- and suffixes are evident.

Let us formulate the problem in the following way. Given a finite alphabet $\Sigma = \{a, c, g, t\}$ and denote the collection of all possible strings made of letters from Σ, including an empty string ϵ, as Σ^*. Suppose $B \in \Sigma^*$ is a given set of "bad" or missing words that are not allowed to appear. For example, $B = \{gcctagg\}$ for *E. coli* or $B = \{gcgcgcg, gcgcgca, agcgcgc, tgcgcgc\}$ for *A. aeolicus*. Denote by a_K the number of "clean" strings of length K in Σ^* that do not contain any word from B as substrings. Our task is to calculate a_K precisely, taking into account all possible overlaps among pre- and suffixes of members of B. It might be difficult to get a_K separately for each and every K. However, if we are able to define and calculate an auxiliary function (a generating function)

$$f(s) = \sum_{K=0}^{\infty} a_K s^K, \tag{1}$$

then the counting problem would be solved once and for all K. There are at least two independent ways[5,6] to solve the problem, using either the Goulden-Jackson cluster method[7] from enumerative combinatorics or so-called factorizable language[8] from formal language theory.

5. Fractal Dimensions Behind the K → ∞ Limit of Two-Dimensional Histograms

A better way to see the K-string composition of a long DNA sequence or a genome is to display the results of string counting as a two-dimensional histogram. In order to count the number of single nucleotides in a genome one needs four counters allocated on the computer screen as a 2 by 2 matrix $M = \begin{pmatrix} g & c \\ a & t \end{pmatrix}$. In order to visualize the distribution of K-strings one takes the direct product of K copies of the matrix M:

$$M^{\{K\}} = M \otimes M \otimes \cdots \otimes M.$$

Each element in this $2^K \times 2^K$ square is a counter for a given type of K-string. We use a 16-color scale to represent the range of counts, e.g., white for missing, red for counts 1 to 4, $\cdots$, and black for any count that is greater than 40. The use of such a crude color code is an effective way of coarse-graining that helps to highlight some regular patterns in the final histogram.

We have implemented a user-friendly program entitled SeeDNA[9] to realize the above visualization algorithm. Taking the genomic sequence of a bacterium as input, SeeDNA displays a colorful two-dimensional histogram (a "portrait") of the K-string composition. Many bacterial portraits show characteristic regular patterns caused mainly by the absence or under-representation of one or more types of substrings[10] ("avoidance signature"[11]). Shown in Fig. 3 is the "portrait" of the human chromosome 22. The fractal-like self-similar and self-overlapping patterns are caused by mutations well-known in almost all mammalian genomes that have made the number of the dinucleotide cg significantly less than gc, though both are under-represented.

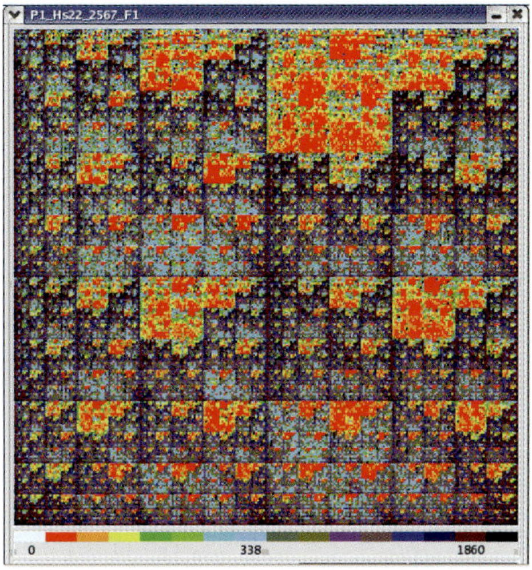

Fig. 3. A two-dimensional histogram of 8-strings in human chromosome 22.

In order to understand the origin of the fractal-like patterns let us look at the location of all counters of strings that contain at least a substring *cg* (*cg*-tagged strings). This is shown in Fig. 4 for four increasing values of K. These patterns are caused by the allocation of counters and have nothing to do with biology. However, they provide a characteristic background in real genomic portraits that may have biological implications.

In the $K \to \infty$, hence non-biological, limit one can define a fractal precisely. In order to avoid any confusion we emphasize that the patterns seen in Fig. 4 are not fractals but ordinary two-dimensional objects. Only the complementary set obtained by excluding all seen regular patterns in the $K \to \infty$ limit makes a fractal. Let us try to calculate the dimension of this complementary set.

We start with $K = 0$. Let the overall scale of the portrait be $\delta_0 = 1$. There is only $a_0 = 1$ empty string ϵ that is "clean", i.e, does not contain the "bad" tag *cg*. At $K = 1$ we have $\delta_1 = 1/2$ and all $a_1 = 4$ single-letter counters are "clean". At $K = 2$ we have $\delta_2 = 1/2^2$ and among the 4^2 dinucleotide counters there is one that represents *cg* and should be excluded. Therefore, we have $a_2 = 15$ "clean" cells. Suppose we know how to calculate a_K at scale $\delta_K = 1/2^K$ for a general K, then the fractal dimension is given by

$$D = \lim_{K\to\infty} \frac{\log a_K}{-\log \delta_K} = \lim_{K\to\infty} \frac{\log {a_K}^{1/K}}{\log 2}. \tag{2}$$

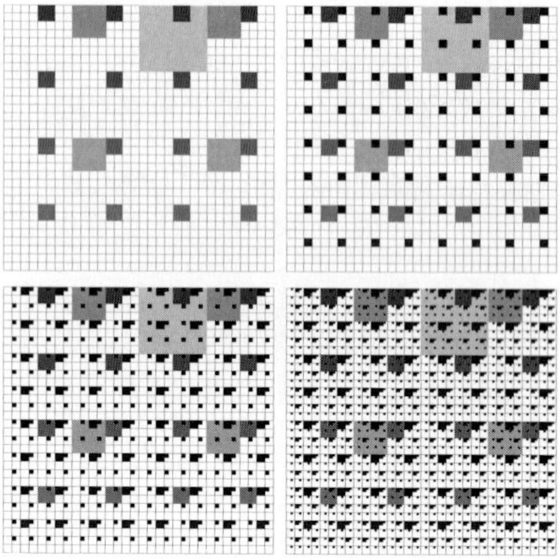

Fig. 4. The locations of counters for *cg*-tagged strings at $K = 4$ to 7. Please note the self-overlapping structure in addition to self-similarity.

The $\lim_{K\to\infty} \log {a_K}^{1/K}$, if exists, is nothing but the Cauchy convergence radius $|\lambda|$ of the generating function Eq. (1) viewed as a series in terms of the auxiliary variable s. It is given by the "minimal module" zero s_0 of $f(s)^{-1} = 0$:

$$\lim_{K\to\infty} \log {a_K}^{1/K} = |\lambda| = \frac{1}{|s_0|}. \tag{3}$$

Hence we have the dimension

$$D = -\frac{\log |s_0|}{\log 2}. \tag{4}$$

In the last two sections we have formulated two problems: the number of redundant missing strings taken away by shorter true missing ones and the dimension of the complementary set of tagged-string counters in the $K \to \infty$ limit. Now it is clear that these two problems happen to be one and the same, the latter being the graphic representation of the former. It turns out that this problem may be solved exactly by using at least two independent methods: the Goulden-Jackson cluster method in enumerative combinatorics and factorizable language in formal language theory. We skip the mathematical details and only give the generating function for number of strings that contain at least one "bad" substring *gcctagg* in the *E. coli* genome:[5]

$$f(s) = \frac{1 + s^6}{1 - 4s + s^6 - 3s^7}.$$

Then the generating function for all "clean" strings is $1/(1 - 4s) - f(s)$.

The language theory approach led to a minimal finite-state automaton whose "transfer function" is closely related to the generating function. For the *A. aeolicus* genome, which has four missing 7-strings, the corresponding automaton consists of 14 nodes. For details please consult a recent chapter written for *Annual Review in Nonlinear Science and Complexity.*[8]

6. Significant Improvement of Bacterial Phylogeny by Simple Markov Subtraction

Bacteria are the most successful species on the Earth. They have been thriving for more than 3 billion years. They have shaped the biochemical and even part of geochemical environment for all living organisms. Yet human knowledge on bacteria has been rather restricted. Even the first thing in any biological study, the taxonomy of bacteria, has been a long-standing problem mainly due to the limited morphological features useful for classification. A great progress has been made in the late 1970s when microbiologists started to use so-called 16S rRNA sequence (a tiny segment of about 1 500 letters which can be extracted and sequenced separately without sequencing the whole genome) to infer relatedness (phylogeny) of bacterial species. It has gone so far that the present classification of bacteria is largely based on 16S rRNA analysis.[12] Can these tiny segments represent the whole species? Are they immune to so-called "lateral gene transfer"? There is an urgent need to develop whole-genome based method to infer bacterial phylogeny and thus to provide an independent check of the present 16S rRNA analysis.

Bacterial genomes differ significantly in their size and gene content. The genome size of sequenced species differs more than 20 times. A small genome has about 400 genes while a large one may possess more than 7000 genes. There is no way to align them letter by letter. A few years ago when we discovered the species-specific "avoidance signature" (see the last section) in bacterial genomes we tried to use these signatures to infer phylogenetic relationship but failed.[11] Eventually we found a K-string *Composition Vector* method: each species is represented by a vector with 20^K components, obtained by simple counting of K-strings in the collection of all proteins. Then a subtraction procedure aimed to diminish the influence of so-called "neutral mutations" and to highlight the shaping role of natural selection is applied. The number of K-strings is first predicted from that of $(K-1)$- and $(K-2)$-strings by making a weakest Markov assumption,[13] then the difference of the predicted number and actual count is taken as a new component of the composition vector.

The Markov assumption is so simple that we sketch the idea here. The probability of a K-string $\alpha_1\alpha_2\cdots\alpha_K$, where α_i is one of the 20 amino acid letters, is expressed via a conditional probability:

$$p(\alpha_1\alpha_2\cdots\alpha_K) = p(\alpha_K|\alpha_1\alpha_2\cdots\alpha_{K-1})p(\alpha_1\alpha_2\cdots\alpha_{K-1}).$$

Now the Markov assumption: the "far-most" α_1 in the unknown conditional probability may be ignored. The also unknown new conditional probability may be excluded by using a similar relation:

$$p(\alpha_2\alpha_3\cdots\alpha_K) = p(\alpha_K|\alpha_2\alpha_3\cdots\alpha_{K-1})p(\alpha_2\alpha_3\cdots\alpha_{K-1}).$$

We are left with probabilities which may all be calculated from the frequency of the corresponding K-, $(K-1)$-, and $(K-2)$-strings. The Markov subtraction has improved the phylogeny so much that now it can be compared with taxonomy in an exhaustive way. The branchings in our phylogenetic tree based on 432 bacterial genomes available by the end of 2006 revealed only 7 "outliers" at the phylum level.[14]

In fact, our new approach was first announced at the conference in honor of Prof. C. N. Yang's 80th birthday[15] and published in a biological journal.[13] A free Web Server named CVTree has been published[16] for easy use by biologists.

7. Decomposition and Reconstruction of Protein Sequences: The Problem of Uniqueness

We have been working on the justification of the composition vector tree approach. In so doing we encountered another mathematical problem.[17] Let us look at the following piece of a winter flounder antifreeze protein:

MALSLFTVGQ LIFLFWTMRI TEASPDPAAK AAPAAAAAPA AAAPDTASDA AAAAALTAAN AKAAAELTAA NAAAAAAATA RG

It is an alanine(A)-rich protein made of 82 amino acids. Let us decompose this sequence into a collection of overlapping 5-strings (penta-peptides): *MALSL*, *ALSLF*, etc. One gets 78 such strings some of which appear several times. Now we ask the converse: given the collection of these 5-strings, if we reconstruct a sequence by using each penta-peptide once and only once, how unique would the reconstruction be? The inverse problem is solvable because at least one can get the original protein. Obviously, when K is big enough, the reconstruction is unique. This problem has a natural connection to a well-studied problem in graph theory, namely, the number of so-called Eulerian loops in a graph. Indeed, by treating 5-string $MALSL$ as a transition from $\boxed{MALS}$ to $\boxed{ALSL}$, etc., and simplifying what we obtained by reducing nodes that do not affect the number of Eulerian loops, we get the graph shown in Figure 5.

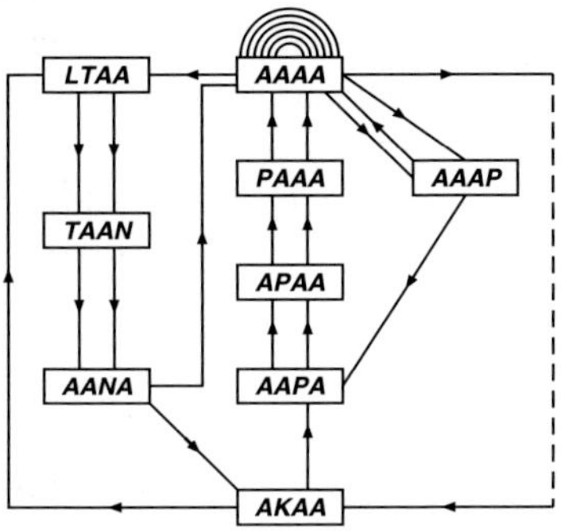

Fig. 5. An Euler graph determined by the winter flounder antifreeze protein.

This protein sequence has 1512 different reconstructions at $K = 5$. At $K = 6$ there are 60 reconstructions, at $K = 7$ there are 2, and there is an unique reconstruction at $K = 8$. We note that most of naturally occurring proteins have unique construction at $K = 5$ or 6.[18] It has been proved recently that there exist finite-state automata which can recognize whether a given sequence has an unique reconstruction at a given K. It is interesting to note that equipped with the automata just mentioned, one could discover a few proteins with a huge number of reconstructions at moderate K from a big protein database without possessing any background biological knowledge.

Acknowledgments

It is a great honor to dedicate this paper to the 85th birthday of Prof. C. N. Yang. In the summer of 1973 Prof. Yang met with a number of Beijing physicists in Lakeside Pavilion of Peking University and asked about research interest of everyone. Scientifically we were in very bad shape then. In April 1980 during my first short visit to USA Prof. Yang introduced me to recent progress of integrable systems and handed me a huge pile of preprints/reprints. Unfortunately, all these papers got lost *en route* mailing back to China. I missed the chance of getting into that exciting field and turned to chaotic dynamics. In December 1995 I arrived at Stony Brook at Prof. Yang's invitation. Prof. Yang talked with great enthusiasm on new discoveries in Bose-Einstein condensation. I did have E-mailed back all the fresh electronic files to a PhD student who later completed his thesis on BEC, but I myself was already engaged in a transition to biological problems. Anyway, during all these 35 years I have felt constant influence and encouragement of Prof. Yang. This work was partially supported by the National Basic Research Program of China (973 Program) Grant No. 2007CB814800. I am also very grateful to Nanyang University of Technology, Singapore, for supporting my attendance of the international conference in honor of Prof. C. N. Yang's 85th Birthday.

References

1. The Genomes On-Line Database (GOLD): `http://www.genomesonline.org/`
2. Claude E. Shannon, *Bell Syst. Tech. J.* **27** 379 (1948).
3. Huimin Xie and Bailin Hao, in *Bioinformatics. CSB2002 Proceedings*, (IEEE Computer Society, Los Alamitos, California, pp.31-42, 2002).
4. Chan Zhou and Huimin Xie, *Ann. Combin.* **8** 499 (2004).
5. Bailin Hao, *Physica A* **282** 225 (2000).
6. Bailin Hao, Huimin Xie, Zuguo Yu and Guoyi Chen, *Ann. Combin.*, **4** 247 (2000).
7. I. Goulden and D. M. Jackson, *J. London Math. Soc.* **20** 567 (1979).
8. Bailin Hao and Huimin Xie, Factorizable Language: from Dyanmics to Biology, chapter in *Annual Reviews of Nonlinear Science and Complexity*, ed. by Heinz G. Schuster, (Wiley-VCH, to appear).
9. Junjie Shen, Shuyu Zhang, Hoong-Chien Lee and Bailin Hao, *Genomics, Proteomics and Bioinformatics* **2** 192 (2004).
10. Hoong-Chien Lee, Shuyu Zhang and Bailin Hao, *Chaos, Solitons and Fractals*, **11** 825 (2000).
11. Bailin Hao, Ji Qi, *J. Bioinf. & Comput, Biol.*, **2** 1 (2004).
12. G. M. Garrity *et al.*, *Taxonomic Outline of the Bacteria and Archaea*, Rel. 7.7, 6 March 2007, Michigan State University.
13. Ji Qi, Bin Wang and Bailin Hao, *J. Mol. Evol.* **58** 1 (2004).
14. Lei Gao, Ji Qi, Jiandong Sun and Bailin Hao, *Science in China* (Series C Life Science) **50** 587 (2007).
15. Bailin Hao, Ji Qi, and Bin Wang, *Mod. Phys. Lett. B* **17**, 91 (2003).
16. Ji Qi, Hong Luo and Bailin Hao, *Nucl. Acids Res.* **32**, Web Server Issue, W45 (2004).
17. Xiaoli Shi, Huimin Xie, Shuyu Zhang and Bailin Hao, *J. Korean Phys. Soc.* **50** 118 (2007).
18. Li Xia and Chan Zhou, *J. Syst. Sci. & Complexity* **20** 18 (2007)

SPIN PRECESSION AND INTERFERENCE IN TWO-DIMENSIONAL ELECTRON GAS

CHING-RAY CHANG
Department of Physics, National Taiwan University, Taipei 10617, Taiwan

JYH-SHINN YANG
Institute of Optoelectronic Science, National Taiwan Ocean University, Keelung 202, Taiwan

We investigate the spatial behavior of spin precession of traversing electrons in a two dimensional electron gas, and, also, the property of spin interference in square loop devices with Rashba and Dresselhaus spin-orbit couplings. Treating the effects due to the two coupling mechanisms by means of a spin rotation operator, we develop a convenient framework for studying the property of spin precession. We first derive analytical expressions for the spin precession which allow a more concrete description of the spatial distribution of the spin orientation. For example, the properties of spin precession, such as the rotation axis, the rotation angle and the cone angle, can be easily determined. We then extend the analytic framework to derive the spin-orbit coupling-induced phase for spin interference in square rings; this procedure makes the optimal control of the interference condition more convenient, and the spin filter more accessible experimentally.

1. Introduction

The spin precession effect due to the spin-orbit (SO) coupling in mesoscopic semiconductor structures has been one of the most important and attractive topics in the emerging field of spintronics from both the academic and the practical perspective [1–4]. In particular, the SO coupling provides a promising means of manipulating electron spin in semiconductor nanostructures purely by external electric fields [5] or gate voltages [6]. The SO coupling in zinc-blende semiconductor heterostructures usually consists of the two dominant mechanisms. One is due to the structure inversion asymmetry (SIA) [7], called the Rashba effect, which can be controlled by the applied gate voltage or by means of specific design heterostructures [8].

The other is the bulk inversion asymmetry (BIA) [9], called the Dresselhaus effect, which induces SO coupling whose strength is either material specific or controlled by the strain in bulk GaAs structures [10]. The SO coupling causes the spin state splitting, equivalent to an effective magnetic field (EMF) acting on the spin [11], and the transport electron then undergoes spin precession within the two-dimensional electron gas (2DEG). This unique property due to inversion asymmetry in 2DEG opens the possibility of semiconductor spintronic devices such as the spin-field-effect transistor [7, 12, 13] and the spin filter [14, 15]. To date a real device has not yet been realized, and a full understanding of the effects due to the SIA and BIA on the behavior of spin precession is thus crucial to manufacturing such devices [1, 13].

In general, the effective magnetic field depends on the electron transport direction, while in a [001]-grown zinc-blende semiconductor with equal coupling constants in the Rashba and Dresselhaus terms, it is independent of the transport direction [16]. Similar behavior is expected for the [110] Dresselhaus (linear) model [17]. As a result, in these two special cases, the spin precession angle depends only on the traveled distance but not on the motion direction of the electron. Accordingly, the electron precesses as a helix and the configuration of the spin orientation is persistent against any momentum dependent (but spin independent) scattering. The so-called persistent spin helix (PSH) thus results [16, 17]. Recently, it has been shown that the interplay of SO couplings can substantially suppress the D'yakonov-Perel' spin-relaxation mechanism or the Elliot-Yafet spin-flip mechanism in the above model system, producing long spin lifetimes. The utilization of this novel property then makes a high performance transistor possible [12, 18].

In spintronics applications, ideal filters capable of producing one type of spin-polarized electron currents are highly desirable and useful. Very recently, a perfect spin filter was theoretically predicted by Hatano *et al.* [15], who showed that the SO coupling can be regarded as non-Abelian or $SU(2)$ gauge fields which impose spin dependent phases on the traveling electron. By further adjusting the Rashba coupling strength and the external magnetic field, they achieve ideal spin-filters, in which one spin component undergoes a fully destructive interference while the other undergoes a fully constructive one. However, the polarization efficiency of these filters, which exploit only the SIA effect, is expected to be low due to the uncertainty in the rotation angle of the spin, which is due to the uncertainties in the detected position, the Rashba coupling strength, and the randomization of spin states caused by the scattering. Several theoretical analyses suggest that the influence of BIA effect on the spin filtering efficiency could be significant in some cases [12, 14].

In this paper, we shall focus on the spin precessional properties of conduction electrons in quantum wells subject to both Rashba and Dresselhaus couplings. Treating the two SO couplings on an equal footing, we develop a convenient framework based on the spin rotation operator to describe in general the behavior of spin precession during the electron transport. Our approach provides also a natural scheme for investigating the spin interference effect in one-dimensional ballistic polygon loops. We also analytically derive the spin-orbit induced phase for the spin interference in a square ring, with which we can optimally tune the related parameters to meet the spin interference condition. This analysis makes the spin filter more accessible experimentally.

2. Theory

Let us suppose that an electron with a definite spin is injected via an ideal point contact into a 2DEG without inversion symmetry. The 2DEG is assumed to be semi-infinite so that any boundary effects may be neglected. Setting the growth direction of the model layer to be [001], and the x and y axes to be [100] and [010] respectively, the single-particle Hamiltonian including both the Rashba and the Dresselhaus terms is given by [2]

$$H = \frac{1}{2m^*}(p_x^2 + p_y^2) + \frac{\alpha_R}{\hbar}(p_y\sigma_x - p_x\sigma_y) + \frac{\alpha_D}{\hbar}(p_x\sigma_x - p_y\sigma_y)\,, \qquad (1)$$

where m^* is an effective band mass, $p_{x,y}$ are the electron's momentum, $\sigma_{x,y}$ are the Pauli spin matrices, and α_R and α_D denote the coupling constants measuring the strengths of the SIA and BIA effects respectively. It is easy to verify that

$$\psi_\pm(\vec{r}) = e^{i\vec{k}_\pm\cdot\vec{r}}\chi_\pm = e^{i\vec{k}_\pm\cdot\vec{r}}\frac{1}{\sqrt{2}}\begin{pmatrix} 1 \\ \mp ie^{i\varphi} \end{pmatrix} \qquad (2)$$

are the eigenstates of Eq. (1), with eigenenergies

$$E_\pm = \frac{1}{2m^*}(\hbar\vec{k}_\pm)^2 \pm \gamma(\phi)k_\pm\,, \qquad (3)$$

where $\chi_\pm$ denote the eigenspinors, the position vector $\vec{r} = (x, y)$, and the in-plane wave vector $\vec{k} = (k_x, k_y)$ with $k_x = k\cos\phi$ and $k_y = k\sin\phi$, where ϕ is the traveling angle of electrons or the azimuthal angle of $\vec{k}$. The symbols

γ and φ are defined as

$$\gamma(\phi) = \sqrt{\alpha_R^2 + \alpha_D^2 + 2\alpha_R\alpha_D \sin(2\phi)}\,, \tag{4a}$$

$$\varphi = \arg[\alpha_R \cos\phi + \alpha_D \sin\phi + i(\alpha_R \sin\phi + \alpha_D \cos\phi)]\,. \tag{4b}$$

Clearly, the double signs in Eqs. (2) and (3) indicate that there are two values of k, or two available states, corresponding to the same energy due to the SO coupling. For later convenience we define the two Fermi wave vectors: $k_+ = k_F - \Delta k$ and $k_- = k_F + \Delta k$, where $k_F = (2m^* E_F)^{1/2}/\hbar$ with E_F being the Fermi energy, and $\Delta k = m^*\gamma/\hbar^2 \ll k_F$ for weak SO couplings.

Our purpose now is to calculate the SO-induced spin precession along the motion direction. We proceed in the following manner: First, we consider an arbitrarily injected spin at $r = 0$, $|\psi(0)\rangle = (a\ b)^\dagger$ with $aa^* + bb^* = 1$, and expand $|\psi(0)\rangle$ in terms of the basis states, Eq. (2), obtaining the coefficients:

$$\langle\chi_+|\psi(0)\rangle = c_+\,; \qquad \langle\chi_-|\psi(0)\rangle = c_-\,, \tag{5}$$

where $c_\pm = a \pm i\exp(-i\varphi)b$. Then, after a displacement $\vec{r}$ along the ϕ direction, the precessing electron will be in the state

$$\psi(\vec{r}) = c_+ e^{i\vec{k}_+\cdot\vec{r}}\chi_+ + c_- e^{i\vec{k}_-\cdot\vec{r}}\chi_-\,. \tag{6}$$

After some algebra, we arrive at

$$\psi(\vec{r}) = e^{i\vec{k}_F\cdot\vec{r}} \begin{pmatrix} \cos\eta & \sin\eta e^{-i\varphi} \\ -\sin\eta e^{i\varphi} & \cos\eta \end{pmatrix} \begin{pmatrix} a \\ b \end{pmatrix}, \tag{7}$$

where $\eta = m^*\gamma|\vec{r}|/\hbar^2$. Obviously, the 2×2 matrix in Eq. (7), denoted by U, stands for the rotation operator of spin 1/2; thus it describes a process in which the spin of the traversing electron undergoes precession around the unit vector $\hat{n} = (\sin\varphi, -\cos\varphi, 0)$ with a rotation angle of 2η. Note that the spin rotation axis $\hat{n}$ just points along the effective in-plane magnetic field due to the SO coupling, spanning the angle $\varphi - \pi/2$ relative to the x axis. As shown below, the formalism of Eq. (7) provides a convenient framework for investigating the properties of spin transport, such as the spin precession and the interference in 2DEG subject to the two SO couplings.

3. Results and Discussion

We first investigate the spatial features of spin precession due to the SO coupling by employing the present scheme. We assume that the electron has a conserved wave vector, and is free to move along any straight line path in the 2DEG. For brevity, the case with $\alpha \geq 0$ and $\beta \geq 0$ is only considered here. The corresponding results for the other cases can be inferred by symmetry.

In order to study the spin configuration, we need to find the expectation value of the spin vector $\langle \vec{S} \rangle = (\hbar/2)\langle \vec{\sigma} \rangle$ along the x, y, and z axes. With Eq. (2), straightforward algebra yields

$$\langle \sigma_x \rangle = \sin\theta_s\{-\sin(\phi_s - \varphi)\sin\varphi + \cos(2\eta)[\cos\phi_s + \sin(\phi_s - \varphi)\sin\varphi]\} \\ - \cos\theta_s \sin(2\eta)\cos\varphi\,, \tag{8a}$$

$$\langle \sigma_y \rangle = \sin\theta_s\{\sin(\phi_s - \varphi)\cos\varphi + \cos(2\eta)[\sin\phi_s - \sin(\phi_s - \varphi)\cos\varphi]\} \\ - \cos\theta_s \sin(2\eta)\sin\varphi\,, \tag{8b}$$

$$\langle \sigma_z \rangle = \sin\theta_s \sin(2\eta)\cos(\phi_s - \varphi) + \cos\theta_s\cos(2\eta) \tag{8c}$$

for injected spin, $\chi_{inj} = \langle \cos(\theta_s/2), \sin(\theta_s/2)\exp(-i\phi_s) \rangle^{\dagger}$ where θ_s and ϕ_s are the directional angles. We recall that identical formulae were presented in [13], but the present ones make the behavior of the spin precession more transparent. Obviously, the last terms are due to the out-of-plane component of injected spin, while the other terms are due to the in-plane one. The first term in the curly bracket, namely the component of the injected spin along the spin rotation axis, is always kept constant along a straight line path, while the other terms are a sinusoidal function of x and y which results from the spin precession. Clearly, Eqs. (8) demonstrate that the spatial behavior of spin precession under SO coupling can be classically treated as a precessional motion of magnets with constant velocity traveling through uniform, in-plane magnetic fields. In other words, moving with the traversing spin we see that the spin is precessing on a cone around the EMF where the cone angle is the smaller one of $|\phi_s - \varphi + \pi/2|$ or $|\phi_s - \varphi - 3\pi/2|$. Note that the spin direction remains unchanged when $\theta_s = \pi/2$ and $\phi_s = \varphi \pm \pi/2$ in Eq. (8); this corresponds to eigenspinor states, Eq. (2), leading to null spin precession. Interestingly, as the spin undergoes the full cycle of spin precession, i.e., half of the spin-precession angle $\eta = n\pi$ with n being an integer, we reach the same spin direction as was the original one. Here the spin of the traversing electron returns exactly to the original orientation after arriving at these specific positions. The special contours can be found to be a family of parallel lines, or concentric circles, or ellipses, for the cases with $\alpha_R/\alpha_D = \pm 1$, $\alpha_R\alpha_D = 0$, or others, respectively (see the figures below).

Using Eqs. (7) and (8) we can understand in depth the effects of two SO coupling mechanisms on the spatial distribution of the spin orientation.

As an explicit illustration, we set $\theta_s = \pi/2$ and $\phi_s = \pi/8$. We now begin the case with pure Rashba coupling or SIA effect ($\alpha_D = 0$). Using Eq. (7), we easily find that the spin precession angle is $2m^*\alpha_R r/\hbar^2$, which is proportional to the traveled length as well as to the coupling strength. The direction of the spin rotation axis is $\hat{p} \times \hat{z}$, always perpendicular to the propagating path of the traversing electron, and spanning an angle $\phi - \pi/2$ with the x axis. This fact shows that the Rashba field is circularly polarized, and invariant under the rotation about the z axis of the 2DEG plane. Figure 1(a) clearly displays the Rashba spin precession (RSP) projected onto the x-y plane, $\langle\sigma_x\rangle$ and $\langle\sigma_y\rangle$. Since the spin rotation axis is always perpendicular to the electron momentum, the RSP along the transport path generally behaves simply like a pendulum swinging about the direction of EMF, except for two special paths, $\phi = -3\pi/8$ and $5\pi/8$ in the present case (see the crossed shaded line with a large slope). Along these two paths the spin polarization of traversing electrons is always perpendicular to its transport direction, or parallel to the Rashba field, and the spin is subject to zero torque, leading to the occurrence of null spin precession. In contrast, as the electron travels along the polarization direction of injected spin (see the crossed shaded line with a small slope), the spin exhibits an upright precession with a maximum cone angle $\pi/2$, since the spin polarization spans a right angle with the Rashba field. Note that the prominent

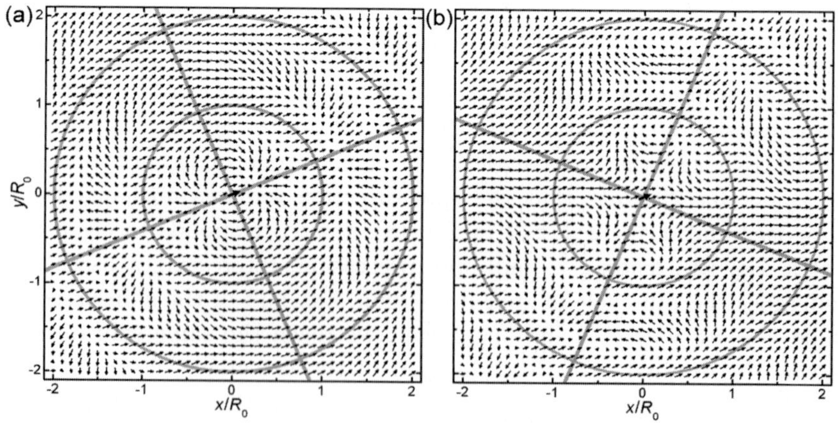

Fig. 1. Spin precession due to pure Rashba (a) and pure Dresselhaus (b) spin-orbit couplings. The length unit $R_0 = \pi\hbar^2/m^*\alpha_R$ (a) or $\pi\hbar^2/m^*\alpha_D$ (b). The injected spin (shown by the bold arrow at the center) is in-plane polarized, spanning an angle $\pi/8$ with the x axis. The meaning of the crossed shaded lines and circular contours is described in the text.

contours corresponding to the full cycles of spin precession are a family of concentric circles (see the shaded curves), due to the rotation symmetry of the Rashba fields.

In the case with only the Dresselhaus coupling ($\alpha_R = 0$), the spin precession angle is $2m^*\alpha_D r/\hbar^2$, suggesting a directional independence of the strength of effective fields, similar to the Rashba case. The spin rotation axis is along $(x/r, -y/r, 0)$, spanning an angle $-\phi$ with the x axis; the direction of EMF is no more perpendicular to the electron's momentum, except for the specific paths at $\phi = \pm\pi/4$ (and $\pm 3\pi/4$). Thus, the injected spin encounters another type of EMFs, and hence the spatial behavior of Dresselhaus spin precession (DSP) on the general path is quite different from the Rashba one [see Fig. 1(b)], while at $\phi = \pm\pi/4$ ($\pm 3\pi/4$), the DSP exhibits the similar spatial behavior with the RSP. Although the DSP appears to be more complicated than the RSP, the main feature of position-dependent spin orientation can be still explained in terms of the relevant spin rotation axis and cone angle. It is worth noting that similar to the Rashba case, there also exist two specific types of contours: the two crossed straight-line paths and the family of concentric circles. As the electron travels along the $-\pi/8$ and $7\pi/8$ paths (see the crossed shaded line with a small slope), the polarization of the injected spin is just along the rotation axis and the spin direction remains unchanged during transport since there is no action on the spin by the EMF. By contrast, as the electron travels along the $3\pi/8$ and $-5\pi/8$ paths (see the crossed shaded line with a large slope), the spin direction is always perpendicular to the EMF, thus the spin exhibits an upright precession with a maximum cone angle $\pi/2$. Interestingly, the special contour corresponding to the full cycles of spin precession is also a family of concentric circles (see the shaded curves), same as in the Rashba case.

In a case with combined Rashba and Dresselhaus couplings, the spin undergoes precession around the resultant of EMFs due to the two couplings, and then the spatial distribution of spin orientation can be obtained by superposing the RSP and the DSP. Since the spin rotation axes due to the two EMFs are generally not collinear, except those along the symmetry-preferred paths, the spin texture cannot be easily determined. However, for cases dominated by the SIA or BIA effects, the pattern of spin precession, as expected, should exhibit the RSP-like or the DSP-like behavior as shown in Figs. 2(a) and 2(b). Notably, the above-mentioned two crossed paths appear again, where the combined effects make the paths with null spin precession turn towards the $-\pi/4$ ($-3\pi/4$) axes. Interestingly, the special contours become ellipses (see the shaded curves). In addition, the short axes

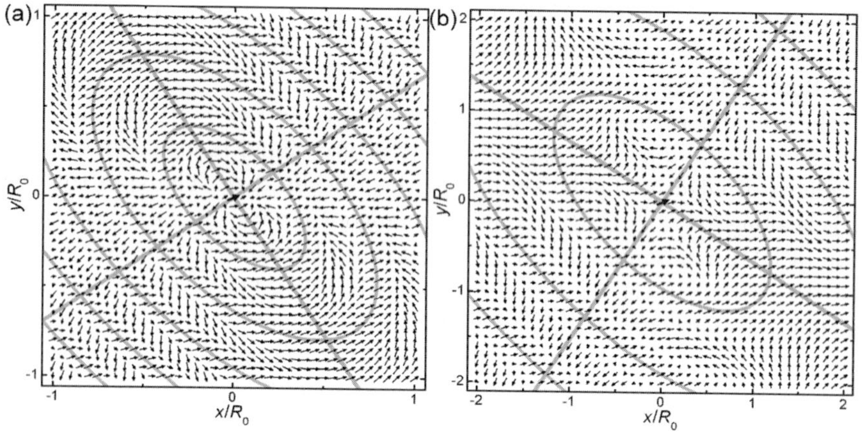

Fig. 2. Patterns of spin precession under the combined Rashba and Dresselhaus spin-orbit couplings. The coupling ratios $\alpha_R/\alpha_D = 3$ (a) and 1/3 (b). The length unit $R_0 = \pi\hbar^2/m^*\alpha_D$. The injected spin (shown by the bold arrow at the center) is in-plane polarized, spanning an angle $\pi/8$ with the x axis. The meaning of the crossed shaded lines and ellipses is described in the text.

of ellipses all align along the [110] direction due to the high spin precession rate along these paths, on which the EMFs, due to that the BIA and SIA effects are parallel to each other and enhanced. In the special case of terms of equal coupling strength ($\alpha_R = \alpha_D > 0$), the spin precession angle is $(2^{3/2}m^*\alpha_R/\hbar^2)|x+y|$, and the directions of spin rotation axis are just $[1-10]$ for $x+y>0$ and $[-110]$ otherwise. Therefore, the fixed direction of the resultant EMFs on either side of the $3\pi/4$ axis reasonably make the spin precession angle be merely proportional to the net displacement along the direction [110], irrespective of the traveled path and the initial spin orientation. As a result, the special pattern of spin precession, called the PSH [16, 17], is developed as shown in Fig. 3(a). This pattern exhibits a translation invariance with a period of the spin-precession length, $\pi\hbar^2/2m^*\alpha_R$, or the distance between the neighboring contours with full spin-precession cycles. A similar pattern of PSH can be also observed in Fig. 3(b) for the case $\alpha_R = -\alpha_D > 0$, where the Dresselhaus field reverses the direction, and the axis of translation invariance rotates 90° counterclockwise, compared with Fig. 3(a).

We now study the spin interference in square rings with corners at $(0,0)$, $(\ell,0)$, (ℓ,ℓ) and $(0,\ell)$; the lower-left and the upper-right corners are in contact with two ideal leads. Assume now that the SO coupling exists only within the loop, but it is absent in the two leads. Here we focus on the SO

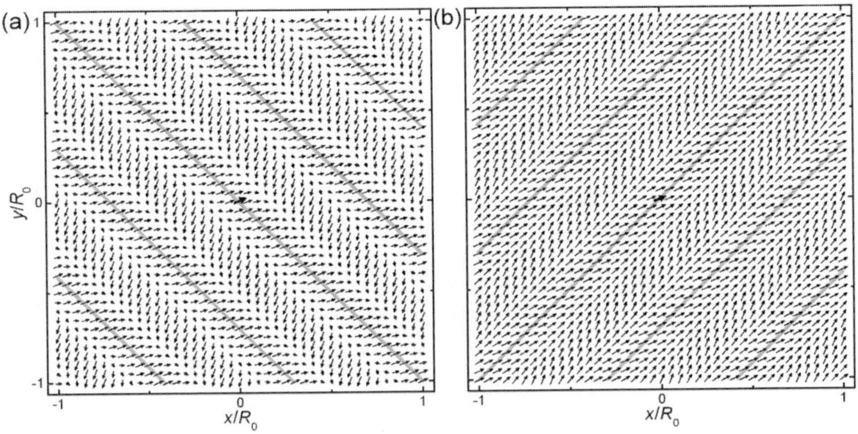

Fig. 3. Persistent spin helix precession under equal Rashba and Dresselhaus spin-orbit coupling strengths. The coupling constants: $\alpha_R = \alpha_D > 0$ (a) and $\alpha_R = -\alpha_D > 0$ (b). The length unit $R_0 = \pi\hbar^2/m^*|\alpha_D|$. The injected spin (shown by the bold arrow at the center) is in-plane polarized, spanning an angle $\pi/8$ with the x axis. The meaning of shaded lines is described in the text.

coupling induced phase, and neglect the effect of external magnetic fields, since it just creates an additional spin-independent phase of the same sign for both up and down spins. Suppose now that an incident spin ψ_{inc} from the left lead is split into a pair of partial waves at the lower-left corner, and then each of them follows the bottom-right path I (counterclockwise) or the left-top path II (clockwise). They finally merge at the upper-right corner, and enter into the right lead, giving rise to spin interference. Using the above spin-rotation operator U, the outgoing spin $\psi_{I,II}$ along the path I (II) can be described by the successive matrix product of spin rotation operators along the traveled path, i.e., $\psi_I = U_I\psi_{inc}$ and $\psi_{II} = U_{II}\psi_{inc}$, where $U_I = U_yU_x$ and $U_{II} = U_xU_y$, with subscripts x and y denoting the x- and y-oriented paths, respectively [19, 20]. Owing to the unitarity of the matrix U, one can readily perform the matrix product. We now seek to find the phase difference between the two paths, or equivalently, the phase factor that the traversing electron acquires upon circling the square ring:

$$U_{\text{phase}} = U_{II}^{+}U_I = \begin{pmatrix} a & b \\ -b^* & a \end{pmatrix}, \tag{9}$$

where $a = 1 - 2u^2 + 2iu[\cos^2\beta + \sin^2\beta\sin(2\varphi_0)]$ and $b = u(\cos\varphi_0 + \sin\varphi_0)(1+i)$, with $\beta = m^*\ell({\alpha_R}^2 + {\alpha_D}^2)^{1/2}/\hbar^2$, $\varphi_0 \equiv \varphi(\phi = 0) = \tan^{-1}(\alpha_D/\alpha_R)$ and $u = \sin^2\beta\cos(2\varphi_0)$. The eigenvalues of the 2×2 matrix

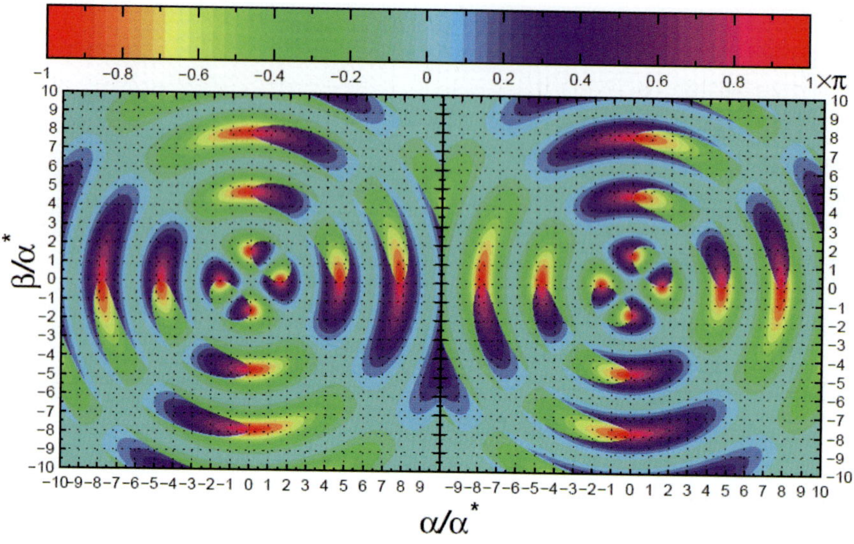

Fig. 4. Spin-orbit phases of the tilted up spin (left) and the tilted down spin (right) for a square ring patterned in the Rashba-Dresselhaus [001] 2DEG. Here $\alpha^* = (\hbar^2/m^*\ell)\sin^{-1}(2^{-1/4})$.

of the phase factor (9) are

$$\lambda = 1 - 2u^2 \pm 2i\sqrt{u^2(1-u^2)}\,. \tag{10}$$

Alternatively, the eigenvalues can be expressed as $\exp(\pm i\phi_{RD})$, since their modulus is unity. Thus the phases acquired by the tilted up and down spins are $\pm\phi_{RD}$. This shift is due to the SO coupling. One can check that for the pure Rashba coupling, the result Eq. (10) agrees with that in [15] by putting $\varphi_0 = 0$. In addition, setting $\phi_{RD} = \pi/2$ and $\alpha_D = 0$, we again obtain the same condition for achieving the ideal filters with Eq. (30) in [15], i.e., $\sin^4\beta = 1/2$, corresponding to the specific Rashba strength, $\alpha^* = (\hbar^2/m^*\ell)\sin^{-1}(2^{-1/4})$.

Figure 4 displays the SO phases of the tilted-up and tilted-down spins plotted as functions of the reduced coupling constants, α_R/α^* and α_D/α^*. Clearly, the SO phases are symmetric under $(\alpha_R, \alpha_D) \to (-\alpha_R, -\alpha_D)$, but anti-symmetric under $(\alpha_R, \alpha_D) \to (\alpha_D, \alpha_R)$. For the special case with $|\alpha_R| = |\alpha_D|$, there is a null SO phase, $\phi_{RD} = 0$ from $\cos(2\varphi_0) = 0$. Physically these results from the fact that the effects of the EMFs on the x- and y-oriented paths are identical, thus the spin which follows either the path I or II acquires the same spin-dependent SO phase. Moreover, in the vicinity of the circles $R_n = ({\alpha_R}^2 + {\alpha_D}^2)^{1/2}/\alpha^* \approx n\pi$, with n being positive integers, we have very small SO phases, $\phi_{RD} \approx 0$. Between these circles are the bands where the SO phases oscillate at $R_{n+1/2}$ with a full period 2π within each quadrant. The fine structure inside the band clearly shows discontinuities of ϕ_{RD} [see, for example, the boundaries between $\pi_{RD} \approx -0.4\pi$ (green/light) and $\phi_{RD} \approx 0.4\pi$ (blue/dark)]. In particular, the parameters proposed for ideal spin filters [15], $\alpha_R = \pm\alpha^*$ and $\alpha_D = 0$, corresponding to $\phi_{RD} = \pm\pi/2$, are quite unstable. Interestingly, the presence of α_D moves ϕ_{RD} away from these discontinuities, and makes the spin filter therefore experimentally more accessible. An example is given by taking $\alpha_R = \pm 4.55\alpha^*$ and $\alpha_D = -1.8\alpha^*$ in the second band ($\approx R_{3/2}$).

To conclude, we have developed a framework based on the spin rotation operator, this formalism allows one to study the behavior of spin precession and spin interference. We then derived analytical formulae for a model system subject to the Rashba and the Dresselhaus SO couplings. In comparison to previously published results, the present expression for spin precession is more transparent and useful for analyzing the spatial distributions of the RSP, the DSP, and of composite cases. In particular, the unique features of spin precession patterns can be easily identified within the spin rotation operator formalism. We have found that the behavior of RSP can be satisfactorily explained by the rotational symmetry of the Rashba field perpendicular to the traveling direction of electrons. In contrast, the DSP exhibits a more complicated behavior due to the anisotropic property of the EMF; however, it preserves the RSP behavior on the $\pm\pi/4$ and $\pm 3\pi/4$ axes. This implies that the spatial behavior of the spin precession due to inversion asymmetry in the 2DEG along these four specific directions is always invariant, and the same as the behavior of the RSP, regardless of the influence of the Dresselhaus term. In particular, the condition for null spin precession, and the special contour where the spins all return to its original orientation are explicitly displayed. Finally, we analyzed the property of the SO phase due to the two couplings in square-ring interferometers, and proposed a practicable condition for spin interference such as might be used to construct the best spin filter. We expect that these findings shall allow us to properly manipulate the behavior of spin precession in spin transport devices employing the SIA and BIA effects.

References

1. R. Winkler, *Phys. Rev.* **B69**, 045317 (2004).
2. M.-H. Liu and Ching-Ray Chang, and S.-H. Chen, *Phys. Rev.* **B71**, 153305 (2005).

3. S. A. Wolf, D. D. Awschalom *et al.*, *Science* **294**, 1488 (2001); Igor Žutić, Jaroslav Fabian, and S. Das Sarma, *Rev. Mod. Phys.* **76**, 323 (2004).
4. *Semiconductor Spintronics and Quantum Computation*, edited by D. D. Awschalom, D. Loss, and N. Samarth (Springer, Berlin, 2002).
5. K. C. Nowack, F. H. Koppens, Yu. V. Nazarov, and L. M. K. Vandersypen, *Science* **318**, 1430 (2007).
6. S. Datta and B. Das, *Appl. Phys. Lett.* **56**, 665 (1990).
7. Yu. A. Bychkov and E. I. Rashba, *J. Phys.* **C17**, 6039 (1984).
8. T. Koga, J. Nitta, T. Akazaki, and H. Takayanagi, *Phys. Rev. Lett.* **89**, 046801 (2002).
9. G. Dresselhaus, *Phys. Rev.* **100**, 580 (1955).
10. Y. Kato, R. C. Myers, A. C. Gossard, and D. D. Awschalom, *Nature* (London) **427**, 50 (2003).
11. E. A. D. E. Silva, *Phys. Rev.* **B46**, 1921 (1992).
12. J. Schliemann, J. C. Egues, and D. Loss, *Phys. Rev. Lett.* **90**, 146801 (2003).
13. M.-H. Liu and Ching-Ray Chang, *Phys. Rev.* **B73**, 205301 (2006).
14. D. Z.-Y. Ting and X. Cartoixà, *Phys. Rev.* **B68**, 235320 (2003).
15. N. Hatano, R. Shirasaki, and H. Nakamura, *Phys. Rev.* **A75**, 032107 (2007).
16. B. A. Bernevig, J. Orenstein, and S. C. Zhang, *Phys. Rev. Lett.* **97**, 236601 (2006).
17. M.-H. Liu, K.-W. Chen, S.-H. Chen, and Ching-Ray Chang, *Phys. Rev.* **B74**, 235322 (2006).
18. X. Cartoixà, D. Z.-Y. Ting, and Y.-C. Chang, Appl. *Phys. Lett.* **83**, 1462 (2003).
19. V. M. Ramaglia, V. Cataudella, G. De Fillipps, and C. A. Perroni, *Phys. Rev.* **B73**, 155328 (2006).
20. Son-Hsien Chen and Ching-Ray Chang, e-print cond-mat/0709.0079v2.

ATOMS AND IONS; UNIVERSALITY, SINGULARITY AND PARTICULARITY: ON BOLTZMANN'S VISION A CENTURY LATER

MICHAEL FISHER

Institute for Physical Science and Technology,
University of Maryland
E-mail: claremon@ipst.umd.edu

Ludwig Boltzmann died by his own hand 101 years ago last September. He was a passionate believer in atoms: underlying thermodynamics, he felt, lay a statistical world governed by the mechanics of individual particles. His struggles against critics — "Have you ever seen an atom?" taunted Ernst Mach — left him pessimistic. Nevertheless, following Maxwell and clarified by Gibbs, he established the science of Statistical Mechanics. But today, especially granted our understanding of critical singularities and their universality, how much do atomic particles and their charged partners, ions, really matter? The answers we have also met opposition. But Boltzmann would have welcomed the insights gained and approved of applications of statistical dynamics to biology, sociology, and other enterprises.

INSIGHTS FROM COMPUTER SIMULATION

E. G. WANG

Institute of Physics,
Chinese Academy of Sciences
E-mail: egwang@aphy.iphy.ac.cn

Pattern formation and decay in the early stage of growth is fundamental to many materials physics and chemistry. Understanding the complex interplay between factors that influence the evolution of surface-based nanostructures can be challenging and so computer simulation can play an important role in providing insight. In this talk, I will first introduce a one-, two-, and three-dimensional Ehrlich-Schwoebel (ES) barrier in kinetics-driven growth. Within this framework, I will show how one can control the island shape, the island instability, and the film roughness efficiently. Furthermore, I will discuss a novel concept: a true upward adatom diffusion on metal surface, which is beyond the traditional Ehrlich-Schwoebel (ES) barrier model. This process offers new indications as how to use ab initio kinetic Monte Carlo simulation can uncover some of the building regulations of the evolution mechanism down to atomic-scale.

FIFTY YEARS OF HARD-SPHERE BOSE GAS: 1957–2007

KERSON HUANG

Physics Department, Massachusetts Institute of Technology, Cambridge, MA 02139, USA
kerson@MIT.EDU

Fifty years ago, Yang and I worked on the dilute hard-sphere Bose gas, which has been experimentally realized only relatively recently. I recount the background of that work, subsequent developments, and fresh understanding. In the original work, we had to rearrange the perturbation series, which was equivalent to the Bogoliubov transformation. A deeper reason for the rearrangement has been a puzzle. I can now explain it as a crossover from ideal gas to interacting gas behavior, a phenonmenon arising from Bose statistics. The crossover region is infinitesimally small for a macroscopic system, and thus unobservable. However, it is experimentally relevant in mesoscopic systems, such as a Bose gas trapped in an external potential, or on an optical lattice.

Keyword: Bose gas; hard sphere.

1. Introduction

On this occasion in celebrating Professor Chen Ning Yang's eighty-fifth birthday, I want to offer a bit of reminiscence of the time Professor Yang and I worked on the dilute hard-sphere Bose gas, and a bit of fresh understanding on this old subject.

I met Professor Yang for the first time in 1956 at the Institute for Advanced Study in Princeton, when I arrived as a post-doctorate from MIT. Long before that, of course, I had known him by reputation. I was enthralled by his beautiful result on the spontaneous magnetization of the Ising model,[1] and his work with T. D. Lee on phase transitions and the "circle theorem".[2,3]

When I met T. D. Lee in 1954, the first thing he told me was that Yang had just proposed a marvelous theory, in "which space was filled with gyroscopes at every point". He was, of course, referring to the Yang-Mills gauge theory of 1954, which has become the foundation of the standard model of elementary particles.[4]

We started working on the quantum-mechanical hard-sphere interaction, and published our first paper in 1957, exactly fifty years ago.[5] That was a memorable year, as Yang and Lee won the Nobel prize later that year for their work on parity violation.[6]

The hard-sphere Bose gas has become relevant experimentally since 1995. On the theoretical front, we have continued to gain new understanding up to now.

2. Hard Spheres

My Ph.D. thesis at MIT was on the saturation of nuclear forces.[9] Victor Weisskopf was my supervisor, but I worked on a daily basis with Sidney Drell, who was then assistant professor. Meson theory[a] predicted that nucleons interact with many-body forces,[10] and we wanted to see whether they lead to nuclear saturation. Drell and I did independent calculations, and bet a nickel whenever we disagreed. One of the problems we struggled with was how to treat the hard core in the nuclear potential. We used what was known as the "Jastrow wave function"[11,12] to do variational calculations, but it was not satisfactory from a basic point of view. I continued to think about this problem when I arrived at Princeton.

As it turned out, Yang also worked on the hard-sphere problem the year before, in collaboration with J. M. Luttinger; but abandoned the project because they got divergent results.

Yang and Luttinger replaced the hard-sphere potential by a delta function, and tried to calculate the energy in perturbation theory. For a quantum gas of N identical particles, the ground-state energy per particle to first order is

$$\frac{E_0}{(\hbar^2/2m)N} = 4\pi an \tag{1}$$

where a is the hard-sphere diameter, m the particle mass, and n the particle density. This result was first obtained by W. Lenz in 1929, by estimating the quantum-mechanical kinetic energy due to excluded volume.[13] However, the second-order result diverges.

[a]The prevailing meson theory was pseudoscalar theory with psudoscalar coupling, soon to be superceded by pseudovector coupling in the Chew-Low theory of the "three-three resonance".

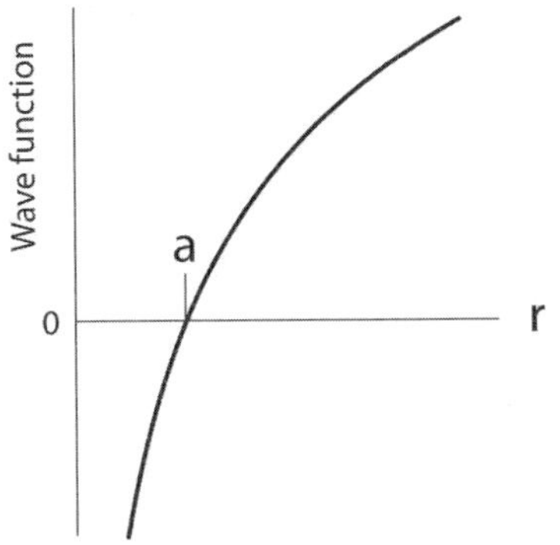

Fig. 1. The wave function of a particle in a delta-function potential vanishes at the scattering length a, which is an effective hard-sphere diameter. When contnued inside the hard core, it develops a $1/r$ singularity, which is spurious.

3. Pseudopotential

After thinking a bit about Yang and Luttinger's calculation, I thought I knew where the divergence came from and how to fix it. As a graduate student at MIT, I had worked with Weisskopf on the electron-neutron interaction. This was treated earlier by Fermi,[14] using a delta-function potential called a pseudopotential, with the proviso that it be used only to first order in perturbation theory. And Weisskopf knew how to improve it.[15]

The point can be illustrated in the two-body problem. As illustrated in Fig. 1, the delta function potential $\delta(r)$ makes the wave function vanish at a, which is what we wanted. But if we continued it to inside the hard sphere, the wave function diverges like $1/r$. We should ignore the wave function inside a, but there is no way to do this in perturbation theory, for the formalism tells us to integrate over all space, and the $1/r$ singularity leads to a spurious divergence. Weisskopf's solution is to modify the potential through the replacement

$$\delta(r) \to \delta(r)\frac{\partial}{\partial r}r\,. \tag{2}$$

The differential operator $\partial/\partial r$ expunges any term proportional to $1/r$, but does nothing if the wave function is regular at $r = 0$. The pseudopotential is really a way to introduce a boundary condition just outside of the potential, to properly describe scattering from the potential. The correct pseudopotential is in fact

$$V(r) = \left(\frac{4\pi a\hbar^2}{m}\right)\delta(r)\frac{\partial}{\partial r}r\,. \tag{3}$$

Here a is the scattering length, the long-wavelength limit of

$$-\frac{\tan \delta(k)}{k} \tag{4}$$

where k is the scattering wave number, and $\delta(k)$ the s-wave scattering phase shift. For hard-sphere scattering, $\delta(k) = -ka$, and the scattering length coincides with the hard-sphere diameter. As a quantum-mechanical operator, the pseudopotential is non-hermitian; but it gives an effective Hamiltonian, and should yield real eigenvalues.

I outlined the argument to Yang, and he was somewhat dubious at first, but left a note on my desk that evening:

"I thought about it. What you said about $\tan \delta$ was correct."

We then calculated the energy levels of a Bose gas and a Fermi gas in perturbation theory, with no divergences.[5] We also calculated the virial expansion of the equation of state of a hard-sphere Bose gas, and studied how the interaction modifies the Bose–Einstein condensation.[16] This was a fulfillment of what Yang and Luttinger had started to do.

4. An Encounter with Pauli

Yang was away when I gave a seminar at the Institute for Advanced Study on our calculations. Besides local luminaries, including Oppenheimer, Wigner, and Dyson, there was a very special guest in the audience, Wolfgang Pauli. When the seminar began, he started to nod and fall sleep, until I mentioned that the effective Hamiltonian was not hermitian.

"Not hermitian?" Pauli opened his eyes.

Oppenheimer, sitting next to Pauli, leaned over to him and said, "That's really OK, because ..."

"It is not hermitian! I do not like this!" said Pauli.

In his more youthful days, I guess, Pauli would have chased me out of the room, or left the room himself. But that day, he nodded back to sleep after some grumbling, and I was able to finish the seminar.

At that time, I was renting a room in a gracious old house in Evelyn Place owned by a European lady, Mrs. Loewy. Her late husband must have been a serious scholar, for the living room was filled to the ceiling with old tomes, mostly in German. When I went back to the house that evening, Mrs. Loewy was waiting for me.

"Guess who came to visit today", she beamed, "Pauli!"
"He said you gave a seminar, and it was '*nicht dumm*'[b]!"

5. Peritization

We calculated the ground state energy per particle to second order in perturbation theory:

$$\frac{E_0}{(\hbar^2/2m)N} = 4\pi an\left[1 + A_2\left(\frac{a}{L}\right)\right] \tag{5}$$

where L is the linear dimension of the box containing the system, and

$$A_2 = 2.837297\cdots \tag{6}$$

In Ref. [5], this constant was cited as 2.73, a poor estimate due to my clumsy numerical work. It was calculated exactly by M. Lüscher[17] years later, and Yang[18] found an elementary derivation of his result. To the same order of approximation, the excited states are separated from the ground state by an energy gap ϵ_0:

$$\frac{\epsilon_0}{\hbar^2/2m} = 8\pi an\,. \tag{7}$$

The second-order correction to the ground state energy per particle vanishes in the macroscopic limit, when the system becomes infinite with finite density. So, we got zero instead of a divergence.

Upon closer examination, however, there is trouble. A graphical analysis shows that the expansion has the general form[19]

$$\begin{aligned}\frac{E_0}{(\hbar^2/2m)N} - 4\pi an = {}& \frac{1}{NL^2}\left[A_2\left(\frac{aN}{L}\right)^2 + A_3\left(\frac{aN}{L}\right)^3 + \cdots\right]\\ &+\frac{1}{N^2L^2}\left[B_2\left(\frac{aN}{L}\right)^2 + B_3\left(\frac{aN}{L}\right)^3 + \cdots\right] + \cdots\end{aligned} \tag{8}$$

We see that the expansion parameter is aN/L, which diverges in the macroscopic limit. Thus, beyond second order, the divergence comes back in a different form.

We left the issue unresolved over the summer of 1957, when I went to Bell Labs and worked on the electron gas with David Bohm and David Pines.[20] I think Yang spent the summer working mainly on weak interactions with T. D. Lee. In their spare time, they toyed with the idea that summing the

[b]'not dumb'

series in (8) horizontally might yield a finite result. That is, one should sum the most divergent terms.

The first horizontal line in (8) has the form $[1/(NL^2)]f(aN/L)$, which in the macroscopic limit must tend to a function of the finite combination $aN/L^3 = an$. Assuming $f(x)$ to be a power, one finds that the power must be 5/2.[21] Thus, instead of a power series, the expansion would become one in fractional powers.

A few years earlier, Lev Landau[22] summed the "leading logarithms" (most divergent terms) in the perturbation series for the renormalized charge of the electron in quantum electrodynamics (QED), and obtained the provocative result now referred to as "triviality": the renormalized charge of the electron is zero in the limit of infinite cutoff. Noticing a pattern, Abraham Pais proposed the name "peritization" for the summing of the most divergent terms.

Back from summer vacation, T. D. Lee joined us to actually sum the series. After some hard work we succeeded, and obtained the well-known result[23]

$$\frac{E_0}{(\hbar^2/2m)N} \to 4\pi an\left[1+\frac{128}{15\sqrt{\pi}}\sqrt{na^3}+O(na^3)\right], \tag{9}$$

where the arrow denotes the macroscopic limit.

In the old power-series scheme, the excited states were separated from the ground state by an energy gap. In the new scheme, the gap closes, and we have phonon excitations with energy ω_k, given by

$$\frac{\omega_k}{\hbar^2/2m} = k\sqrt{k^2+16\pi an}\,. \tag{10}$$

This identifies the sound velocity to be $(\hbar/2m)\sqrt{16\pi an}$, which agrees with that calculated independently from the compressibility of the ground state.

After the work was done, we learned that N. N. Bogoliubov[24] had derived the same spectrum through a transformation of the Hamiltonian that now bears his name. The Bogoliubov transformation, which has become a standard tool, may be looked upon as an easy and elegant way to "peritize" the perturbation series.[c]

Landau's peritization in QED can also be done more elegantly – through the renormalization group.[25]

[c]There is a technical difference, however. In the Bogoliubov transformation, one assumes that the ground-state occupation is a number instead of an operator. We did not make that assumption in our explicit summation.

6. Higher-Order Corrections

Wu[35] calculated the ground-state energy per particle to the order beyond $a^{5/2}$:

$$4\pi a n \left[8\left(\frac{4\pi}{3} - \sqrt{3}\right) n a^3 \ln(12\pi n a^3) \right].$$

The logarithmic dependence was interesting and unexpected. However, a variational calculation[36] disagrees strongly with this correction, while agreeing very well with the lower-order terms. A more recent Monte Carlo calculation[37] obtained a similar result. In particular, agreement with the above correction took a dramatic turn for the worse when $na^3 > 10^{-2}$. What this means remains a mystery.

7. Superfluidity

All particles have zero momentum in the ground state, in the ideal gas,. However, the occupation of the zero-momentum level is less than 100% in the interacting gas. The deficit is known as the "depletion fraction", and was calculated to be $8/3\sqrt{na/\pi}$. However, at zero temperature, the entire system is a superfluid, i.e., $n_s = n$, where n_s is the superfluid density. We can find n_s through the linear response of the system to an infinitesimal hypothetical velocity field.[36] A calculation at zero temperature, using the pseudopotential, was carried out by Huang and Meng[37] in the presence of a quenched random potential. As expected, $n_s = n$ in the absence of randomness, but n_s decreases as the strength of the random potential increases, and there is a critical strength above which $n_s = 0$. This indicates that the random potential pins the condensate, preventing it from moving. The result has been verified by an independent calculation.[38] Thus, we can have condensate without superfluidity.

8. Two-Dimensional Bose Gas

In spatial dimension $D = 2$, a Bose gas does not have a condensate, interacting or not. This is due to the absence of long-range correlations, due to strong local fluctuations of the quantum phase of the ground-state wave function. A local condensate does exist, but it contains topological defects, which are vortices. In a low-temperature phase, when vortices and antivortices form bound states, the system is a superfluid. Thus, we can have superfluidity without condensate. At a higher temperature, the Kosterlitz–Thouless phase transition occurs, whereby the bound vortex pairs become ionized, and supefluidity ceases.[39]

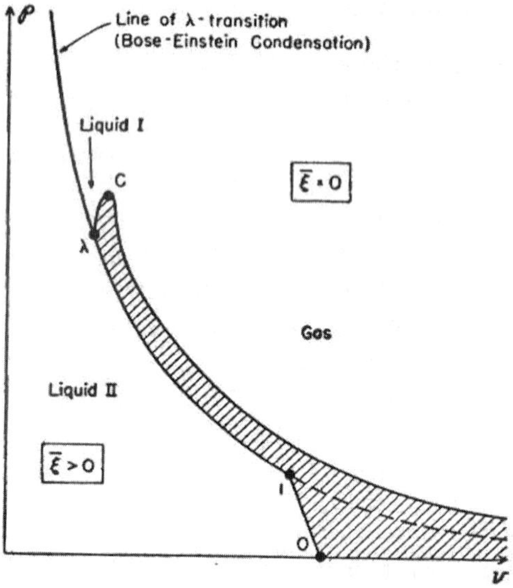

Fig. 2. Equation of state of dilute Bose gas with hard-sphere repulsion plus short-range attractive interaction, as represented in a P-V diagram. the shaded region is the modification to the transition line of the ideal gas, showing the formation of a liquid phase. The Bose–Einstein condensation occurs in the liquid phase, just as in liquid helium. The whole shaded region shrinks about the ideal gas line and slides off to the left, when interactions are turned off.

9. Liquid Helium and Atomic Trap

The physical motivation to study the hard-sphere Bose gas was the understanding of the superfluid transition in liquid helium. However, that transition in liquid helium happens not in the gas phase, but in the liquid phase. To reproduce the phase diagram, at least qualitatively, one must produce a first-order gas-liquid phase transition, and that requires the addition of an attractive potential. I was able to do that, and show how the phase diagram emerges from that of the ideal Bose gas, as one turns on the hard-sphere and attractive interactions.[26,27] This is shown in Fig. 2.

There, the matter lay for almost forty years until 1995, when a dilute Bose gas was experimentally realized in an atomic trap, and Bose–Einstein condensation was observed.[7,8] This inspired a fury of theoretical activity,[28] all based on an interaction described by an effective hard-sphere diameter, the s-wave scattering length.

Yang[29] returned to the subject to treat the trapped atomic gas, whose density is not uniform in space, through a local density approximation. I participated in formulating a generalized Thomas-Fermi approximation.[30]

10. Crossover between Ideal and Interacting Gas

It has been a puzzle why the expansion parameter changes from aN/L to $a^{1/3}n$. I recently realize that this is an expression of a crossover from ideal gas to interacting gas.[31]

The idea of crossover originates in the study of the renormalization group pioneered by Kenneth Wilson and Michael Fisher.[32] Generally, a system is governed by different effective Hamiltonians at different scales, each corresponding to a "fixed point" in the space of Hamiltonians. When the scale changes, the system "crosses over" from one fixed point to the other. For example, a system confined between two plates may look like a 2D system, but when the plate separation increases, it will eventually cross over to 3D behavior.

The expansion parameter in an extremely dilute Bose gas is

$$\frac{aN}{L} \sim \sqrt{\frac{\text{System size}}{\text{Mean-free-path in condensate}}}\,. \tag{11}$$

The mean-free-path in the condensate is of order $(n\sigma)^{-1}$, with scattering cross section $\sigma \sim Na^2$, where the factor N comes from Bose enhancement. For small aN/L, the system is a Knudsen gas, in which collisions are infrequent. In another region with small mean-free-path, one goes over to the hydrodynamic regime. This is illustrated in Fig. 3 for a gas coming out of a hole in the wall.

The crossover happens because of the factor N in aN/L, which makes the parameter diverge in the macroscopic limit. It is thus a property of Bose statistics, and there is no corresponding phenomenon in the Fermi gas.

Can we observe the crossover experimentally? To answer this question, let me retrace the origin of the idea.

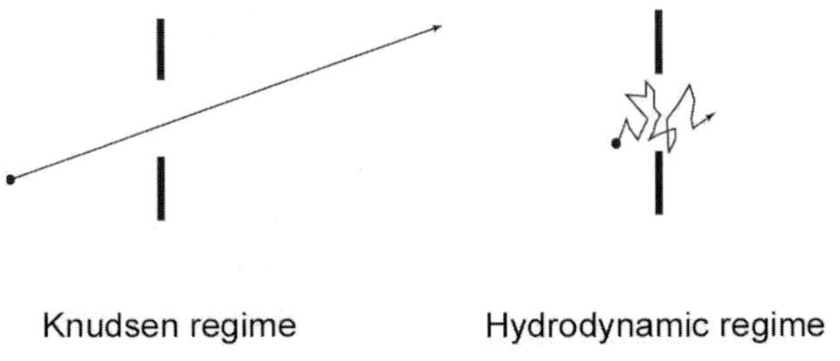

Fig. 3. In the Knudsen (or collisionless) regime, the mean-free-path is much greater than relevant lengths in the system. The opposite is true in the hydrodynamic regime. There is a crossover between these regimes in an interacting Bose gas as a consequence of Bose statistics, as explained in the text.

In the ideal Bose gas, Bose–Einstein condensation occurs in 3D because the state of nonzero momentum cannot accommodate more than a certain number of particles. This number depends on temperature, and when the temperature falls below a critical value, these states can no longer accommodate all the particles present, and the excess is forced into the state of zero momentum, forming the Bose–Einstein condensate. The critical temperature $T_c^{(0)}$ for an ideal Bose gas in 3D, of density n, is

$$k_B T_c^{(0)} = C n^{-2/3} \tag{12}$$

where $C = 2\pi\hbar^2/m[\zeta(3/2)]^{-2/3}$.

I calculated the critical number for the hard-sphere gas, using the virial series we calculated forty years ago, and obtained a shift in transition temperature T_c due to the interaction[33]

$$\Delta_0 \equiv \frac{T_c - T_c^{(0)}}{T_c^{(0)}} = c_0 \sqrt{a n^{1/3}} \tag{13}$$

where $c_0 = 8\sqrt{2\pi}/3[\zeta(3/2)]^{2/3} \approx 3.527$. This disagrees strongly with numerical computations based on a scheme due to Gordon Baym,[34] which give a linear dependence on a:

$$\Delta_1 = c_1 a n^{1/3} . \tag{14}$$

The constant c_1 is of order unity, and varies according to the computation scheme.

The two results differ because they were calculated in different regions. My calculation was done in the gas phase, approaching the transition from the high-temperature side, whereas the other calculations approach it from within the condensed phase. These give different answers because of the crossover.

11. Observation of Crossover

The Knudsen regime is relevant for kinetic processes, such as effusion through a hole, when the mean-free-path is much larger than the dimension L of the hole, as illustrated earlier in Fig. 3. For a macroscopic system in thermal equilibrium, L is the size of the whole system, and the Knudsen region consists of an infinitesimally small neighborhood of zero density, and is of little experimental interest.

For a Bose gas condensed in a potential or on an optical lattice, however, the Knudsen regime spans a region accessible to experimentation. In this case, one can vary the system size, particle number, and even the scattering length through Feshbach resonances. The relevant dimension L in a harmonic trap of frequency ω_0 is the harmonic length

$$L = \sqrt{\frac{\hbar}{m\omega_0}}\,. \tag{15}$$

The trapped gas behaves like an ideal gas when $aN/L \ll 1$, and crosses over to the Thomas-Fermi regime when $aN/L \gg 1$. During the crossover, the size of the condensate increases from the ideal-gas radius

$$R_0 \sim L \tag{16}$$

to the Thomas–Fermi radius[30]

$$R_{\mathrm{TF}} \sim \left(\frac{aN}{L}\right)^{1/5} L\,. \tag{17}$$

The excitations change from ideal-gas states in the trap to collective oscillations, the lowest ones being the "breathing" (monopole) and quadrupole modes.[28]

As illustration, take $a = 1$ nm, $L = 100$ μm. The critical particle number is

$$N_c = \frac{L}{a} = 10^5\,. \tag{18}$$

The transition temperature depends on N, and interpolates between two fractional shifts:

$$\Delta(N) = \begin{cases} \Delta_0 & (N \ll N_c) \\ \Delta_1 & (N \gg N_c) \end{cases}\,. \tag{19}$$

The values given in Eqs. (13) and (14) are for the uniform gas, but may be used for order-of-magnitude estimates. The two curves corresponding to Δ_0 and Δ_1 are plotted in Fig. 4. By interpolating between them, we get an estimate of how the actual shift might behave. The fractional shift is more than ten times greater in the Knudsen gas, indicating that the condensate is more stable against excitation owing to the energy gap.

Currently, experiments involve millions of atoms, and are well in the hydrodynamic regime. It would be interesting to explore the Knudsen regime on an optical lattice, not so much for measuring the temperature shift, but for kinetic properties of adding atom to and extracting from the lattice.

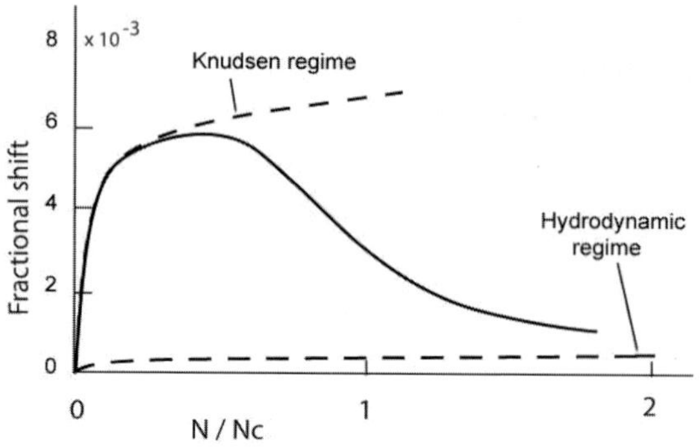

Fig. 4. Fractional shift of the Bose–Einstein transition temperature for the Knudsen and hydrodynamic regimes are shown as dotted curves. The solid line is a visual interpolation between the two.

12. Conclusion

After fifty years, the dilute hard-sphere Bose gas still enjoys a vigorous life, and we wish the same for Professor Yang.

13. Questions and Answers

C. N. Yang:

I want to make several remarks. The first one was: actually around 1956 to 1957, I wrote more than 10 papers more or less about the subject. But later on since there were no experimental verification and, theory-wise we hit the stone wall. So I tend to forget it. In the 1990's there was a Prof. B. A. Li of Kentucky who wrote an article about my contributions to physics and he asked me to name something like 10-areas. I recently checked, I did not mention these 10 papers. Of course, I'm very happy now that with this beautiful and the unbelievably exciting new developments about Bose–Einstein condensation, and the cold atom physics that Prof. Cohen-Tannoudji had told us about, the subject is revived and the people are now measuring sound velocities in both Fermi and Bose gases. And I checked the science citation — my papers of 1956, 1957 and around that time — netted something like 600 references. It is a bit like a revival of an old friend who has been dead for many years!

My second remark is that the subject is extremely subtle and reading some of the new papers, I'm afraid a number of theorists working on it did not completely appreciate the subtlety. I would advise them to read the paper of Lee, Huang and me, ... which by the way was discussed at the same time, in a different area, by Bardeen, Cooper and Schrieffer, in particular, its pairing, the elimination of the singularity that Kerson just mentioned and the excitation spectrum, etc. are very non-trivial and very subtle things.

The third remark I want to make is that if you read the paper, the three of us calculated the number of particles in the momentum zero state. We made a series expansion of that and clearly demonstrated that it is less than 100%, when you have any interaction. And yet because of this beautiful experiment by Andronikashvili the percentage is 100%. So did we make a mistake? I don't think so. In fact also Onsager and Penrose had in another way made an estimate, and they said the momentum zero state for zero temperature in liquid helium is only 8%. And not 100%. And yet this beautiful Russian experiment showed that the condensate had to be 100%. OK, I know the secret. I believe I know the secret. It's because of the shape of the excitation spectrum near the ground state. What happens is that although you have less than 100% condensed from a mathematical calculation, the part which is not condensed does not form new states. There is a gap. Now I think that is a very deep subject and I wish I have the energy that Kerson Huang wished me to have to explore this. I think this is the most important subject. Thank you.

David Thouless:

I think I would like to take the chair's privilege to answer at least part of Prof. Yang's question. Yes, we've known for a very long time that there's a difference between the proportion of particles in the condensate and the superfluid density. We've known what it is. And part of our understanding comes from work in the 1960's by Brian Josephson who discussed the fact that the critical experiment relating to the superfluid fraction appeared to be thermal coefficients and were not related to the magnetization in the magnetic analogy.[42] So that's something which is water under the bridge a very long time ago. I want to make another comment which is of course at the same time as the beautiful work on the Bose system came out, there were a lot of papers on the Fermi system where, in a sense, the experimental challenge is much greater because people knew an awful lot about plasma frequency and such things and they had to get it right. And there was a lot done; I won't mention all the names but the work of Bohm and Pines, two of the pioneers on resumming terms, was a major part of getting that straight, even if it was less elegant than the work on the Bose–Einstein system.

Michael Fisher:

I would just like to add a little more to the history because one of the aims of my own talk was to bring out what is wrong with the BCS theory: it was such a successful theory that you couldn't see where the critical fluctuations were buried! To reveal that, was one of the great things David Thouless did which, for me, removed my own blinkers entirely. So I'm very glad that he has now stressed that we do have a good understanding of the superfluid density, ρ_s. Again the issue really concerns another oversight of the simple mean field theory or of the Landau approach. One has to understand the difference between the condensate fraction, $n_0(T)$, and $\rho_s(T)$. The distinction is crucial in what David would be too modest to mention, namely, the Kosterlitz–Thouless superfluid transition in two dimensions where n_0 is rigorously zero but ρ_s is not! The point is not unrelated to the question I was asked about $dv = 2 - \alpha$. One has to understand the correlation functions in quite a lot of detail; and, even in a simple situation, putting it all together is more subtle than one's first thoughts would suggest. Indeed, one has to look much more carefully at what an experiment like Andronikashvili's actually measures.

Kerson Huang:

I believe the difference is that the condensate fraction is an equilibrium property where ρ_s is a transport coefficient.

Michael Fisher:

No, for me that is really not the best interpretation. Rather, as Josephson showed, ρ_s is closer to an equilibrium elastic modulus or, even, an interfacial free energy. Once more this is where finite-size effects come in. The issue, indeed, goes back to Onsager because one can ask: "What is the interfacial tension?" And answer to say: "Well I have two phases and I have to work out the excess free energy of the interface between them." But what Onsager did in his original paper was to say: "Well let me just change the boundary conditions." So he considered an antiferromagnet and chose the number of layers in the system. When one went around below T_c one found $+, -, +, -, +, -, \ldots$ layers. If the whole system has an even number of layers, the alternation fits together as one pure phase.

But Onsager said: "Well, let's take it as an odd number." Then the sequence of layers goes $+, -, +, -, \ldots$, until one comes to $+, +$, or to $-, -$, that is, a *mismatch seam* must arise: consequently the free energy is not going to be as low. So there is a difference in free energy between odd and even for the full system and Onsager identified that as simply the *interfacial tension.* Now if you go to an XY-type spin system, the spin direction can represent the local quantum-mechanical phase of the equilibrium state of a Bose system. Then, in analogy to the antiferromagnet one can impose boundary conditions twisted, say, by an angle of π or less. That adds to the free energy; the change is proportional to ρ_s and decreases as $1/L^2$. Thus $\rho_s(T)$ is defined via boundary conditions in a fully equilibrium situation.

Kerson Huang:

But that's just one way to calculate the transport coefficient!

Michael Fisher:

Well, not really! The question is must one think of ρ_s as a dynamical effect? And the answer is "No!" Rather it describes a fully equilibrium free energy induced by boundary conditions [see e.g., M. E. Fisher, M. N. Barber and D. Jasnow, *Phys. Rev. A* **8**, 1111 (1973)]. And one might recall that if one took periodic boundary conditions and actually set up a flow, that would then decay by spontaneously nucleating vortices! So just as Professor Yang emphasized, the more you look into the details at each level, the greater the subtleties! It can be difficult to keep abreast: but I think its most appropriate that you, Kerson, reminded us of these fascinating but tricky issues of crossover and finite-size effects[d] and that this discussion was stimulated by Professor Yang's question.

David Thouless:

We can have one more question.

Anonymous audience:

You told that there's no Bose condensate in 2D system but can we conclude that there's no superfluidity for 2D system?

Kerson Huang:

Well Michael just said that in spite of the fact that there's no Bose condensate there is superfluidity. There is. The Thouless' transition.

[d]A related crossover from ideal Bose-gas to interacting lambda-point behavior has been studied experimentally by J. D. Reppy and coworkers [*Physica* **126B**, 335 (1984)] and studied theoretically by P. B. Weichmann *et al.*, *Phys. Rev. B* **33**, 4632 (1986).

References

1. C. N. Yang, *Phys. Rev.* **85**, 809 (1952).
2. C. N. Yang and T. D. Lee, *Phys. Rev.* **87**, 404 (1952).
3. T. D. Lee and C. N. Yang, *Phys. Rev.* **87**, 410 (1952).
4. C. N. Yang and R. L. Mills, *Phys. Rev.* **96**, 191 (1954).
5. K. Huang and C. N. Yang, *Phys. Rev.* **105**, 767 (1957).
6. T. D. Lee and C. N. Yang, *Phys. Rev.* **104**, 245 (1956).
7. M. H. Anderson, J. R. Ensher, M. R. Mathews, C. E. Wieman and E. A. Cornell, *Science* **269**, 198 (1995).
8. K. B. Davis, M.-O. Mewes, M. R. Andrews, N. J. van Druten, D. S. Durfee, D. M. Kurn and W. Ketterle, *Phys. Rev. Lett.* **75**, 3969 (1995).
9. S. D. Drell and K. Huang, *Phys. Rev.* **91**, 1527 (1953).
10. M. Lévy, *Phys. Rev.* **88**, 25 (1952).
11. R. Jastrow, *Phys. Rev.* **79**, 389 (1950).
12. R. Jastrow, *Phys. Rev.* **81**, 165 (1951).
13. W. Lenz, *Z. Phyzik* **56**, 778 (1929).
14. E. Fermi, *Ricerca Sci.* **7**, 13 (1936).
15. J. M. Blatt and V. F. Weisskopf, *Theoretical Nuclear Physics* (Wiley, New York, 1952), p. 74.
16. K. Huang, C. N. Yang and J. M. Luttinger, *Phys. Rev.* **105**, 776 (1957).
17. M. Lüscher, *Comm. Math. Phys.* **105**, 153 (1986).
18. C. N. Yang, *Chin. J. Phys.* **25**, 80 (1987).
19. K. Huang, in Hard-sphere Bose gas: The method of pseudopotentials, *The Many-Body Problem: Les Houches Summer School in Physics* (Dounod, Paris, 1958), p. 602.
20. D. Bohm, D. Pines and K. Huang, *Phys. Rev.* **107**, 71 (1957).
21. T. D. Lee and C. N. Yang, *Phys. Rev.* **105**, 1119 (1957).
22. L. D. Landau, in *Niels Bohr and the Development of Physics*, ed. W. Pauli (McGraw-Hill, New York, 1955).
23. T. D. Lee, K. Huang and C. N. Yang, *Phys. Rev.* **106**, 1135 (1957).
24. N. N. Bogoliubov, *J. Phys. USSR* **II**, 23 (1947).
25. K. Huang, *Quarks, Leptons, and Gauge Fields*, 2nd edn. (World Scientific, Singapore, 1992), p. 192.
26. K. Huang, *Phys. Rev.* **115**, 765 (1959).
27. K. Huang, *Phys. Rev.* **119**, 1129 (1960).
28. F. Dalfovo, S. Giorgini, L. Pitaeveskii and S. Stringari, *Rev. Mod. Phys.* **71**, 463 (1999).
29. T. T. Chou, C. N. Yang and L. H. Yu, *Phys. Rev. A* **53**, 4259 (1996).
30. E. Timmermans, P. Tommasini and K. Huang, *Phys. Rev. A* **55**, 3645 (1997).
31. K. Huang, Bose-Einstein condensation of a Knudsen gas, arXiv:0707:2234 [cond-mat.stat-mech] (2007).
32. K. Huang, *Fundamental Forces of Nature: the Story of Gauge Fields* (World Scientific, Singapore, 2007) gives a popular account of the renormalization group and crossover in Chap. 22.
33. K. Huang, *Phys. Rev. Lett.* **83**, 3770 (1999).

34. M. Holtzmann, G. Baym, J. P. Blaizot and F. Laloë, *Phys. Rev. Lett.* **87**, 129403 (2001), and references therein.
35. T. T. Wu, *Phys. Rev.* **115**, 1390 (1959).
36. R. Drachman, *Phys. Rev.* **131**, 1881 (1963).
37. S. Giorgini, J. Boronat and J. Casulleras, *Phys. Rev. A* **60**, 5129 (1999).
38. See, for example, K. Huang, Bose-Einstein condensation and superfluity in *Bose-Einstein Condensation*, eds. A. Griffin, D. W. Snow and S. Stringari (Cambridge University Press, Cambridge, England, 1995).
39. K. Huang and H. F. Meng, *Phys. Rev. Lett.* **69**, 644 (1992).
40. S. Giorgini, L. Pitaevskii and S. Stringari, *Phys. Rev. B* **49**, 12398 (1994).
41. K. Huang, Quantum Field Theory, from Operators to Path Integrals (Wiley, New York, 1998), Chap. 18, gives a general discussion of two-dimensional systems.
42. B. D. Josephson, *Phys. Lett.* **21**, 608 (1966).

Quantum Physics

QUANTUM PHENOMENA VISUALIZED BY ELECTRON WAVES

AKIRA TONOMURA

Hitachi, Ltd., Hatoyama, Saitama 350-0395, Japan
Frontier Research System, RIKEN, Wako, Saitama 351-0198, Japan
Okinawa Institute of Science and Technology, 7542, Onna, Okinawa 904-0411, Japan
akira.tonomura.zk@hitachi.com

Quantum phenomena have become directly observable with the development of advanced techniques such as coherent field-emission electron beams, sensitive detectors, and microlithography. Examples are the single-electron build-up of an interference pattern, which contains, as R. Feynman describes in his textbook, the heart of quantum mechanics, and the Aharonov–Bohm (AB) effect, which indicates the physical significance of gauge fields. Using the AB effect, i.e., the fundamental principle behind the interaction of an electron wave with electromagnetic fields, new ways to directly observe previously unobservable microscopic quantum objects and phenomena were developed by detecting the phase of electrons.

Keyword: Aharonov–Bohm effect; field-emission; electron waves; phases.

1. Introduction

In 1897, electrons were identified as particles that are constituents of atoms.[1] Thirty years later, to explain the stability of atoms and other unexplicable data obtained until that time, electrons were also considered to have wave properties. The direct evidence for the wave properties of electrons was given by C. Davisson and L. H. Germer,[2] J. P. Thomson and A. Reid,[3] and S. Kikuchi *et al.*[4] They irradiated parallel electron beams onto materials and observed electron beams diffracted from arrays of atoms that comprise the materials; scattered electrons from the atoms were found to interfere with each other to form diffracted beams.

The wave nature of electrons is not limited to microscopic regions in atomic dimensions. In 1956, interference patterns formed by two electron beams were observed using an electron biprism.[5] Furthermore, electron interferometry has progressed greatly[6] because of the development of highly collimated, monochromatic field-emission electron beams,[7] just like optical interferometry progressed because of the advent of lasers. Our recently developed 1-MV field-emission electron beam[8] has the highest brightness (2×10^{10} A/cm^2·sr) and narrowest monochromaticity ($\Delta E/E = 5\times 10^{-7}$ min^{-1}) ever obtained. This paper reports both quantum-mechanical experiments made possible by bright electron beams, and the observation of phase objects by using the wave properties of electrons.

2. Development of Coherent Electron Beams

D. Gabor invented electron holography in 1949[9] to break through the resolution limit of electron microscopes, by using the wave nature of electrons. In the first step, a hologram, which is an interference pattern between an object wave and a reference wave, is formed with electrons, and in the second step, the electron wavefronts are reconstructed by illuminating a light reference wave onto the hologram. In the optical reconstruction stage, the effect of aberrations of electron lenses is optically compensated for, so as to improve the resolution.

We demonstrated the possibility of electron holography in 1968.[10] However, we were convinced from the experiments that bright electron beams, just like laser beams in optics, would be needed to make electron holography practically useful.

We soon began developing bright, yet monochromatic electron beams that are field-emitted from a pointed tip, and we have continued to do so. This electron source is extremely small, typically 50 Å in diameter, so it has to be immobile within a fraction of the source diameter. Therefore, we had to overcome these technical difficulties in order to prevent even the slightest mechanical vibration of the tip, the accelerating tube, the microscope column, or the slightest deflection of the fine beam by stray AC magnetic fields. Otherwise, the inherent high brightness of the electron beam would deteriorate.

After ten years of work, we developed an 80-kV electron beam[11] which was two orders of magnitude brighter than that of the thermal beams used then. Electron interference patterns became directly observable on a fluorescent screen, and as many as 3,000 interference fringes were recorded on film compared to 300 fringes till then. By using electron holography and bright electron beams, new information that could not be obtained by conventional electron microscopy became obtainable. For example, magnetic lines of force inside and outside ferromagnetic samples were directly and quantitatively observed in h/e flux units[12,13] in interference micrographs, which can be obtained optically in the reconstruction stage of electron holography.

After that, we continued to develop even brighter electron beams. A series of experiments to study the AB effect[14–17] were carried out with a 250-kV electron microscope. Magnetic vortices in metal superconductors[18] were observed with a 350-kV microscope,[19] and unusual behaviors of vortices peculiar to high-T_c superconductors[20–22] were observed using a 1-MV microscope.[23] These observations became possible by precisely detecting the phase of an electron wave using bright electron beams.

3. Electron Phase and Aharonov-Bohm Effect

Phase shifts of an electron wave transmitted through electromagnetic potentials, V and $\mathbf{A}$, can be derived from the Schrödinger equation. The phase S of an electron wave, especially when electromagnetic fields are sufficiently weak for the WKB approximation to be valid, can simply be expressed as follows:

$$S = \frac{1}{\hbar} \int (m\mathbf{v} - e\mathbf{A}) ds \,, \qquad (1)$$

where the line integral is carried out along an electron path. The effect of electrostatic potentials V is included in $\mathbf{v}$. Conversely, Eq. (1) can also give us the definition of electromagnetic potentials.

From Eq. (1), we can understand how electromagnetic potentials physically influence the electron phase. The first term, $\int m\mathbf{v}\, d\mathbf{s}/\hbar$, in this equation corresponds to the optical path length. The effect of vector potentials $\mathbf{A}$ is given by the second term.

By comparing these two terms, $-e\mathbf{A}$ can be interpreted as a kind of electron momentum. In exact terms, $-e\mathbf{A}$ is the momentum exchanged between the sources of the fields and an electron, and that occurs with the electron only because an electron has an electric charge, irrespective of whether it is at rest or moving. If a unit charge is placed in a magnetic field, the integral value of the field momentum (the vector product between electric and magnetic fields, $\mathbf{E} \times \mathbf{B}$) over all the space becomes equal to vector potentials $\mathbf{A}$ at the point of the charge.[24]

In 1959, Y. Aharonov and D. Bohm[25] theoretically predicted that a relative phase shift can exist even when electron beams pass only through spaces free of $\mathbf{E}$ and $\mathbf{B}$.[26] This effect was later called "the Aharonov–Bohm effect". They attributed this effect to potentials, V and $\mathbf{A}$, which are considered to have no physical meaning in classical physics.

The significance of the AB effect increased during the 1970s, in relation to the unified theories of all fundamental interactions in nature, where potentials are extended to "gauge fields" and regarded as the most fundamental physical quantity. T. T. Wu and C. N. Yang[27] stressed the significance of the AB effect in relation to the physical reality of gauge fields (potentials) as follows:

> "The concept of an SU_2 gauge field was first discussed in 1954. In recent years, many theorists, perhaps a majority, believe that SU_2 gauge fields do exist. However, so far there is no experimental proof of this theoretical idea, since conservation of isotopic spin only suggests, and does not require, the existence of an isotopic spin gauge field. What kind of experiment would be a definitive test of the existence of an isotopic spin gauge field? A generalized Bohm-Aharonov experiment would be."

However, potentials have long been regarded as mathematical auxiliaries, so some people questioned the existence of the AB effect, thus causing a controversy.[28,29] Although the AB effect had been experimentally tested[30–33] to exist for the magnetic case soon after its prediction, these results were attributed to the effect of magnetic fields leaking from both ends of finite solenoids or ferromagnets used in the experiments.[34,35]

We made a series of experiments[14–17] using tiny leakage-free magnetic samples of toroidal geometry, and by quantitatively measuring precision in leakage-free conditions using holographic interference microscopy.[36] By the last experiment[16] using a toroidal magnet covered with superconductors, which C. G Kuper[37] and C. N. Yang[38] proposed, the existence of the AB effect was conclusively confirmed.

We used the underlying principle of the AB effect to observe magnetic lines of force[12,13] and quantized vortices in superconductors[39] quantitatively, as electron interference micrographs, and to observe the movements of vortices dynamically[18] by Lorentz microscopy (out-of-focus transmission electron microscopy). Next, we report our experimental results on the AB effect and then its applications using coherent electron beams we developed over the past 40 years.

4. Confirmation Experiments on AB Effect

We carried out a series of experiments to clarify any ambiguities raised in the controversy, and we introduce the last experiment here,[16] which is considered to be the most conclusive up to the present. We used a toroidal ferromagnet instead of a straight solenoid, which has inevitable leakage fluxes from both ends of the solenoid. An infinite solenoid is experimentally unattainable, but an ideal geometry with no flux leakage can be achieved by the finite system of a toroidal magnetic field.[37] Furthermore, the toroidal ferromagnet was covered with a superconducting niobium layer to completely confine the magnetic field.

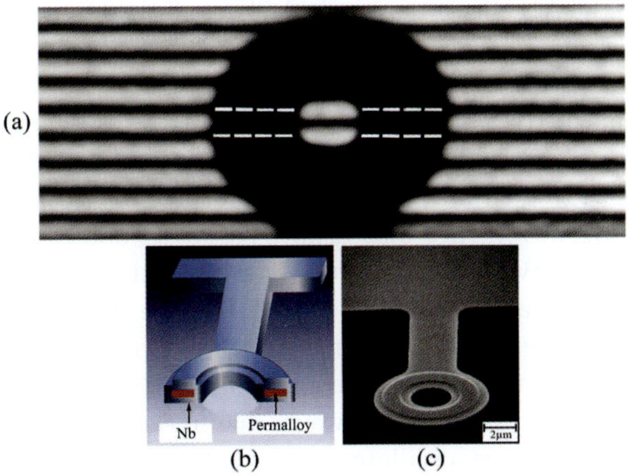

Fig. 1. Photographic evidence for AB effect. (a) Interference pattern. (b) Schematic of sample. (c) Scanning electron micrograph.

An electron wave was incident on a tiny toroidal sample fabricated using the most advanced lithography techniques, and the relative phase shift ΔS between two waves passing through the hole and around the toroid was measured as an interferogram.

Although samples that had various magnetic flux values were measured, the ΔS was either 0 or π. The conclusion is now obvious. The photograph in Fig. 1 indicates that a relative phase shift of π is produced, indicating the existence of the AB effect even when the magnetic fields are confined within the superconductor and shielded from the electron wave. An electron wave must be physically influenced by the vector potentials.

In this experiment, a quantization of the relative phase shift, either 0 or π, assured that the niobium layer surrounding the magnet actually became superconductive. When a superconductor completely surrounds a magnetic flux, then the flux is quantized to an integral multiple of quantized flux, $h/(2e)$. When an odd number of vortices are enclosed inside the superconductor, the relative phase shift becomes π (mod. 2π). For an even number of vortices, the phase shift is 0. Therefore, the occurrence of flux quantization can be used to confirm that the niobium layer actually became superconductive, that the superconductor completely surrounded the magnetic flux, and that the Meissner effect prevented any flux from leaking out. Therefore, we can conclude that electron waves passing through the field-free regions inside and outside the toroidal magnet are phase-shifted by π, although the waves never touch the magnetic fields.

Soon after the AB effect was conclusively confirmed by a series of experiments using electron beams, electrons inside metals were also found to exhibit the AB effect.

R. A. Webb of IBM used a tiny ring circuit to demonstrate that electrons inside metals also exhibit interference and the AB effect.[40] When 100 electrons enter a ring circuit, as a matter of course in classical physics, 100 electrons exit. However, they also behave as waves, and even a single electron can split into two partial waves. Therefore, the number of electrons that exit the ring can become 90 or 110 because of constructive or destructive interference, depending on their relative phases. Therefore, when magnetic flux passes through the ring circuit and changes the relative electron phase between the two partial waves due to the AB effect, the electron current, or the resistance, oscillates.

The AB effect was also detected in carbon nanotubes.[41] In a cylinder, electrons can take many different paths to get from one point to another along the axis of the cylinder: a direct route or a left or right-handed path. If magnetic flux passes through this cylinder, their relative phase changes due to the AB effect, thus changing the resistance. Ohm's law is no longer valid in this microscopic world, and the AB effect now plays an essential role in understanding the performance of ultra-microscopic devices. As recently reported,[42] metal carbon nanotubes can be changed to semiconductors due to the AB effect because a phase factor is added to the wavefunction, thus even changing the band structures.

As these examples demonstrate, the AB effect is encroaching upon the more macroscopic and practical world, although the AB effect has not yet been confirmed in an exact sense except for the electron-beam experiments.

5. Observation of Magnetic-Flux-Quantization Process

Using the technique developed for the confirmation of the AB effect, we can observe the process of magnetic flux being quantized, which C. N. Yang mentioned in 1983. He proposed a new experiment in his comments on my talk at ISQM '83,[38]

> "Your beautiful recent experiment has shown that, in a ring with minimum flux leakage, flux is not quantized. Can you also fabricate a small superconducting solenoid in an overall shape of a ring, with flux inside? If you can, then you can dramatically demonstrate flux quantization in the interference fringes' line-up matching."

When we carried out this experiment in 1986 and reported the result at ISQM '86, Prof. Yang asked the following:

> "Can you take a series of holograms as the temperature is lowered? This will allow you to see whether flux is expelled or sucked in, as the Nb becomes superconducting. Fairbank and his collaborator have done this in their experiment of 1961."

The result of the temperature dependence observation is as follows. When the temperature T of this sample is 300 K [Fig. 2(a)], the phase difference is 0.5 π. The fringe position inside the hole is a little bit above the outside fringes.

When T decreases to 15 K [Fig. 2(b)], the fringe inside the hole moves up to a level slightly lower than the middle line between the two outside fringes.

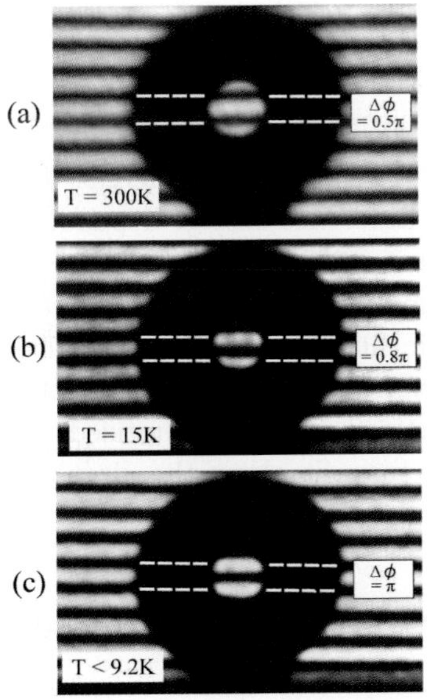

Fig. 2. Temperature dependence of relative phase shifts between two electron waves passing inside the hole and outside the toroidal magnet. (a) $T = 300$ K. (b) $T = 15$ K. (c) $T = 5$ K.

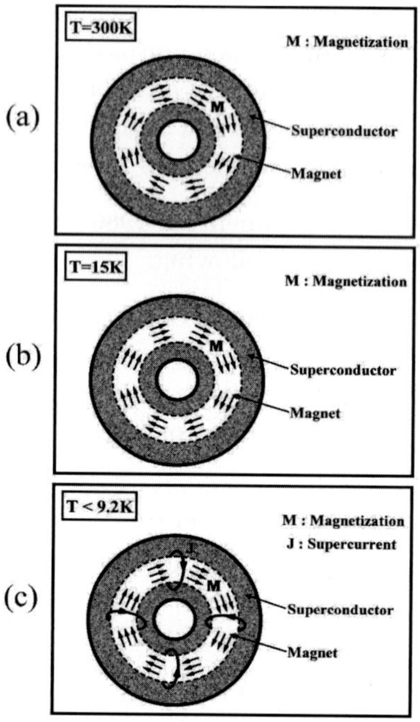

Fig. 3. What happens to the troidal magnet covered with superconductors when T decreases? (a) $T = 300$ K. (b) $T = 15$ K. (c) $T = 5$ K.

The phase difference is 0.8 π. What happens when the temperature further decreases below 9.2 K? The phase shift increased to exactly π [Fig. 2(c)].

Why did the phase difference change depending on the temperature?

When T decreases from 300 K to 15 K (see Fig. 3), the spin directions in the magnet are aligned due to the decrease in thermal fluctuations, thus increasing the magnetic flux by 5%. When T decreases below T_c, 9.2 K [Fig. 3(c)], a supercurrent circulating around the magnet is induced to adjust the total magnetic flux to be quantized, so an integral number of the Cooper-pair waves may fit along the circulating orbit. These results exactly confirm Yang's predictions.

6. Applications of AB Effect in Electromagnetic-Field Observation

The AB effect principle can be used to observe microscopic distributions of electromagnetic fields by detecting the phase of the transmitted electron beam. More specifically, the thickness of a specimen with a uniform material distribution can be observed as the thickness contours in the interference micrograph obtained through an electron holography process.[43] That is because the phase of an electron wave is shifted by the inner potential of the specimen when the wave passes through it.

Relative phase shifts can be detected in the conventional interference pattern with a $2\pi/4$ precision, but the precision increases up to $2\pi/100$ by using a phase-amplification technique peculiar to holography. This technique has helped detect thickness changes due to monatomic steps[44] and carbon nanotubes.[45]

6.1. *Magnetic lines of force*

In the case of pure magnetic fields, the phase shift is produced by vector potentials. When the phase distribution is displayed as a contour map, the micrograph can be interpreted in the following straightforward way.[12]

(1) Contour fringes in the interference micrograph indicate magnetic lines of force because no relative phase shift is produced between two beams passing through two points along a magnetic line.
(2) Contour fringes exhibit magnetic flux in units of h/e because the relative phase shift between two beams enclosing a magnetic flux of h/e is 2π.

An observation of magnetic lines of force inside a ferromagnetic fine particle is shown in Fig. 4. Only the triangular outline of this particle can be observed by electron microscopy. In the interference micrograph, two kinds of contour fringes appear: narrow fringes parallel to the edges indicate thickness contours in 200 Å units, and circular fringes in the inner region indicate in-plane magnetic lines of force in $h/(2e)$ flux units because the micrograph is phase-amplified two times and the specimen thickness is uniform in the inner region.

6.2. *Vortices in superconductors*

Interference microscopy is not the only technique that can be used to visualize the phase distribution. For example, some kind of phase object can be observed in an out-of-focus image called "Lorentz microscopy" because the phase change is transformed into an intensity change when the image is defocused. This method is convenient for real-time observation.

A quantized vortex in a superconductor, which acts as a pure weak-phase object under an illuminating electron beam, has actually been visualized as a spot in a defocused image, or a Lorentz micrograph.[18]

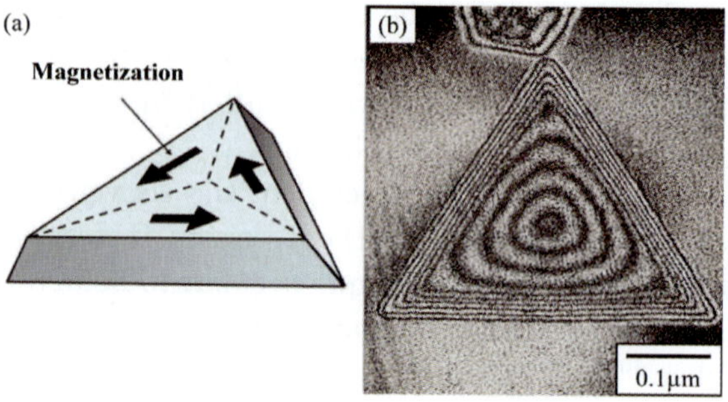

Fig. 4. Cobalt fine particle. (a) Schematic diagram. (b) Interference micrograph.

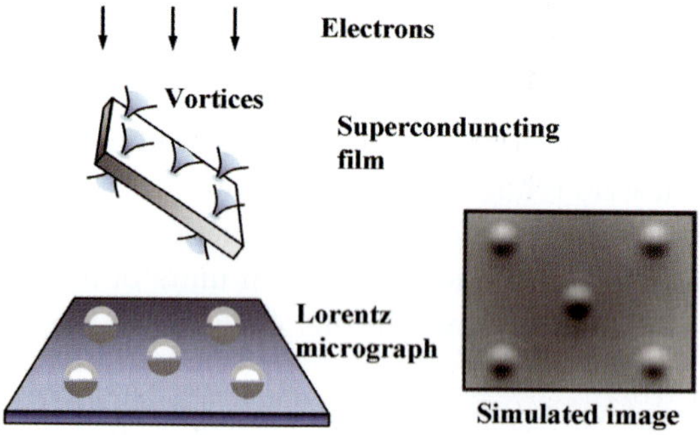

Fig. 5. Principle behind vortex observation.

The experimental arrangement for observing vortices in a superconducting thin film is shown in Fig. 5. When a magnetic field is applied to the tilted film, vortices are produced. Electrons passing through the film are phase shifted by magnetic fluxes of vortices due to the AB effect. Vortices can be observed by simply defocusing the electron microscope image. That is, when the intensity of electrons is observed in an out-of-focus plane, a vortex appears as a pair of bright and dark contrast features (Fig. 5).

We can thus observe the dynamics of vortices in real time by applying Lorentz microscopy. Those dynamics include behaviors of vortices at pinning centers and surface steps under various conditions of sample temperatures and applied magnetic fields. Vortices move in interesting ways as if they were living organisms.

An interesting example[46] is shown in Fig. 6, where two kinds of vortex images appear in a single field of view. They are vortices and antivortices produced in a niobium thin film when the 100 G magnetic field applied

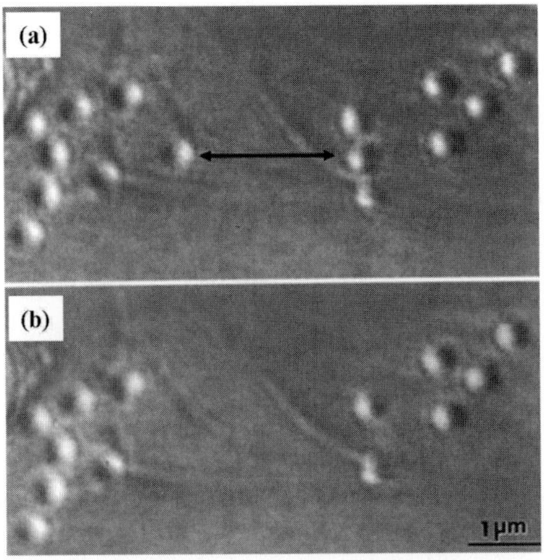

Fig. 6. Annihilation of vortices and antivortices in thin film of niobium. (a) Before annihilation. (b) After annihilation.

to the film is suddenly reversed and its magnitude increases. The original vortices are leaving the film, but cannot instantly do so because they are pinned down by defects, while the oppositely oriented vortices begin to penetrate the film from its edges. Where two streams of vortices and antivortices collide head-on, the vortex-antivortex pairs of the heads of those two streams annihilate each other. The directly observed pair annihilation appears to simulate that of particles and antiparticles.

6.3. *Manipulation of vortices*

We artificially controlled the motion of individual vortices using the "ratchet mechanism" aiming at vortex devices.[47–49] The ratchet system is well-known as a kind of wheel, which permits only "unidirectional rotation". Such a movement occurs under a (1) spatially-asymmetric periodic potential, and (2) symmetric AC driving force.

The asymmetric distribution of artificial defects that trap vortices was used as the ratchet system of vortices. As shown in Fig. 7, the vortices form an array of arrowheads, and the calculated potential was confirmed to be asymmetric along the vertical axis.

In Fig. 8, vortices are mostly trapped at defects, and some are untrapped, as shown in a Lorentz micrograph.[50] When a weak driving force is applied to vortices, only untrapped vortices began to move. The left

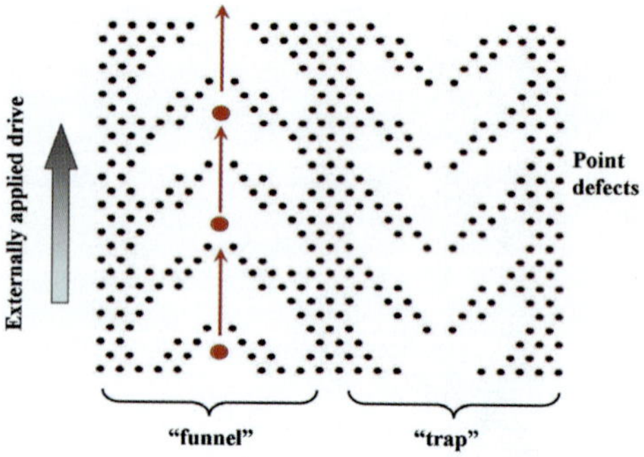

Fig. 7. Vortex ratchet system.

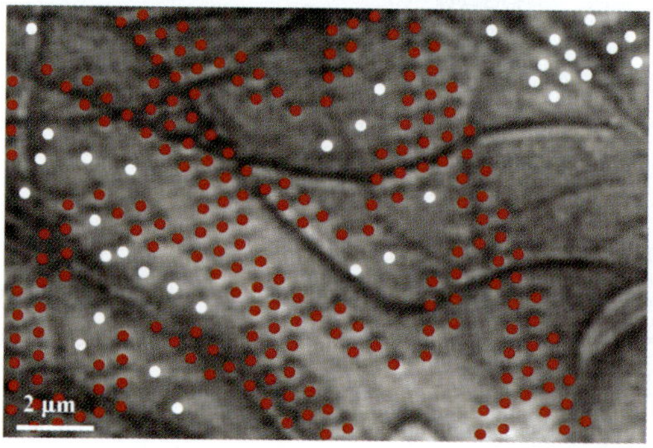

Fig. 8. Lorentz micrograph of vortices in ratchets. White dots are trapped vortices, red dots are untrapped vortices.

arrowhead acted as a "funnel", while the right one acted as a "trap". When the driving force increased, the vortices in the funnel went further upwards, but the trapped vortices in the right arrowhead remained trapped, which demonstrates the ratchet principle.

We confirmed that even when the driving force changed from upward to downward such as in AC driving, vortices only went upward in the left arrowhead, confirming the rectification of vortex flow. Thus, the ratchet motion of vortices was observed in units of individual vortices, which demonstrated the motion control of vortices using asymmetric pinning.

7. 1-MV Microscope

We developed a 1-MV field-emission electron microscope[23] (Fig. 9) to observe unconventional behaviors of vortices in high-T_c superconductors. One

Fig. 9. 1-MV field-emission electron microscope.

MV electrons were needed because electrons had to penetrate a film thicker than the magnetic radius (penetration depth) of vortices in high-T_c superconductors. With this microscope, we first observed the internal behavior of vortices inside high-T_c Bi-2212 thin films.[20]

The columnar defects, which are considered optimal pinning traps for vortices in layered structure materials, were produced in Bi-2212 films in a tilted direction by the irradiation of high-energy heavy ions. The tilted columns can be seen as tiny lines in the electron micrographs. When these images are defocused, they are blurred, and eventually, they completely disappear by spreading out. However, when they are defocused even further, vortex images appear because they are produced by phase contrast.

The resultant Lorentz micrograph of vortices is shown in Fig. 10. Two kinds of vortex images can be observed: circular images and "red" elongated images. Circular images correspond to vortices perpendicularly penetrating the film, while the elongated images are trapped at tilted columnar defects. Two different images are formed because distributions of magnetic fields are different, thus forming different phase shifts due to the AB effect. We

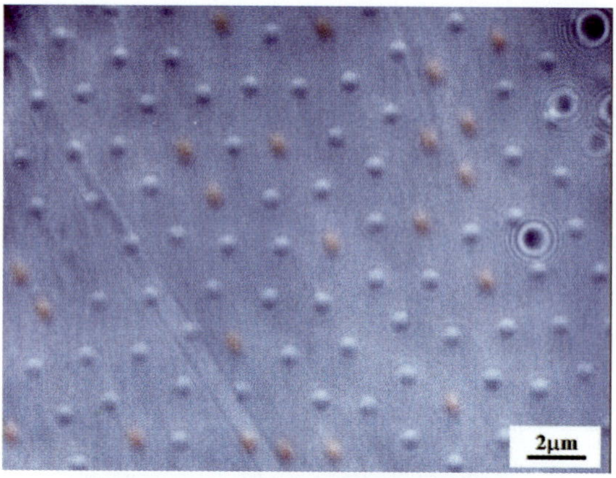

Fig. 10. Lorentz micrograph of superconducting high-T_c layered Bi-2212 thin film. "Red" elongated images correspond to vortices trapped along tilted columnar defects and circular images to vortices penetrating the film perpendicularly.

also confirmed the two different images by simulation.

When a driving force is applied, the difference in pinning forces becomes evident. Untrapped vortices soon begin to move, but trapped vortices do not. This has enabled us to use these different vortex images to investigate whether vortices are trapped or not under various conditions[21] even when they are moving.

8. Single Electron Build-Up of an Interference Pattern

Until now, only the "wave nature" of electrons has been used to observe quantum phenomena and electromagnetic fields. However, electrons have a "particle nature" also. An experiment demonstrating the dual nature of electrons described by R. Feynman in his famous textbook[51] has become feasible to carry out.[52] This experiment is Young's double-slit interference experiment using electrons (Fig. 11).

Although Feynman remarked that this was a thought experiment, it has become feasible, due to advanced equipment such as a sensitive detector that can detect individual electrons (Fig. 12). Electrons are emitted from the source, pass through an electron biprism, and arrive at the detector. A positive voltage is applied to the filament, so that electrons are attracted toward the filament, which acts as an interferometer. Electrons that arrive at the detector are displayed on a monitor. When a coherent beam is intense, interference fringes can be observed.

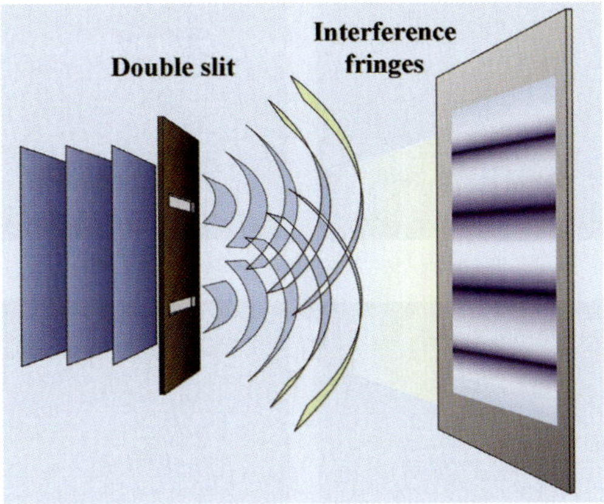

Fig. 11. Double-slit experiment.

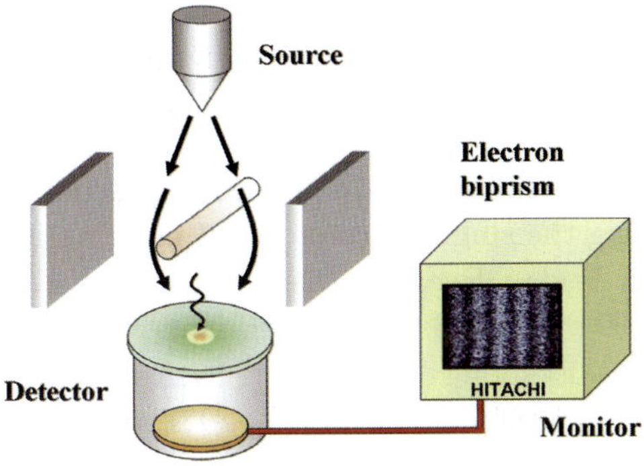

Fig. 12. Two-beam interference experiment for electrons.

What happens when the beam intensity becomes extremely weak so that no more than one electron may exist in the microscope? In this case, at first, electrons appear one by one to be here and there at random (Fig. 13). Electrons are surely "particles". However, if we wait for an hour for many electrons to accumulate, something like interference fringes begin to appear (Fig. 14). The interference fringes are formed only when two electron waves pass through on both sides of the biprism and overlap in the lower plane.

The photographs in Figs. 13 and 14(a) are really strange. No theory can predict the distributions of electrons in these photographs. We have to wait for one hour to obtain the experimental result [Fig. 14(c)] that can be theoretically predicted by quantum mechanics.

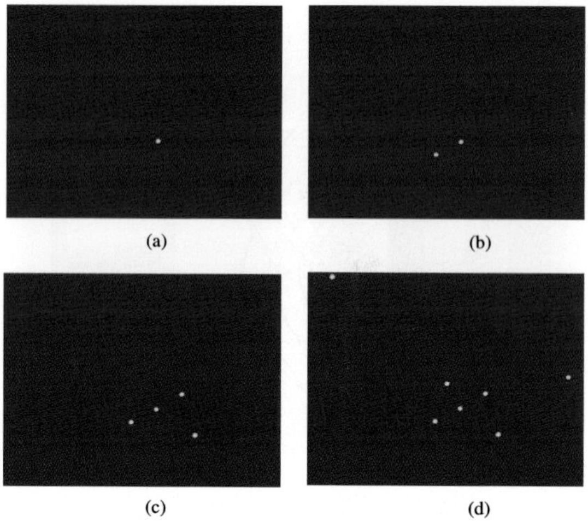

Fig. 13. Video scenes of the electron interference experiment in the central part of the monitor. Numbers of electrons are 1(a), 2(b), 4(c), and 9(b). The electron density is so sparse that only one electron exists at one time in the apparatus.

We carried out the experiments in a field-emission electron microscope in 1989,[53] but the experiment has gradually become the focus of attention. The video demonstrating the single-electron build up of an interference pattern was shown at my Friday Evening Discourse at the Royal Institution in Great Britain in 1994.[54] In 1999, our experiment was described in the Nobel Archives[a] by Gösta Ekspong. In 2002, "Physics World" readers selected the most beautiful experiments of all time.[55] The first was our double-slit experiment applied to the interference of single electrons, together with C. Jönsson's experiments in 1961.[56] I was humbled to learn that historical experiments by Galileo and Millikan were ranked lower. R. P. Crease, who wrote this article, later published a book with the same title.[57]

He saw my video, which I showed at the Friday Evening Discourse, and wrote in his book,

> "Tonomura showed this movie at a talk at the Royal Institution. He sped up the video to show the interference pattern materialize — hauntingly — out of individual, apparently random specks, the way a galaxy might form before your eyes at dusk out of tiny stars, a pattern that is undeniable and that hints at the existence of deeper universal structures."

[a]http://nobelprize.org/nobel_prizes/physics/articles/ekspong/index.html

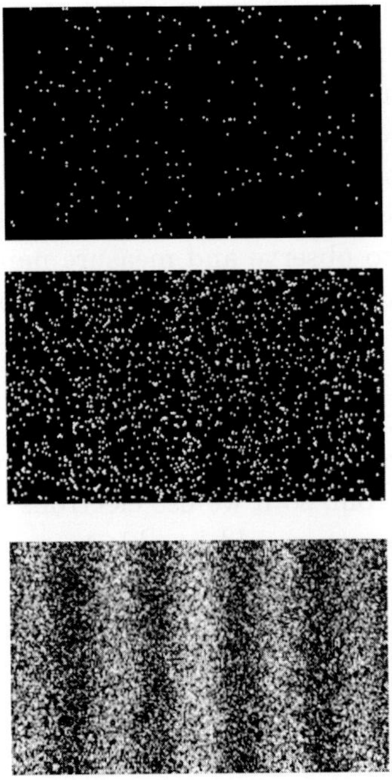

(a)

(b)

(c)

Fig. 14. Video scenes of the electron interface experiment in the whole view of the monitor. Numbers of electrons are (a) 230, (b) 6,000, (c) 140,000. When electrons are accumulated, the biprism interference pattern, which is formed between two electron waves passing through on both sides of the biprism, can be observed. It is as if a single electron splits into two, and passes through on both sides of the biprism.

Amazingly, this most strange, most basic nature of quantum mechanics is no longer imaginary, but can actually be used for practical purposes such as quantum computers.

9. Challenge of New Technology

I recall looking up at our 1-MV electron microscope standing 7 meters tall and weighing 40 tons (Fig. 9). When I first began my experiments on the AB effect in 1981, it never occurred to me in my wildest dreams that one day I would be involved in building such an enormous piece of equipment to make the quantum world visible to the naked eye.

Looking back now, a great deal of effort was required to pursue this line of basic research. I was employed by a company, Hitachi, so some effort was required to persuade those around me that doing such basic research was so interesting and significant. C. N. Yang helped me greatly in that respect.

Fortunately, Japan's level of sophistication with the electron microscope in both industry and academia has remained extremely high, right up to the present. However, this situation is entirely based on the legacy that the previous generation of researchers have built up and left behind, and I am a bit concerned about the future. The decisive factor in making further advances in nanotechnology and biotechnology to control atoms and molecules in the nanorealm is the ability to observe and measure nanostructures and their states at atomic dimensions. The electron microscope is an indispensable tool permitting us to see atomic structures of micro-objects and quantum phenomena.

However, I am not yet satisfied with the present performance of electron microscopes. Electrons accelerated to 1-MV have a wavelength only 1/1,000th the size of an atom, so if we use electrons scattered by a sample in the electron microscope, we would be able to see figures of the sample atoms in greater detail, and more clearly. We would be remiss if we did not thoroughly extract all of the information contained in the electrons.

I would certainly like to see up-and-coming younger researchers pursue long-term objectives, including the above ultra-high resolution based on a clear vision, and take on bigger challenges in choosing their research objectives. We, of the older generation, have always dreamed of the day 50 or 100 years down the road when Japan will become the recognized leader in some areas of science and technology, and I believe that we are truly obligated to provide the necessary funding and environment enabling younger researchers to pursue their dreams and fully immerse themselves in their research.

References

1. J. J. Thompson, *Proc. Roy. Inst.* **15**, 419 (1897).
2. C. Davisson and L. H. Germer, *Nature* **119**, 558 (1927).
3. G. P. Thomson and A. Reid, *Nature* **119**, 890 (1927).
4. S. Kikuchi, *Jpn. J. Phys.* **5**, 83 (1928).
5. G. Möllenstedt and H. Düker, *Z. Phys.* **145**, 377 (1956).
6. A. Tonomura, *The Quantum World Unveiled by Electron Waves* (World Scientific, Singapore, 1998).
7. A. Tonomura, T. Matsuda, J. Endo, H. Todokoro and T. Komoda, *J. Electron Microsc.* **28**(1), 1 (1979).
8. T. Kawasaki, T. Yoshida, T. Matsuda, N. Osakabe, A. Tonomura, I. Matsui and K. Kitazawa, *Appl. Phys. Lett.* **76**(9), 1342 (2000).
9. D. Gabor, *Proc. R. Soc. Lond. A* **197**, 454 (1949).
10. A. Tonomura, A. Fukuhara, H. Watanabe and T. Komoda, *Jpn. J. Appl. Phys.* **7**, 295 (1968).

11. A. Tonomura, T. Matsuda, J. Endo, H. Todokoro and T. Komoda, *J. Electron Microsc.* **28**, 1 (1979).
12. A. Tonomura, T. Matsuda, J. Endo, T. Arii and K. Mihama, *Phys. Rev. Lett.* **44**, 1430 (1980).
13. A. Tonomura, T. Matsuda, J. Endo, T. Arii and K. Mihama, *Phys. Rev. B* **34**, 3397 (1986).
14. A. Tonomura, T. Matsuda, R. Suzuki, A. Fukuhara, N. Osakabe, H. Umezaki, J. Endo, K. Shinagawa, Y. Sugita and H. Fujiwara, *Phys. Rev. Lett.* **48**, 1443 (1982).
15. A. Tonomura, H. Umezaki, T. Matsuda, N. Osakabe, J. Endo and Y. Sugita, *Phys. Rev. Lett.* **51**, 331 (1983).
16. A. Tonomura, N. Osakabe, T. Matsuda, T. Kawasaki, J. Endo, S. Yano and H. Yamada, *Phys. Rev. Lett.* **56**, 792 (1986).
17. N. Osakabe, T. Matsuda, T. Kawasaki, J. Endo, A. Tonomura, S. Yano and H. Yamada, *Phys. Rev. A* **34**, 815 (1986).
18. K. Harada, T. Matsuda, J. Bonevich, M. Igarashi, S. Kondo, G. Pozzi, U. Kawabe and A. Tonomura, *Nature* **360**, 51 (1992).
19. T. Kawasaki, T. Matsuda, J. Endo and A. Tonomura, *Jpn. J. Appl. Phys.* **29**, 508 (1990).
20. A. Tonomura, H. Kasai, O. Kamimura, T. Matsuda, K. Harada, Y. Nakayama, J. Shimoyama, K. Kishio, T. Hanaguri, K. Kitazawa, M. Sasase and S. Okayasu, *Nature* **412**, 620 (2001).
21. T. Matsuda, O. Kamimura, H. Kasai, K. Harada, T. Yoshida, T. Akashi, A. Tonomura, Y. Nakayama, J. Shimoyama, K. Kishio, T. Hanaguri and K. Kitazawa, *Science* **294**, 2136 (2001).
22. A. Tonomura, K. Kasai, O. Kamimura, T. Matsuda, K. Harada, T. Yoshida, T. Akashi, J. Shimoyama, K. Kishio, T. Hanaguri, K. Kitazawa, T. Masui, S. Tajima, N. Koshizuka, P. L. Gammel, D. Bishop, M. Sasase and S. Okayasu, *Phys. Rev. Lett.* **88**, 237001 (2002).
23. T. Kawasaki, T. Yoshida, T. Matsuda, N. Osakabe, A. Tonomura, I. Matsui and K. Kitazawa, *Appl. Phys. Lett.* **76**, 1342 (2000).
24. K. J. Konopinski, *Am. J. Phys.* **46**, 499 (1978).
25. Y. Aharonov and D. Bohm, *Phys. Rev.* **115**, 485 (1959).
26. W. Ehrenberg and R. W. Siday, *Proc. Phys. Soc. London B* **62**, 8 (1949).
27. T. T. Wu and C. N. Yang, *Phys. Rev. D* **12**, 3845 (1975).
28. P. Bocchieri and A. Loinger, *Nuovo Cimento* **47A**, 475 (1978).
29. M. Peshkin and A. Tonomura, in The Aharonov–Bohm effect, *Lecture Notes in Physics*, Vol. 340 (Springer, Heidelberg, 1989).
30. R. G. Chambers, *Phys. Rev. Lett.* **5**, 3 (1960).
31. H. A. Fowler, L. Marton, J. A. Simpson and J. A. Suddeth, *J. Appl. Phys.* **32**, 1153 (1961).
32. G. Möllenstedt and W. Bayh, *Phys. Bl.* **18**, 299 (1962).
33. H. Boersch, H. Hamisch and K. Grohmann, *Z. Phys.* **169**, 263 (1962).
34. S. M. Roy, *Phys. Rev. Lett.* **44**, 111 (1980).
35. P. Bocchieri and A. Loinger, *Lett. Nuovo. Cimento* **30**, 449 (1981).
36. A. Tonomura, *Electron Holography*, 2nd edn. (Springer, Heidelberg, 1999).

37. C. G. Kuper, *Phys. Lett.* **794**, 413 (1980).
38. C. N. Yang, Remarks to the talk "Electron Halography, Aharonov–Bohm Quantization Effect and Flux" by A. Tonomura *et al.*, in *Proc. Int. Symp. on Foundations of Quantum Mechanics*, Tokyo, 1983, eds. S. Kamefuchi *et al.* (Phys. Society of Japan, Tokyo, 1984), p. 27.
39. J. E. Bonevich, K. Harada, T. Matsuda, H. Kasai, T. Yoshida, G. Pozzi and A. Tonomura, *Phys. Rev. Lett.* **70**, 2952 (1993).
40. R. A. Webb, S. Washburn, C. P. Umbach and R. B. Laibowitz, *Phys. Rev. Lett.* **54**, 2696 (1985).
41. A. Bachtold, C. Strunk, J. P. Salvetat, J. M. Bonard, L. Forro, T. Nussbaumer and C. Schönenberger, *Nature* **397**, 673 (1999).
42. E. D. Minot, Y. Yalsh, V. Sazonova and P. L. McEuen, *Nature* **428**, 536 (2004).
43. A. Tonomura, J. Endo and T. Matsuda, *Optik* **53**, 143 (1979).
44. A. Tonomura, T. Matsuda, T. Kawasaki, J. Endo and N. Osakabe, *Phys. Rev. Lett.* **54**, 60 (1985).
45. A. Tonomura, T. Matsuda, J. Endo, T. Arii and K. Mihama, *Phys. Rev. Lett.* **44**, 1430 (1980).
46. K. Harada, H. Kasai, T. Matsuda, M. Yamasaki and A. Tonomura, *J. Electron Microsc.* **46**(3), 227 (1997).
47. J. F. Wambaugh *et al.*, *Phys. Rev. Lett.* **83**, 5106 (1999).
48. C. J. Olson *et al.*, *Phys. Rev. Lett.* **87**, 177002 (2001).
49. J. E. Villegas *et al.*, *Science* **302**, 1188 (2003).
50. Y. Togawa *et al.*, *Phys. Rev. Lett.* **95**, 087002 (2005).
51. R. P. Feynman, R. B. Leighton and M. Sands, *The Feynman Lecture on Physics*, Vol. III (Addison-Wesley, Reading, Mass, 1965), pp. 1.1–1.5.
52. A. Tonomura, J. Endo, T. Matsuda, T. Kawasaki and H. Ezawa, *Am. J. Phys.* **57**, 117 (1989).
53. A. Tonomura, J. Endo, T. Matsuda, T. Kawasaki and H. Ezawa, *Am. J. Phys.* **57**(2), 117 (1989).
54. A. Tonomura, Electron waves unveil the microcosmos, in *Unveiling the Microcosmos — Essays on Science and Technology from the Royal Institution*, ed. P. Day (Oxford University Press, Oxford, 1996), p. 1.
55. R. P. Crease, The most beautiful experiment, *Physics World September*, 19 (2002).
56. C. Jönsson, *Zeitschrift für Physik* **161**, 454 (1961).
57. R. P. Crease, *The Prism and the Pendulum — The Ten Most Beautiful Experiments in Science* (Random House, Inc., New York, 2002).

PHASE SEPARATION OF ATOMS IN OPTICAL LATTICES

HAI-QING LIN

Department of Physics and Institute of Theoretical Physics,
The Chinese University of Hong Kong, Hong Kong, China
E-mail: hqlin@phy.cuhk.edu.hk

We study dynamics of two species of fermionic atoms in optical lattices in the framework of the asymmetric Hubbard model. A common phenomenon, called phase separation is predicted to occur. We provide arguments on the existence of phase separation, accompanied by a rigorous proof that, even for a single hole case, the density wave state is unstable to the phase separation in the strong interaction limit. Using the state-of-the-art numerical techniques, we obtain the ground state phase diagram and investigate the quantum phase transition from the density wave to phase separation by studying both the corresponding charge order parameter and quantum entanglement. We also discuss experimental realization of phase separation in optical lattices.

ULTRACOLD ATOMS ACHIEVEMENTS AND PERSPECTIVES

CLAUDE COHEN-TANNOUDJI

Collège de France and Ecole Normale Supérieure,
Laboratoire Kastler Brossel, France
E-mail: cct@lkb.ens.fr

Our ability to control and to manipulate atomic systems has considerably increased during the last few years. Very precise measurements with ultracold atoms provide now severe new tests of fundamental theories like general relativity. The possibility to control all experimental parameters of an ultracold atomic sample, like the temperature, the density, the strength of the interactions, allows one to realize simple models of more complex systems found in other fields of physics and to get a better understanding of their behavior. A few of these developments will be briefly reviewed.

Ultracold Atoms
Achievements and Perspectives

Claude Cohen-Tannoudji

Conference in honor of
Professor CN Yang's 85th Birthday
31 October-3 November 2007, Singapore

1530
Collège de France

CENTRE NATIONAL DE LA RECHERCHE SCIENTIFIQUE

IFRAF

1

Evolution of Atomic Physics

Characterized by spectacular advances in our ability to manipulate the various degrees of freedom of an atom

- Spin polarization (optical pumping)
- Velocity (laser cooling, evaporative cooling)
- Position (trapping)
- Atom-Atom interactions (Feshbach resonances)

Purpose of this lecture

1 – Briefly describe the basic methods used for producing and manipulating ultracold atoms

2 - Review a few examples showing how ultracold atoms, which have become a very powerful tool in atomic physics, are allowing one to

- perform new severe tests of basic physical laws
- achieve new situations where all parameters can be carefully controlled, providing in this way simple models for understanding more complex problems appearing in other fields.

2

PRODUCING AND MANIPULATING ULTRACOLD ATOMS

- **Cooling**
- **Trapping**
- **Feshbach resonances**

Forces exerted by light on atoms

A simple example

Target C bombarded by projectiles *p* coming all along the same direction

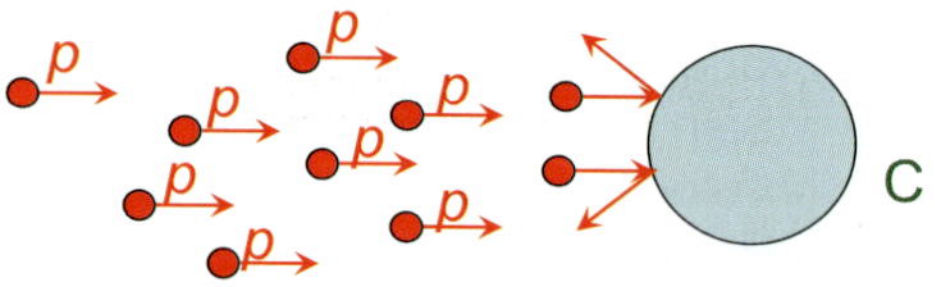

As a result of the transfer of momentum from the projectiles to the target C, the target C is pushed

Atom in a light beam

Analogous situation, the incoming photons, scattered by the atom C playing the role of the projectiles p

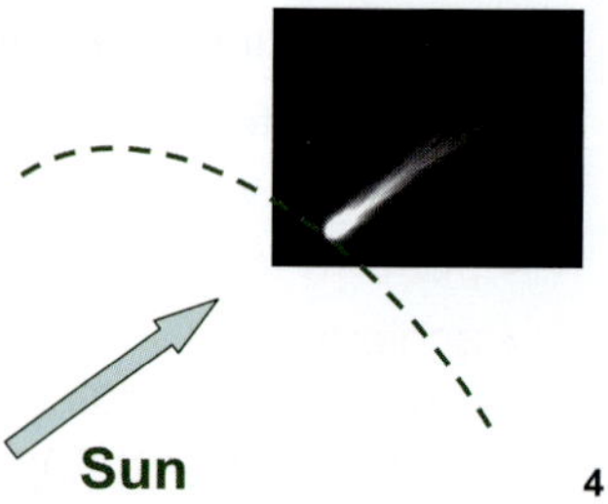

Explanation of the tail of the comets

In a resonant laser beam, the radiation pressure force can be very large

Atom in a resonant laser beam

Fluorescence cycles (absorption + spontaneous emission) lasting a mean time τ (radiative lifetime of the excited state) of the order of 10^{-8} s

Mean number of fluorescence cycles per sec : $W \sim 1/\tau \sim 10^8$ sec^{-1}

In each cycle, the mean velocity change of the atom is equal to:

$$\delta v = v_{rec} = h\nu/Mc \sim 10^{-2} \text{ m/s}$$

Mean acceleration a (or deceleration) of the atom

$$\begin{aligned} a &= \text{velocity change /sec} \\ &= \text{velocity change } \delta v \text{ / cycle x number of cycles / sec } W \\ &= v_{rec} \times (1/\tau_R) = 10^{-2} \times 10^8 \text{ m/s}^2 = 10^6 \text{ m/s}^2 = 10^5 \text{ g} \end{aligned}$$

Huge radiation pressure force!

Stopping an atomic beam

Atomic beam

Laser beam

J. Prodan
W. Phillips
H. Metcalf

Tapered solenoid

Zeeman slower

5

Laser Doppler cooling

T. Hansch, A. Schawlow, D. Wineland, H. Dehmelt
Theory : V. Letokhov, V. Minogin, D. Wineland, W. Itano

2 counterpropagating laser beams

Same intensity Same frequency ν_L ($\nu_L < \nu_A$)

$\nu_L < \nu_A$ v $\nu_L < \nu_A$

Atom at rest (v=0)

The two radiation pressure forces cancel each other out

Atom moving with a velocity v

Because of the Doppler effect, the counterpropagating wave gets closer to resonance and exerts a stronger force than the copropagating wave which gets farther

Net force opposite to v and proportional to v for v small

Friction force "Optical molasses"

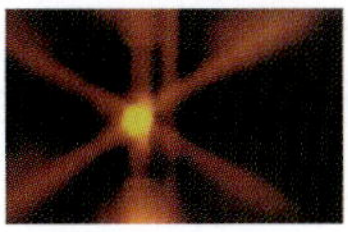

6

"Sisyphus" cooling

J. Dalibard
C. Cohen-Tannoudji

Several ground state sublevels

Spin up Spin down

In a laser standing wave, spatial modulation of the laser intensity and of the laser polarization

- Spatially modulated light shifts of g↑ and g↓ due to the laser light
- Correlated spatial modulations of optical pumping rates g↑ ↔ g↓

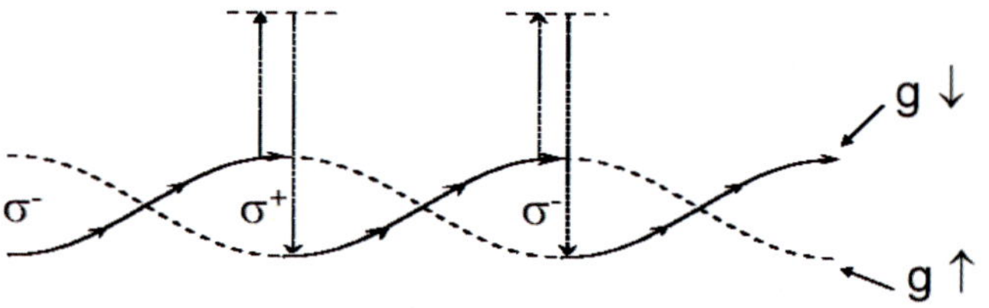

The moving atom is always running up potential hills (like Sisyphus)!

Very efficient cooling scheme leading to temperatures in the μK range

7

Evaporative cooling

H. Hess, J.M. Doyle MIT

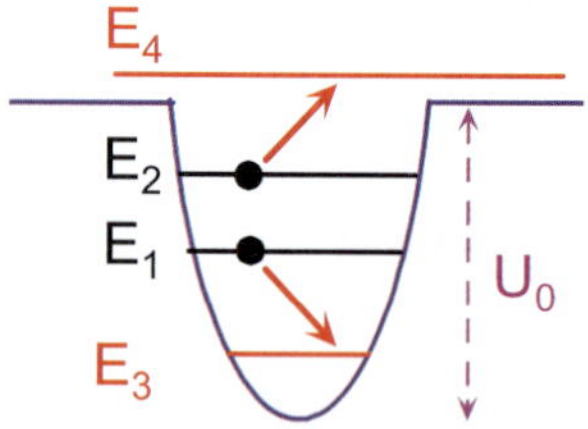

Atoms trapped in a potential well with a finite depth U_0

2 atoms with energies E_1 et E_2 undergo an elastic collision

After the collision, the 2 atoms have energies E_3 et E_4, with

$$E_1 + E_2 = E_3 + E_4$$

If $E_4 > U_0$, the atom with energy E_4 leaves the well

The remaining atom has a much lower energy E_3. After rethermalization of the atoms remaining trapped, the temperature decreases

8

Temperature scale (in Kelvin units)

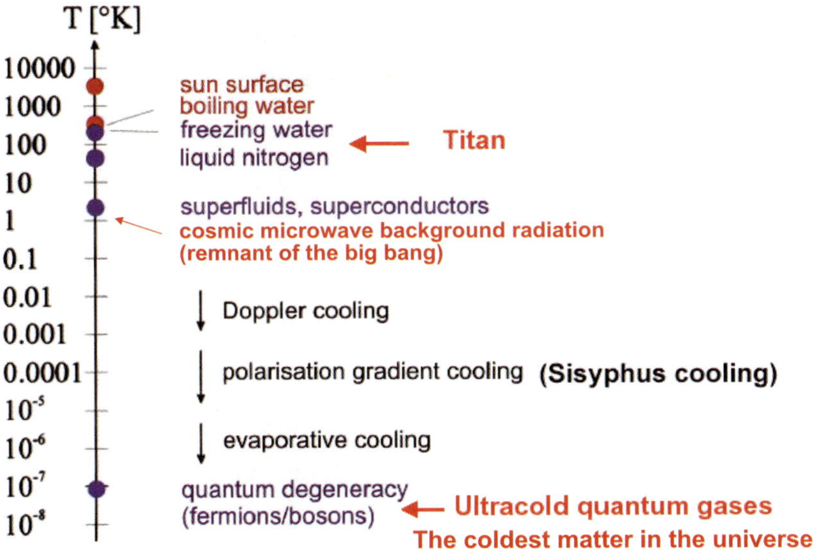

9

Traps for neutral atoms

"Optical Tweezers"

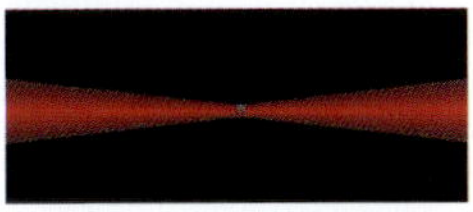

Spatial gradients of laser intensity

Focused laser beam. Red detuning ($\omega_L < \omega_A$)

The light shift δE_g of the ground state g is negative and reaches its largest value at the focus. Attractive potential well in which neutral atoms can be trapped if they are slow enough

"Optical lattice"

Spatially periodic array of potential wells associated with the light shifts of a detuned laser standing wave

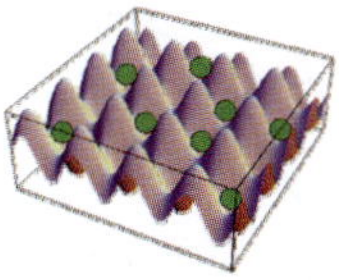

Other types of traps using magnetic field gradients or magnetic field gradients combined with the radiation pressure of properly polarized laser beams ("Magneto Optical Traps")

10

Optical lattices

The dynamics of an atom in a periodic optical potential, also called "optical lattice", shares many features with the dynamics of an electron in a crystal. But it also offers new possibilities!

New possibilities offered by optical lattices

They can be easily manipulated, much more than the periodic potential inside a crystal

- Possibility to switch off suddenly the optical potential by switching off the laser light
- Possibility to vary the depth of the periodic potential well by changing the laser intensity
- Possibility to change the spatial period of the potential by changing the angle between the 2 running laser waves
- Possibility to change the frequency of one of the 2 couterpropagating waves and to obtain a moving standing wave

Furthermore, possibility to control atom-atom interactions, both in magnitude and sign, by using "Feshbach resonances"

Defect free artificial "crystal of light"

11

Feshbach Resonances

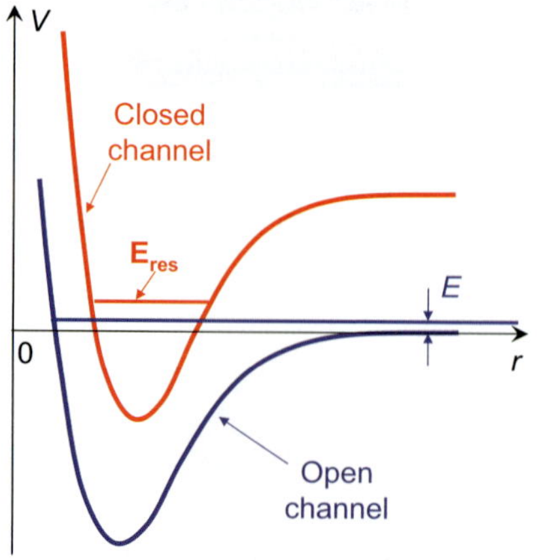

The 2 atoms collide with a very small positive energy E in a channel which is called "open"

The energy of the dissociation threshold of the open channel is taken as the zero of energy

There is another channel above the open channel where scattering states with energy E cannot exist because E is below the dissociation threshold of this channel which is called "closed"

There is a bound state in the closed channel whose energy E_{res} is close to the collision energy E in the open channel

12

Physical mechanism of the Feshbach resonance

The incoming state with energy E of the 2 colliding atoms in the open channel is coupled by the interaction to the bound state φ_{res} in the closed channel.

The pair of colliding atoms can make a virtual transition to the bound state and come back to the colliding state. The duration of this virtual transition scales as $\hbar$ / I E_{res}-E I, i.e. as the inverse of the detuning between the collision energy E and the energy E_{res} of the bound state.

When E is close to E_{res}, the virtual transition can last a very long time and this enhances the scattering amplitude

Analogy with resonant light scattering when an impinging photon of energy hν can be absorbed by an atom which is brought to an excited discrete state with an energy hν_0 above the initial atomic state and then reemitted. There is a resonance in the scattering amplitude when ν is close to ν_0

13

Sweeping the Feshbach resonance

The total magnetic moment of the atoms are not the same in the 2 channels (different spin configurations). The energy difference between the 2 channels can thus be varied by sweeping a magnetic field

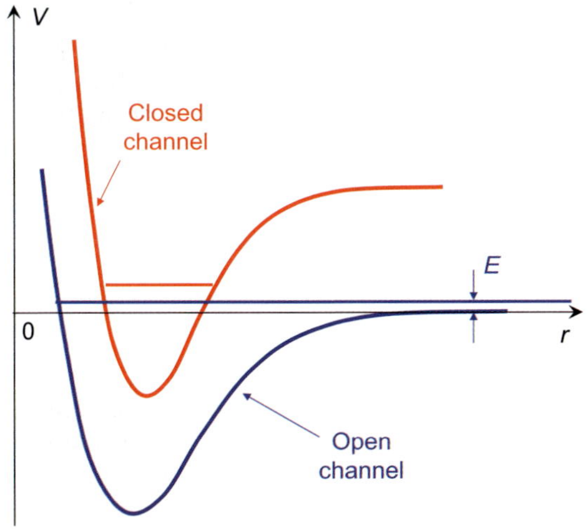

14

Scattering length versus magnetic field

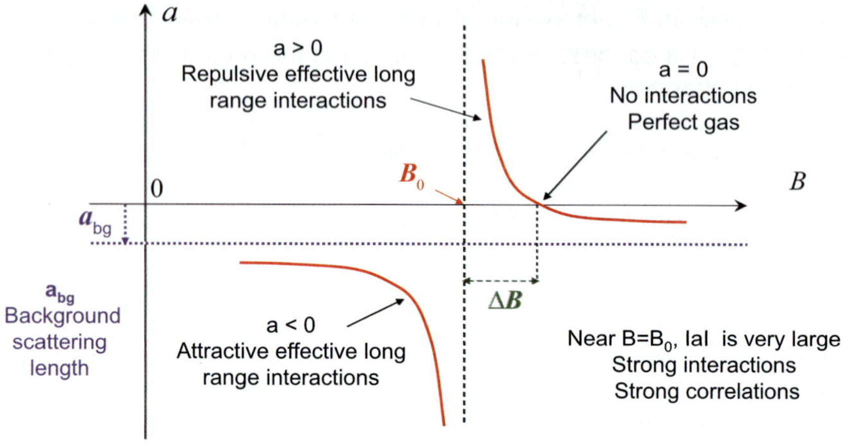

B_0 : value of the field for which the energy of the bound state, in the closed channel (shifted by its interaction with the continuum of collision states in the open channel) coincides with the energy E~0 of the colliding pair of atoms 15

Bound state of the two-channel Hamiltonian

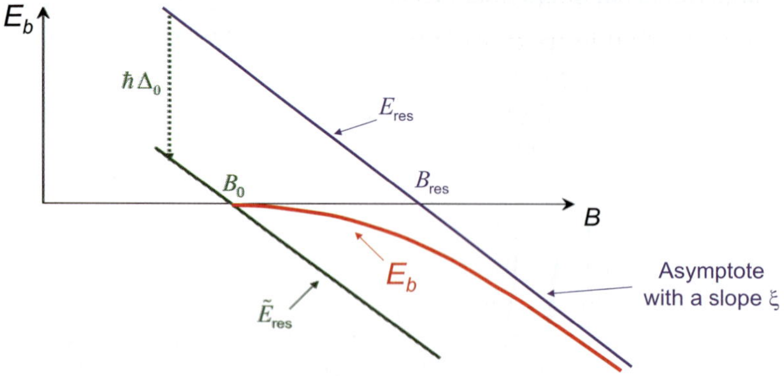

- The bound state of $H_{2\text{-channel}}$ appears for $B > B_0$, in the region a>0.
- E_b first decreases quadratically with B-B_0 and then tends to the unperturbed energy E_{res} of the bound state φ_{res} of the closed channel
- If B_0 is swept through the Feshbach resonance from the region a<0 to the region a>0, a pair of ultracold atoms can be transformed into a molecule 16

TESTING FUNDAMENTAL LAWS WITH ULTRACOLD ATOMS

Ultraprecise Atomic Clocks

17

Measuring time with atomic clocks

Principle of an atomic clock

The correction loop locks the frequency of the oscillator to the frequency ω_A of the hyperfine transition of ^{133}Cs used for defining the second

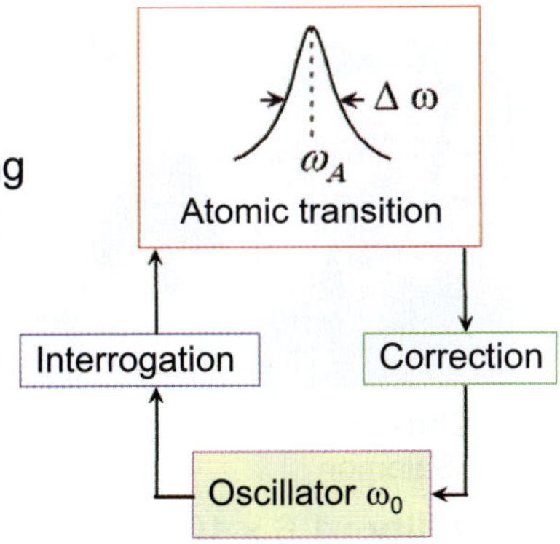

The narrower the atomic line, *i.e.* the smaller $\Delta\omega$, the better the locking of the frequency of the oscillator to ω_A.

$$\Delta \nu \approx 1 / T$$

T : Observation time

It is therefore interesting to use slow atoms in order to increase T, and thus to decrease $\Delta \omega$

18

Improving atomic clocks with ultracold atoms

Usual clocks using thermal Cs atoms

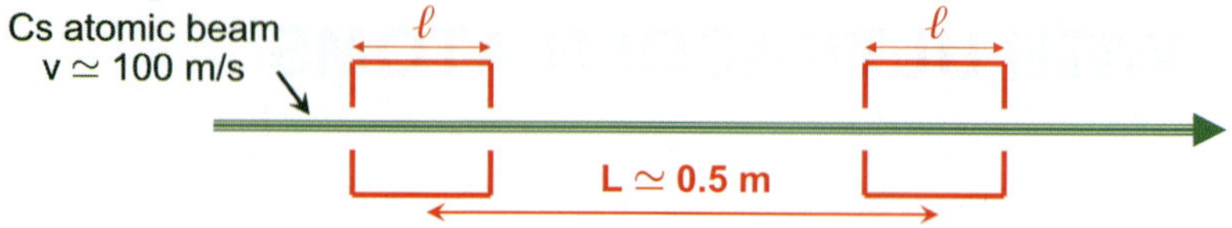

Appearance in the resonance of Ramsey fringes having a width determined by the time $T = L/v \simeq 0.005$ s

Fountains of ultracold atoms

Throwing a cloud of ultracold atoms upwards with a laser pulse to have them crossing the <u>same</u> cavity twice, once in the way up, once in the way down, and obtaining in this way 2 interactions separated by a time interval T

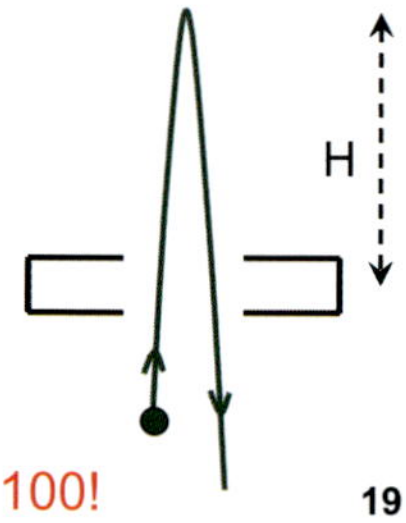

$$g(T/2)^2/2 = H \quad \rightarrow \quad T = 2\sqrt{2H/g}$$

H = 30 cm → T = 0.5 s   Improvement by a factor 100!

19

Examples of atomic fountains

- Sodium fountains : Stanford S. Chu
- Cesium fountains : BNM/SYRTE C. Salomon, A. Clairon

Christophe Salomon

André Clairon

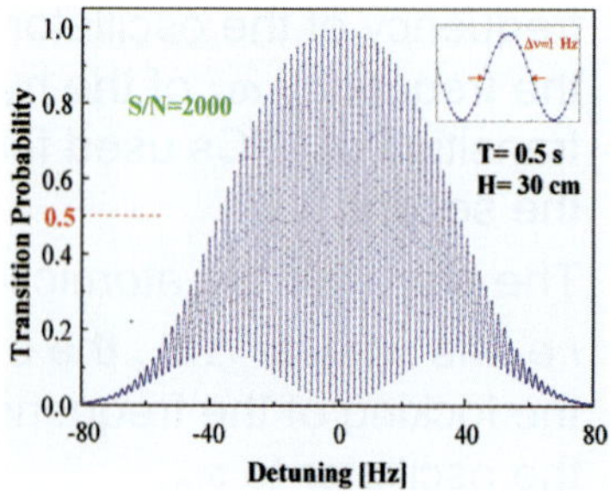

Stability : 1.6 x 10^{-16} for an integration time 5 x 10^{4} s
Accuracy : 3 x 10^{-16}

A stability of 10^{-16} corresponds to an error smaller than 1 second in 300 millions years

20

From terrestrial clocks to space clocks

Parabolic flights

Atomic Clock Ensemble in Space

esa

cnes
CENTRE NATIONAL D'ETUDES SPATIALES

- A cold atom Cs standard in space
- Fundamental physics tests
- Worldwide access

21

Gravitational shift of the frequency of a clock

An observer at an altitude *z* receives the signal of a clock located at the altitude $z+\delta z$ and measures a frequency $\omega_A(z+\delta z)$ different from the frequency, $\omega_A(z)$, of his own clock

$$\frac{\omega_A(z+\delta z)-\omega_A(z)}{\omega_A(z)}=\frac{\delta\,\omega_A}{\omega_A}=\frac{g\,\delta z}{c^2}$$

$$\mathrm{g}=10\,\mathrm{m/s}^2 \quad \delta z=1\mathrm{m} \quad \mathrm{c}=3\times 10^8\,\mathrm{m/s} \;\rightarrow\; \delta\omega/\omega=10^{-16}$$

2 clocks at altitudes differing by 1 meter have apparent frequencies which differ in relative value by 10^{-16}.
A space clock at an altitude of 400 kms differs from a terrestrial clock by 4 x 10^{-11} . Possibility to check this effect with a precision 25 times better than all previous tests

Another possible application : determination of the "geoid", surface where the gravitational potential has a given value

22

Laboratory tests of a possible variation of fundamental constants

Because of relativistic corrections, the hyperfine structure of an alkali atom depends on the fine structure constant and on the atomic number Z.

J. Prestage, R. Tjoelker, L. Maleki, Phys. Rev. Lett. 74, 3511 (1995)

By comparing the hyperfine frequencies of cesium and rubidium measured with 2 fountains, and by following the evolution of their ratio over several years, on can put an upper bound on

$$\frac{\dot{\alpha}}{\alpha} = \frac{1}{\alpha}\frac{d\alpha}{dt}$$

Other (non laboratory) tests

- Natural nuclear reactor of Oklo (Gabon)
- Absorption spectroscopy of the light emitted by distant quasars

23

FROM ULTRACOLD ATOMS TO MORE COMPLEX SYSTEMS

- **From ultracold atoms to ultracold molecules**
- **Crossing a Feshbach resonance with fermionic atoms**
  BEC – BCS crossover
- **Bosonic atoms in periodic optical lattices**
  Superfluid Mott-insulator transition
- **Low dimensional systems**
  Berezinski-Kosterlitz-Thouless transition
- **Ultracold atoms as "quantum simulators"**

24

Ultracold Molecules

How do ultracold atoms interact? Can they form molecules? Can one extend to molecules the results obtained for atoms? Trying to answer to these questions has stimulated the emergence of new fields:

- Cold collisions
- Atomic Feshbach resonances
- Ultracold molecules. Molecular BEC

<u>From ultracold atoms to ultracold molecules</u>

Gluing 2 ultracold atoms with one or two-photon photoassociation

Starting from ultracold atoms and sweeping a Feshbach resonance

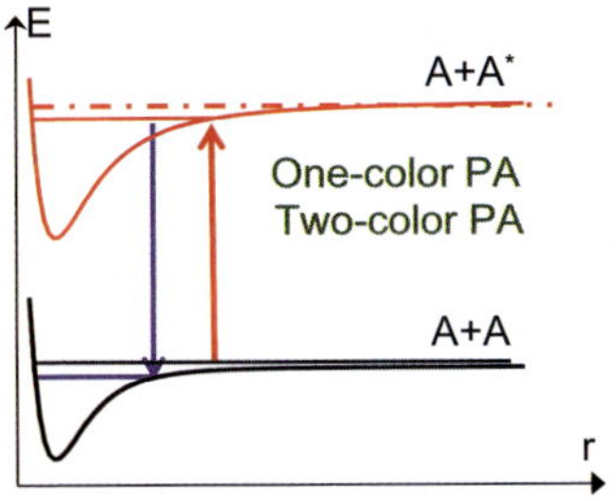

25

BEC-BCS crossover observed with ultracold fermions

By varying the magnetic field around a Feshbach resonance, one can explore 3 regions

- Region a>0 (strong interactions). There is a bound state in the interaction potential where 2 fermions with different spin states can form molecules which can condense in a BEC
- Region a<0 (weak interactions). No molecular state, but long range attractive interactions giving rise to weakly bound Cooper pairs which can condense in a BCS superfluid phase
- Region a=∞ (center of the resonance). Strong interactions Strongly correlated systems with universal properties.

Recent observation at MIT (W. Ketterle et al) of quantized vortices in all these 3 zones demonstrating the superfluid character of the 3 phases

Science, <u>435</u>, 1047 (2005)

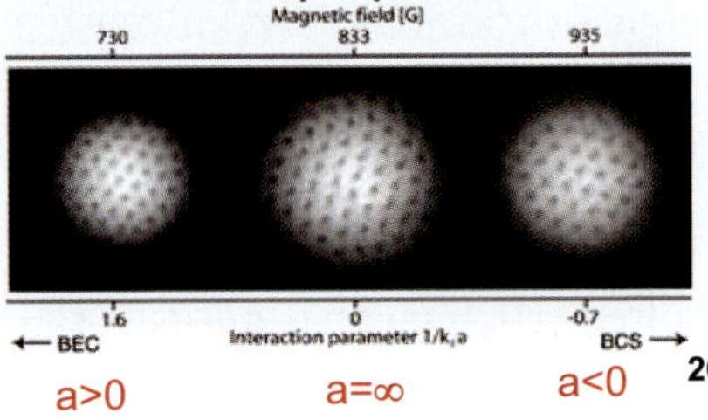

26

Fermions with 2 different spin states in an optical lattice

Investigation of the transition towards a state where fermions in different spin states arrange themselves in an antiferromagnetic order (Néel transition)

Other open problems with fermionic atoms

- Lattices with frustration
- p-wave pairing in BCS
- Imbalance between the 2 spin state populations
- Bose-Fermi mixtures
- Fermi Hubbard model

27

BEC in a periodic optical potential Superfluid – Mott insulator transition

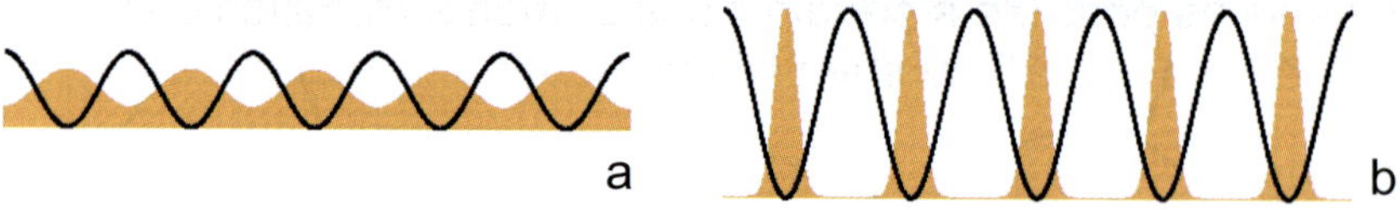

a – Small depth of the wells. Delocalized matter waves. Superfluid phase

b - Large depth of the wells. Localized waves. Insulator phase

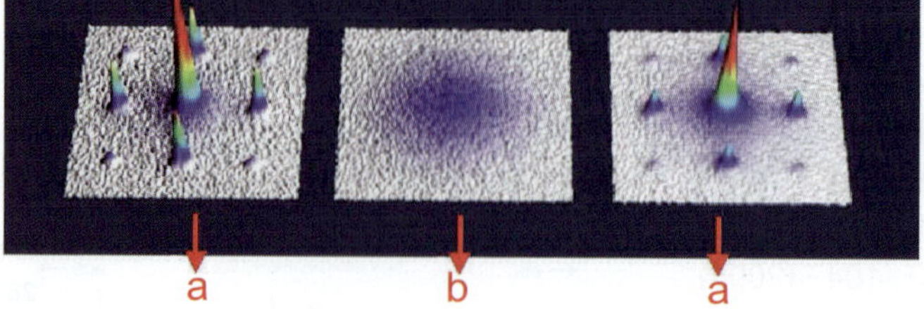

I. Bloch group in Munich Nature, 415, 39 (2002)

28

BKT crossover in a trapped 2D atomic gas

J. Dalibard group, ENS, Paris, Nature, 441,1053 (2006)

How to prepare the 2D gas

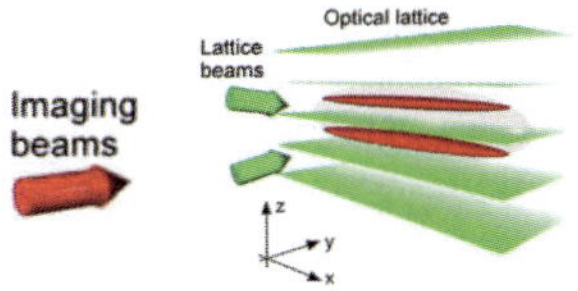

How to detect phase coherence

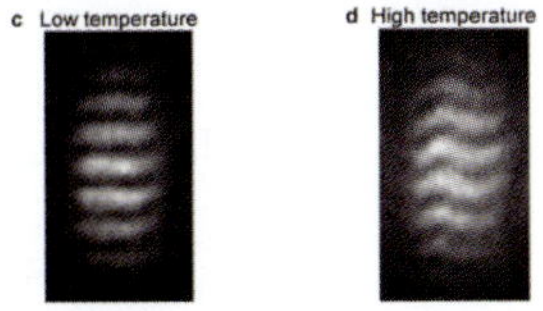

Interference fringes changing at high T (lower contrast, waviness)
Quasi-long-range order (vortex-antivortex pairs) lost at high T

Detection of the appearance of free vortices

Onset of sharp dislocations in the interference pattern coinciding with the loss of long-range order

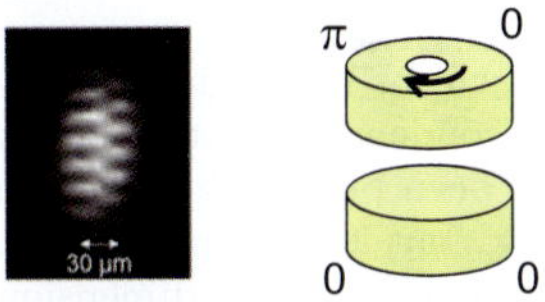

Conclusion : the mechanism of the BKT crossover is the unbinding of vortex-antivortex pairs with the appearance of free vortices

29

Quantum Simulators

Realizing an experimental system whose behavior reproduces as close as possible a certain class of model Hamiltonians.

Feynman's idea

Requirements for a "quantum simulator"

- Tailoring the potential in which particles are moving
- Controlling the interactions between particles
- Controlling the temperature, the density
- Ability to measure various properties of the system

Possibilities offered by ultracold atomic gases

- Very flexible optical potentials, with all dimensionalities, with all possible shapes (periodic, single well,...)
- Tuning the interactions with Feshbach resonances
- Various cooling schemes and measurement methods

Hope to answer in this way questions unreachable for classical computers because of memory, speed and size limitations

30

QUANTUM SPIN HALL EFFECT

SHOUCHENG ZHANG

Stanford University
E-mail: sczhang@stanford.edu

Search for topologically non-trivial states of matter has become a prime goal for condensed matter physics. Recently, a new class of topological insulators has been proposed. These topological insulators have an insulating gap in the bulk, but have topologically protected edge states due to the time reversal symmetry. In two dimensions the edge states give rise to the quantum spin Hall (QSH) effect, in the absence of any external magnetic field. We show that the QSH state can be realized in HgTe/CdTe semiconductor quantum wells. By varying the thickness of the quantum well, the electronic state changes at a critical thickness. This is a topological quantum phase transition between a conventional insulating phase and a phase exhibiting the QSH effect with a single pair of helical edge states. This theoretical proposal has been tested in a recent experiment carried out at University of Wuerzburg, and the distinct signatures of the QSH state have been experimentally observed.

BERRY PHASE IN YANG–BAXTER SYSTEMS AND BOGOLIUBOV HAMILTONIAN AS A DERIVATIVE OF DIRAC HAMILTONIAN VIA BRAID RELATION

MO-LIN GE

Theoretical Physics Division, Chern Institute of Mathematics, Nankai University, Tianjin 300071, People's Republic of China
E-mail: geml@nankai.edu.cn

KANG XUE

Department of Physics, Northeast Normal University, Changchun, Jilin 130024, People's Republic of China

JING-LING CHEN

Theoretical Physics Division, Chern Institute of Mathematics, Nankai University, Tianjin 300071, People's Republic of China
E-mail: chenjl@nankai.edu.cn

Spin interaction Hamiltonians are obtained from the unitary Yang–Baxter $\breve{R}$-matrix. Based on which, we study Berry phase and quantum criticality in the Yang–Baxter systems. We also indicate that Bogoliubov Hamiltonian is a derivative of Dirac Hamiltonian via braid relation.

Keyword: Berry phase; Yang–Baxter system; Dirac Hamiltonian; Braid relation.

1. Introduction

Yang–Baxter equation (YBE) is a kind of fruitful resource that provides multi-spin interaction Hamiltonians.[1,2] A braiding matrix B is an asymptotic limit of matrix $\breve{R}(u)$, which is the solution to the following YBE

$$\breve{R}_{12}(u)\breve{R}_{23}(u+v)\breve{R}_{12}(v) = \breve{R}_{23}(v)\breve{R}_{12}(u+v)\breve{R}_{23}(u). \quad (1)$$

Generally a solution of $\breve{R}(u)$ depends on two parameters, say, ϕ (the q-deformation parameter with $q = e^{i\phi}$, or it may be originated from other parameter such as $\eta = e^{i\phi}$) and θ (the spectral parameter or the one dimensional linear momentum). In physics the parameter ϕ is the flux that can be dependent on time t. If we define a quantum state

$$|\Phi(\theta, \phi(t))\rangle = \breve{R}(\theta, \phi(t))|\Phi(0)\rangle, \tag{2}$$

where $|\Phi(0)\rangle$ is the initial state independent of t. The normalization condition of the quantum states $\langle\Phi(\theta, \phi(t))|\Phi(\theta, \phi(t))\rangle = \langle\Phi(0)|\Phi(0)\rangle = 1$ requires the unitary condition $\breve{R}^\dagger(\theta, \phi(t)) = \breve{R}^{-1}(\theta, \phi(t))$. It follows from Eq. (2) that

$$\begin{aligned} &i\hbar\tfrac{\partial|\Phi(\theta,\phi(t))\rangle}{\partial t} \\ &= i\hbar\left[\tfrac{\partial\breve{R}(\theta,\phi(t))}{\partial t}\breve{R}^\dagger(\theta, \phi(t))\right]\breve{R}(\theta, \phi(t))|\Phi(0)\rangle \\ &= H(t)|\Phi(\theta, \phi(t))\rangle, \end{aligned} \tag{3}$$

where the Hamiltonian

$$H(t) = i\hbar\frac{\partial\breve{R}(\theta, \phi(t))}{\partial t}\breve{R}^\dagger(\theta, \phi(t)). \tag{4}$$

The $\breve{R}(\theta, \phi)$-matrix is related to the braiding matrix. Suppose $|\Phi(0)\rangle$ represents a direct-product state without entangling, and $B(\phi)|\Phi(0)\rangle$ yields an entangled state (in general, a maximally entangled state). For the case with two distinct eigenvalues, through the Yang-Baxterization $B(\phi) \to \breve{R}(\theta, \phi)$, Eq. (4) defines the Hamiltonian for the Yang–Baxter systems.

This paper has two purposes: (i) we investigate the Berry phases (BP)[8] and quantum criticality (QC) phenomenon[3–7] in Yang–Baxter systems; (ii) we indicate that Bogoliubov Hamiltonian is a derivative of Dirac Hamiltonian via braid relation. The paper is organized as follows. In Sec. 2 we study the Berry phase and the quantum criticality for a kind of Yang–Baxter Hamiltonian related to the extra-special two-group. In Sec. 3, we study Berry phase and the quantum criticality for a Yang–Baxter Hamiltonian related to the well-known six-vertex model and the Temperley-Lieb (TL) algebra. In Sec. 4, we connect the Bogoliubov Hamiltonian (BH) with the Dirac Hamiltonian (DH) via the braid relation. Conclusion and discussion are made in the end.

2. BP and QC for Hamiltonian $H_1(\theta, \phi(t))$

The braiding operators satisfy the following braid relations:

$$\begin{aligned} b_i b_{i+1} b_i &= b_{i+1} b_i b_{i+1}, \qquad && 1 \le i \le n-1, \\ b_i b_j &= b_j b_i, && |i-j| \ge 2, \end{aligned} \tag{5}$$

where the notation $b_i \equiv b_{i,i+1}$ has been used. Let us consider the following type braiding matrix for two spin-1/2 particle[9,10]

$$B^{\frac{1}{2}\frac{1}{2}} = \frac{1}{\sqrt{2}}(I + M^{\frac{1}{2}\frac{1}{2}}), \tag{6}$$

where I is the 4×4 identity matrix and

$$M^{\frac{1}{2}\frac{1}{2}} = \begin{pmatrix} & & & e^{i\phi} \\ & & \epsilon & \\ & -\epsilon & & \\ -e^{-i\phi} & & & \end{pmatrix}, \tag{7}$$

with $\epsilon = \pm 1$, and $\phi = \phi(t)$ represents the arbitrary flux. The matrix $M^{\frac{1}{2}\frac{1}{2}}$ satisfies the algebraic relations of the extra-special two-group.[11,12]

The trigonometric Yang-Baxterization approach[13–15] $\breve{R}(x) = \rho(xB - x^{-1}B^{-1})$ (here ρ is a normalization factor) gives

$$\breve{R}(x) = [2(x^2 + x^{-2})]^{-1/2}[(x + x^{-1})I^{\frac{1}{2}\frac{1}{2}} + (x - x^{-1})M^{\frac{1}{2}\frac{1}{2}}],$$

$$[\breve{R}(x)]^{-1} = [2(x^2 + x^{-2})]^{-1/2}[(x + x^{-1})I^{\frac{1}{2}\frac{1}{2}} - (x - x^{-1})M^{\frac{1}{2}\frac{1}{2}}]. \tag{8}$$

The unitary condition $[\breve{R}(x)]^{-1} = \breve{R}(x^{-1})$ leads to $\phi(t)$ =real.

Equation (7) can be rewritten as

$$M^{\frac{1}{2}\frac{1}{2}} = e^{i\phi(t)} S_1^+ S_2^+ - e^{-i\phi} S_1^- S_2^- + \epsilon(S_1^+ S_2^- - S_1^- S_2^+), \tag{9}$$

where $S_i^{\pm} = S_i^1 \pm i S_i^2$ are raising and lowering operators of spin-1/2 angular momentum for the i-th lattice. We then have from Eq. (4) that

$$\begin{aligned} H_1(x, \phi(t)) &= -\hbar\dot{\phi}[2(x^2 + x^{-2})]^{-1}(x - x^{-1}) \times \\ &\quad \{(x - x^{-1})(S_1^3 + S_2^3) + \\ &\quad (x + x^{-1})(e^{i\phi} S_1^+ S_2^+ + e^{-i\phi} S_1^- S_2^-)\}. \end{aligned} \tag{10}$$

By using

$$\begin{aligned} x &= [-\cos 2\theta]^{-1/2}(\cos\theta + \sin\theta) \\ x^{-1} &= [-\cos 2\theta]^{-1/2}(\sin\theta - \cos\theta), \end{aligned} \tag{11}$$

Eq. (10) can be recast to

$$H_1(\theta, \phi(t)) = -\hbar\dot{\phi}\cos\theta[\cos\theta(S_1^3 + S_2^3) + \sin\theta(e^{i\phi} S_1^+ S_2^+ + e^{-i\phi} S_1^- S_2^-)]. \tag{12}$$

More precisely, the Hamiltonian reads

$$H_1(\theta, \phi(t)) = -\hbar\dot{\phi}(t)\cos\theta \times \begin{pmatrix} \cos\theta & 0\;0 & e^{-i\phi(t)}\sin\theta \\ 0 & 0\;0 & 0 \\ 0 & 0\;0 & 0 \\ e^{i\phi(t)}\sin\theta & 0\;0 & -\cos\theta \end{pmatrix}.$$

The eigen-problem of Eq. (12) under adiabatic approximation is

$$H_1(\theta, \phi(t))|\Phi_\pm(\theta, \phi(t))\rangle_1 = E^1_\pm(t)|\Phi_\pm(\theta, \phi(t))\rangle_1, \tag{13}$$

where the two non-zero eigenvalues are

$$\begin{aligned} E^1_\pm &= \mp\hbar\dot{\phi}\cos\theta \\ &= \mp\hbar\omega\cos\theta \quad \text{for } \phi = \omega t, \end{aligned} \tag{14}$$

and the corresponding eigenstates are

$$\begin{aligned} |\Phi_+(\theta, \phi)\rangle &= \cos\tfrac{\theta}{2}|\uparrow\uparrow\rangle + \sin\tfrac{\theta}{2}e^{-i\phi}|\downarrow\downarrow\rangle, \\ |\Phi_-(\theta, \phi)\rangle &= -\sin\tfrac{\theta}{2}e^{i\phi}|\uparrow\uparrow\rangle + \cos\tfrac{\theta}{2}|\downarrow\downarrow\rangle. \end{aligned} \tag{15}$$

The physical consequence of Berry phase for the above Yang–Baxter Hamiltonian system, i.e., $H_1(\theta, \phi(t))$, has been discussed in.[16] Namely, from the definition of Berry phase

$$\begin{aligned} \gamma(c) &= i\int_0^T dt\langle n(\vec{R})|\frac{\partial}{\partial t}|n(\vec{R})\rangle = i\int_0^T dt A(t) \\ &= i\int_0^{2\pi} dt\dot{\phi}^{-1}\langle n(\vec{R})|\frac{\partial}{\partial\phi}|n(\vec{R})\rangle, \end{aligned} \tag{16}$$

here $\vec{R} = (\sin\theta\cos\phi, \sin\theta\sin\phi, \cos\theta)$ and $|n(\vec{R})\rangle = |\Phi_\pm(\theta, \phi)\rangle$, one then obtains the Berry phases for the Yang–Baxter system as

$$\gamma^1_\pm = (\pm\int_0^{2\pi} d\phi)\sin^2\frac{\theta}{2} = \pm\pi(1-\cos\theta) = \pm\frac{\Omega}{2}, \tag{17}$$

where $\Omega = 2\pi(1-\cos\theta)$ is the familiar solid angle enclosed by the loop on the Bloch sphere.

Observing the Hamiltonian (12), which is obtained through the Schrödinger evolution of the Bell state with $\phi = \omega t$, it does have a nice physical interpretation. Since $\mathbf{S}_1$ and $\mathbf{S}_2$ are two-dimensional representation operators of $SU(2)$ for particles 1 and 2 respectively, we have $(S_i^\pm)^2 = 0$ $(i = 1, 2)$. It is convenient to introduce the following fermionic operators:

$$S_i^- = \begin{pmatrix} 0\;0 \\ 1\;0 \end{pmatrix}_i = \hat{f}_i, \qquad S_i^+ = \begin{pmatrix} 0\;1 \\ 0\;0 \end{pmatrix}_i = \hat{f}_i^\dagger \tag{18}$$

Then by means of $[S_i^3, S_j^{\pm}] = \pm S_i^{\pm}\delta_{ij}$ and $[S_i^+, S_j^-] = 2\delta_{ij}S_i^3$ we have

$$\begin{aligned}&\{\hat{f}_i, \hat{f}_i^\dagger\} = \hat{f}_i\hat{f}_i^\dagger + \hat{f}_i^\dagger\hat{f}_i = 1,\\ &[\hat{f}_i, \hat{f}_j] = [\hat{f}_i, \hat{f}_j^\dagger] = 0, \quad (\text{for } \ i \neq j),\\ &S_i^3 = \hat{f}_i^\dagger\hat{f}_i - \frac{1}{2} = \hat{n}_i - \frac{1}{2},\end{aligned} \tag{19}$$

i.e., $\hat{f}_i$'s satisfy the fermonic anticommutator for the same i-th lattice and satisfy the bonsonic commutator for different sites of the lattices, and $\hat{n}_i = \hat{f}_i^\dagger\hat{f}_i$ is the number operator that can be 0 and 1. It is easy to check that

$$\begin{aligned}&S^+ = \prod_{i=1}^{2}\hat{f}_i^\dagger, \quad S^- = \prod_{i=1}^{2}\hat{f}_i,\\ &S^3 = \frac{1}{2}(S_1^3 + S_2^3),\end{aligned} \tag{20}$$

form an $SU(2)$ group satisfying $[S^3, S^{\pm}] = \pm S^{\pm}$ and $[S^+, S^-] = 2S^3$.

In terms of Eqs. (18)-(20) the Hamiltonian (12) can be recast to the form

$$\begin{aligned}&H_1(\theta, \phi(t)) = -\hbar\omega\cos\theta[\cos\theta\cdot(\hat{n}_1 + \hat{n}_2 - 1)\\ &+\sin\theta(e^{i\phi(t)}S^+ + e^{-i\phi(t)}S^-)],\end{aligned} \tag{21}$$

or

$$H_1(\theta, \phi(t)) = -\hbar\omega\eta(\theta)H_0(\theta, \phi(t)) \tag{22}$$

where

$$H_0(\theta, \phi(t)) = 2\eta(\theta)S^3 + \Delta(t)S^+ + \Delta(t)^*S^- \tag{23}$$

$$\eta(\theta) = \cos\theta, \quad \Delta(t) = \sin\theta e^{i\phi(t)} \tag{24}$$

The standard procedure making $H_0(\theta, \phi(t))$ diagonal is

$$W^\dagger H_0 W = 2\varepsilon S_3, \quad \varepsilon = \sqrt{(\varepsilon(\theta))^2 + |\Delta(t)|^2} = 1, \tag{25}$$

and the eigenstate is

$$\begin{aligned}&|\xi(\theta)\rangle = W|\text{vacuum}\rangle = \exp(\xi S_+ - \xi^* S_-)|\text{vacuum}\rangle,\\ &S_-|\text{vacuum}\rangle = 0,\end{aligned} \tag{26}$$

with

$$\xi = re^{i\phi(t)}, \quad \cot(2r) = -\frac{\varepsilon(\theta)}{|\Delta(t)|} \tag{27}$$

Substituting Eq. (24) into Eq. (25) and Eq. (27) we obtain

$$r = -\frac{\theta}{2},$$

in other words, we have

$$\begin{aligned} W^\dagger H_1 W|\xi(\theta)\rangle &= -\hbar\omega \cdot 2\cos\theta S_3|\xi(\theta)\rangle \\ &= -\hbar\omega\cos\theta(\hat{n}_1 + \hat{n}_2 - 1)|\xi(\theta)\rangle. \end{aligned} \tag{28}$$

It is nothing but an oscillator Hamiltonian formed by two fermions with the frequency $\omega\cos\theta$. When $\theta = 0$ Eq. (23) reduces to the standard oscillator for $\Delta(t) = 0$. However, When $\theta \neq 0$, $\Delta(t)$ plays a role of the "energy gap" and the wave function takes the form of spin coherent state.[17,18] We know that the eigenfunction of oscillator is the Hermitian polynomial, whereas the wave function of Eq. (23) with $\Delta(t) \neq 0$ is the spin coherent state shown by

$$\begin{aligned} |\xi\rangle &= \frac{1}{(1+|\tau|^2)^{\frac{1}{2}}}\exp(\tau S_+)|\text{vacuum}\rangle \\ &= \frac{1}{\sqrt{1+|\tau|^2}}\{|0,0\rangle + \tau|1,1\rangle\}, \end{aligned} \tag{29}$$

where

$$\begin{aligned} \tau &= -e^{i\phi}\tan\frac{\theta}{2}, \\ |1,1\rangle &= |n_1 = 1, n_2 = 1\rangle, \\ |0,0\rangle &= |n_1 = 0, n_2 = 0\rangle. \end{aligned} \tag{30}$$

We thus conclude that $\theta = 0$ is a point of quantum criticality. It is not caused by temperature, but by the degree of entanglement related to the parameter θ.

The degree of entanglement (or the concurrence[19]) for an arbitrary two-qubit state $|\psi\rangle = a|\uparrow\uparrow\rangle + b|\uparrow\downarrow\rangle + c|\downarrow\uparrow\rangle + d|\downarrow\downarrow\rangle$ is $\mathcal{C} = 2|ad - bc|$. The Berry phases in Eq. (17) can be expressed in terms of the concurrence of the states $|\Phi_\pm(\theta,\varphi)\rangle$ as

$$\gamma_\pm^1 = \mp\pi(1 - \sqrt{1 - \mathcal{C}^2}), \tag{31}$$

where $\mathcal{C} = |\sin\theta|$ is the concurrence of $|\Phi_\pm(\theta,\varphi)\rangle$. It is worthy of mentioning that (i) the parameter space that Berry phases occur is an entanglement space specified by θ and ϕ; (ii) in a noisy environment, the states $|\Phi_\pm(\theta,\phi)\rangle$ may become mixed states as $\rho_\pm(r,\theta,\phi) = g\,|\Phi_\pm\rangle\langle\Phi_\pm| + (1-g)\rho_{\text{noise}}$, where $0 \le g \le 1$. In this case, one can have a mixed-state geometric phase as

$\gamma_{\pm}^{\text{mixed}} = \mp\arctan(g\tan\frac{\Omega}{2})$, for more details one may refer to Ref. [16]. Moreover, one may observe that when $\theta = 0$ or $\mathcal{C} = 0$, the quantum criticality occurs in the Hamiltonian system $H_1(\theta, \phi(t))$ and at the same time Berry phases vanish correspondingly.

3. BP and QC for Hamiltonian $H_2(\theta, \phi(t))$

In this section, we come to study the Berry phase and also the quantum criticality for a kind of Yang–Baxter Hamiltonian related to the well-known six-vertex model[2] and the Temperley-Lieb algebra.

For the well-known six-vertex model, the braiding matrix reads

$$\begin{aligned} B = S^{\frac{1}{2}\frac{1}{2}} &= \begin{bmatrix} q & 0 & 0 & 0 \\ 0 & 0 & -\eta & 0 \\ 0 & -\eta & q-q^{-1} & 0 \\ 0 & 0 & 0 & q \end{bmatrix} \\ &= q(I - q^{-1}U^{\frac{1}{2}\frac{1}{2}}), \end{aligned} \tag{32}$$

where

$$U^{\frac{1}{2}\frac{1}{2}} = \begin{pmatrix} 0 & 0 & 0 & 0 \\ 0 & q & \eta & 0 \\ 0 & \eta^{-1} & q^{-1} & 0 \\ 0 & 0 & 0 & 0 \end{pmatrix}. \tag{33}$$

The matrix $U^{\frac{1}{2}\frac{1}{2}}$ satisfies the Temperley-Lieb algebra, i.e., $U_iU_{i\pm1}U_i = U_i$, $U_i^2 = d\ U_i$ (for the above matrix $U^{\frac{1}{2}\frac{1}{2}}$, $d = q + q^{-1}$). Similarly, the trigonometric Yang-Baxterization approach gives

$$\begin{aligned} \breve{R}(x) = [q^2 + q^{-2} - (x^2 + x^{-2})]^{-1/2}[(qx - q^{-1}x^{-1})I \\ -(x - x^{-1})U^{\frac{1}{2}\frac{1}{2}}], \end{aligned} \tag{34}$$

$$\begin{aligned} [\breve{R}(x)]^{-1} = [q^2 + q^{-2} - (x^2 + x^{-2})]^{-1/2}[(qx^{-1} - q^{-1}x)I \\ +(x - x^{-1})U^{\frac{1}{2}\frac{1}{2}}]. \end{aligned} \tag{35}$$

It is easy to check that $[\breve{R}(x)]^{\dagger}=[\breve{R}(x)]^{-1}=\breve{R}(-x)$ for $x = e^{i\vartheta}$, $\eta = e^{i\varphi(t)}$, and $\theta, \varphi(t), q \in$ real.

One may symmetrize the matrix $\breve{R}(x)$ given by Eq. (34) (i.e., to make the matrix elements $\left[\breve{R}(x)\right]_{1/2,-1/2}^{1/2,-1/2} = \left[\breve{R}(x)\right]_{-1/2,1/2}^{-1/2,1/2}$) through the following unitary transformation

$$\breve{R}_{i\ i+1}(V(x)) = V(x)\breve{R}_{i\ i+1}(x)V(x)^{\dagger}, \tag{36}$$

where $V(x) = V_i(x) \otimes [V_{i+1}(x)]^{-1}$ and

$$V_i(x) = \begin{pmatrix} 0 & x^{-\frac{1}{4}} \\ x^{\frac{1}{4}} & 0 \end{pmatrix}. \tag{37}$$

The resultant $\breve{R}_{i\ i+1}(V(x))$ is still a solution of YBE. Let only the parameter $\eta = e^{i\varphi(t)}$ be time-dependent, it yields from Eq. (4) and Eq. (36) that

$$\begin{aligned} H_2(x, \varphi(t)) = \hbar\dot{\varphi}\left[q^2 + q^{-2} - (x^2 + x^{-2})\right]^{-1}(x - x^{-1}) \\ \times[(x - x^{-1})(S_1^3 - S_2^3) + \\ (q - q^{-1})(e^{i\varphi}S_1^+S_2^- - e^{-i\varphi}S_1^-S_2^+)]. \end{aligned} \tag{38}$$

Putting $x = e^{i\vartheta}$, $\vartheta = \pi/2 - \theta$ and $\varphi(t) = \phi(t) - \pi/2 = \omega t$, we have

$$\begin{aligned} H_2(\theta, \phi(t)) = -4\hbar\omega\left[q^2 + q^{-2} + 2\cos 2\theta\right]^{-1}\cos\theta \\ \times[\cos\theta(S_1^3 - S_2^3) + \\ \tfrac{1}{2}(q - q^{-1})(e^{i\phi}S_1^+S_2^- + e^{-i\phi}S_1^-S_2^+)], \end{aligned} \tag{39}$$

whose two nonzero eigenvalues are

$$\begin{aligned} E_\pm^2 &= -4\hbar\dot{\phi}(q^2 + q^{-2} + 2\cos 2\theta)^{-1}\cos\theta\lambda_\pm \\ &= -\frac{4\hbar\dot{\phi}\cos\theta}{\lambda_\pm}, \end{aligned} \tag{40}$$

with

$$\lambda_\pm = \pm\sqrt{\cos^2\theta + (q - q^{-1})^2/4}. \tag{41}$$

Under the adiabatic approximation the corresponding eigenstates are

$$\begin{aligned} |\Phi_+(\theta, \phi)\rangle &= \tfrac{1}{\sqrt{2\lambda_+}}[(\lambda_+ - \cos\theta)^{-1/2}\left(\tfrac{q-q^{-1}}{2}\right)|\uparrow\downarrow\rangle \\ &\quad +i(\lambda_+ - \cos\theta)^{1/2}e^{-i\phi}|\downarrow\uparrow\rangle], \\ |\Phi_-(\theta, \phi)\rangle &= \tfrac{1}{\sqrt{2\lambda_-}}[i(\lambda_- - \cos\theta)^{-1/2}\left(\tfrac{q-q^{-1}}{2}\right)e^{i\phi}|\uparrow\downarrow\rangle \\ &\quad -(\lambda_- - \cos\theta)^{1/2}e^{-i\phi}|\downarrow\uparrow\rangle]. \end{aligned} \tag{42}$$

The corresponding Berry phases for the Yang–Baxter system are

$$\begin{aligned} \gamma_\pm^2 &= \pm\pi\left(1 - \frac{1}{\lambda_+}\cos\theta\right) \\ &= \pm\pi\left[1 - \frac{\cos\theta}{[\cos^2\theta + (q - q^{-1})^2/4]^{1/2}}\right]. \end{aligned} \tag{43}$$

The above Berry phases have been "q-deformed", when $\lambda_+ = 1$, or $q = \sqrt{1 + \sin^2\theta} \pm \sin\theta$, Eq. (43) reduces to Eq. (17). Remarkably the Berry phases in Eq. (43) can still be expressed in terms of the concurrence of the states $|\Phi_\pm(\theta,\varphi)\rangle$ in Eq. (42) as $\gamma_\pm^2 = \mp\pi(1-\sqrt{1-\mathcal{C}^2})$, where $\mathcal{C} = (q-q^{-1})/(2\lambda_+)$.

Similarly, the Hamiltonian $H_2(\theta,\phi(t))$ can be rewritten in terms of $SU(2)$ generators $J^+ = S_1^+S_2^- = \hat{f}_1^\dagger \hat{f}_2$, $J^- = S_1^-S_2^+ = \hat{f}_1\hat{f}_2^\dagger$, $J^3 = (S_1^3 - S_2^3) = (\hat{n}_1 - \hat{n}_2)/2$ as

$$H_2(\theta,\phi(t)) = -4\hbar\omega\frac{\cos\theta}{q^2+q^{-2}+2\cos 2\theta}H_0'(\theta,\phi(t)), \tag{44}$$

where

$$\begin{aligned} &H_0'(\theta,\phi(t)) = 2\varepsilon(\theta)J^3 + \Delta(t)J^+ + \Delta(t)^*J^-,\\ &\varepsilon(\theta) = \cos\theta, \quad \Delta(t) = e^{i\phi(t)}(q-q^{-1})/2. \end{aligned} \tag{45}$$

When $q - q^{-1} = 0$, or $q = \pm 1$, the Hamiltonian $H_0'(\theta,\phi(t))$ contracts to $H_0'(\theta,\phi(t)) = \varepsilon(\theta)(\hat{n}_1 - \hat{n}_2)$, thus the quantum criticality occurs. Correspondingly, one may easily see that the Berry phases in Eq. (43) vanish.

4. BH as a Derivative of DH via Braid Relation

The Bogoliubov Hamiltonian for quasiparticles in ^{3}He-B is compared to Dirac Hamiltonian for electrons in Ref. [21]. It can be expressed as(we have employed $\hbar = c = 1$):

$$H = m(\vec{k})\beta + \vec{p}\cdot\vec{\alpha} \tag{46}$$

where the mass $m(\vec{k})$ and momentum $\vec{p}$ are

$$m(\vec{k}) = \frac{\vec{k}^2}{2m} - \mu,\ \vec{p} = \vec{k}\Delta_B/k_F, \tag{47}$$

Δ_B is equilibrium order parameter, k_F is Fermi momentum, β and $\vec{\alpha}$ are 4×4 Dirac matrices:

$$\beta = \begin{pmatrix} \mathbf{1_2} & 0 \\ 0 & -\mathbf{1_2} \end{pmatrix},\ \vec{\alpha} = \begin{pmatrix} 0 & \vec{\sigma} \\ \vec{\sigma} & 0 \end{pmatrix}. \tag{48}$$

It is similar to the Dirac Hamiltonian for the electrons with mass m:

$$H = m\beta + \vec{p}\cdot\vec{\alpha} \tag{49}$$

The difference is the mass term in the Bogoliubov Hamiltonian essentially depends on k, and the Bogoliubov equation is not invariant under Lorentz group while the Dirac equation is. By using the braid relation, in this section our result surprisingly shows that the Bogoliubov Hamiltonian can be induced from Dirac Hamiltonian through braid relation, which we would like to illustrate as follows.

For a free electron the Dirac Hamiltonian is given in standard notation by

$$H_D = \frac{1}{E}(m\beta + \vec{p}\cdot\vec{\alpha}) \tag{50}$$

with the normalization $E = \sqrt{p^2+m^2}$, $p = |\vec{p}|$, We introduce the q-deformation of H_D, $R_1(\theta) = e^{i\frac{\theta}{2}H_D}$. Then we can get a q-deformation of another Hamiltonian $\mathcal{H}$, i.e. $R_2(\theta) = e^{i\frac{\theta}{2}\mathcal{H}}$. We expect R_1 and R_2 satisfy the following braid relation

$$R_1R_2R_1 = R_2R_1R_2. \tag{51}$$

After an unitary transformation on both R_1 and R_2, they of course still satisfy the same braid relation. One special unitary matrix

$$V = \frac{1}{\sqrt{2E}}\begin{pmatrix} p_+u_-^{-1} & -p_3u_-^{-1} & 0 & u_- \\ -p_+u_+^{-1} & p_3u_+^{-1} & 0 & u_+ \\ p_3u_-^{-1} & p_-u_-^{-1} & u_- & 0 \\ -p_3u_+^{-1} & -p_-u_+^{-1} & u_+ & 0 \end{pmatrix} \tag{52}$$

with $p_\pm = p_1 \pm ip_2$, $u_\pm = \sqrt{E\pm m}$, transform the H_D diagonalized:

$$VH_DV^\dagger = \begin{pmatrix} \sigma_3 & 0 \\ 0 & \sigma_3 \end{pmatrix}. \tag{53}$$

In terms of such a transformation, $R_1(\theta)$ becomes diagonal as well and is composed of two 2×2 matrices.

$$R_1'(\theta) = Ve^{i\frac{\theta}{2}H_D}V^\dagger = \begin{pmatrix} a(\theta) & 0 \\ 0 & a(\theta) \end{pmatrix}, \tag{54}$$

where $a(\theta) = \begin{pmatrix} e^{i\frac{\theta}{2}} & 0 \\ 0 & e^{-i\frac{\theta}{2}} \end{pmatrix} = e^{i\frac{\theta}{2}}\begin{pmatrix} 1 & 0 \\ 0 & e^{-i\theta} \end{pmatrix}$.

Until now, θ acts as a free deformation parameter. One practical choice is related to the anyon description for FQHE with $\nu = 1/2$.[22] Let $\theta = -\frac{\pi}{2}$, then

$$a = a(\theta = -\frac{\pi}{2}) = e^{-i\pi/4}\begin{pmatrix} 1 & 0 \\ 0 & i \end{pmatrix}. \tag{55}$$

Following the braid relation there should be a matrix b satisfying

$$aba = bab. \tag{56}$$

One typical solution of such matrix b is given by

$$b = \frac{1}{\sqrt{2}}\begin{pmatrix} 1 & i \\ i & 1 \end{pmatrix}. \tag{57}$$

In fact, both matrices a and b emergence exactly at the anyon description for FQHE with $\nu = 1/2$.[22]

Under the action of unitary transformation V, we expect $R_2^{'}(\theta) = Ve^{i\frac{\theta}{2}\mathcal{H}}V^{\dagger}$ take a block representation. Therefore, the matrix

$$R_2^{'} = \begin{pmatrix} b & 0 \\ 0 & b \end{pmatrix} = \frac{1}{\sqrt{2}} \begin{pmatrix} 1 & i & 0 & 0 \\ i & 1 & 0 & 0 \\ 0 & 0 & 1 & i \\ 0 & 0 & i & 1 \end{pmatrix} \tag{58}$$

naturally satisfies the relation

$$R_1^{'} R_2^{'} R_1^{'} = R_2^{'} R_1^{'} R_2^{'}. \tag{59}$$

Utilizing the same transformation V to switch back, one will get

$$R_2 = V^{\dagger} R_2^{'} V = \frac{1}{\sqrt{2}}(1 - i\frac{1}{\frac{Ep}{m}}(\frac{p^2}{m}\beta - \vec{p}\cdot\vec{\alpha})). \tag{60}$$

Combining with the definition of R_2,

$$R_2 = e^{i\frac{\theta}{2}\mathcal{H}} = e^{-i\frac{\pi}{4}\mathcal{H}} = \frac{1}{\sqrt{2}}(1 - i\mathcal{H}), \tag{61}$$

we finally get

$$\mathcal{H} = \frac{1}{\frac{Ep}{m}}(\frac{p^2}{m}\beta - \vec{p}\cdot\vec{\alpha}) = \frac{1}{\mathcal{E}}(\frac{p^2}{m}\beta - \vec{p}\cdot\vec{\alpha}),\ \mathcal{E} = \frac{p}{m}E. \tag{62}$$

It is easy to check that $\mathcal{H}$ is normalized, $\mathcal{H}^2 = 1$. The obtained Hamiltonian has a natural explanation: it is the Bogoliubov Hamiltonian for quasiparticles in ^{3}He-B with the free energy $\mu = 0$ and mass being $m/2$.

5. Conclusion and Discussion

In summary, we have obtained some spin-1/2 interaction Hamiltonians from the unitary Yang–Baxter $\breve{R}$-matrix. Based on which, Berry phases and quantum criticality in the Yang–Baxter systems have been studied. We has also shown that the Dirac Hamiltonian and Bogoliubov Hamiltonian can be surprisingly related with each other with the braid relation. It is known that the braid matrices $R_1^{'}$ and $R_2^{'}$ are related to the anyon description for FQHE with $\nu = 1/2$.[22] If there is evidence of physical consequence of associative Hamiltonian $\mathcal{H}$ based on H_D through braid relation, we may expect that Dirac particle could be decomposed to anyons on the basis of the braid relation. Howerver, The nature of our result is still not clear.

Let us make three more discussions to end this paper.

(i) The Yang–Baxter Hamiltonians are induced from the $\breve{R}(\theta,\phi)$-matrix via Eq. (4). It is natural to ask whether the same Hamiltonian, e.g., $H_1(\theta,\phi(t))$, can be induced from different matrices of $\breve{R}(\theta,\phi)$. The answer is yes. Here we would like to provide such an example.

Considering the following braiding matrix

$$B = S^{\frac{1}{2}\frac{1}{2}} = I + fU^{\frac{1}{2}\frac{1}{2}}, \tag{63}$$

where

$$U^{\frac{1}{2}\frac{1}{2}} = \begin{pmatrix} \epsilon & 0 & 0 & e^{i\varphi} \\ 0 & 0 & 0 & 0 \\ 0 & 0 & 0 & 0 \\ e^{-i\varphi} & 0 & 0 & \epsilon \end{pmatrix} \tag{64}$$

satisfies the TL algebra, and $d = 2\epsilon$, $\beta = -d/2 = -\epsilon$, $\epsilon = \pm 1$, $f = (-d \pm \sqrt{d^2-4})/2 = \beta$. After making the rational Yang–Baxterization,[20] from Eq. (63) one obtains the $\breve{R}(u)$ as

$$\breve{R}(u) = I + G(u)U, \quad G(u) = \frac{u}{\alpha + \beta u}. \tag{65}$$

The unitary condition $[\breve{R}(u)]^\dagger = [\breve{R}(u)]^{-1} = \breve{R}(-u)$ leads to $G(-u) = G(u)^*$, or $(\alpha^{-1}u)^* = -\alpha^{-1}u$. We choose

$$\alpha u^{-1} = i\tan\theta, \tag{66}$$

then it is easy to have from Eq. (4) and Eq. (65) that

$$H(u,\phi) = i\hbar G(u)G(-u)\frac{\partial U}{\partial t}[G(-u)^{-1}I + U]. \tag{67}$$

Substituting Eq. (65) into Eq. (67) one obtains

$$\begin{aligned} H_3(\theta,\varphi(t)) = -\hbar\dot{\varphi}\cos\theta\{&\cos\theta(S_1^3 + S_2^3) \\ &-i\sin\theta[e^{i\varphi}S_1^+S_2^+ - e^{-i\varphi}S_1^-S_2^-]\}. \end{aligned} \tag{68}$$

After redefining the parameter $\varphi(t) = \phi(t) - \pi/2$, one may find that the Hamiltonian $H_3(\theta,\varphi(t))$ is identical to the Hamiltonian $H_1(\theta,\phi(t))$ as shown in Eq. (12).

(ii) The same matrix $U^{\frac{1}{2}\frac{1}{2}}$ may yield inequivalent Yang–Baxter Hamiltonians. For instance, Let

$$U^{\frac{1}{2}\frac{1}{2}} = \begin{pmatrix} 0 & 0 & 0 & 0 \\ 0 & \epsilon & e^{i\varphi} & 0 \\ 0 & e^{-i\varphi} & \epsilon & 0 \\ 0 & 0 & 0 & 0 \end{pmatrix}, \tag{69}$$

which is a special form of Eq. (33) by taking $q = \epsilon = \pm 1$. Based on the same matrix $U^{\frac{1}{2}\frac{1}{2}}$, one can have two kinds of inequivalent braiding matrices, one is $B_1 = q(I - q^{-1}U^{\frac{1}{2}\frac{1}{2}})$ in Eq. (32), the other is $B_2 = I + fU^{\frac{1}{2}\frac{1}{2}}$ in Eq. (63). After making the trigonometric Yang-Baxterization for the former one B_1, it yields the Yang–Baxter Hamiltonian $H_2(\theta, \phi(t))$ as in Eq. (39); similarly, after making the rational Yang-Baxterization for the latter one B_2, it yields the following Yang–Baxter Hamiltonian:

$$\begin{aligned} H_4(\theta, \phi(t)) &= -\hbar\dot{\phi}\cos\theta\{\cos\theta(S_1^3 - S_2^3) \\ &\quad + \sin\theta[e^{i\phi}S_1^+S_2^- + e^{-i\phi}S_1^-S_2^+]\}. \end{aligned} \tag{70}$$

The Hamiltonian $H_4(\theta, \phi(t))$ is inequivalent to Hamiltonian $H_2(\theta, \phi(t))$, and it has the same eigenvalues and Berry phases as those of the Hamiltonian $H_1(\theta, \phi(t))$.

(iii) For spins at N-lattices one may define the following $SU(2)$ generators as

$$\begin{aligned} S^+ &= \prod_{i=1}^N S_i^+ = \prod_{i=1}^N \hat{f}_i^\dagger, \quad S^- = \prod_{i=1}^N S_i^- = \prod_{i=1}^N \hat{f}_i, \\ S^3 &= [S^+, S^-]/2 = \left(\prod_{i=1}^N \hat{n}_i + \prod_{i=1}^N (1 - \hat{n}_i)\right)/2. \end{aligned} \tag{71}$$

When S^+ acts on the vacuum state, it produces the state with all spins up $|\uparrow\uparrow\cdots\uparrow\rangle$, similarly, S^- produces the state with all spins down $|\downarrow\downarrow\cdots\downarrow\rangle$. The states $|\uparrow\uparrow\cdots\uparrow\rangle$ and $|\downarrow\downarrow\cdots\downarrow\rangle$ are the chiral spin states, or the "chiral photons". All the similar discussion on quantum criticality and Berry phase can be extended to multipartite spin-1/2 systems, which we shall investigate subsequently.

Acknowledgments

This work is supported in part by NSF of China (Grant No. 10575053 and No. 10605013) and Program for New Century Excellent Talents in University.

References

1. R. J. Baxter, *Exactly Solved Models in Statistical Mechanics* (Academic Press, London, 1982).
2. *Yang–Baxter Equations in Integrable Systems*, edited by M. Jimbo (World Scientific, Singapore, 1990).

3. A. Osterloh, L. Amico, G. Falci, and R. Fazio, Nature **416**, 608 (2002).
4. G. Vidal, J. I. Latorre, E. Rico, and A. Kitaev, Phys. Rev. Lett. **90**, 227902 (2003); J. I. Latorre, E. Rico, and G. Vidal, Quant. Inform. Comput. **4**, 48 (2004).
5. L. A. Wu, S. Bandyopadhyay, M. S. Sarandy, and D. A. Lidar, Phys. Rev. A **72**, 032309 (2005);
6. J. Anders and V. Vedral, e-print quant-ph/0610268.
7. J. L. Birman and S. Q Zhou, Quantum phase transitions and contraction of dynamical symmetry, to be published.
8. *Geometric Phases in Physics*, edited by A. Shapere and F. Wilczek (World Scientific, Singapore, 1989).
9. L. H. Kauffman and S. J. Lomonaco Jr., New J. Phys. **6**, 134 (2004);
10. Y. Zhang, L. H. Kauffman and M. L. Ge, Int. J. Quant. Inform. **3**, 669 (2005).
11. J. M. Franko, E. C. Rowell, and Z. Wang, J. Knot Theory Ramifications **15**, 413 (2006).
12. Y. Zhang, E. C. Rowell, Y. S. Wu, Z. Wang, and M. L. Ge, e-print arXiv:0706.1761.
13. V. F. R. Jones, Commun. Math. Phys. **125**, 459 (1987).
14. M. L. Ge, Y. S. Wu, and K. Xue, Int. J. Mod. Phys. A **6**, 3735 (1991); M. L. Ge and K. Xue, J. Phys. A **26**, 281 (1993).
15. M. L. Ge, K. Xue, and Y. S. Wu, Yang–Baxterization and algebraic structures, *Braid Group, Knot Theory and Statistical Mechanics II*, edited by C. N. Yang and M. L. Ge (World Scientific, Singapore, 1994) p.130.
16. J. L. Chen, K. Xue, and M. L. Ge, Phys. Rev. A **76**, 042324 (2007).
17. W. Zhang, D. Feng, and R. Gilmore, Rev. Mod. Phys. **62** 867 (1990).
18. S. Chaturvedi, M. S. Sriram, and V. Srinivasan, J. Phys. A **20**, L1091 (1987).
19. W. Wootters, Phys. Rev. Lett. **80**, 2245 (1998).
20. M. L. Ge and K. Xue, Phys. Lett. A **152**, 266 (1991).
21. G.E. Volovik, *Exotic Properties of superfluid* ^3He,(World Scientific, Singapore,1992), p77.
22. S. D. Sarma, M. Freedman, C. Nayak, S. H. Simon, and Ady Stern, e-print Arxiv: 0707.1889.

HIERARCHICAL QUANTUM SEARCH

VLADIMIR E. KOREPIN[1] and YING XU[2]

C.N. Yang Institute for Theoretical Physics,
State University of New York at Stony Brook,
Stony Brook, NY 11794-3840, USA
[1]*E-mail: korepin@max2.physics.sunysb.edu*
[2]*E-mail: yixu@ic.sunysb.edu*

Database search has wide applications and is used as a subroutine in many important algorithms. In this paper we will consider a database with a single target item. Quantum algorithm [Grover] locates the target item faster than any classical algorithm. In addition to a full [Grover] search, it frequently occurs that one is looking for a group of items [a block] containing the target item, rather than the target item itself. This problem is known as partial search. As a generalization of the full search, partial search is of particular importance in practice. Partial search trades accuracy for speed, i.e. it works faster than a full search. There exists different versions of partial search. We will study the optimized version of the algorithm discovered by Grover and Radhakrishnan and call it GRK. GRK can be applied successively [in a sequence]. First the database is partitioned into blocks and GRK is applied to find the target block. Then this target block is partitioned into sub-blocks and GRK is used again to find the target sub-block. This procedure can be repeated if the database is large enough. [This sequence of GRK's is called a hierarchy.] Another possibility is to partition the database into sub-blocks directly and use GRK to find the target sub-block once. In this paper we will prove that the latter is faster [makes less queries to the oracle].

Keyword: Quantum Search; Grover Algorithm; Partial Search; hierarchical Search.

1. Introduction

Database search has many applications. Search algorithm enters as a subroutine in many important algorithms in computer sciences.[1–3] Grover discovered a quantum algorithm which searches a database faster than any classical algorithm.[4] Let's consider a database with one target item. We use number of queries to the oracle as complexity measure. The Grover algorithm finds the target item [with probability 1] in

$$j_{\text{full}} = \frac{\pi}{4}\sqrt{N}, \qquad N \to \infty \tag{1}$$

iterations [queries to the oracle]. We shall call it a full search.

It occurs frequently in practice that less information is needed. For example, the address of the target item in binary form is $|t\rangle = |b_1 b_2 b_3 ... b_n\rangle$, and we want to find only the first 3 bits $b_1 b_2 b_3$. This means that the database is partitioned into 8 blocks. All items in a block share the common feature such that the first 3 bits being the same. We want to find the block containing the target item. This is an example of partial search. The general problem of partial search considers the following: An N item database is partitioned into K blocks, each of the same size

$$b = \frac{N}{K}. \tag{2}$$

A user wants to find the block containing the target item, instead of the target item itself. The block with the target item is called the target block; others non-target blocks. Partial search naturally arises in list matching.[5] Partial search is not only a compromise on accuracy for speed, but also has it own significance. Partial search can find all items in the database which share some features with the target item. This can be considered as a special case of sorting problem. The GRK algorithm of partial search was suggested by Grover and Radhakrishnan,[6] and optimized in 7. It takes $\sim \frac{\pi}{4}(1 - \text{coeff}(K))\sqrt{N}$ number of queries to find the target block. Here coeff(K) is a finite positive number, which depends on K and has a limit when blocks are large $b \to \infty$. GRK is the most efficient partial search algorithm known in literature.[6–10]

GRK can be applied in a sequence [one after another], i.e. after the first GRK, the target block found can be further partitioned into sub-blocks. Then a second GRK can be applied to find the sub-block containing the target item [called the target sub-block]. We shall call the sequence of GRK's a partial search hierarchy. In hierarchical search we iterate GRK. A practical example would be: In order to find a hotel, we first look at a State map and then a town map. We shall see that the second GRK works faster than the first one. Actually, GRK can be conducted repeatedly until we find the smallest target sub-sub-block interested. The total number of queries is the sum of queries of each GRK in the hierarchy. [We use number of queries as measure of complexity.]

Alternative to a partial search hierarchy which finds the target sub-sub-block, we could partition the database directly into sub-sub-blocks and use GRK once: We shall call it direct partial search. Although each GRK works faster than the previous one in the hierarchy, it is not guaranteed that the total number of queries in the hierarchy [sequence of GRK's] is less than that of a direct partial search. On the contrary, we will prove that *direct partial search works faster*, which is the main result of the paper. For example, consider a database partitioned into 2 blocks. Each block is partitioned into 2 sub-blocks, so totally 4 sub-blocks. One could first find the target block using GRK, then the target sub-block using sequential GRK. However, it is faster to run a GRK directly over the 4 sub-blocks, which finds the target sub-block once.

The paper consists of two parts: In the first part, we start with the Grover algorithm and the GRK algorithm. Then we study the partial search hierarchy in detail. The second part proceeds to a comparison of the hierarchical partial search with direct partial search. Then we prove our main result that direct partial search works faster.

2. The Grover Search Algorithm

In our paper, we consider different methods of partial search. They are all built on the original idea of the full Grover search.[2,4,11] Let's formulate the problem. Consider a database of N items with one target item. [a] The database is associated with a Hilbert space with N normalized basis vectors. The basis vector corresponding to item x is denoted by $|x\rangle$. The Grover search is a quantum algorithm which starts from the uniform superposition of all basis vectors in the whole database:

$$|s_1\rangle = \frac{1}{\sqrt{N}} \sum_{x=0}^{N-1} |x\rangle, \qquad \langle s_1|s_1\rangle = 1. \tag{3}$$

The algorithm searches for a single target item $|t\rangle$ iteratively. The Grover iteration is a unitary transform:

$$G_1 = -I_{s_1} I_t. \tag{4}$$

Later we shall call it a global iteration in GRK. Here I_t and I_{s_1} are two inversions about the target item $|t\rangle$ and the uniform superposition $|s_1\rangle$ defined in (3), respectively:

$$I_t = \hat{I} - 2|t\rangle\langle t|, \tag{5}$$

$$I_{s_1} = \hat{I} - 2|s_1\rangle\langle s_1|, \tag{6}$$

[a] Target item also called in literature marked item or solution.

where $\hat{I}$ is the identical operator. The Grover iteration G_1 is a rotation in the Hilbert space from $|s_1\rangle$ towards the target $|t\rangle$ by an angle θ_1 defined by:[11]

$$\sin^2 \theta_1 = \frac{1}{N}. \tag{7}$$

After j_1 iterations the state of the database becomes:[2,11]

$$G_1^{j_1}|s_1\rangle = \sin((2j_1+1)\theta_1)|t\rangle + \frac{\cos((2j_1+1)\theta_1)}{\sqrt{N-1}} \sum_{x \neq t}^{\text{N-1 items}} |x\rangle. \tag{8}$$

Therefore after $j_{\text{full}} = \pi/(4\theta_1) - 1/2$ iterations the probability amplitude of $|t\rangle$ becomes unity and amplitudes of other items all vanish. i.e.

$$G_1^{j_{\text{full}}}|s_1\rangle = |t\rangle. \tag{9}$$

As N becomes large $j_{\text{full}} = \pi/(4\theta_1) - 1/2$ approaches (1). More details on Grover search can be found in 2.

3. Algorithms for Partial Search

Before introducing the GRK partial search algorithm [see next section], we look at a few other algorithms for comparison:

(i) Naive Search

Pick a block randomly and make a full Grover search in it [which makes $\frac{\pi}{4}\sqrt{\frac{N}{K}}$ queries to the oracle]. If we find the target item then we understand that this is the target block. If not, then we discard this block and pick another randomly. Make a full Grover search in it and repeat this procedure till we find the target block. In the worst case the target block will be the last one. So with probability 1 we have to use

$$r(N,K) = \frac{(K-1)}{\sqrt{K}} \frac{\pi}{4} \sqrt{N} \tag{10}$$

iterations [queries] to find the target block. [b]

A full Grover search finds the target item in $(\pi/4)\sqrt{N}$ queries. If we know the exact address of the target item then we also know the target block. Comparing $(\pi/4)\sqrt{N}$ with $r(N,K)$ in (10), we see that the naive version is faster only for two blocks $K = 2$. [If $K \geq 3$ a full search is faster].

---

[b] $(\sqrt{K}/2)(\pi/4)\sqrt{N}$ queries on average.

(ii) Binary Search

Assume that $K = 2^k$ with k being a positive integer. Divide the database into two blocks and make a full Grover search in one block. If the target item is not found, then take the remaining block and divide it into two sub-blocks. Pick a sub-block randomly and make a full search again in it. Repeat the procedure until we are left with the last block. In the worst case, the number of queries necessary to find the target block is

$$\frac{\pi}{4}\sqrt{N}\left\{\frac{1}{\sqrt{2}}+\frac{1}{\sqrt{4}}+\ldots+\frac{1}{2^{k/2}}\right\}, \qquad k=\log_2 K. \tag{11}$$

The first two terms in the braces of (11) are greater than 1 for $K \geq 3$,

$$\frac{1}{\sqrt{2}}+\frac{1}{2}=\frac{\sqrt{2}+1}{2}>1. \tag{12}$$

So this algorithm is less efficient than a full Grover search, when $K > 2$.

(iii) Grover and Radhakrishnan Version

A faster version was found in 6. Pick randomly a block and make a full Grover search in the compliment [all items in the rest of the database]. Either the target item [and block] is found after the search or the picked block is the target block. This requires $\frac{\pi}{4}\sqrt{b(K-1)} = \frac{\pi}{4}\sqrt{N}\sqrt{\frac{K-1}{K}}$ queries. It is faster than a full search.

4. The GRK Partial Search Algorithm

Grover and Radhakrishnan also discovered a faster quantum algorithm for partial search,[6] which uses the same oracle as the main Grover algorithm. [See Summary and Appendix D.] Partial search also starts from the uniform superposition of all basis states (3). A general structure of the algorithm is:[6–10,12]

Step 1. Global iterations: j_1 standard Grover iterations (4). After this step the state of database is $G_1^{j_1}|s_1\rangle$.

Step 2. Simultaneous local iterations in each block: j_2 local Grover iterations defined in (13) below. After step 2 the state of database is $G_2^{j_2}G_1^{j_1}|s_1\rangle$.

Local iteration is defined by

$$G_2 = \bigoplus_{\text{blocks}}^{K} G_2^{\text{one block}} = -\left(\bigoplus_{\text{blocks}}^{K} I_{s_2}\right) I_t. \tag{13}$$

It is a direct sum of Grover iterations [called local queries] defined in each block

$$G_2^{\substack{\text{one}\\\text{block}}} = -I_{s_2} I_t. \tag{14}$$

In the expression I_t is the same inversion (5), i.e. query to the oracle. I_{s_2} is a local inversion

$$I_{s_2} = \hat{I} - 2|s_2\rangle\langle s_2|. \tag{15}$$

Here $|s_2\rangle$ is the uniform superposition of items in one block

$$|s_2\rangle = \frac{1}{\sqrt{b}} \sum_{\text{one block}}^{\text{b items}} |x\rangle. \tag{16}$$

Local iteration G_2 is a the Grover iteration in each block done simultaneously in all blocks. G_2 acts trivially on non-target blocks. A non-trivial operation [rotation] is present only in the target block with new rotation angle θ_2 defined by

$$\sin^2 \theta_2 = \frac{K}{N} = \frac{1}{b}. \tag{17}$$

Note that amplitudes of all items in non-target blocks remain intact.

Step 3. Location of the target block with a final global iteration:[7,8,10] We have to vanish amplitudes of all items in non-target blocks. We can do it by application of one more global iteration. The resulting state is

$$|d\rangle \equiv G_1 G_2^{j_2} G_1^{j_1} |s_1\rangle = \sin\omega |t\rangle + \frac{\cos\omega}{\sqrt{b-1}} \sum_{\substack{x\neq t\\ \text{target block}}}^{\substack{\text{b-1}\\\text{items}}} |x\rangle. \tag{18}$$

The final state (18) is expressed as a superposition over items in the target block only. This is realized by requiring that the amplitude of any non-target block vanishes after the partial search, i.e.

$$\langle x|d\rangle = 0. \tag{19}$$

Here x is an arbitrary item in any non-target block. This vanishing condition can be written explicitly as an equality for j_1 and j_2, see 7. We shall call it a cancellation condition.

This partial search algorithm was further optimized in 7. In the large block limit $b \to \infty$, the total number of items also large $N \to \infty$, while the ratio $K = N/b$ kept finite. Then the expression for rotation angles (7) and (17) simplifies

$$\theta_1 \to \frac{1}{\sqrt{N}}, \qquad \theta_2 \to \frac{1}{\sqrt{b}}. \tag{20}$$

It turns out convenient to rewrite numbers of iterations in a scale form:[6]

$$j_1 = \left(\frac{\pi}{4} - \frac{\eta}{\sqrt{K}}\right)\sqrt{N}, \qquad j_2 = \frac{\alpha}{\sqrt{K}}\sqrt{N}. \tag{21}$$

Here η and α are parameters of order 1 [they have a limit]. The ranges of these parameters are discussed in Appendix B. The vanishing condition (19) in terms of these parameters simplifies in the large b limit:[7,9]

$$\tan\left(\frac{2\eta}{\sqrt{K}}\right) = \frac{2\sqrt{K}\sin 2\alpha}{K - 4\sin^2\alpha}. \tag{22}$$

The total number of queries is

$$S(K) \equiv j_1 + j_2 + 1 \overset{b\to\infty}{\longrightarrow} \left(\frac{\pi}{4} + \frac{\alpha - \eta}{\sqrt{K}}\right)\sqrt{N}. \tag{23}$$

It was minimized [subject to the constraint (22)] in 7. The minimum number of queries is achieved at

$$\eta\,(K) = \frac{1}{2}\sqrt{K}\arctan\left(\frac{\sqrt{3K-4}}{K-2}\right), \qquad \alpha\,(K) = \frac{1}{2}\arccos\left(\frac{K-2}{2(K-1)}\right). \tag{24}$$

Thus the minimized number of queries of GRK partial search [as a function of K] is

$$S\,(K) \overset{b\to\infty}{\longrightarrow} \left(\frac{\pi}{4} + \frac{\alpha\,(K) - \eta\,(K)}{\sqrt{K}}\right)\sqrt{N}. \tag{25}$$

A proof of (24) being the minimum is given in Appendix C. Note that $\alpha - \eta$ is negative and number of blocks $K \geq 2$ in a non-trivial situation.

In the large block limit, the ω appeared in (18) is

$$\omega = \alpha(K), \tag{26}$$

see 7. As a consequence, the state of database after GRK (18) is the following: The amplitudes of items in non-target blocks all vanish and the state of the target block is

$$|d\rangle = \sin\alpha(K)|t\rangle + \frac{\cos\alpha(K)}{\sqrt{b-1}} \sum_{\substack{x\neq t \\ \text{target block}}}^{\substack{\text{b-1} \\ \text{items}}} |x\rangle. \tag{27}$$

5. The Partial Search Hierarchy

A partial search hierarchy is a sequence of GRK's. After location of the target block, we may consider a subsequent GRK partial search: The target block is further partitioned into $\tilde{K}$ sub-blocks and we search for the sub-block containing the target item [target sub-block]. For example we can use Google Earth to find the State of New York first on the map of USA and then make a sequential search for Stony Brook in the State map.

We shall show below that a sequential GRK can be done faster than the first GRK. The coefficient $\pi/4$ in (25) is replaced by a smaller number:

$$\frac{\pi}{4} \to \frac{\pi}{4} - \frac{1}{4}\arccos\left(\frac{K-2}{2(K-1)}\right). \tag{28}$$

Each successive GRK works faster than the previous one for two reasons. First, the new database is smaller [only one block of the previous one]. Second, the initial state of the new database (27) can be represented in different forms (30) and (40) below. We see that for sequential GRK, the initial state is no longer a uniform superposition of basis vectors of the new database. It is an unevenly weighted superposition with emphasis on the target $|t\rangle$, see (30) and (40). In other words, the new initial state of the database is equivalent to a partially searched [though not fully searched] one. This fact was studied in 7. It was shown that after the first GRK the state of the target block [new database] can be written as [(27) rewritten]

$$|d\rangle = G_1 G_2^{j_2} G_1^{j_1}|s_1\rangle = \sin\alpha(K)|t\rangle + \frac{\cos\alpha(K)}{\sqrt{b-1}} \sum_{\substack{x\neq t \\ \text{target block}}}^{\substack{\text{b-1} \\ \text{items}}} |x\rangle. \tag{29}$$

We have used relation (26). Compared with (8), we see that the state after the first GRK (29) takes the form

$$|d\rangle = G_1 G_2^{j_2} G_1^{j_1}|s_1\rangle = G_2^{\frac{\alpha(K)}{2}\sqrt{b}}|s_2\rangle, \tag{30}$$

which serves as the initial state of the sequential GRK.

For notational convenience, we use a "$\sim$" to indicate variables in sequential GRK and make the following definitions:

$$\begin{aligned}
\text{Number of items in new database :} \quad & \tilde{N} = b = \tilde{K}\tilde{b}, && (31)\\
\text{Uniform superposition of new database :} \quad & |\tilde{s_1}\rangle = |s_2\rangle, && (32)\\
\text{New global inversion :} \quad & I_{\tilde{s}_1} = I_{s_2}, && (33)\\
\text{New global iteration :} \quad & \tilde{G_1} = G_2, && (34)\\
\text{New global rotation angle :} \quad & \tilde{\theta_1} = \theta_2, && (35)\\
\text{Uniform superposition of one sub-block :} \quad & |\tilde{s_2}\rangle = \frac{\sum_{\text{one sub-block}}^{\tilde{b}\text{ items}} |x\rangle}{\sqrt{\tilde{b}}}, && (36)\\
\text{New local inversion :} \quad & I_{\tilde{s_2}} = I - 2|\tilde{s}_2\rangle\langle\tilde{s}_2|, && (37)\\
\text{New local iteration :} \quad & \tilde{G_2} = -I_{\tilde{s}_2} I_t, && (38)\\
\text{New local rotation angle :} \quad & \sin^2\tilde{\theta}_2 = \frac{1}{\tilde{b}}. && (39)
\end{aligned}$$

Written in these notations, the initial state of new database (30) is equivalent to a partially searched one with $\frac{\alpha(K)}{2}\sqrt{\tilde{N}}$ new global queries, i.e.

$$|d\rangle = G_1 G_2^{j_2} G_1^{j_1}|s_1\rangle = \tilde{G_1}^{\frac{\alpha(K)}{2}\sqrt{\tilde{N}}}|\tilde{s_1}\rangle. \tag{40}$$

Steps of sequential GRK can be written similarly to the first GRK using new notations (31)-(39). The resultant state of target sub-block is

$$\begin{aligned}
|\tilde{d}\rangle &\equiv \tilde{G_1}\tilde{G_2}^{\tilde{j}_2}\tilde{G_1}^{\tilde{j}_1}\left(\tilde{G_1}^{\frac{\alpha(K)}{2}\sqrt{\tilde{N}}}|\tilde{s_1}\rangle\right)\\
&= \sin\tilde{\omega}|t\rangle + \frac{\cos\tilde{\omega}}{\sqrt{\tilde{b}-1}} \sum_{\substack{x\neq t\\ \text{target sub-block}}}^{\tilde{b}-1\text{ items}} |x\rangle.
\end{aligned} \tag{41}$$

Note that the vector in the parentheses is $|d\rangle$ of (27). We also have [similar to (19)]

$$\langle x|\tilde{d}\rangle = 0, \qquad \forall x \in \{\text{items of non-target sub-blocks}\}. \tag{42}$$

This yields cancellation condition relating $\tilde{j}_1$ and $\tilde{j}_2$, see 7. We introduce parameters $\tilde{\eta}$ and $\tilde{\alpha}$ defined by

$$\tilde{j}_1 = \left(\frac{\pi}{4} - \frac{\alpha(K)}{2} - \frac{\tilde{\eta}}{\sqrt{\tilde{K}}}\right)\sqrt{\tilde{N}}, \qquad \tilde{j}_2 = \tilde{\alpha}\sqrt{\tilde{b}}. \tag{43}$$

The algorithm is also optimized 7 in the large sub-block limit: $\tilde{b} \to \infty$, $\tilde{N} \equiv \tilde{K}\tilde{b} \to \infty$. In the limit, the angles (35) and (39) simplify

$$\tilde{\theta_1} = \frac{1}{\sqrt{\tilde{N}}}, \qquad \tilde{\theta_2} = \frac{1}{\sqrt{\tilde{b}}}. \tag{44}$$

The minimum is achieved at

$$\tilde{\eta}\left(\tilde{K}\right) = \tfrac{1}{2}\sqrt{\tilde{K}}\arctan\left(\tfrac{\sqrt{3\tilde{K}-4}}{\tilde{K}-2}\right) = \eta\left(\tilde{K}\right),$$
$$\tilde{\alpha}\left(\tilde{K}\right) = \tfrac{1}{2}\arccos\left(\tfrac{\tilde{K}-2}{2(\tilde{K}-1)}\right) = \alpha\left(\tilde{K}\right). \tag{45}$$

Similar to (26), we have in the large sub-block limit

$$\tilde{\omega} = \alpha(\tilde{K}). \tag{46}$$

As a result the number of queries of the sequential GRK is

$$\bar{S}\left(K,\tilde{K}\right) \equiv \tilde{j}_1 + \tilde{j}_2 + 1 \overset{\tilde{b}\to\infty}{\longrightarrow} \left(\frac{\pi}{4} - \frac{\alpha(K)}{2} + \frac{\alpha\left(\tilde{K}\right) - \eta\left(\tilde{K}\right)}{\sqrt{\tilde{K}}}\right)\sqrt{\tilde{N}}. \tag{47}$$

In principle, sequential GRK's can be conducted successively until the smallest target sub-sub-block is found. Here arises a question on the efficiency of hierarchical partial search, i.e. whether or not is a sequence of GRK's works faster than a direct GRK partial search of the smallest sub-sub-blocks. As will be shown in the following section, direct GRK partial search makes less queries in the quantum case.

6. Comparison of Hierarchical Partial Search with Direct Partial Search

The partial search hierarchy forms a sequence of GRK's. It starts from searching for the largest target block and ends with searching for the smallest target sub-sub-block. On the other hand, it is also possible to partition the database directly into the smallest sub-sub-blocks and use a GRK to find the target sub-sub-block in one time. One question of significance is **whether the hierarchical search works faster than the direct search or not**. This question is of practical importance and the answer turns out to be *negative*. We prove the statement by studying the first two successive GRK's in the hierarchy.

We have already derived the optimized number of queries of the first two GRK's in (25) and (47), respectively. So that the total number of queries is the sum:

$$T\left(K,\tilde{K}\right) \equiv S(K) + \bar{S}\left(K,\tilde{K}\right)$$
$$= \left\{\tfrac{\pi}{4} + \tfrac{\frac{\pi}{4}+\frac{1}{2}\alpha(K)-\eta(K)}{\sqrt{K}} + \tfrac{\alpha(\tilde{K})-\eta(\tilde{K})}{\sqrt{K\tilde{K}}}\right\}\sqrt{N}. \tag{48}$$

On the other hand, if the database is partitioned directly into $K\tilde{K}$ blocks, a direct GRK algorithm would require

$$S\left(K\tilde{K}\right) = \left[\frac{\pi}{4} + \frac{\alpha\left(K\tilde{K}\right) - \eta\left(K\tilde{K}\right)}{\sqrt{K\tilde{K}}}\right]\sqrt{N} \tag{49}$$

queries instead. Let us compare $T(K,\tilde{K})$ and $S(K\tilde{K})$, assuming that both $K \geq 2$ and $\tilde{K} \geq 2$.

6.1. *Numerical Comparison of Query Numbers and Asymptotic Analysis*

Before giving the complete proof, we illustrate this fact by looking at a few concrete examples. Here in Table 1 we give a few numerical examples of query numbers $S(K\tilde{K})$ and $T(K,\tilde{K})$ as well as their difference, for a better understanding. It is clear that each $T - S$ is positive in the last column.

Table 1. Numerical Examples of Query Numbers.

K	$\tilde{K}$	$S(K\tilde{K})/\sqrt{N}$	$T(K,\tilde{K})/\sqrt{N}$	$(T(K,\tilde{K}) - S(K\tilde{K}))/\sqrt{N}$
2	2	0.61548	0.670379	0.054899
2	3	0.646015	0.695421	0.049406
3	2	0.646015	0.721158	0.075143
2	4	0.664521	0.71289	0.048369
4	2	0.664521	0.73929	0.074769
3	3	0.671394	0.741605	0.070211

Independently, when number of blocks and sub-blocks both being large, i.e. $K \to \infty$, $\tilde{K} \to \infty$, asymptotic forms of $\alpha(x)$ and $\eta(x)$ when $x \to \infty$ are obtained from (24) as

$$\alpha(x) \sim \frac{\pi}{6} + \frac{1}{2\sqrt{3}x} + \frac{5\sqrt{3}}{(6x)^2}, \qquad \eta(x) \sim \frac{\sqrt{3}}{2} + \frac{1}{2\sqrt{3}x} + \frac{11\sqrt{3}}{90x^2}. \tag{50}$$

Then the query numbers (48) and (49) take asymptotic forms using (50)

$$S(K\tilde{K}) \sim \left\{\frac{\pi}{4} + \left[\frac{\pi}{6} - \frac{\sqrt{3}}{2} + \frac{1}{5\sqrt{3}(2K\tilde{K})^2}\right]\frac{1}{\sqrt{K\tilde{K}}}\right\}\sqrt{N} \tag{51}$$

$$T(K,\tilde{K}) \tag{52}$$

$$\sim \left\{\frac{\pi}{4} + \left[\left(\frac{\pi}{3} - \frac{\sqrt{3}}{2}\right) - \frac{1}{4\sqrt{3}K} - \frac{19\sqrt{3}}{10(6K)^2}\right]\frac{1}{\sqrt{K}} + \left[\frac{\pi}{6} - \frac{\sqrt{3}}{2} + \frac{1}{5\sqrt{3}(2\tilde{K})^2}\right]\frac{1}{\sqrt{K\tilde{K}}}\right\}\sqrt{N}.$$

As for the difference (57) of query numbers, the ratio $K/\tilde{K}$ becomes relevant in determining the asymptotic behavior. There are 3 possibilities:
If $K/\tilde{K} \to 0$, then $1/K$ is dominating, and

$$T(K,\tilde{K}) - S(K\tilde{K}) \sim \left[\left(\tfrac{\pi}{3} - \tfrac{\sqrt{3}}{2}\right) K^{-\frac{1}{2}}\right] \tfrac{1}{\sqrt{N}}. \tag{53}$$

If $K/\tilde{K} \to \infty$, then $1/\tilde{K}$ is dominating, and

$$T(K,\tilde{K}) - S(K\tilde{K}) \sim \left(\tfrac{1}{20\sqrt{3}} K^{-\frac{1}{2}} \tilde{K}^{-\frac{5}{2}}\right) \tfrac{1}{\sqrt{N}}. \tag{54}$$

If $K/\tilde{K} \to$ finite number, then we have the same result as (53). In both the expressions (53) and (54) the coefficients of $1/\sqrt{N}$ are positive. Up to now we saw that $T > S$. Now let us formally prove as a theorem (58) that $T > S$ in general, when $K \geq 2$ and $\tilde{K} \geq 2$.

6.2. *General Proof that* $T(K,\tilde{K}) > S(K\tilde{K})$

Now we prove that $T\left(K,\tilde{K}\right) - S\left(K\tilde{K}\right)$ is always positive in the region $K,\ \tilde{K} \in [2,+\infty)$. In order to complete the proof we need the following two lemmas.

Lemma 6.1.

$$\frac{\pi}{4} + \left(\frac{1}{2}\alpha - \eta\right)(x) > 0, \qquad \forall x \in [2,+\infty)\,. \tag{55}$$

Proof. $\forall x \in [2,+\infty)$
The derivative $\left[\frac{\pi}{4} + \left(\frac{1}{2}\alpha - \eta\right)\right]'(x) = \frac{1}{4\sqrt{x}} f(x)$ with $f(x) \equiv \frac{3}{\sqrt{3x-4}} - \arctan\frac{\sqrt{3x-4}}{x-2}$. While $f'(x) = \frac{-9x+8}{2x(x-1)(3x-4)^{\frac{3}{2}}} < 0$, so that $f(x)$ monotonically decreasing. Further, since that $f(2) = \frac{3}{\sqrt{2}} - \frac{\pi}{2} > 0$, $f(x) \overset{x\to+\infty}{\longrightarrow} 0$, then continuous function $f(x) > 0$ in the region. [$f(x)$ is positive at one point $x = 2$ and tends to zero as x tends to infinity. As a continuous and monotonic function, $f(x)$ can never become negative nor zero in the region.] Therefore $\left[\frac{\pi}{4} + \left(\frac{1}{2}\alpha - \eta\right)\right]'(x) > 0$, so that $\frac{\pi}{4} + \left(\frac{1}{2}\alpha - \eta\right)(x)$ is a monotonically increasing function of x. With $\frac{\pi}{4} + \left(\frac{1}{2}\alpha - \eta\right)(2) = \frac{3-2\sqrt{2}}{8}\pi > 0$, we conclude that $\frac{\pi}{4} + \left(\frac{1}{2}\alpha - \eta\right)(2) > 0$ in the region. □

Lemma 6.2.

$$(\alpha - \eta)(x) \ \ \textit{monotonically decreasing}, \qquad \forall x \in [2,+\infty)\,. \tag{56}$$

Proof. $\forall x \in [2, +\infty)$
The derivative $(\alpha - \eta)'(x) = \frac{1}{4\sqrt{x}} g(x)$ with $g(x) \equiv \frac{\sqrt{3x-4}}{x-1} - \arctan \frac{\sqrt{3x-4}}{x-2}$. While $g'(x) = \frac{1}{x(x-1)^2\sqrt{3x-4}} > 0$, so that $g(x)$ monotonically increasing. Further, since that $g(2) = \sqrt{2} - \frac{\pi}{2} < 0$, $g(x) \stackrel{x \to +\infty}{\longrightarrow} 0$, then continuous function $g(x) < 0$ in the region. [$g(x)$ is negative at one point $x = 2$ and tends to zero as x tends to infinity. As a continuous and monotonic function, $g(x)$ can never become positive nor zero in the region.] Therefore $(\alpha - \eta)'(x) < 0$, we conclude that $(\alpha - \eta)(x)$ is a monotonically decreasing function of x in the region. □

Having proved these two lemmas, we look at the structure of $T\left(K, \tilde{K}\right) - S\left(K\tilde{K}\right)$ using (48) and (49):

$$T\left(K, \tilde{K}\right) - S\left(K\tilde{K}\right) \tag{57}$$
$$= \left\{ \frac{\frac{\pi}{4} + \frac{1}{2}\alpha(K) - \eta(K)}{\sqrt{K}} + \frac{\left[\alpha(\tilde{K}) - \eta(\tilde{K})\right] - \left[\alpha(K\tilde{K}) - \eta(K\tilde{K})\right]}{\sqrt{K\tilde{K}}} \right\} \sqrt{N}\,.$$

Making use of **Lemma 1** (55), we see that the numerator $\frac{\pi}{4} + \frac{1}{2}\alpha(K) - \eta(K)$ of the first term appearing in the brace of (57) is positive for $K \geq 2$. Making use of **Lemma 2** (56) and since $K\tilde{K} > \tilde{K}$, the monotonic property of $\alpha - \eta$ ensures that $\alpha\left(\tilde{K}\right) - \eta\left(\tilde{K}\right) > \alpha\left(K\tilde{K}\right) - \eta\left(K\tilde{K}\right)$. So that the second term in the brace of (57) is also positive for both $K \geq 2$ and $\tilde{K} \geq 2$. Therefore the whole brace of (57) is positive. As a consequence, we conclude our result in the following theorem:

Theorem 6.1.

$$T\left(K, \tilde{K}\right) > S\left(K\tilde{K}\right), \qquad \forall K, \tilde{K} \in [2, \infty)\,. \tag{58}$$

i.e. Hierarchical partial search makes more queries to the oracle than direct partial search. *Direct GRK partial search works faster.*

6.3. *Hierarchy with Many GRK's*

Theorem (58) can be extended to the case of hierarchical search with an arbitrary number of GRK's. The direct GRK always works faster. We prove the statement as follows.

Consider a hierarchy with m GRK's. Assume that $m \geq 2$. We denote the whole operations $G_1 G_2^{j_2} G_1^{j_1}$ of each GRK by one symbol and define an operator

$$\mathcal{G} \equiv G_1 G_2^{j_2} G_1^{j_1}. \tag{59}$$

The hierarchical search works on the initial state $|s_1\rangle$ as

$$\mathcal{G}_m \dots \mathcal{G}_3 \mathcal{G}_2 \mathcal{G}_1 |s_1\rangle, \tag{60}$$

where the sub-index denotes position of the GRK in the hierarchy [sequence]. The proof can be written formally in the following way. Define the total number of queries of the hierarchy

$$T(K_1, K_2, \dots, K_m) \equiv S(K_1) + \sum_{i=2}^{m} \bar{S}(K_{i-1}, K_i). \tag{61}$$

Here K_i is number of "sub"-blocks in the ith partition of database. [We denoted K_1 and K_2 by K and $\tilde{K}$ respectively in previous sections.] $S(K_1)$ is number of queries of the first GRK, and $\bar{S}(K_{i-1}, K_1)$ that of the ith GRK in the hierarchy. Note that S and $\bar{S}$ are not of the same function form. S takes the form corresponding to a direct GRK (25):

$$S(K_1) = \left(\frac{\pi}{4} + \frac{\alpha(K_1) - \eta(K_1)}{\sqrt{K_1}} \right) \sqrt{N}. \tag{62}$$

While (for $i \geq 2$) $\bar{S}$ takes a form of sequential GRK similar to (47):

$$\bar{S}(K_{i-1}, K_i) = \left(\frac{\pi}{4} - \frac{\alpha(K_{i-1})}{2} + \frac{\alpha(K_i) - \eta(K_i)}{\sqrt{K_i}} \right) \frac{\sqrt{N}}{\sqrt{\prod_{j=1}^{i-1} K_j}}. \tag{63}$$

[We denoted $S(K_1)$ and $\bar{S}(K_1, K_2)$ by $S(K)$ and $\bar{S}(K, \tilde{K})$ respectively in previous sections.] Let us substitute these expressions into (61):

$$\begin{aligned} &T(K_1, K_2, \dots, K_m) \\ &= \left\{ \frac{\pi}{4} + \sum_{i=1}^{m-1} \frac{\frac{\pi}{4} + \frac{1}{2}\alpha(K_i) - \eta(K_i)}{\sqrt{\prod_{j=1}^{i} K_j}} + \frac{\alpha(K_m) - \eta(K_m)}{\sqrt{\prod_{i=1}^{m} K_i}} \right\} \sqrt{N}. \end{aligned} \tag{64}$$

On the other hand, if we partition the database directly into the smallest sub-blocks, then the number of these sub-blocks would be $\prod_{i=1}^{m} K_i$. A direct GRK will locate the smallest target sub-block. This would require

$$S\left(\prod_{i=1}^{m} K_i\right) = \left\{ \frac{\pi}{4} + \frac{\alpha(\prod_{i=1}^{m} K_i) - \eta(\prod_{i=1}^{m} K_i)}{\sqrt{\prod_{i=1}^{m} K_i}} \right\} \sqrt{N} \tag{65}$$

queries to the oracle. Therefore the difference of (64) and (65) is

$$T(K_1, K_2, \ldots, K_m) - S\left(\prod_{i=1}^{m} K_i\right) \tag{66}$$

$$= \left\{ \left(\sum_{i=1}^{m-1} \frac{\frac{\pi}{4} + (\frac{1}{2}\alpha - \eta)(K_i)}{\sqrt{\prod_{j=1}^{i} K_j}} \right) + \frac{(\alpha - \eta)(K_m) - (\alpha - \eta)(\prod_{i=1}^{m} K_i)}{\sqrt{\prod_{i=1}^{m} K_i}} \right\} \sqrt{N}.$$

We will show that this expression is always positive when each $K_i \geq 2$. Using **Lemma 1** (55), we see that each term under the summation of (66) is positive. Using **Lemma 2** (56), $\alpha - \eta$ is a monotonically decreasing function. Note that product of all K_i's is larger than K_m, we see that the remaining term in the brace of (66) is also positive. Consequently, we conclude our result in the following corollary:

Corollary 6.1.

$$T(K_1, K_2, \ldots, K_m) > S\left(\prod_{i=1}^{m} K_i\right), \qquad \forall K_i \in [2, +\infty). \tag{67}$$

i.e. Hierarchy of *arbitrary* number of GRK's makes more queries to the oracle than a direct GRK. *Direct GRK partial search always works faster.*

7. Summary

The present paper studied quantum search. Partial search algorithm is called GRK. We studied partial search hierarchy and compared it with direct partial search [GRK]. Consider database of N items with a single target item [target item also called marked item or solution]. The database is partitioned into K blocks, each block further partitioned into $\tilde{K}$ sub-blocks. Hierarchical search is: use GRK and sequential GRK to find the target block and target sub-block, respectively. Successive GRK's can be made if the database is further partitioned. Each sequential GRK in the hierarchy works faster than the previous one. However, the total number of queries to the oracle adds up. The main conclusion is that *a partial search hierarchy works slower than a direct partial search*, see theorem (58) and corollary (67). For example, consider a database partitioned into 3 blocks. Each block is further partitioned into 3 sub-blocks, so totally there are 9 sub-blocks. One could first find the target block using GRK, then the target sub-block by a sequential GRK. Nevertheless, it is faster to run a GRK partial search directly over the 9 sub-blocks and finds the target sub-block once.

Note: Only the class of algorithms using the standard Grover oracle was considered in the paper. This means that if one has already built the main Grover algorithm experimentally, then we do not need any new hardware to run the GRK algorithm. Another advantage of using the same oracle I_t as the main Grover algorithm is more subtle: We can use ancilla [additional or auxiliary] q-bits to label different partitions of the database into blocks of equal size $b = N/K$. Then we are able to run GRK algorithm simultaneously for different partitions. [See Appendix D for more details.] Later a user can measure the ancilla q-bits and choose his or her favorite partition, by that time the target block already will be found.

Acknowledgments

The work is supported by NSF Grant DMS-0503712.

Appendix A. Differences of the Last Operation of GRK in Literature

Different versions of the last operation in **Step 3** of GRK appeared in literature.[6,7,9] People have finalized [after steps 1 and 2] the state $|v\rangle \equiv G_2^{j_2} G_1^{j_1} |s_1\rangle$ with different operations I_{s_1}, $-I_t I_{s_1}$, or $G_1 \equiv -I_{s_1} I_t$. Grover and Radhakrishnan used I_{s_1}.[6] This makes one less query to the oracle but the amplitude of the target item is negative in the final state $I_{s_1}|v\rangle$. Paper 7 used $-I_t I_{s_1}$ but paper 10 used G_1. The last two version become the same in the large block limit. This means that final states $-I_t I_{s_1}|v\rangle$ and $G_1|v\rangle$ are equivalent [of the same form] when $b \to \infty$, though I_{s_1} and I_t do not commute in general. We choose G_1 in our paper because it uses the same Grover iteration.

Appendix B. Ranges of Parameters α and η

We are going to specify ranges of parameters α and η introduced in (21). Because of the constraint (22) relating the two parameters, it is sufficient to specify the range of α. It was shown in 7 that amplitudes [of items in the database after GRK] depend on $\sin(2j_2\theta_2) \sim \sin(2\alpha)$ and $\cos(2j_2\theta_2) \sim \cos(2\alpha)$. So that it is sufficient to take values of α within one period: $\alpha \in [a, a + \pi]$, with a some real number determined later. We are looking for the exact boundaries of α set by physical considerations.

Query numbers (21) are non-negative:

$$j_1 = \left(\frac{\pi}{4} - \frac{\eta}{\sqrt{K}}\right)\sqrt{N} \geq 0, \tag{B.1}$$

$$j_2 = \frac{\alpha}{\sqrt{K}}\sqrt{N} \geq 0. \tag{B.2}$$

Total query number (23) should be less than that of a full Grover search:

$$j_1 + j_2 = \left(\frac{\pi}{4} + \frac{\alpha - \eta}{\sqrt{K}}\right)\sqrt{N} \leq \frac{\pi}{4}\sqrt{N}. \tag{B.3}$$

These three inequalities (B.1), (B.2) and (B.3) yield that

$$0 \leq \alpha \leq \eta \leq \frac{\pi}{4}\sqrt{K}. \tag{B.4}$$

We use constraint (22) to express η as a function of α

$$\eta(\alpha) = \frac{1}{2}\sqrt{K}\,\mathrm{Arctan}\left(\frac{2\sqrt{K}\sin 2\alpha}{K - 4\sin^2\alpha}\right) \tag{B.5}$$

with function Arctan(x) multi-valued. But according to (B.4), we have

$$0 \leq \mathrm{Arctan}\left(\frac{2\sqrt{K}\sin 2\alpha}{K - 4\sin^2\alpha}\right) \leq \frac{\pi}{2}. \tag{B.6}$$

Therefore we could take the principal branch $\arctan(x)$. Now inequality (B.4) becomes

$$0 \leq \alpha \leq \frac{1}{2}\sqrt{K}\arctan\left(\frac{2\sqrt{K}\sin 2\alpha}{K - 4\sin^2\alpha}\right) \leq \frac{\pi}{4}\sqrt{K}. \tag{B.7}$$

This inequality determines range of α.

The solution of (B.4) can be written in the following form:

$$0 \leq \alpha \leq \alpha_B(K). \tag{B.8}$$

Here the upper bound $\alpha_B(K)$ is a function of K. When $K = 2$, 3 or 4, $\alpha_B(K)$ coincide with the singularities of $\eta(\alpha)$. [$K - 4\sin^2\alpha = 0$ at these singularities.] When $K \geq 5$, values of $\alpha_B(K)$ can be solved numerically. As K increases, $\alpha_B(K)$ approaches a certain positive number $\alpha_B(\infty)$. This limit $\alpha_B(\infty) = 0.947747\ldots$ is the solution of $\alpha = \sin(2\alpha)$. [Inequality $\alpha \leq \eta(\alpha)$ becomes $\alpha \leq \sin(2\alpha)$ as $K \to \infty$.] The value of $\alpha_B(K)$ always lies in between $\alpha_B(\infty)$ and $\frac{\pi}{2}$ when $K \geq 5$. We list these results in Table B1.

Table B1. Upper Bound of α.

K	2	3	4	5	6	100	∞
$\alpha_B(K)$	$\frac{\pi}{4}$	$\frac{\pi}{3}$	$\frac{\pi}{2}$	1.22683	1.15100	0.956221	0.947747

Appendix C. Minimization of the Total Number of Queries of GRK

Here we give a proof that (24) is the global minimum of $\alpha - \eta$ under constraint (22). In Appendix B we used (22) to express η as a function of α

$$\eta(\alpha) = \frac{\sqrt{K}}{2} \arctan\left(\frac{2\sqrt{K}\sin 2\alpha}{K - 4\sin^2\alpha}\right). \tag{C.1}$$

Now we define a function

$$f(\alpha) \equiv \alpha - \eta(\alpha) \tag{C.2}$$

which we want to minimize within the range $0 \leq \alpha \leq \alpha_B(K)$. We first prove that (24) is a local minimum of $f(\alpha)$.

Appendix C.1. *Case* $K \geq 3$

The first derivative of $f(\alpha)$ is

$$f'(\alpha) = \frac{16(K-1)\sin^4\alpha - 4K^2\sin^2\alpha + K^2}{16(K-1)\sin^4\alpha - 8K\sin^2\alpha - K^2}. \tag{C.3}$$

It vanishes at (24) with $\sin^2\alpha = \frac{K}{4(K-1)}$. We calculate next the second derivative

$$f''(\alpha) = \frac{4K\sin 2\alpha[4(K-1)(K-2)\cos^2 2\alpha + 16(K-1)\cos 2\alpha + (K-2)^2(K+2)]}{[16(K-1)\sin^4\alpha - 8K\sin^2\alpha - K^2]^2}. \tag{C.4}$$

Note that the value of the denominator at (24) is $\frac{K^6}{(K-1)^2}$, which is strictly positive as $K \geq 3$. The numerator is also positive because both $\sin 2\alpha$ and $\cos 2\alpha$ are positive at (24) with $K \geq 3$. [See 7 for the range of $\alpha(K)$.] Therefore $f'(\alpha) = 0$ and $f''(\alpha) > 0$ at the solution (24), so that (24) is a local minimum for $K \geq 3$.

Appendix C.2. *Case* $K = 2$

The case that $K = 2$ is more subtle. Expression (24) yields that $\alpha = \frac{\pi}{4}$ and $\eta = \frac{\pi}{2\sqrt{2}}$. However, both first (C.3) and second (C.4) derivatives of

$\alpha - \eta(\alpha)$ vanish at this critical point. The third derivative is non-zero: $f'''(\alpha = \frac{\pi}{4}) = -4$. So we expand function $\alpha - \eta(\alpha)$ about the critical point

$$\alpha - \eta(\alpha)|_{K=2} = -\frac{4}{3!}\left(\alpha - \frac{\pi}{4}\right)^3 + \mathcal{O}\left(\left(\alpha - \frac{\pi}{4}\right)^4\right). \qquad \text{(C.5)}$$

We see that $\alpha = \frac{\pi}{4}$ is actually a saddle point due to the non-vanishing cubic term. The form (C.5) suggests that if α goes greater than $\frac{\pi}{4}$, value of function $\alpha - \eta(\alpha)$ could be further reduced than the value at the saddle point. However, $\alpha = \frac{\pi}{4}$ is a boundary set by physical considerations [see Table B1]. Definition of α and η in (21) involves query numbers j_1 and j_2, which are non-negative. Therefore $j_1 \equiv \left(\frac{\pi}{4} - \frac{\eta}{\sqrt{2}}\right)\sqrt{N} \geq 0$, i.e. $\eta \leq \frac{\pi}{2\sqrt{2}}$. Now we allow α to go beyond $\frac{\pi}{4}$ and write

$$\alpha = \frac{\pi}{4} + \delta, \qquad \eta = \frac{\pi}{2\sqrt{2}} + \epsilon. \qquad \text{(C.6)}$$

Here δ and ϵ are infinitesimals, $\delta > 0$. Then constraint (22) requires that

$$\epsilon = \delta. \qquad \text{(C.7)}$$

So that η would be greater than the physically allowed maximal value $\frac{\pi}{2\sqrt{2}}$ and j_1 would be negative $j_1 = -\frac{\delta}{\sqrt{2}}\sqrt{N}$. This analysis showed that α can never go beyond $\frac{\pi}{4}$ and function $\alpha - \eta(\alpha)$ is minimized at this boundary. Therefore, expression (24) as a local minimum is also valid in the case that $K = 2$.

Now we have proved that the critical point (24)

$$\alpha\,(K) = \frac{1}{2}\arccos\left(\frac{K-2}{2(K-1)}\right) \qquad \text{(C.8)}$$

is a local minimum of $f(\alpha)$. Note that $f(\alpha)$ is analytical as $0 \leq \alpha \leq \alpha_B(K)$ and there is no singularity in this range any more. Therefore we can show that this local minimum (C.8) is also global by comparing the value of $f(\alpha)$ at (C.8) with those at the boundaries. [We always have $0 < \alpha(K) \leq \alpha_B(K)$ and equality holds only for $K = 2$.] We list the comparison results for $K = 2$, 3 and 4 in Table C1.

When $K \geq 5$, $f(0) = f(\alpha_B(K)) = 0$, while $f(\alpha(K)) < 0$. Therefore, we conclude that the critical point (24) or (C.8) is always the global minimum.

Appendix D. Different Partitions of a Database

A data base of N items can be partitioned into blocks in different ways. For example, items in one block may have the first 3 bits of their addresses the same for one partition or the last 3 bits the same for another partition.

Table C1. Comparison of values of $f(\alpha)$ at different points.

K	$f(0)$	$f(\alpha(K))$	$f(\alpha_B(K))$
2	0	$\frac{\pi}{4}\left(1-\sqrt{2}\right) \approx -0.325323$	$\frac{\pi}{4}\left(1-\sqrt{2}\right) \approx -0.325323$
3	0	-0.337098	-0.313152
4	0	-0.339837	0

For a database partitioned into K blocks of equal size $b = N/K$, there are totally

$$P(N,K) = \frac{N!}{(b!)^K \, K!} \tag{D.1}$$

different ways of partition. We could use ancilla q-bits [also called additional or auxiliary q-bits] to label these partitions. As N and b both being large, we shall need

$$\log_2 P(N,K) \sim N \log_2 K - \log_2 K! \tag{D.2}$$

ancilla q-bits. For example, if we have $N = 4$ items and $K = 2$ blocks, then the number of partitions is $P(4,2) = 3$ and we shall need $\log_2 3 \approx 2$ ancillas. In practice, The number (D.2) can be further reduced if we only label the partitions commonly used, not all partitions. Then we can run GRK simultaneously for those selected partitions. When a user measures ancilla q-bit in his/her favorite partition, the target block will already be found by that time.

References

1. T. H. Cormen, C. E. Leiserson, R. E. Leiserson, R. L. Rivest and C. Stein, *Introduction to Algorithms* (MIT Press, 2003), 2nd ed.
2. M. A. Nielsen and I. L. Chuang, *Quantum Computation and Quantum Information* (Cambridge Univ. Press, 2000).
3. R. Cleve, A. Ekert, L. Henderson, C. Macchiavello and M. Mosca, *Complexity* **4**, 33 (1998), quant-ph/9903061.
4. L. K. Grover, Proceedings, *28th Annual ACM Symposium on the Theory of Computing (STOC)* (1996), 212, quant-ph/9605043.
5. M. Heiligman, quant-ph/0006136.
6. L. K. Grover and J. Radhakrishnan, *ACM Symp. on Parallel Algorithms and Architectures, Las Vegas, Nevada, USA* (2005), 186, quant-ph/0407122.
7. V. E. Korepin, *J. of Phys. A* **38**, L731 (2005), quant-ph/0503238.
8. V. E. Korepin and J. Liao, *Quantum Information Processing* **5**, 209 (2006), quant-ph/0510179.
9. B.-S. Choi and V. E. Korepin, *Quantum Information Processing* **6**, 97 (2007), quant-ph/0608106.

10. V. E. Korepin and B. C. Vallilo, *Prog. Theor. Phys.* **116**, 783 (2006), quant-ph/0609205.
11. M. Mosca, *Theoretical Computer Science* **264**, 139 (2001).
12. V. E. Korepin and L. K. Grover, *Quantum Information Processing* **5**, 5 (2006), quant-ph/0504157.

SPIN HALL EFFECT IN ULTRACOLD ATOMIC GAS

XIONG-JUN LIU*,† and C. H. OH*,‡

*Department of Physics, National University of Singapore,
2 Science Drive 3, Singapore 117542
†Department of Physics, Texas A&M University,
College Station, Texas 77843-4242, USA
‡phyohch@nus.edu.sg

In this paper we discuss the spin Hall effect (SHE) in cold atoms through optical coherent control means. We first briefly introduce the concepts of spin-orbit coupling and SHE in solid systems and then discuss the optical means in creating adiabatic gauge field for neutral atomic system. The optically induced SHE are then presented.

1. Introduction

The intrinsic spin Hall effect (SHE) has attracted much attention since its recent prediction in solid systems with spin-orbit coupled structures [1–4]. The physics of SHE in semiconductors can be described in the following way: In the presence of spin-orbit coupling, the applied electric field leads to a transverse motion (perpendicular to the electric field) of spins, with the spin-up and spin-down carriers transporting oppositely to each other, thus a spin current is created perpendicular to the electric field. SHE has potential applications in spintronics, for it provides an underlying theory in the controllable electric injection of spin currents. Furthermore, motivated by the connection between SHE and quantum Hall physics [5, 6], quantized SHE was studied in several materials such as metallic graphene [7, 8], mercury telluride-cadmium telluride semiconductor systems [9] and in semiconductors with a strain gradient structure [10]. In the latter cases quantum SHE is essentially the phenomenon in which particles in different spin directions experience opposite effective gauge fields formed through spin-orbit couplings, resulting in a non-vanishing quantized spin hall conductance and a vanishing charge hall conductance.

Compared with solid matters, ultra-cold atomic system provides a unique access to the study of complex many-body dynamics with its very clean environment and remarkable controllability [11]. The remarkable controllability in the parameters of these systems allows a clean study of complex physics in a controllable fashion. For example, one of the most exciting advances in ultracold atomic gases is that rotating Bose-Einstein condensates (BECs) provide a conceptual link between the physics of trapped gases and the physics of condensed matter systems such as superfluids, type-II superconductors, and quantum-Hall effect (QHE) materials [12–15]. In the rotating BEC systems, the trap rotation provides an effective (spin-independent) magnetic field for the electrically neutral atoms, for which the analog of the fractional quantum Hall effect (FQHE) can be studied. However, since the rotation does not depend on internal degrees (atomic angular-momentum states or atomic spins), the SHE cannot be achieved in such systems.

On the other hand, recent advances in the optical coherent manipulation of ulracold atoms suggest an alternative to create adiabatic gauge fields for the atomic systems by coupling the atoms to external optical fields which are spatially dependent in the amplitude and phase [16–19], e.g. optical fields with with orbital angular-momentum (OAM) [20]. Note the fact that the adiabatic gauge fields are fully controlled by the laser induced transitions between different atomic states. Properly adjusting the parameters of external laser fields, one can arrive at the situation that different atomic spin states experience opposite gauge fields, leading to SHE in atomic systems [21, 22].

The paper is organized as follow. In the following section, we briefly introduce the spin-orbit coupling interaction in the solid systems and the related SHE. Then, in Sec. 3, the adiabatic gauge field for the atomic system is discussed; in Sec. 4, we discuss the realization of SHE in the ultracold fermi gas. Conclusions are given in Sec. 5.

2. Spin-Orbit Coupling in Solids

A familiar example for the spin-orbit coupling is that the reduction from Dirac equation to non-relativistic case can result in the coupling between the orbital and spin degree of electrons, in the presence of some external potential. However, since the spin-orbit coupling Hamiltonian for such case

$$H_{so} = \frac{e^2\hbar^2}{4m_0c^2}\vec{\sigma} \cdot (\nabla V \times \frac{\mathbf{P}}{m_0}) \tag{1}$$

depends inversely on the light speed c, this effect is normally very weak for the non-relativistic momentum $\mathbf{P}$ of the electron. Here V is the applied external potential. Actually, for the vacuum case the weak spin-orbit coupling is a result of the large energy gap between electrons and positrons ($\Delta E \sim 2m_0c^2 \sim 2.0$MeV). Such a situation, however, can be changed in the semiconductor quantum wells, whose band structure shows many similarities to the situation of free relativistic electrons. In the III-V semiconductor quantum wells, for example, the energy gap between conduction and valence band is of order 1eV or smaller [23], which makes it possible to obtain a relatively large spin-orbit coupling effect in the semiconductor materials.

By now the SHE in solids are mainly studied with spin-orbit coupling Hamiltonian for conduction band electrons and valence band holes. Specifically, for the conduction band electrons, we have two types of spin-orbit couplings known by the Dresselhaus terms [24] and Rashba terms [25], respectively. The former spin-orbit coupling Hamiltonian is of the bulk-inversion asymmetry:

$$H_D^{so} = \frac{\gamma}{\hbar^3}\left(\sigma_x P_x (P_y^2 - P_z^2) + \sigma_y P_y (P_z^2 - P_x^2) + \sigma_z P_z (P_x^2 - P_y^2)\right), \quad (2)$$

which is trilinear in the momentum. For sufficiently low temperatures, and considering a very narrow quantum well grown along the [001] direction, one can approximately neglect the nonlinear terms of P_x and P_y and replace P_z^2 by $\langle P_z^2\rangle$. Thus H_D^{so} can be recast into

$$H_D^{so} = -\frac{\alpha}{\hbar}(\sigma_x P_x - \sigma_y P_y^2), \quad (3)$$

with $\alpha = \gamma\langle P_z^2\rangle$. On the other hand, the Rashba Hamiltonian is of the form

$$H_R^{so} = -\frac{\lambda}{\hbar}(\sigma_x P_y - \sigma_y P_x). \quad (4)$$

The typical values of the Dresselhaus coefficient and Rashba coefficient are about $\alpha \sim 10^{-11}$eVm and $\lambda \leq 10^{-11}$eVm for GaAs quantum well. The spin-orbit coupling Hamiltonian can be equivalently understood as that the electrons experience a non-Abelian gauge field for the orbital motion. For example, the gauge potential for the Rashba model (4) reads $A_x = \frac{\lambda}{\hbar}\sigma_y, A_y = -\frac{\lambda}{\hbar}\sigma_x$, which is associated with a spin-dependent magnetic field along the z direction:

$$\mathbf{B} = \frac{e\lambda^2}{\hbar^2 c}\sigma_z \hat{e}_z. \quad (5)$$

Semiclassically, from this result one can judge that electrons with spin pointing to $+z$ direction experience an effective B field in $+z$ direction, while electrons with spin pointing to $-z$ direction experience an effective B field in $-z$ direction. As a result, if we apply an electric field in the $+x$ direction, the electrons in the $+z$ directional spin-polarization will move in the $-y$ direction, while the electrons in the $-z$ directional spin-polarization will move in the $+y$ direction, leading to a spin current in the y direction. This result was firstly found by J. Sinova *et al.* [2] in another way.

For the valence-band hole systems around Γ point, the effective Hamiltonian are known as Luttinger model [26]

$$H_L^{so} = \frac{\hbar^2}{2m}\left((\gamma_1 + \frac{5}{2}\gamma_2)k^2 - 2\gamma_2(\mathbf{k}\cdot\mathbf{S})^2\right), \tag{6}$$

where $\mathbf{S}$ is the spin-3/2 operator. It was shown by S. Murakami *et al.*[1] that the above Hamiltonian describes a spin-dependent monopole field in the k-space after diagonalizing the term $\gamma_2(\mathbf{k}\cdot\mathbf{S})^2$ by a $\mathbf{k}$-dependent unitary transformation $U(\mathbf{k})$, which makes the electrons with spin-up polarizations and spin-up polarizations move in the opposite directions, respectively. Thus the Lorentz force also accounts for a transverse pure spin current obtained in the Luttinger model when applying an external electric field.

3. Adiabatic Gauge Field in Atoms: Effective Spin-Orbit Couplings

3.1. *Adiabatic gauge field*

In this subsection we consider the general theory of adiabatic gauge field in the quantum many-level systems, which results in the effective spin-orbit coupling interactions for such systems. The idea that adiabatic gauge fields can be generated in quantum systems was firstly proposed by Wilczek and Zee more than twenty years ago [27]. Here we shall review the general theory of the adiabatic gauge fields created in the quantum N-level systems. For convenience, we denote $H_0(r)$ as the unperturbed Hamiltonian that describes the N internal levels $|\chi_k\rangle (k = 1, 2, ..., N)$ for the system, and $\mathcal{E}_k$ as the eigenvalue of corresponding energy level. In the presence of perturbation Hamiltonian H_I that can be created through e.g. external trap potentials, laser induced coupling for atomic systems, etc, the dynamics of the quantum system can be studied via perturbation theory. Without loss of the generality, in this paper we consider the case of a weak perturbation so that H_I does not induce the energy shift to the energy levels of internal Hamiltonian, but causes the coupling between different eigenstates $|\chi_k\rangle$. For this we can generally expand the wave function for the system according to $\tilde{\Psi}(\mathbf{r}, t) = \sum_{k=1}^{N} \Psi_k(\mathbf{r}) \exp(-i\mathcal{E}_k t/\hbar)|\chi_k\rangle$, where $\mathcal{E}_k$ is the eigenvalue of corresponding internal energy level. Denoting by $\Psi = (\Psi_1, \Psi_2, ..., \Psi_N)^T$, the coupling between states Ψ_k by the interaction Hamiltonian can be given by

$$i\hbar\frac{\partial\Psi}{\partial t} = [-\frac{\hbar^2}{2m}\nabla^2 + V(\mathbf{r})]\Psi + H_I(\mathbf{r})\Psi, \tag{7}$$

where $V(\mathbf{r})$ is the external trap potential, and the interacting Hamiltonian H_I depends on the local parameters, e.g. the spatial position $\mathbf{r}$. The diagonalization of the interacting Hamiltonian can be done through a local unitary transformation $U(\mathbf{r})$, $\Psi = U(\mathbf{r})\Phi$, $H_I^d = U^\dagger H_I U$, where $\Phi = (\Phi_1, \Phi_2, ..., \Phi_N)$ is the diagonalized basis of the interacting Hamiltonian, and H_I^d is diagonal on this basis. The diagonal elements $(H_I^d)_{kk} = E_k$ represents the eigenvalue of Φ_k. Note that the specific form of the unitary transformation $U(\mathbf{r})$ shall be solely determined by the interacting Hamiltonian. Substituting the transformation $U(\mathbf{r})$ into eq. (7) and by a straightforward calculation we can finally rewrite the equation of evolution for Φ as

$$i\hbar\frac{\partial\Phi}{\partial t} = \frac{1}{2m}\left(i\hbar\nabla + \frac{e}{c}\mathbf{A}(\mathbf{r})\right)^2\Phi + V'(\mathbf{r})\Phi + H_I^d(\mathbf{r})\Phi, \tag{8}$$

where we have introduced the gauge potential

$$\mathbf{A}(\mathbf{r}) = i\frac{\hbar c}{e}U^\dagger(\mathbf{r})\nabla U(\mathbf{r}) \tag{9}$$

with e the electric charge constant, and $V' = U^\dagger V(\mathbf{r})U$. Since $U(\mathbf{r})$ considered here is a local unitary transformation on the N-dimensional Hilbert space, it is a $N \times N$ parameter-dependent matrix. As a result, the gauge defined based on eq. (9) is non-Abelian and of the $N \times N$ matrix form (of the SU(N) symmetry). The non-Abelian gauge field is associated with the curvature (effective magnetic field) that is given by

$$F_{\mu\nu} = \partial_\mu A_\nu - \partial_\nu A_\mu - i\frac{e}{c}[A_\mu, A_\nu] \tag{10}$$

with the effective magnetic field $B_j = 1/2\epsilon_{jkl}F_{kl}$. One should keep in mind that to this step we do not apply the adiabatic condition in the equation of motion. The gauge field defined in eq. (9) is a pure gauge and it is easy to verify that $F_{\mu\nu} = 0$. It is noteworthy that the non-Abelian pure gauge may have nontrivial significance since it can result in a phase to the wave function Φ

$$\Phi(\mathbf{r}, t) \to \mathcal{P}e^{i\frac{e}{\hbar c}\int \mathbf{A}\cdot d\mathbf{l}}\Phi(\mathbf{r}, t), \tag{11}$$

where $\mathcal{P}$ is the operator of chronological ordering. The above phase factor is known as Wilson loop integral [28]. This phase factor can be non-zero even for a pure gauge, since for the non-Abelian gauge field we have $\nabla \times \mathbf{A} \neq \mathbf{B} = 0$. To find the possible observable consequence for this phase factor is an interesting issue in gauge field theory.

Now we consider the adiabatic condition that leads to the adiabatic gauge field for the present system [29]. Under the adiabatic condition, the adiabatic gauge potential is generally associated with a non-zero curvature $F_{\mu\nu} \neq 0$. Specifically, we shall consider the case in the following that the system is non-degenerate, i.e. E_n is different for different eigenstates Φ_n.

When each eigenstate is separated from others by a sufficiently large difference in the eigenvalues, the coupling between each two states, say Φ_j and Φ_k, induced by the off-diagonal element $\mathbf{A}_{jk}(\mathbf{r})$ satisfies $|\frac{\mathbf{P}\cdot\mathbf{A}_{jk}}{m(E_j - E_k)}| << 1$. In this way, the adiabatic condition is then fulfilled and the transition between different states in eq.(8) can be ignored. The non-Abelian gauge field is reduced from SU(N) symmetry to N independent U(1) Abelian gauge fields with

$$\mathbf{A}_j(\mathbf{r}) = i\frac{\hbar c}{e}\langle \Phi_j|\nabla|\Phi_j\rangle, \qquad j = 1, 2, ..., N, \tag{12}$$

and the corresponding scalar potential after applying the adiabatic condition $\phi_j(\mathbf{r}) = \frac{1}{2m}\sum_{k\neq j} \mathbf{A}_{jk} \cdot \mathbf{A}_{kj}, j = 1, 2, ..., N$. In the dynamical equation of evolution (8), all components of the column vector Φ become decoupled to each other and we have

$$i\hbar\frac{\partial \Phi_j}{\partial t} = \frac{1}{2m}\big(i\hbar\nabla + \frac{e}{c}\mathbf{A}_j(\mathbf{r})\big)^2\Phi_j + \bar{V}_j'(\mathbf{r})\Phi_j, \tag{13}$$

with $\bar{V}_j'(\mathbf{r}) = V'(\mathbf{r}) + \phi_j + E_j$. The curvature of U(1) gauge fields is given by $F_{\mu\nu}^{(j)} = \partial_\mu A_\nu^{(j)} - \partial_\nu A_\mu^{(j)}$, which is generally non-zero.

3.2. *Two-level atomic system*

As a concrete example, here we consider the adiabatic gauge field created in a two-level I-type atomic system shown in Fig. 1(a). The transition $|b\rangle \to |a\rangle$ is coupled by an optical field with Rabi-frequency $\Omega(r) = \Omega_0(r)e^{i\phi(\mathbf{r})}$, where $\Omega_0(r)$ is the amplitude and $\phi(\mathbf{r})$ is the spatially-dependent phase. Δ is the detuning of the optical dipole transition. Under the rotating-wave approximation the Hamiltonian in the interacting picture of the present system reads

$$\begin{aligned} H &= -\frac{\hbar^2}{2m_e}\nabla^2 + V(\mathbf{r}) + H_I \\ H_I &= \hbar\Delta|a\rangle\langle a| - (\hbar\Omega|a\rangle\langle b| + h.c.), \end{aligned} \tag{14}$$

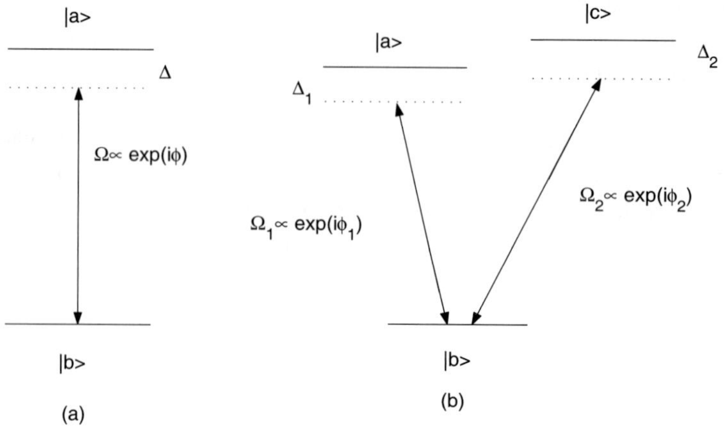

Fig. 1. (a) Two-level system coupled to one laser field. (b) Three-level V-type system coupled to two external laser fields.

where $V(\mathbf{r})$ is the external trap potential. The interaction Hamiltonian H_I can be diagonalized with the local unitary transformation: $\tilde{H}_I = U(\mathbf{r})H_I U^\dagger(\mathbf{r})$ with

$$U(\mathbf{r}) = \begin{bmatrix} \cos\theta & \sin\theta e^{-i\phi(\mathbf{r})} \\ -\sin\theta e^{i\phi(\mathbf{r})} & \cos\theta \end{bmatrix}. \tag{15}$$

Here the mixing angle θ is defined according to $\tan\theta = E_-/E_+$, where the eigenvalues of $|\Phi_+\rangle$ and $|\Phi_-\rangle$ can be calculated by: $E_\pm = (\Delta \pm \sqrt{\Delta^2 + 4\Omega_0^2})/2$. Accordingly, the eigenstates of interaction Hamiltonian are given by $|\Phi_+\rangle = \cos\theta|a\rangle + \sin\theta e^{-i\phi(\mathbf{r})}|b\rangle$, $|\Phi_-\rangle = -\sin\theta e^{i\phi(\mathbf{r})}|a\rangle + \cos\theta|b\rangle$. Now we consider the large detuning case, i.e. $\Delta^2 \gg \Omega_0^2$. By calculating all values up to the order of Ω_0^2/Δ^2, one can easily verify that $\tan\theta \approx -\Omega_0/\Delta$, and $E_+ \approx \Delta \gg E_- \approx -\Omega_0^2/\Delta$. According to the discussion in the former subsection, atomic motion may lead to the transitions between ground eigenstate and excited one as follow $\tau_\pm \sim |\bar{\mathbf{v}}_f \cdot \nabla\theta(\mathbf{r}) + \frac{1}{2}\sin 2\theta\bar{\mathbf{v}}_f \cdot \nabla\phi(\mathbf{r})|$, where $\bar{\mathbf{v}}_f$ is the fermi velocity of the atomic system. For our purpose we require $\tau_\pm \ll |E_+ - E_-|$ so that the coupling between ground state and excited state can be neglected. Then, we introduce the adiabatic condition that the system is restricted in the ground state. The adiabatic gauge field in the present case is

$$\mathbf{A}(\mathbf{r}) = -\frac{\hbar c}{e}\sin^2\theta\nabla\phi(\mathbf{r}), \tag{16}$$

which is Abelian, and the effective trap potential given by

$$V_{eff}(\mathbf{r}) = V(\mathbf{r}) - \hbar\Omega_0^2/\Delta + \frac{\hbar^2}{2m}\left[(\nabla\theta)^2 + \sin^2\theta\cos^2\theta(\nabla\phi)^2\right]. \quad (17)$$

This result can be easily extended to three-level V-type system (Fig. 1(b)), which is coupled to two optical fields $\Omega_1 = \Omega_1^{(0)}\exp[i\phi_1(\mathbf{r})]$ and $\Omega_2 = \Omega_2^{(0)}\exp[i\phi_2(\mathbf{r})]$. For the large detuning case we can also apply the adiabatic condition and the atoms are fully trapped at the ground state. By a similar calculation we obtain the adiabatic gauge field for the ground state

$$\mathbf{A}_g(\mathbf{r}) = -\frac{\hbar c}{e}\frac{\Omega_{10}^2}{\Delta_1^2}\nabla\phi_1(\mathbf{r}) - \frac{\hbar c}{e}\frac{\Omega_{20}^2}{\Delta_2^2}\nabla\phi_2(\mathbf{r}), \quad (18)$$

which is Abelian, and the effective trap potential given by

$$\begin{aligned} V_{eff}(\mathbf{r}) = {} & V(\mathbf{r}) - \hbar\Omega_{10}^2/\Delta_1 - \hbar\Omega_{20}^2/\Delta_2 \\ & + \frac{\hbar^2}{2m}\left[(\nabla\theta_1)^2 + \sin^2\theta_1\cos^2\theta_1(\nabla\phi_1)^2\right] \\ & + \frac{\hbar^2}{2m}\left[(\nabla\theta_2)^2 + \sin^2\theta_2\cos^2\theta_2(\nabla\phi_2)^2\right]. \end{aligned} \quad (19)$$

4. Spin Hall Effect in Atoms

By now, we know the effective gauge field on atoms can be generated through atom-light couplings. Properly adjusting the parameters of external laser fields, one can arrive at the situation that different atomic spin states (angular-momentum states) experience opposite gauge fields, leading to SHE in atomic systems. Such an idea was firstly proposed in the Refs. [21] and [22].

We consider an ensemble of cold Fermi atoms interacting with two external light fields (Fig. 2). The ground ($|g_\pm, \pm\frac{1}{2}\rangle$) and excited ($|e_\pm, \pm\frac{1}{2}\rangle$) states are hyperfine angular momentum states (atomic spins) with their total angular momenta $F_g = F_e = 1/2$. The laser fields with orbital angular-momenta are characterized by the Rabi-frequencies $\Omega_1 = \Omega_{10}\exp(i(\mathbf{k}_1\cdot\mathbf{r} + l_1\vartheta))$ and $\Omega_2 = \Omega_{20}\exp(i(\mathbf{k}_2\cdot\mathbf{r} + l_2\vartheta))$. Here the phase $\vartheta = \tan^{-1}(y/x)$. For simplicity, in the following discussion we replace the notations $|\alpha_\pm, \pm\frac{1}{2}\rangle$ by $|\alpha_\pm\rangle$ ($\alpha = e, g$). The $\mathbf{r}$-representation atomic wave function is denoted by $\Phi_\alpha(\mathbf{r}, t)$. Similar to former discussions, we introduce the slowly-varying amplitudes of atomic wave-functions by (setting $\omega_{g\pm} = 0$): $\Psi_{g\pm} = \Phi_{g\pm}, \Psi_{e+} = \Phi_{e+}(\mathbf{r}, t)e^{-i(\mathbf{k}_2\cdot\mathbf{r} - (\omega_{e+} - \Delta_1)t)}, \Psi_{e-} = \Phi_{e-}(\mathbf{r}, t)e^{-i(\mathbf{k}_1\cdot\mathbf{r} - (\omega_{e-} - \Delta_2)t)}$, where $\hbar\omega_\alpha$ is the energy of the state $|\alpha\rangle$, Δ_1

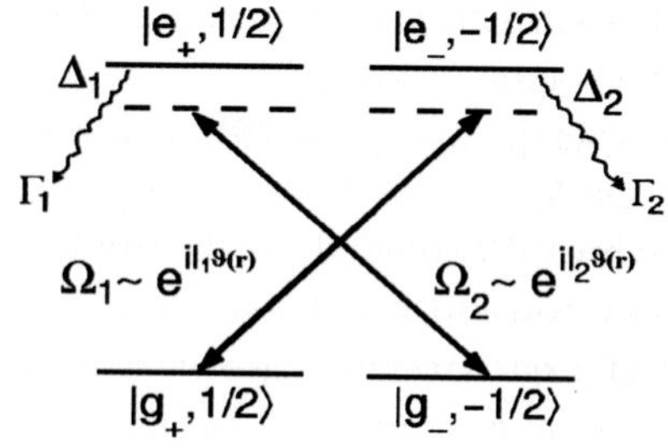

Fig. 2. Fermi atoms with four-level internal hyperfine spin states interacting with two light fields. This can be experimentally realized with alkali atoms, such as ^6Li atoms $(2^2S_{1/2}(F=1/2) \longleftrightarrow 2^2P_{1/2}(F'=1/2))$ [30].

and Δ_2 are detunings. The Hamiltonian of the present system can be written as $H = H_0 + H_1 + H_2$, where

$$H_0 = \sum_{\alpha=e\pm g\pm} \int d^3r\Psi^*_\alpha\Big(-\frac{\hbar^2}{2m}\nabla^2 + V(\mathbf{r})\Big)\Psi_\alpha,$$

$$H_1 = \hbar\Delta_1 \int d^3r\Psi^*_{e+}S_{e_+e_+}\Psi_{e_+} + \hbar\int d^3r(\Psi^*_{e_+}\Omega_{10}e^{il_1\vartheta}S_{1+}\Psi_{g_-} + h.a.),$$

$$H_2 = \hbar\Delta_2 \int d^3r\Psi^*_{e_-}S_{e_-e_-}\Psi_{e_-} + \hbar\int d^3r(\Psi^*_{e_-}\Omega_{20}e^{il_2\vartheta}S_{2+}\Psi_{g_+} + h.a.) \tag{20}$$

with atomic operators $S_{e_\pm e_\pm} = |e_\pm\rangle\langle e_\pm|$, $S_{1+} = |e_+\rangle\langle g_-|$, $S_{2+} = |e_-\rangle\langle g_+|$, $S^\dagger_{1+} = S_{1-}$ and $S^\dagger_{2+} = S_{2-}$. $V(\mathbf{r})$ is the external trap potential. The collisions (s-wave scattering) between cold Fermi atoms are negligible. Through the local unitary transformation the interaction part of the Hamiltonian can be diagonalized, and the four eigenstates can be obtained as $\big[|\psi_+\rangle, |\psi_-\rangle\big]^T = U_1\big[|e_+\rangle, |g_-\rangle\big]^T$ and $\big[|\phi_+\rangle, |\phi_-\rangle\big]^T = U_2\big[|e_-\rangle, |g_+\rangle\big]^T$ with

$$U_j = \begin{bmatrix} \cos\theta_j & \sin\theta_j e^{il_j\vartheta(\mathbf{r})} \\ -\sin\theta_j e^{-il_j\vartheta(\mathbf{r})} & \cos\theta_j \end{bmatrix}, \quad j = 1, 2.$$

The mixing angles θ_j are defined according to $\tan\theta_1 = E^-_\psi/E^+_\psi$ and $\tan\theta_2 = E^-_\phi/E^+_\phi$, where the eigenvalues of $|\psi_\pm\rangle$ and $|\phi_\pm\rangle$ are $E^\pm_\psi = (\Delta_1 \pm \sqrt{\Delta_1^2 + 4\Omega_{10}^2})/2$ and $E^\pm_\phi = (\Delta_2 \pm \sqrt{\Delta_2^2 + 4\Omega_{20}^2})/2$.

In order to suppress the spontaneous emission of the excited states, we consider also the large detuning case so that the coupling between ground states and excited states can be neglected. Then, we introduce the adiabatic condition that the population of the higher levels $|\psi_+\rangle$ and $|\phi_+\rangle$ is adiabatically eliminated, whilst the total system is restricted to the ground eigenstates $|\Psi\rangle = \cos\gamma|S_\downarrow\rangle + \sin\gamma|S_\uparrow\rangle$, where, to facilitate further discussions, we have put the effective spin states: $|S_\downarrow\rangle = |\psi_-\rangle$, $|S_\uparrow\rangle = |\phi_-\rangle$ with their z-component effective spin polarizations $S_z^\uparrow = \frac{\hbar}{2}(\cos^2\theta_2 - \sin^2\theta_2) \approx \hbar/2$ and $S_z^\downarrow = \frac{\hbar}{2}(\sin^2\theta_1 - \cos^2\theta_1) \approx -\hbar/2$. The parameter γ describes the possibility of an atom in states $|S_\downarrow\rangle$ and $|S_\uparrow\rangle$, and can be determined by initial condition. Atomic decay can be safely neglected in the present situation, since the probability of excited states is much smaller than that of the ground states for $|S_{\uparrow,\downarrow}\rangle$ ($\sin^2\theta_j \ll \cos^2\theta_j$). The adiabatic gauge potentials and scalar potentials for the spin-up and spin-down particles are given by

$$\mathbf{A}_\downarrow = -\mathbf{A}_\uparrow = \hbar l c e^{-1}\frac{\Omega_0^2}{\Delta^2}(x\hat{e}_y - y\hat{e}_x)/\rho^2 \tag{21}$$

and $V_{eff}(\mathbf{r}) = V_{\uparrow,\downarrow}(\mathbf{r}) = V(\mathbf{r}) - \hbar\Omega_0^2/\Delta - \frac{\hbar^2 l^2\Omega_0^4}{(2m\Delta^4\rho^2)}$ with $\rho = \sqrt{x^2+y^2}$, when we choose $\Delta_1 = \Delta_2 = \Delta$, $\Omega_{10} = \Omega_{20} = \Omega_0$ and $l_1 = -l_2 = l$, i.e. the angular momenta of the two light fields are opposite in direction. This result is intrinsically interesting: by coupling the atomic spin states to radiation, we find the atomic system can be described as an ensemble of charged particles with opposite spins experiencing opposite magnetic fields but the same electric field. With the help of gauge and trap potentials, we can rewrite the Hamiltonian in the following effective form:

$$\begin{aligned} H = &\int d^3r\Psi_{s_\downarrow}^*\left[\frac{1}{2m}(\hbar\partial_k + i\frac{e}{c}A_k)^2\right]\Psi_{s_\downarrow} \\ &+ \int d^3r\Psi_{s_\uparrow}^*\left[\frac{1}{2m}(\hbar\partial_k - i\frac{e}{c}A_k)^2\right]\Psi_{s_\uparrow} \\ &+ \int d^3r(V_\downarrow|\Psi_{s_\downarrow}|^2 + V_\uparrow|\Psi_{s_\uparrow}|^2)), \end{aligned} \tag{22}$$

where $\Psi_{s_\downarrow}(\mathbf{r}) = \cos\gamma\langle\mathbf{r}|S_\downarrow\rangle$ and $\Psi_{s_\uparrow}(\mathbf{r}) = \sin\gamma\langle\mathbf{r}|S_\uparrow\rangle$ are spin wave functions in $\mathbf{r}$-representation.

The spin current in the present system has several important properties. Firstly, it satisfies the conservation law. The spin density in the present system is calculated according to $\vec{S}(\mathbf{r},t) = \Psi_{s_\downarrow}^*\vec{S}_\downarrow\Psi_{s_\downarrow} + \Psi_{s_\uparrow}^*\vec{S}_\uparrow\Psi_{s_\uparrow}$ with $\vec{S}_{\uparrow\downarrow} = (S_x^{\uparrow\downarrow}, S_y^{\uparrow\downarrow}, S_z^{\uparrow\downarrow})$. On the other hand, the spin current density $\vec{J}_k(\mathbf{r},t) = (J_k^{s_x}, J_k^{s_y}, J_k^{s_z})$ is defined by

$$\begin{aligned} \vec{J}_k = &-\frac{i\hbar}{m}\vec{S}_\downarrow(\Psi_{s_\downarrow}^* D_{1k}\Psi_{s_\downarrow} - \Psi_{s_\downarrow}D_{1k}^*\Psi_{s_\downarrow}^*) \\ &-\frac{i\hbar}{m}\vec{S}_\uparrow(\Psi_{s_\uparrow}^* D_{2k}\Psi_{s_\uparrow} - \Psi_{s_\uparrow}D_{2k}^*\Psi_{s_\uparrow}^*), \end{aligned} \tag{23}$$

where $D_{1k} = \partial_k + i\frac{e}{c}A_k$ and $D_{2k} = \partial_k - i\frac{e}{c}A_k$ are the covariant derivative operators. By a straightforward calculation we can verify the following continuity equation

$$\partial_t \vec{S}(\mathbf{r},t) + \partial_k \vec{J}_k = A_k \sigma_z \hat{\mathbf{e}}_z \times \vec{J}_k(\mathbf{r},t), \tag{24}$$

where σ_z is the pauli matrix. It is easy to see for the s_z-component spin current, the right hand side of Eq. (24) equals zero. Thus the spin current $J_k^{s_z}$ is conserved. Also, it can be verified that the orbit-angular momentum current is also conserved. These results confirm the conservation law for s_z-component of total angular momentum current of atoms. Secondly, the spin current in the present system may exhibit interesting topological properties. For convenience, we denote $\Psi_{s_\alpha} = \sqrt{2m}\eta\zeta_\alpha$, $\alpha =\uparrow,\downarrow$, where the complex $\zeta_\alpha = |\zeta_\alpha|e^{-i\varphi_\alpha}$ with $|\zeta_\downarrow|^2 + |\zeta_\uparrow|^2 = 1$. The modulus η is related to the total atomic density which is assumed to be constant in our case. Then the s_z-component of eq. (23) can be recast into $J_k^{s_z} = \hbar^2\eta^2[\partial_k(\varphi_1 - \varphi_2) - (|\zeta_\downarrow|^2 - |\zeta_\uparrow|^2)\partial_k(\varphi_1 + \varphi_2)] - 2\hbar^2\eta^2\frac{e}{c}A_k$. Furthermore, we introduce the unit vector field by $\vec{\lambda} = (\bar{\zeta}, \vec{\sigma}\zeta)$, where $\zeta = (\zeta_\downarrow, \zeta_\uparrow^*)^T$ and $\vec{\sigma} = (\sigma_x, \sigma_y, \sigma_z)$. It then follows that $\lambda_1 = \zeta_\downarrow^*\zeta_\uparrow^* + \zeta_\downarrow\zeta_\uparrow, \lambda_2 = i(\zeta_\downarrow\zeta_\uparrow - \zeta_\downarrow^*\zeta_\uparrow^*)$ and $\lambda_3 = |\zeta_\downarrow|^2 - |\zeta_\uparrow|^2$. Based on these variables we can finally get the following result:

$$(\nabla \times \mathbf{J}^{\mathbf{s_z}})_{\mathbf{m}} = -2\hbar^2\eta^2\frac{\mathbf{e}}{\mathbf{c}}\mathbf{B}_{\mathbf{m}} - \frac{1}{2}\hbar^2\eta^2\epsilon_{\mathbf{mkl}}\vec{\lambda}\cdot(\partial_{\mathbf{k}}\vec{\lambda} \times \partial_{\mathbf{l}}\vec{\lambda}). \tag{25}$$

The contribution $\vec{\lambda}\cdot(\partial_k\vec{\lambda} \times \partial_l\vec{\lambda}$ indicates a topological term in the spin current induced by optical fields. Note each value of $\vec{\lambda}$ represents a point in the two-dimensional sphere S^2 (see Fig. 3). The variation of $\vec{\lambda}$ depends on the relative density of the two spin components $\gamma(\mathbf{r})$ and the sum of the phase spreadings $\varphi_\uparrow(\mathbf{r}) + \varphi_\downarrow(\mathbf{r})$. It is easy to verify that $\vec{\lambda}$ can cover the entire surface S^2 for the parameter distributions in the interaction region: $\varphi_1 + \varphi_2 : 0 \to 2n_1\pi$ and $\gamma : 0 \to n_2\pi/2$ with integers $n_1, n_2 \geq 1$. We then obtain a map between the unit vector field $\vec{\lambda}$ and the spatial vectors

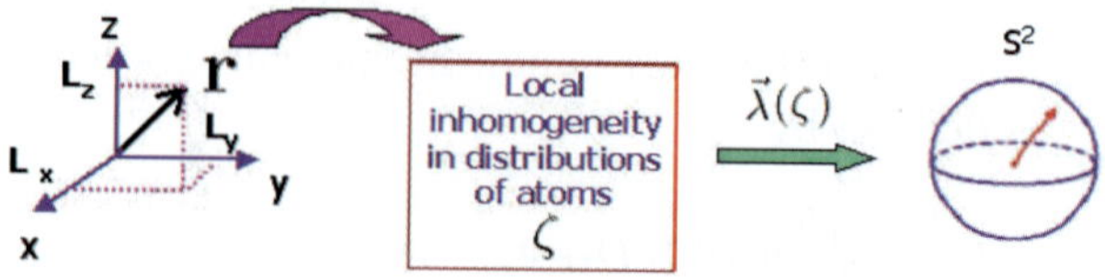

Fig. 3. Possible parameter choice for generating a spin current with topological properties.

$F : \vec{\lambda} \rightarrow \mathbf{r}/r$ so that the closed-surface integral

$$\oint_s (\nabla \times \mathbf{J}^{\mathbf{s_z}}) \cdot \mathbf{dPS} \sim \oint_{\mathbf{s}} \mathbf{ds_m} \epsilon_{\mathbf{mkl}} \frac{\mathbf{r}}{\mathbf{r}} \cdot (\partial_{\mathbf{k}} \frac{\mathbf{r}}{\mathbf{r}} \times \partial_{\mathbf{l}} \frac{\mathbf{r}}{\mathbf{r}}) \sim \mathbf{4\pi n},$$

where n is the winding number. In physics, such mapping degree corresponds to the formation of a local inhomogeneity in the densities of the spin-up and spin-down atoms. The Hamiltonian (22) can also be understood as a system with *two-flavor* oppositely charged particles interacting with *one* external effective magnetic field. Such model has a wide range of applications in, e.g. two-band superconductivity [31], etc. For instance, under certain conditions liquid metallic hydrogen might allow for the coexistence of superconductivity with both electronic and protonic Cooper pairs [32]. Faddeev et al. have discovered a series of nontrivial topological properties in these systems [33]. Based on the present technique, one can develop an optical way to realize *two-flavor* artificially charged system, which may allow for a deeper understanding of the basic physical mechanisms in cold atomic systems.

For practical application, an important goal is to create a pure spin current injection, i.e. there is no massive current accompanied. For this we use two columnar spreading light fields that [34] $\Omega_{01}(\mathbf{r}) = \Omega_{02}(\mathbf{r}) = f\rho$ with $f > 0$. Furthermore, we set the potential $V(\mathbf{r}) = \frac{1}{2}\nu(\vec{\rho} + x_0\hat{\mathbf{e}}_x)^2$ with the coefficient $\nu = (1 + \frac{\hbar l^2 f^2}{2m\Delta^3})\frac{2\hbar f^2}{\Delta}$, say, we use a two-dimensional harmonic potential centered at $\vec{\rho} = -x_0\hat{\mathbf{e}}_x$ [35]. In the parameter regime we reach a set of uniform magnetic field $\mathbf{B}_\downarrow(-\mathbf{B}_\uparrow) = \nabla \times \mathbf{A}_\downarrow = \frac{\hbar l c}{e}\frac{f^2}{\Delta^2}\hat{e}_z$, and electric fields $\mathbf{E} = -\nabla V_{eff}(\mathbf{r}) = -(1 + \frac{\hbar l^2 f^2}{4m\Delta^3})\frac{2\hbar f^2 x_0}{e\Delta}\hat{e}_x$, respectively in z $(-z)$ and x directions. Atoms in different spin states $|S_\alpha\rangle$ experience the opposite magnetic fields $\mathbf{B}_\alpha$ but the same electric field $\mathbf{E}$. This may lead to a Landau level structure for each spin orientation. Together with the applied in-plane electric field, one can obtain a pure spin current in the y direction, while the massive current is zero. To exactly calculate the spin currents, we need to calculate the eigenstates of the present system. The Hamiltonian for a single atom in state $|S_\alpha\rangle$ can be written as:

$$\begin{aligned} H^\alpha &= H_0^\alpha + H', \quad \alpha = \downarrow, \uparrow \\ H' &= -eEx \\ H_0^\alpha &= \frac{\hbar^2 eB}{2mc}(R_\alpha^2 + P_\alpha^2) + \frac{1}{2m}p_z^2 \end{aligned} \tag{26}$$

with $R_{\uparrow\downarrow} = (\frac{c}{eB})^{1/2}(p_x - \frac{e}{c}A_x^{\uparrow\downarrow})$, $P_{\uparrow\downarrow} = (\frac{c}{eB})^{1/2}(p_y - \frac{e}{c}A_y^{\uparrow\downarrow})$ and $B = |\mathbf{B}_{\uparrow,\downarrow}|$. One can verify that $[R_\alpha, P_\beta] = i\hbar\delta_{\alpha\beta}$, so the eigenfunction of H_0^α is Hermite polynomial, written as $\mu^{\alpha}_{nk}(R)$ with the Landau level $E_{nk,\alpha} = (n+1/2)\hbar\omega + \hbar k^2/2m$ and $\omega = eB/mc$. In the weak field case, the spin/massive current carried by an atom can be calculated to the first order correction with perturbation theory on the state $|\mu^{\alpha}_{nk}\rangle$ (linear response)

$$J^y_{s_z,m} = \frac{1}{L_x L_z}\int dE\big((j^{(0)}_{s_z,m})_{nk,\alpha} + (j^{(1)}_{s_z,m})_{nk,\alpha}\big)f(E) \tag{27}$$

with $f(E)$ the Fermi distribution function, and the space-spreading lengths of the atoms in x and z directions are L_x and L_z. Here the spin current in the zeroth and first order is given by

$$\begin{aligned}(j^{(0)}_{s_z,m})_{nk,\alpha} &= \langle\mu^{\alpha}_{nk}|j^y_{s_z,m}|\mu^{\alpha}_{nk}\rangle\\ (j^{(1)}_{s_z,m})_{nk,\alpha} &= \sum_{n'\alpha'}\frac{\langle\mu^{\alpha'}_{n'k}|H'|\mu^{\alpha}_{nk}\rangle\langle\mu^{\alpha}_{nk}|j^y_{s_z,m}|\mu^{\alpha'}_{n'k}\rangle}{E_{n,\alpha} - E_{n',\alpha'}} + h.a.,\end{aligned} \tag{28}$$

and the spin current operator is $(j^y_{s_z})_{\alpha} = \frac{\hbar}{2}(S^{\alpha}_z v^{\alpha}_y + v^{\alpha}_y S^{\alpha}_z)$ with $v^{\alpha}_y = [y, H^{\alpha}]/i\hbar = (eB/m^2c)^{1/2}P_{\alpha}$. The massive current operator $j_m = mv_y$. For ^{6}Li atoms we consider the initial condition that $\sin^2\gamma = \cos^2\gamma = 1/2$, i.e. the atoms have the equal possibility in state $|S_{\downarrow}\rangle$ and $|S_{\uparrow}\rangle$, which is familiar in optical trap [36]. Note that under the present definitions the perturbation part can be rewritten as $H' = i\hbar E(\frac{ec}{B})^{1/2}\frac{\partial}{\partial R}$. Substituting this result into Eq (28) and (27) we get

$$J^y_{s_z} = n_a(\hbar + \frac{\hbar^2 l^2 f^2}{4m\Delta^3})\frac{\Delta x_0}{l}, \tag{29}$$

where $n_a = 2m\eta^2$ is the density of atoms, whereas the massive current $J_m = 0$. In fact, under present interaction of external effective electric and magnetic fields, the atoms with opposite velocities in y direction have opposite spin polarizations. Thus the massive current vanishes whereas a pure spin current is obtained. This result allows us to create conserved spin currents without using atomic beams.

To observe SHE in present cold atomic system, a spin-sensitive measurement [37] can be used. Experimentally, for example, one can detect the spatially separated spin accumulation due to the spin current via, e.g. fluorescence measurement. For ^{6}Li atoms, one could perform resonant Raman transitions from the accumulated spin state $S_z = -1/2$ to $2^2P_{1/2}(F = 3/2, m = -3/2)$ and from $S_z = 1/2$ to $2^2P_{1/2}(F = 3/2, m = 3/2)$ and then observe fluorescence. As long as the spin accumulations are separated to say larger than $5 - 10$ microns, the two separated images can be resolved in the experiment.

Finally, one can give some numerical evaluations for the present system. The typical values of the parameters can be set $\Delta \sim 10^7 s^{-1}, l < 10^4, f = 10^{7\sim 8}(s \cdot m)^{-1}, x_0 \approx 0.2mm$. The velocity then satisfies $\bar{v} < 1.0m/s$ and $\tau_\pm/|E_\psi^+ - E_\psi^-| < 10^{-3} \ll 1$, which guarantees the validity of the adiabatic condition. Furthermore, if we employ an atomic system with the atomic density $n_a \approx 1.0 \times 10^{10} cm^{-3}$, one can find the spin current $J_{s_z}^y \approx 1.322 \times 10^{-5} eV/cm^2$. On the other hand, practically the light fields have a finite cross section S_{xy}, whose effect is ignored in the calculation of spin currents. For our purpose, this effect can be neglected when $lf^2 S_{xy}/\Delta^2 \gg 1$, which means the effective magnetic flux induced by the light fields can support an enough large number of quantum states for each Landau level. Since the decoherence rate of atomic spin currents is generally much smaller than that of the counterparts in semiconductors, the created atomic spin currents could be applicable in quantum information devices.

5. Summary

To summarize, we have discussed the technique of realizing spin Hall effect (SHE) in ultracold atomic gas. Firstly, we briefly reviewed the the concepts of spin-orbit coupling and SHE in solid systems and then discuss the optical means in creating adiabatic gauge field for neutral atomic system. In this way, we can obtain the effective spin-orbit coupling interaction for neutral atomic system. Especially, we discussed how to generate the spin-dependent effective magnetic field for the ultracold fermi atomic gas, and then obtain the SHE in atoms. The properties of spin current in the present atomic system are also discussed. Further investigations on the applications of the adiabatic gauge field in atoms and experimental demonstration of the theoretically predicted results will be the next focus in this intriguing topic.

Acknowledgments

This work is supported by NUS academic research Grant No. WBS: R-144-000-189-305, US NSF under the grant DMR-0547875.

References

1. S. Murakami, N. Nagaosa, S.-C. Zhang, Science 301, 1348 (2003).
2. J. Sinova et al., Phys. Rev. Lett. 92, 126603 (2004).
3. Y. K. Kato et al., Science, 306, 1910 (2004); J. Wunderlich B. Kaestner, J. Sinova, and T. Jungwirth, Phys. Rev. Lett. 94, 047204 (2005).

4. B.K. Nikolic, S. Souma, L.P. Zârbo and J. Sinova, Phys. Rev. Lett. 95, 046601 (2005); S. Murakami, N. Nagaosa and S.-C. Zhang, Phys. Rev. Lett. 93, 156804 (2004).
5. K. v. Klitzing, G. Dorda, and M. Pepper, Phys. Rev. Lett. 45, 494 (1980).
6. D. C. Tsui, H. L. Stormer, and A. C. Gossard, Phys. Rev. Lett. 48, 1559 (1982).
7. C.L. Kane and E.J. Mele, Phys. Rev. Lett. 95, 226801 (2005).
8. N.A. Sinitsyn, J.E. Hill, Hongki Min, J. Sinova, and A.H. MacDonald, cond-mat/0602598.
9. B.A. Bernevig, Taylor L. Hughes and S.-C. Zhang, Science, **314**, 1757, (2006).
10. B.A. Bernevig and S.-C. Zhang, Phys. Rev. Lett. 96, 106802 (2006).
11. M. Lewenstein, A. Sanpera, V. Ahufinger, B. Damski, A. Sen De, and U. Sen, Adv. Phys. 56, 243 (2007).
12. P. Engels, I. Coddington, P. C. Haljan, and E. A. Cornell, Phys. Rev. Lett. 89, 100403 (2002).
13. B. Paredes, P. Fedichev, J.I. Cirac, and P. Zoller, Phys. Rev. Lett. 87, 010402 (2001); Phys. Rev. A, 66, 033609 (2002).
14. J. R. Abo-Shaeer et al., Science 292, 476 (2001); T. L. Ho, Phys. Rev. Lett. 87, 060403 (2001); U. R. Fischer and G. Baym, Phys. Rev. Lett. 90, 140402 (2003).
15. A. S. Søensen, E. Demler, and M. D. Lukin, Phys. Rev. Lett. 94, 086803 (2005).
16. G. Juzeliunas and P. Ohberg, Phys. Rev. Lett. 93, 033602 (2004).
17. D. Jaksch and P. Zoller, New J. Phys. 5, 56 (2003).
18. G. Juzeliunas, P. Ohberg, J. Ruseckas, and A. Klein, Phys. Rev. A 71, 053614 (2005).
19. J. Ruseckas, G. Juzeliunas, P. Ohberg, and M. Fleischhauer, Phys. Rev. Lett. 95, 010404 (2005).
20. M. T. Gabriel, P. Tores Juan, T. Lluis, Nature Physics, 3, 305 (2007).
21. X. -J. Liu, X. Liu, L. C. Kwek and C. H. Oh, Phys. Rev. Lett. 98, 026602 (2007), cond-mat/0603083.
22. S.-L. Zhu, H. Fu, C.-J. Wu, S.-C. Zhang and L.-M. Duan, Phys. Rev. Lett. 97, 240401 (2006).
23. W. Zawadzki, in Optical Properties of Solids, E. D. Haidemenakis (ed.), Gordon and Breach, New York, 1970.
24. G. Dresselhaus, Phys. Rev. 100, 580 (1955).
25. E.I. Rashba, Sov. Phys. - Solid State 2 (1960) 1109.
26. J. M. Luttinger, Phys. Rev. 102, 1030 (1956).
27. F. Wilczek and A. Zee, Phys. Rev. Lett. 52, 2111 (1984).
28. K. G. Wilson, Phys. Rev. D, 10, 2445 (1974).
29. C. P. Sun, M. L. Ge, Phys. Rev. D 41, 1349 (1990).
30. B. DeMarco and D. S. Jin, Science, 285, 1703 (1999); M. E. Gehm, Preparation of an optically-trapped degenerate Fermi gas of ^{6}Li: Finding the route to degeneracy, Doctor Thesis, Duke University, 2003; J Fuchs, et al., physics/0604026 (2006).

31. L. Shen, N. Senozan, and N. Phillips, Phys. Rev. Lett. 14, 1025 (1965); F. Bouquet et al., Phys. Rev. Lett. 87, 047001 (2001).
32. K. Moulopoulos and N.W. Ashcroft Phys. Rev. B 59, 12309 (1999).
33. L. Faddeev and A. J. Niemi, Phys. Rev. Lett. 85, 3416(2000); E. Babaev, L. D. Faddeev, and A. J. Niemi, Phys. Rev. B 65, 100512 (2002).
34. L. Allen, S. M. Barnett and M. J. Padgett, Optical Angular Momentum, 2003 (Bristol: Institute of Physics); E. M. Wright et al, PRA 63, 013608 (2001)).
35. R. Grimm et al., Advances in Atomic, Molecular and Optical Physics Vol. 42, 95 (2000).
36. Z. Hadzibabic et al., Phys. Rev. Lett. 91, 160401 (2003).
37. Artem M. Dudarev et al., Phys. Rev. Lett. 92, 153005 (2004).

MAGNETIC COUPLING AND QUANTUM WELL STATES

Z. Q. QIU

Department of Physics, University of California at Berkeley, USA
E-mail: qiu@socrates.berkeley.edu

Electrons in ultrathin films are confined to form quantum well states (QWS). Photoemission provides the most direct observation of QWS in k-space. The unique capabilities now available at the Advanced Light Source (ALS) at Berkeley make it possible to measure the QW states with atomic layer resolution. In this talk, the photoemission result from the ALS on the Cu/Co(100) QW system is presented. Firstly, using a Ni monolayer to probe the Cu QWS at different positions, it is shown that the QWS in metallic thin films can be described by the envelope function of the Bloch wave. Secondly, by measuring the magnetic coupling using the magnetic dichroism, the relationship is demonstrated between the oscillatory magnetic coupling and the oscillations of the density of states at the Fermi level.

QUANTUM INFORMATION PROCESSING: PRESENT STATUS AND PERSPECTIVES

IGNACIO CIRAC

Max-Planck-Institut fr Quantenoptik
E-mail: Ignacio.Cirac@mpq.mpg.de

In this talk I will describe the present situation of a young field of research: quantum information processing. In particular, I will explain how by using basic concepts coming from Quantum Mechanics it should be possible to build devices that process information in a very efficient way, and allow us to transmit information in a completely new way. Furthermore, during the last few years it has been shown that several of the ideas developed in the context of quantum information can be used to describe complicated many-body quantum systems. In particular, I will explain how new powerful algorithms can be built that should allow us to simulate important problems in the fields of atomic and condensed matter physics.

DEGRADABLE CHANNELS IN QUANTUM INFORMATION THEORY

MARY BETH RUSKAI

Tufts University
E-mail: Marybeth.Ruskai@tufts.edu

Degradable channels are one of the very few for which the quantum capacity can be computed exactly in terms of a one-step formula. We describe several important classes of degradable channels, as well as some differences between qubits and channels on spaces of higher dimension.

Other Topics

PROF. YANG AND PROF. STEINBERGER

WEIMIN WU
Fermi Lab
E-mail: weimin@fnal.gov

There is a photo of Prof Yang and Prof Steinberger together in May 1986, Beijing. The talk will tell the story behind the picture.

Prof.Yang and Prof.Steinberger

Weimin Wu, Fermilab
Oct. 2007, Singapore

1

Prof.Yang and Prof.Steinberger together

This is a photograph that I took in May 1986, Beijing, with Prof.Yang and Prof.Steinberger together. I will briefly talk about the story behind this photo: how these two great scientists came to respect each other personally, and finally met again after a long absence in 1986.

2

Prof. Yang

The first time I met Prof.Yang in person was in 1983 at Stony Brook, when he invited me to visit, introduced me to physicists from the D0 collaboration so I could discuss a possible collaboration between Stony Brook and IHEP in Beijing. In 1981, Prof.Yang was raising funds from Hong Kong and the US to create a CEEC fellowship at Stony Brook, so there were several visiting scholars from China at Stony Brook when I was there.

My second contact with Prof. Yang was in July 1984, Leipzig, Germany at the 22nd International Conference on High Energy Physics.

3

Prof. Yang (cont.)

The 3rd time was in 1992, Washington DC for his 70th birthday.

4

Prof. Yang (cont.2)

We had long discussions during these three visit, especially during the Leipzig conference. We visited the art museum and had dinner together so we had time to discuss many topics, from physics, to politics, to art and photography. He knew I worked in Prof.Steinberger's CDHS/ALEPH group, so Prof. Yang talked about his friendship with Prof.Steinberger during their time together at the University of Chicago.

Although we have not met in person since that time, Prof. Yang and I have always kept in close contact via e-mail.

5

Prof.Steinberger

I went to CERN to participate in Prof.Steinberger's CDHS experiment in Nov. 1979 from IHEP, through a program setup by Prof. TD Lee. Soon after, I created an IHEP group that joined the ALEPH collaboration from 1982 onward. Since I was resident at CERN for almost 7 years starting in 1979, I had many opportunities to talk to Prof.Steinberger. Since then, I have continued to have very close communication with him up to the present.

Prof.Yang once told me, "You are very lucky to have the opportunity to work with Steinberger. Jack is one of the very few experimental physicist who has a rich knowledge of theoretical physics, and is a great master, with a whole view on fundamental physics. You really should make use of the opportunity to learn from Jack".
Furthermore, Prof.Yang told me that "I am not good in experimental physics. When I was student whenever I tried to do anything in laboratory, something always went to wrong... that is why I become a theoretical physicist instead. But Jack is different. He is not only a great experimental physicist, but he knows the theory too... so he does the experiment with vision".

From Prof.Yang's words, I feel he has great respect for Prof.Steinberger.

So does Prof.Steinberger to Prof.Yang.

6

Prof.Steinberger (cont.)

I remember when I met Jack the first time, he told me that, "The reason I accepted visitor-scholars from China to my CDHS group, is that I had great respect for Prof. TD Lee, CN Yang, and Madam Wu."

7

Prof.Steinberger (cont.2)

As you know, the year 1979, it was only a few years after Mao's "Cultural revolution" in China. All scientists and students there were completely isolated from the world. It was first time we heard a words of QCD or quark model in 1978, our knowledges on High Energy Physics at that time, was far less than a student just graduated from a western University. However, as Jack said, "From CN Yang, TD Lee, and Madam Wu, I fully believe that the Chinese people are very intelligent, you will catch up very quick".
Jack especially mentioned that when he was student in University of Chicago, he had met CN Yang, TD Lee, etc.. Jack said, "Even though they had on recently come to the US from China, where there was still war going on, Yang seemed already to know about everything that was going on in Physics." Prof.Steinberger not only told me this personally but he wrote about Yang as a classmate in 1989.

He wrote: *"For me, particle physics began in 1947, when I was a graduate student at the University of Chicago. It was a great privilege to be a student in that department at that time. There were excellent teachers: Fermi, Edward Teller, Willie Zachariasen and Maria Mayer, to name a few. My fellow students also had a great impact on my formation. There were a remarkable group: Chen Ning Yang, Tsung Dao Lee, Marvin L. Goldberger, Marshal Rosenbluth, Lincoln Wolfenstein, Owen Chamberlain, Geoffrey Chew, Richard Garwin and others . The most impressive student-teacher in that group was Yang, who had come from China after the war and at the age of twenty-four, despite the limited wartime possibilities in China, knew all of modern physics fluently by the time he entered graduate school in Chicago".* (Pions to quarks, Cambridge 1989,307).

8

Lost Contact

Jack told me that he feel very sorry that he has no contact to Prof.Yang for long time, after he went to Columbia, where he had more contact with TD Lee. He said it was a pity I made one friend and lost another. I told him that I didn't think that he had lost a friend, may be just lost contact for awhile. He said: "I hope so!".

These words by Steinberger was deeply remembered in my heart. I wanted to do something about it.

The opportunity came in May 1986. The Institute of High Energy Physics, Beijing, invited Steinberger to visit IHEP, on the occasion of his 65 birthday, and also to promote the ALEPH collaboration, particularly to promote to set up the first computer network connection between China and CERN. After Jack's visit, on August 25, 2006, the first Chinese e-mail was sent by me from Bejing to Steinberger at CERN. This was really the first momentous step for China to enter the Internet age.

(And look how far we have come since then!)

9

Contact Restored

When Jack was in Beijing, I learned Prof. CN Yang was in Shanghai at that time. I discussed with the Chinese Academy of Science that, if both agreed, CN Yang come to Beijing to celebrate Jack's 65th birthday. My idea got support from Prof. Zhou, the president of CAS at that time. Prof. Zhou invited both of them for dinner, and the two great scientists finally met each other after long time without any contact. I took a photo of both of them after dinner was finished. I got permission from both of them to publish this photo in my recently published book "The Beauty of Physics".

Jack send me a e-mail recently, saying: "Dear Wei Min, Your book of photographs arrived, many thanks. Thank you also for including one of myself together with C.N. Yang. You were certainly very gifted with the camera, one of your pictures has been up on our walls since many years."

10

"ALMOST EVERY PROBLEM HE TOUCHED EVENTUALLY TURNED INTO GOLD"

LING-LIE CHAU

Physics Department, University of California, Davis, CA 95616, USA
E-mail: chau@physics.ucdavis.edu

The title is a quote from my "Yang's Pyramid" — published in the festschrift celebrating Professor Yang's 70th birthday, ≪Chen Ning Yang, a Great Physicist of the Twentieth Century≫. I likened the collection of his scientific works to a pyramid. As time goes on, hidden treasures continue to be discovered, adding to the well-known brilliant ones. This observation of mine continues to bear out in an impressive way. For the celebration of his 85th birthday, I will highlight recent developments of such remarkable achievements and make an attempt to identify the reasons for them, with the backdrop of my own perspectives.

1. Introduction

On this happy occasion of celebrating the 85th birthday of Professor Yang, I would like to wish him, and us, many, many more.[a]

The title is a quote from my "Yang's Pyramid" — published in the festschrift celebrating Professor Yang's 70th birthday, ≪Chen Ning Yang, a Great Physicist of the Twentieth Century≫, 47–48. Many recent new developments continue to confirm this observation. Also, happily I found that Professor Yang likes it. Among all that has been written about Professor Yang's accomplishments, this quote from my two-page essay was the one used by Professor John Toll, the then President of Washington College, on the occasion of Professor Yang being honored with the Honorary Degree in Science, see Fig. 1, and printed in their Magazine, Summer 1999 issue. I feel like having hit a jackpot.

[a]The day before this conference, at the check-in counter of the conference hotel, Swissotel Merchant Court, Singapore, I saw Lay-Nam Chang and his wife. I greeted them, "Great to see you! Long time no see." He said, "Yah, a long time, it must be more than 10 years." I replied, "Weren't you at Yang 80 in Beijin?" He said, "You are right, I was there." So I said, "Well, it was 5 years since we had seen each other." We all laughed and agreed that we would be lost without the get-togethers for Professor Yang's birthday celebrations.

It is my extraordinary good luck and privilege to know and to be mentored by a great physicist like Professor Yang. During 1969–1986 I was fortunate to be close to his institute, now the C.N. Yang Institute of Theoretical Physics, SUNY Stony Brook. I learned a lot from him through osmosis around his institute and from direct contacts. He was friendly and generous with discussions. I have three copies of his ≪Selected Papers, 1945–1980, with Commentary≫, one at my office and two at home — one of which is the original hardbound copy with his handwritten comments, Fig. 2. These comments reveal more about his friendliness than the worth of my "valuable advice and criticism".

Professor Yang also initiated my life-transforming move from pure research institution experience at the Institute for Advanced Study (1967–1969) and Brookhaven National Laboratory (1969–1986), to university experience at UC Davis. (On the occasion of the 75th Anniversary in 1983, UC Davis invited Professor Yang to visit and asked him for a list of people to consider for hiring in the Physics Department. He suggested two names, of which mine was one. Thanks to my colleagues at Davis, they chose to go after me and obtained thirteen reference letters for my hiring.) My life has been enriched by the two experiences, in an ideal order in spacetime.

I hope that what I write here is helpful to those who would like to learn a successful way of doing physics from Professor Yang. I thank the organizers for this enjoyable and intellectually stimulating conference.[b]

[b]2007 is the 50th anniversaries of many important events. It is the 50th anniversary of Professor Yang's Nobel Prize with Professor Lee, and of the publications of several of his important papers, some of which I will discuss. 2007 is also the 50th anniversary of the Bardeen-Cooper-Schrieffer (BCS) paper for superconductivity, of the launching of Sputnik by the Soviet Union, of the Broadway premier production of Bernstein's West Side Story, etc. On a much more minute scale, 2007 is also the 50th anniversary of my graduation from the Tainan Girls School, which I had completely forgotten till my classmates organized our reunion. It was at the Tainan Girls School where our physics teacher, Mr. Jin-Chuen Shen, got me hooked on physics. I still remember how happy I was when I thought that I understood the centrifugal force. 2007 also marks the first year of my new life as a professor emerita. Coming to this conference and giving this talk gave me a valuable opportunity to reflect and to plan ahead as a physicist. I hope that you all will forgive my indulgence in chatting more than probably I should.
2007 happens also to be the year that the Berkeley Center for Theoretical Physics (BCTP) held its official opening symposium. The funding of it was initiated by the donation from Jane and Robert Wilson in memory of their friendship with Oppenheimer. Many interesting talks were given. It is particularly heart-warming to see that the photos of my other two mentors, Professor Chew (my Ph.D. thesis adviser) and Professor Mandelstam, put together with that of Oppenheimer in a small gallery at BCTP. My deep gratitude to Chew and to Mandelstam will be written about in the future.

2. Reasons for "Almost Every Problem He Touched Eventually Turned Into Gold"

Based upon what I know about his papers and him as a person, I would give two main reasons for why "Almost Every Problem He Touched Eventually Turned Into Gold": one, insightful choice of problems to work on; two, thoroughness in the investigation and getting to the primordial sources.[c] The former comes from in-depth understanding of physics and keeping alert about current experiments. The latter requires dedication and perseverance. Both involve being gifted and making best use of it. Some of his papers were recognized quickly. Some were ahead of their times; however, they will eventually shine when their relations to experiments are made. I will illustrate these qualities of his papers next.

3. Examples of Yang's Papers that Continue to Shine or Have Newly Turned into Gold

Here I list examples of Yang's papers in the three major fields of the physical sciences. Any one of these papers could be a proud accomplishment of a whole career.

3.1. *Particle physics*

Yang-Mills 1954, Phys. Rev. 96, 191–195, "Conservation of Isotopic Spin and Isotopic Spin Invariance"

The paper established local gauge symmetry as principle for interactions in the quantum description. (Yang as a graduate student at Chicago already was thinking about big problems and asked the question of how gauge transform was related to interactions. See Attachment 1.) The demand of covariance of equations of motion under a local phase factors multiplication — called local gauge symmetry — led Yang and Mills derive the equations that now bear their name. This expanded Einstein's using general coordinate transformation symmetry to derive equations in physics.[d]

---

[c] A primordial source or truth — e.g., the Pythagorean theorem, Newton's universal law of gravity, the Heisenberg algebra, the Yang-Baxter relation, or a symmetry law — can be present in many problems, simple or complex. Being able to get to the primordial source of a well chosen problem (relevant and solvable in a reasonable time) is of crucial importance. This quality is pervasive in Yang's papers.

[d] Pedagogical expositions of these will be given in the two books that I have been writing: ≪Basic Mathematics for the Physical Sciences≫ and ≪Group Theory for the Fundamental Laws of Physics≫, based upon the material I developed for my teaching.

It took more than twenty years before the true important nature of this paper was realized in physics. After contributions by many distinguished physicists, the theory now has been fully quantized and developed to become the theory for electroweak and strong interactions, called the Standard Model. See the book, ≪50 Years of Yang-Mills Theory≫, edited by G. 't Hooft. With the verifications in experiments/observations, it has been established that the local gauge symmetry and Einstein's general coordinate transformation symmetry form the bases (primordial sources) of all major interactions: electroweak, strong, and gravity. These physics developments also have a profound influence on mathematics. (How to quantize gravity theory fully is still under intense research.)

The formal citation number of the paper is about 1450, which averages about 27/yr. It peaked to about 80/year during the late '70s to early '80s and now stays about 20/year. There are probably as many, if not more, informal citations using the term Yang-Mills without citing the paper.[e]

Lee-Yang 1956, Phys. Rev. 104, 254–258, "Question of Parity Conservation in Weak Interactions"

Responding to the so called θ-τ puzzle from weak decay experiments, the paper questioned the validity of the parity symmetry (a global symmetry), P, and stimulated experimenters to look into it. Within a year of its publication, experiments confirmed the violation of P. In about fourteen months after the publication of their paper, Lee and Yang were awarded the Nobel Prize. This paper is among the top few that has had the shortest time between the publication and the award of a Nobel Prize to its authors.

Currently this paper has about 1000 citations, averaging about 20/year and in recent years staying around 20–30/year.

Lee-Oheme-Yang 1957, Phys. Rev. 106, 340–345, "Remarks on Possible Noninvariance Under Time Reversal and Charge Conjugation"

The paper raised questions about the violation of other global symmetries, C (charge conjugation) and T (time reversal). Although this paper missed the discussion of CP violation, it set up the parametrizations which are useful for the now called the mass-matrix CP violation.

Currently this paper has about 300 citations, averaging about 6/year and staying steadily at that.

Wu-Yang 1964, Phys. Rev. Lett. 13, 380–385, "Phenomenological Analysis of Violations of CP Invariance in Decays of $K^0$and $\overline{K}^0$"

[e]The citation data given in this section were as of March, 2008.

Responding quickly to the observation of CP violation in 1964, this paper analyzed the CP violation and became the standard reference for the analysis of the now called mass-matrix CP violations.

Later, the Kobayash-Maskawa framework for CP violation allows direct (or amplitude) CP violations as well as mass-matrix CP violations. So this Wu-Yang paper is still relevant. My recent paper showed, among other results, how the Wu-Yang phase convention is always possible in general.[f]

Currently this paper has about 259 formal citations, averaging about 6/year and staying about a few per year. There are many citations of Wu-Yang phase convention without formally citing the paper.

Eventually global symmetries of P, CP, and T have all been established to be violated in weak interactions. However, these violations observed in the laboratories are found not sufficient to explain the lack of anti-matter in the Universe. Whatever the future resolution will be, the generic analysis made in these papers by Yang and coworkers will be relevant.

Yang 1974, Phys. Rev. Lett. 33, 445–447, "Integral Formalism for Gauge Fields"

This paper introduced and emphasized the loop formulation.

The loop formulation has been widely used; however, it is commonly called the Wilson-loop. Based upon my research on the contents and the times of the publications of this paper and those of the paper by K. Wilson, "Confinement of Quarks", Phys. Rev. D 10 (1974) 2445–2459, I have

[f]My recent paper, L.-L. Chau, "Phase Direct CP Violations and General Mixing Matrices", Phys. Lett. B 651 (2007) 293–298, was prompted by the beautiful experimental work of the Belle collaboration, A. Garmash *et al.*, "Evidence for Large Direct CP Violation in $B^{\pm} \to \rho^0 K^{\pm}$ from Analysis of Three-Body Charmless $B^{\pm} \to K^{\pm}\pi^{\pm}\pi^{\mp}$ Decays", Phys. Rev. Lett. 96, 251803 (2006). It was the first experiment of its kind. It confirmed what I wrote and advocated in the 1980s: an observation of CP violation in charged $B^{\pm}$, rather than B^0, would give the cleanest evidence for the direct CP violation in the amplitude; and the reaction $B^{\pm} \to \rho^0 K^{\pm}$ had been the number one on my list; see my overview paper, "In Search of CP Noninvariance in Heavy Quark Systems", 249–268 of ≪ CP Violation≫ edited by C. Jarlskog, 1989, World Scientific.
Here I highlight the new results in this paper: formulating expressions for amplitudes suitable for quantifying both modulus and phase direct CP violations and discovering Möbius transformation (MT) relations, which provide encouraging information for the search of direct CP violations in general; then applying the formulation and calculating the measurements of phase direct CP violations and strong amplitudes by the Belle collaboration; showing a versatile construction procedure for $N \times N$ Cabibbo-Kobayashi-Maskawa (CKM) matrices, Pontecorvo-Maki-Nakagawa-Sakata (PMNS) matrices, and general unitary matrices, thus giving a full derivation of the parametrization for the $N \times N$ cases and a comprehensive understanding of the 3×3 case, which was originally constructed through trials by Chau-Keung, Phys. Rev. Lett. 53 (1984) 1802–1806 and has since been used by the Particle Data Group as the standard parametrization.

formed the opinion that the loop ought to be called the Yang-Wilson-loop. In celebrating Professor Yang's 80th birthday, I wrote it up in the article, "A Forgotten Treasure in Yang's Pyramid", which was distributed but unpublished. I attach it here as Attachment 2, so you can see the details and an interesting encounter with E.P. Wigner.

Currently this paper has about 350 formal citations, averaging about 10/year and staying about a few per year. (I am sure that the citations would have been more if people understand what I have said about it.)

Wu-Yang 1975, Phys. Rev. D 12, 3845–3857, "Concept of Nonintegrable Phase Factors and Global Formulation of Gauge Fields"

This paper extended the previous one, and in particular included the presence of the Dirac magnetic monopole.

Currently this paper has about 800 citations (some overlapping with Yang 1974), averaging about 25/year and recently staying about 20/year.

For additional important papers by Yang in particle physics, see the talk by T.T. Wu at this conference.

3.2. *Many-body physics (which includes statistical mechanics and condensed matter physics)*

Yang-Lee 1952, Phys. Rev. 87, 404–409, "Statistical Theory of Equation of State and Phase Transitions. I. Theory of Condensation"

Lee-Yang 1952, Phys. Rev. 87, 410–419, "Statistical Theory of Equation of State and Phase Transitions. II. Lattice Gas and Ising Model"

These two papers clarified the mathematical description of liquid-vapor phase transition and discovered the famous Lee-Yang circle theorem.

Paper I has about 1000 citations, averaging about 18/year. Impressively, in recent years its citations have stayed about 30/year. Paper II has about 1200 citations (with some overlap between I and II), so averaging about 22/year. Recent citations stay about 30/year, above the average.

Lee-Huang-Yang 1957, Phys. Rev. 106, 1135–1145, "Eigenvalues and Eigenfunctions of a Bose System of Hard Spheres and Its Low-Temperature Properties"

Impressively, the beyond mean-field-theory-approximation results of this paper have been confirmed on the BEC side of the BCS-BEC crossover experiment by A. Almeyer *et al.,* Phys. Rev. Lett. 98, 040401 (2008).[g]

[g]BEC stands for Bose-Einstein condensation. For BCS, see Footnote b. For information about the breath-taking recent developments of ultracold atomic experiments, see the talk by C. Cohen-Tannoudji at this conference.

Currently this paper has about 430 citations, averaging about 9/year and in recent years staying about 15/year, above the average. In 2004, the number of citations surged to 25.

For other aspects of this paper, see K. Huang's talk, in the forthcoming Proceedings or Int. J. Mod. Phys. B 21, 5059–5073 (2007).

Yang 1962, Rev. Mod. Phys. 34, 694–704, "Concept of Off-Diagonal Long-Range Order and the Quantum Phase of Liquid He and of Superconductors"

The paper gave a comprehensive and exact mathematical description of superfluidity/superconductivity. It coined the now standard term ODLRO (off-diagonal-long-range-order).

Currently this paper has about 730 citations, averaging about 16/year and in recent years staying about 25/year, above the average. In 2007 the number of citations surged to 30.

Yang 1967, Phys. Rev. Lett. 19, 1312–1315, "Some Exact Results for the Many-body Problem in One Dimension with Repulsive Delta-Function Interaction"

This paper discovered the now called Yang-Baxter relation and Yangian.

With my coworkers, I had derived generalized Yang-Baxter relations in theories in higher dimensions. I pointed out that the Yang-Baxter relations to the quantum exchange algebras are as basic as Jacobi identities are for algebras, thus as ubiquitous.[h] Not surprisingly, there is a recent surge of interest in this topic related to the string theory.

Currently this paper has about 1020 citations, averaging about 26/year and in recent years staying over 30/year, above the average. In 2007, the number of citations surged to 43. There are many citations of Yang-Baxter or Yangian without formally citing the paper.

For other information about this paper, see the talk by M.L. Ge.

Yang-Yang 1969, J. Math. Phys. 10, 1115–1122, "Thermodynamics of a One-Dimensional System of Bosons with Repulsive Delta-Function Interaction"

Impressively, a recent experimental paper appeared with the title "Yang-Yang Thermodynamics on an Atom Chip", Phys. Rev. Lett. 100, 090402 (2008) by van Amerongen *et al.* It found agreement with the numerical

---

[h] For an overview and references to original journal publications, see L.-L. Chau, "Yang-Mills, Yang-Baxter, and Yangian Algebra: Quantum-Group Symmetry, Quantum Bi-Module Field, and Quanta Bi-Modulons", Chinese J. Phys. **32**, 535–548 (1994), in honor of the retirement of Professor Ta-You Wu from the presidency of Academia Sinica.

calculations — by K.V. Kheruntsyan *et al.*, Phys. Rev. Lett. 91, 040403 (2003) — of the equations given in this 1969 Yang-Yang paper.

Currently this paper has about 600 citations, averaging about 18/year and in recent years staying about 25/year, above the average. In 2007, the number of citations surged to 28.

For additional important papers by Yang in many-body physics, see the talk by F.Y. Wu at this conference.

3.3. *Astrophysics/cosmology*

Yang did not have any paper directly in this field. However, because astrophysics/cosmology uses particle physics and many-body physics together with Einstein's general relativity, it is impacted by Yang's papers.

4. Outlook

The 20th Century had seen fantastic progress in the physical sciences: establishment of quantum phenomena and description; rational understanding of many-body physics and observations of many striking phenomena; development of the Standard Model of electroweak and strong interactions and its experimental confirmation; profound advancement of our understanding of spacetime and the observation of the expansion of the Universe and the imprint of the Universe at around 300,000 years after the Big Bang which occurred about 14 billion years ago; etc.

The 21st Century began with continued confirmations of the striking observation made toward the end of the 20th century: the matter that the stars and you and I are made of constitutes only about 5% of the mass/energy of the Universe, while dark matter takes up 20% and dark energy takes up 75% and is responsible for the acceleration of the expanding Universe! We are facing many exciting challenges.

What is quantum gravity?
How do particles get their masses?
How does the lack of anti-matter in the universe come about?
What is the microscopic description for high Tc superconductivity?
What is dark matter?
What is dark energy?

Whatever the future developments will be, Yang's papers will continue to be relevant and contributing because they have brought about discoveries of the foundations in physics.

Nature, in Her precise language of mathematics, constructs interactions according to symmetry. However, symmetry does not dictate interaction uniquely. Interestingly, Nature always chooses the simplest possibility allowed by symmetry, e.g., the actions for electroweak, strong, and gravity are the simplest possible ones; the groups established in the Standard Model are the simplest Lie groups, $U(1) \otimes SU(2) \otimes SU(3)$. Nature uses symmetry breaking to make our world and the Universe more interesting: e.g., the violation of the global symmetries of P, CP, T in weak interactions; the violations of local gauge symmetries in the phenomena of superfluidity/superconductivity and in the mass generations in the Standard Model. I would like to show my favorite artist's painting which I consider to have wonderfully captured Nature's use of symmetry and symmetry breaking to make our world beautiful and interesting, Fig. 3.

To describe the situation, I modify a quote from my Nobel Laureate classmate, David Gross (mentored by Geoffrey Chew as thesis adviser), in celebrating Yang 70, "Symmetry dictates interactions and Yang dictates symmetry", to "Symmetry plus simplicity dictates interactions, Einstein and Yang dictate symmetry, but who/what dictate symmetry breaking?" A (permuted) quotation of Einstein says it well, "Malicious the Lord is not, but subtle He is!"

We eagerly want to find out more about how exactly Nature does all these. I like to hold the philosophical view that all are out there in Nature: mathematics, music, etc., as well as whatever the human brain can come up with. I concocted a picture for this point of view, Fig. 4. This point of view resonates with what Michelangelo had said about his marble sculpturing (which I am paraphrasing), "The figures are already in the marble, all I have to do is to chip away the unnecessary." Let us follow the role models of Einstein and Yang, be all each of us can be, chip it away and find out more of Nature's beautiful simplicities, primordial sources, and subtleties.

Acknowledgements

I would like to give my heart-felt thanks to Karen Andrews, William and Gloria Rogers, and Weiben Wang (my son) for reading the manuscript and giving valuable and encouraging comments.

Attachment 1:

Transcript of Notes by C.N. Yang,

Written in March 1947 at University of Chicago

Ling-Lie Chau
Physics Department
University of California, Davis, CA95616

Chau's note: I made the transcript of the Notes for the convenience of the reader. I referred to it when I discussed the 1954 Yang-Mills paper in my talk at the Conference celebrating Professor Yang's 85th birthday in Singapore. I thank Professor Yang for proof reading the transcript and acknowledging that he did not find any mistakes in the transcript given here, March 2008.

The photocopy of these Notes was published in ≪50 Years of Yang-Mills Theory≫, edited by G. 't Hooft. In the introduction to the photo copy of these notes, Professor Yang gave the following references:

[1] W. Pauli, *Handbuch der Physik* **24** (1933),
[2] W. Pauli, *Rev. Mod. Phys.* **13**, 203 (1941),
[3] H. Weyl, *Zeitschr. f. Phys.* **56**, 330 (1929).

Gauge Invariance and Interaction

I. The gauge invariance in electromagnetic theory serves for two purposes:

(i) Give rise to the definition of a charge and current density,

(ii) Fix the interaction between an arbitrary field and the photon field by the requirement of the gauge invariance.

By the second purpose, e.g. a real field cannot interact with the photon. (But notice that a complex field that does not undergo a change under a gauge transformation will not interact with the photon either, e.g. the neutron field belongs to this class.)

II. We can easily formulate a theory for meson gauge transformations by requiring that $U_k \to U_k e^{i\alpha\varepsilon_k}$, $U_k^* \to U_k^* e^{-i\alpha\varepsilon_k}$ and get a "meson charge and current" density, where α is a constant, (ε_k = meson charge number of the particle U_k).

In order to fix the interaction between an arbitrary field & the meson field a gauge transformation of the second kind for the meson field is necessary. This can be done in the case of neutral mesons by writing

$$\mathcal{L} = -\frac{1}{2}F_{\mu\nu}F_{\mu\nu} - K^2 U_\nu U_\nu + 2A_\nu M_\nu - 2\frac{1}{K}BM_{\nu,\nu} \quad \swarrow??$$

where A_ν is Stuckelberg's potential (c.f. Pauli, R.M.P.) & regarding A_ν & B as independent variables. Gauge transformation:

$$A_\nu \to A_\nu + \frac{1}{K}f_{,\nu} \quad .$$
$$B \to B - Kf \quad .$$

For charged mesons, the theory perhaps works too, with $\alpha =$ an operator not commutable with τ_3 (isotopic spin).

III. Notice that for photon field, $\varepsilon_k = 0$ or 1 because all particles are either uncharged or have charge e in photon field. But for meson field

$$\frac{G}{e} = \varepsilon_k \cong 3 \text{ if } U_k = \text{ heavy particle field,}$$
$$\frac{g}{e} = \varepsilon_k \cong 10^{-8} \text{ if } U_k = \text{ light particle field.}$$

IV. The gauge transformation described in II will not lead to any fixation of the interaction because the field part of the $\mathcal{L}$ and the particle part are respectively invariant under the transformation, while in the e.m. case they both vary.

This constitutes perhaps a fundamental difference between a field with $K = 0$ & with $K \neq 0$.

Masses of Particles

V. The present theory cannot give a satisfactory account of the masses of the different particles because of infinite difficulties and incompleteness of the theory, e.g. we cannot expect to be able to derive the mass ratio of the proton and the electron, of the proton to the neutron in the present theory.

In a completely satisfactory theory there should be only three universal constants: $\hbar$, c, G (gravitational constant) which in natural units are all 1. The mass or Compton wave lengths of the particles can be derived.

We see that a theory of such kind is necessarily a merge-together of the Q. theory & general relativity (in which the masses of the particles are defined).

Point worth noticing: if electron radius = range of meson forces, $\mathrm{M}_{\mathrm{meson}} = 137$ m.

Gauge Invariance, Charge Density & Neutral Particles

* If we separate each complex wave function into its real and imaginary parts,

$$U = v + iw,$$

the gauge transformation becomes

$$v \to v + \alpha w$$

$$w \to w - \alpha v.$$

$\frac{\delta \mathcal{L}}{\delta \alpha} = 0$ represents gauge invariance & leads to a current S_μ with divergence $= 0$,

$$S_\mu = \text{ const. } [\frac{\partial \mathcal{L}}{\partial v_K} w - \frac{\partial \mathcal{L}}{\partial w_K} v].$$

* For particles of spin 0 & 1, L has terms $v \cdot v$, $w \cdot w$ but no vw.

$\therefore$ $S_\mu =$ term vw.

$\therefore$ $S_\mu = 0$ when $v = 0$ or $w = 0$. which gives Neutral particles

, " , " " " $\frac{1}{2}$, L has the therm $v \cdot w$, no $v \cdot v$, $w \cdot w$

$\therefore$ $S_\mu =$ term vv and ww.

$\therefore$ $S_\mu \neq 0$ when $v = 0$ or $w = 0$. Thus we has no neutral particle.

If, however, we adapt Heisenberg's recipe for passing to q-no. theory S_μ have $v \cdot w$ term again & we can have neutral particles. But this seems unsatisfactory because if put $w = 0$ from the beginning we don't get an $\mathcal{L}$. (See Pauli p.225, 228)

Attachment 2: *This is an article I wrote for the celebration of Professor Yang's 80th birthday. In it I thanked Professor Yang for the important physics I learned from him, related an interesting encounter with Eugene Wigner, and pointed out that the commonly called Wilson-loop ought to be called Yang-Wilson-loop, based upon my research on the contents and the timing of the publications of their papers. This article was distributed but not published. I attach it here as a reference for my discussion of the Yang 1974 paper in the main text.*

A Forgotten Treasure of Yang's Pyramid

Ling-Lie Chau
Physics Department
University of California, Davis, CA 95616

On the occasion of celebrating Professor Yang's 80th birthday, I would like to express my deep gratitude to him.

During seventeen years (1969–1986) of working at Brookhaven National Laboratory and living at Setauket, which is closer to Yang's institute in the State University of New York at Stony Brook than the Lab, I benefited tremendously from Professor Yang and from his institute (which was designated to be "C.N. Yang Institute for Theoretical Physics" since his retirement in 1999). Besides welcoming me to attend seminars and to interact with visitors from all over the world at his institute, he generously allowed me to learn physics directly from him. He would spend hours discussing physics with me in his always superbly organized and elegant office, answering questions and giving guidance. The amazing thing was that he could always find the references, the books or the papers he was looking for. What he provided at his institute helped to broaden and deepen my appreciation and understanding of physics and the related mathematics. Actually, it was not my unique experience. Many other scientists had a similar experience. The environment he provided at his institute was truly unique in the world of frontier physics research. I still do not know how he managed to do it.

An anecdote can demonstrate the importance of what I learned through osmosis from his institute. In 1980 I went to the Erice School as one of the speakers. When I first arrived and went to the top part of the building of the school, which had a beautiful view of the blue Mediterranean Sea, for some reason only Professor Wigner was there. He recognized me from my Princeton days at the Institute of Advanced Study (1967–1969). We started to chat. He asked what I was going to talk about. After hearing the titles of my two scheduled talks, he asked me in his usual overtly polite manner, "Why can't electromagnetism be completely described by the fields? Why do we need to use the gauge potentials?" The answer to his question seemed to me obvious because of what I had learned from the discussions at Yang's institute. However, before I gave the answer to Wigner, my mind quickly turned up the story told by Professor Sam B. Treiman at one of the parties at his home in Princeton. Wigner was famous for going around to ask professors at Princeton his questions in his famous superpolite manner. However, people learned that it was dangerous to answer Wigner's questions. Often, one could easily make a fool of oneself, because Wigner usually had already thought very deeply about his questions. So Treiman said that he never answered Wigner's question about dispersion relations. He just kept on returning Wigner's almost-90-degree bows and asked, "What would you like to know about dispersion relations, Professor Wigner?"

Despite my memory of Treiman's warning, I ventured to give my answer: "Professor Wigner, it is because of the Aharonov-Bohm effect." [Y. Aharonov and D. Bohm, Phys. Rev. 115 (1959) 485–491.] He abruptly straightened his back and gave me a 90-degree bow, "Thank you, I have had my lesson today." He then walked away. I did not quite know what to make of it and we never talked about it again during the conference. To my total surprise, in Wigner's summary talk for the conference he acknowledged me for reminding him of the Aharanov-Bohm effect. Indeed, he had thought more about the problem than I had. When I returned to Brookhaven, I mentioned the story to Dr. Maurice Goldhaber. Maurice said, "Wigner will not acknowledge you in writing." I did not ask what prompted Maurice to say that. A long, long time afterwards, in 1983, the proceedings of the conference arrived and I just automatically put it on the shelf without looking inside, as I often did with conference proceedings. Quite sometime later, Maurice's comment reappeared in my mind and I curiously looked up Wigner's written version of his talk in the proceedings: International School of Subnuclear Physics 1980, Erice, Italy, "The High Energy Limit," Ed. A. Zichichi, Plenum Press. There it was, Wigner acknowledged me in writing! What Wigner said in his summary talk "What We Have Learned," 1065–1073 of the Proceedings, about electromagnetism was very insightful. I would like to quote it here:

"Let me recall that when I learned about the description of the electromagnetic field by a gauge quantity, the electromagnetic potential, I did not like it. The gauge-like quantities cannot be measured and have, therefore, no 'reality.' Thus, for instance, one can add an arbitrary gradient to the electromagnetic potential without changing any of its effects. This shows that at least in classical theory, it cannot be an observable quantity and this remains true in quantum theory. Since I had an aversion to gauge theories, it was good that Dr. Chau reminded me of the article of Aharonov and Bohm which showed the usefulness of the gauge concept. It showed that if the electromagnetic field is described by a gauge-field, its effect on the charged particle traveling through it is local — if it is described by the electromagnetic field, electric and magnetic, no local description is possible. The interference of two electron beams which originated from a single beam is not uniquely given by the electric and magnetic fields through which they travel — the fields between the two beams also have an effect. However, if the field is described by the electromagnetic potential, the description is local — the potential difference between two points contains implicitly the electromagnetic field between those two points. Thus in quantum mechanics we have to choose between a non-local [effect] by an observable field on the one side, and a local description of the effect by a non-observable field. Experience shows that the second description is much easier and it is reasonable, therefore, to adopt it. This has been done."

This year, 2002, is Wigner's 100-year anniversary. On this occasion of the remembrance of Wigner, I am thankful to Wigner for his kindness and for giving me that valuable experience.

As insightful as Wigner was, clearly he did not read or think of what Yang wrote in his 1974 paper, "Integrable Formalism for Gauge Fields," Phys. Rev. Lett. 33 (1974) 445–447, in which Yang clarified and emphasized what are the essential quantities that describe the local gauge theories — not the fields, not the gauge potentials, they are the non-integrable phase factors, which manifest the Aharonov-Bohm effect in electromagnetism.

In the late 1970's, through seminars and talks, I learned from Yang's institute about the Aharonov-Bohm effect and its importance. It became so obvious to me that I could answer Wigner's question. However, not only did I not learn about the effect in my student days, most graduate students even now do not learn it. Most quantum mechanics and electromagnetism textbooks do not mention the effect. I have taken what I learned from Yang and Yang's institute by heart and have always taught it when I teach quantum mechanics and mathematical physics. (I have yet to teach a course on electromagnetism. When I do, I will teach it too.) It is so easy and beautiful for students to understand, and it is so important as Yang has argued and advocated.

As it is well known, the importance of these phase factors was also independently noticed by K.G. Wilson, "Confinement of Quarks," Phys. Rev. D 10 (1974) 2445–2459. I checked the days of submission and publication of the two papers: Yang's was submitted on June 10, 1974 and published on August 12, the same year; Wilson's was submitted on June 12, 1974, and published on October 15, the same year. They were amazingly close. Neither of them knew each other's paper. (I recently checked with Professor Yang and he confirmed it and indicated that he never had any communications on the topic with Wilson, before or after the publication of the two papers.)

Wilson's paper has been universally quoted. For some reason Yang's paper and point of view have not been mentioned when people quote Wilson's paper. So I have started to mention to people that the "Wilson loop" ought to be called "Yang-Wilson loop," in that name order because Yang's paper clearly stated and emphasized that the "loops," i.e., the non-integrable phase factors, are the defining quantities for local gauge theories; while Wilson's paper mainly discussed the utilities of the loops.

"Yang's Pyramid" was the title of my paper contributing to the proceedings of Professor Yang's 70th birthday celebration: "Chen Ning Yang, A Great Physicist of the Twentieth Century," Ed. C.S. Liu and S.-T. Yau, International Press. Thus originates the title of this essay.

I have many other things to thank Professor Yang for. They will require some length to write. I think I will save them for the celebrations of his 90th, 100th, $\cdots$, birthdays.

Fig. 1. The photo is from the Washington College Magazine, Summer 1999 issue, taken by Jim Graham of Washington College '81 on the occasion of Professor Yang (6th from the right) being honored with the Honorary Degree in Science and Mr. J.F. Kennedy Jr. (2nd from the right) giving the commencement speech. Fourth from the right is Professor John Toll, the then President of Washington College. His presidency (1965-1978) at SUNY, Stony Brook had much to do with Professor Yang's move in 1965 from Institute for Advanced Study at Princeton to become the Albert Einstein Professor of Physics of SUNY, Stony Brook and the first director of the now C.N. Yang Institute for Theoretical Physics.

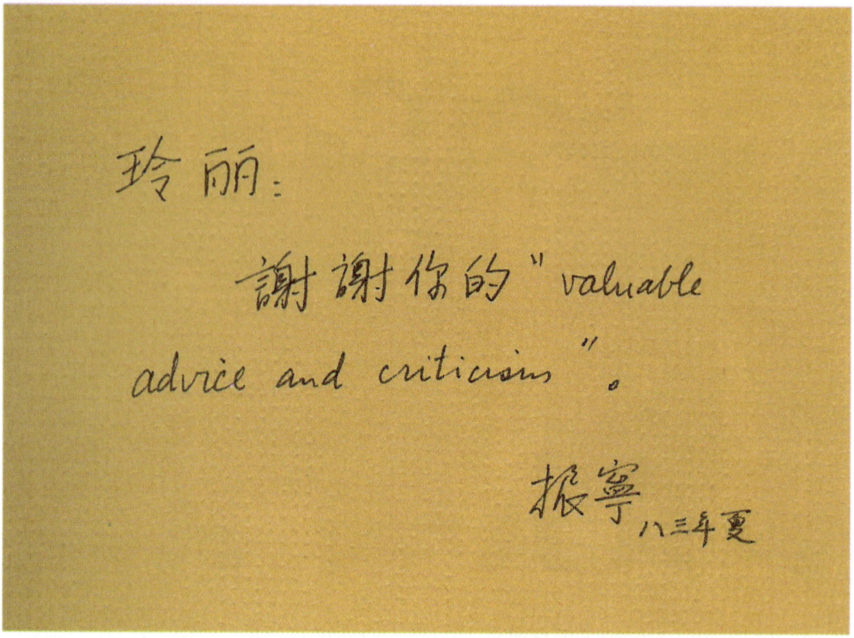

Fig. 2. Professor Yang's handwritten comments on Chau's copy of his ≪Selected Papers, 1945-1980, With Commentary≫. English translation of the comments: <Ling-Lie: Thank you for your "valuable advice and criticism". Chen Ning, '83 summer.>

Fig. 3. Georgia O'Keeffe's 1935 painting, "Ram's Head, White Hollybock – Hills", image from the web page of Brooklyn Museum (Bequest of Edith and Milton Lowenthal). I consider it to have wonderfully captured Nature's use of symmetry and symmetry breaking to make our world beautiful and interesting.

Fig. 4. I concocted this picture to convey my philosophical view that all are out there in Nature: mathematics, music, etc., as well as whatever the human brain can come up with. The background image is from the web page "Inside the Eagle Nebula" of T.A. Rector and B.A. Wolpa, NOAO/AURA.

KIDNEY-BOOJUM-LIKE SOLUTIONS AND EXACT SHAPE EQUATION OF LIPID MONOLAYER DOMAINS

MITSUMASA IWAMOTO

Department of Physical Electronics, Tokyo Institute of Technology, Tokyo, 2-12-1 O-okayama, Meguro-ku, 152-8552, Japan

FEI LIU

Center for Advanced Study, Tsinghua University, Beijing, 100084, China

ZHONG-CAN OU-YANG†

Institute of Theoretical Physics, The Chinese Academy of Sciences, Beijing, P.O. Box 2735, 100080, China
† oy@itp.ac.cn

The shape of the solid lipid monolayer domain surrounded by a fluid phase is of considerable interest from both physical and mathematical points of view. Here, we report two new results about this topic. First, we find an exact analytical solution to an approximated shape equation that we derived recently. This solution can well-describe the kidney- and boojum-like domains that abound in lipid monolayer. Second, we derive an exact domain shape equation by a direct variation of domain energy without any artificial cutoff. We find no continuous solutions that satisfy this shape equation due to the divergence of its coefficients, which is rooted in the continuous description of electrostatic dipoles.

Keyword: Solid lipid monolayer domain; energy variation; shape equation; Neumann's formula.

1. Introduction

In the past two decades, due to the techniques of fluorescence microscopy [FM][1,2] and Brewster angle microscopy (BAM),[3] the experimental study of shape transition in lipid monolayer domains at the air-water interface has revealed a diversity of domain shapes, from round structures to complex labyrinthine patterns,[4–10] and nowadays, becomes the main project of shape

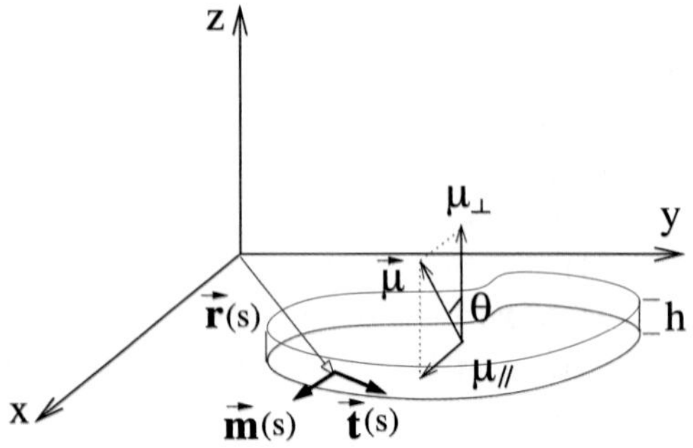

Fig. 1. Schematic illustration of the shape of a monolayer domain, whose dipoles $\vec{\mu}$ uniformly tilt along the x-direction with the same tilted angle θ. The geometric quantities describing its boundary curve are also shown, and h is the thickness of the monolayer.

science. Theoretical understanding of lipid monolayer domain formation is based on competition between electrostatic repulsion and line tension proposed by Andelman *et al.*,[11,12] and is investigated extensively by the McConnel group[13–15] and many others.[16–20] Line tension γ (>0) acts to minimize the domain boundary $\vec{r}(s)$ (s being the arc-length element) and in contrast, the repulsive dipole-dipole interaction tends to separate the dipoles in maximal. Therefore, the considerable task of the mentioned works is the calculation of the energy

$$F = \gamma \oint dl + F_\perp + F_\| \tag{1}$$

for comparison of the energies of various domain shapes, or for evaluation of the stability of a given shape. $F_\perp$ and $F_\|$ are the components of dipole-dipole interaction energy normal and parallel to the water surface, respectively, and can be written as the following double-line integrals:

$$F_\perp = -\frac{\mu_\perp^2}{2} \oint \oint \frac{\mathbf{t}(l) \cdot \mathbf{t}(s)}{|\mathbf{r}(l) - \mathbf{r}(s)|} dl ds\,, \tag{2}$$

and

$$F_\| = \frac{\mu_\|^2}{2} \oint \oint \frac{(\mathbf{t}(l) \cdot \hat{y}_0)(\mathbf{t}(s) \cdot \hat{y}_0)}{|\mathbf{r}(l) - \mathbf{r}(s)|}\,, \tag{3}$$

where $\vec{t} = d\vec{r}/ds = (\cos\phi(s), \sin\phi(s))$ is the unit tangential vector, $\hat{y}_0$ is the unit vector of the y axis and μ_1 and μ_2 are the dipolar density components normal and parallel to the monolayer. Here, all dipoles are supposed to tilt uniformly along the x axis with constant angle θ, i.e., $\vec{\mu} = (\mu_\perp, 0, \mu_\|) = \mu_0(\sin\theta, 0, \cos\theta)$; (see Fig. 1).

Recently, our works[21,22] showed that the double-line integrals in Eq. (1) could indeed be well-approximated by a sum of an additionally negative line tension and a curvature-elastic energy of the domain boundary. An approximated shape equation was derived by a variation of the energy. Many important physical properties and experimental phenomena about the two-dimension (2D) monolayer domain were understood analytically and quantitatively. Although significant progress was made, we still face the core problems in the domain shape studies i.e., Are there analytical solutions to the approximated shape equation? Do the solutions really predict the shapes observed in experiments[6–8,10,23,24] and can we obtain the exact shape equation and its solutions without any approximation? Here, we present our latest efforts to answer these questions.

2. The Approximated Shape Equation and its Exact Solution

We first briefly review the approximation of Eq. (1) and its corresponding shape equation. The details can be found in our previous works.[21,22]

In order to take into account the difference in Gibbs free energy density of g_0 between the outer (fluid) and inner (solid) phases and the domain shape transition induced by surface pressure of compression, Π, we introduced an additional term $\Delta P \oint dA$ in Eq. (1); here $\Delta P = \Pi - g_0$ and $g_0 > 0$ for the solid phase is more stable than the fluid one. Employing the Frenet formulas in the plane and the Taylor's expansion of $\vec{r}$ with respect to the arclength, we obtained the approximation of the two energy terms:

$$F_\perp \approx -\tfrac{\mu_\perp^2}{2} \ln \tfrac{L}{h} \oint ds + \tfrac{11}{96}\mu_\perp^2 L^2 \oint \kappa(s)^2 ds\,, \tag{4}$$

and

$$F_\parallel \approx \frac{\mu_\parallel^2}{2} \ln \frac{L}{h} \oint \sin^2 \phi(s) ds - \frac{1}{192}\mu_\parallel^2 L^2 \oint [11 + 13\cos 2\phi(s)]\kappa(s)^2 ds\,, \tag{5}$$

where $\kappa(s) = -d\phi(s)/ds$ is the curvature of the boundary at s, L and h represent the boundary length and the nonzero monolayer thickness of the domain as a cutoff, respectively; (see Fig. 1). To obtain the shape equation, we derived the first variation of the approximated domain energy and had

$$\Delta P - \Lambda\kappa + \alpha\kappa^3 + \beta\kappa_{ss} + \sigma\kappa\kappa_s = 0\,, \tag{6}$$

where $\kappa_{ss} = d^2\kappa(s)/ds^2$, and the definitions of the coefficients are

$$\begin{aligned}
\Lambda &= \gamma - \frac{\mu_\perp^2}{2}\ln\frac{Le}{h} + \frac{\mu_\parallel^2(1+3\cos 2\phi)}{4}\ln\frac{L}{h} + \frac{11}{48}\mu_\perp^2 L\oint \kappa(s)^2 ds \\
&\quad + \frac{\mu_\parallel^2}{2L}\oint \sin^2\phi(s)ds - \frac{1}{96}\mu_\parallel^2 L\oint(11+13\cos 2\phi)\kappa(s)^2 ds\,, \\
\alpha &= \frac{11}{96}\mu_\perp^2 L^2 + \mu_\parallel^2 L^2\frac{39\cos 2\phi - 11}{192}\,, \\
\beta &= \frac{11}{48}\mu_\perp^2 L^2 - \mu_\parallel^2 L^2\frac{11+13\cos 2\phi}{96}\,, \\
\sigma &= -\frac{13}{24}\mu_\parallel^2 L^2 \sin 2\phi\,,
\end{aligned} \tag{7}$$

respectively. Equation (6) and its coefficients can be further simplified if we are only concerned about the case of $\mu_\parallel = 0$:

$$\Delta P - \Lambda_0\kappa + \alpha_0\kappa^3 + 2\alpha_0\kappa_{ss} = 0\,, \tag{8}$$

where

$$\begin{aligned}
\alpha_0 &= \frac{11}{96}\mu_\perp^2 L^2, \\
\Lambda_0 &= \gamma - \frac{\mu_\perp^2}{2}\ln\frac{eL}{h} + \frac{11}{48}L\mu_\perp^2\oint\kappa(s)^2 ds\,.
\end{aligned} \tag{9}$$

At first sight, both Eqs. (6) and (8) are highly nonlinear and difficult to solve analytically. We have studied them only by substituting a circular domain and investigating the stable and unstable conditions between the radius of a circle and the pressure ΔP and line tension γ. Here, we will show that Eq. (8) has an exact solution if $\Delta P = 0$. In physics, the vanishing ΔP means that the solid and fluid phases of the lipid monolayer coexist and the π_c-plateau is present in the $\Pi - A$ isotherm experiment.[6,7]

Substituting geometrical relations of $\kappa_s = d\kappa/ds = -(1/2)d\kappa^2/d\phi$ and $\kappa_{ss} = (1/2)\kappa d^2\kappa^2/d\phi^2$ into Eq. (8), and multiplying both sides of the equation with $d\kappa^2/d\phi$, we obtain its first integral

$$\left(\frac{d\kappa^2}{d\phi}\right)^2 = \left(\frac{2\Lambda_0}{\alpha_0}\right)\kappa^2 - \kappa^4 + C\,, \tag{10}$$

where C is a constant of integration. For a domain, $\vec{r}(s)$ is a finite closed curve and κ may take extreme values at some points at which $d\kappa^2/d\phi \propto$

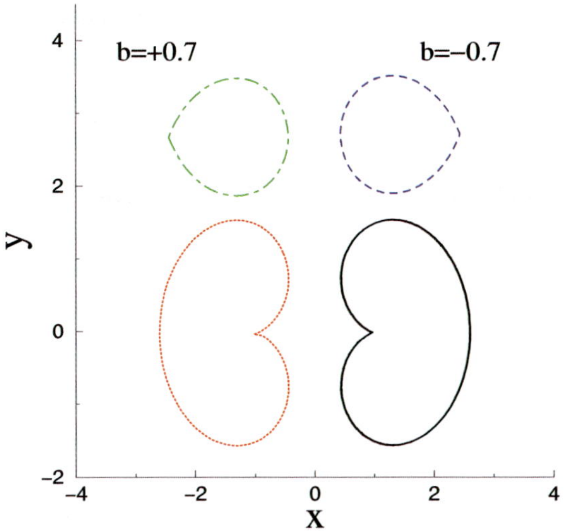

Fig. 2. Illustration of boojum- (upper) and kidney-like (low) shapes calculation by Eq. (11) with $a = 1$, $b = \pm 0.7$ and $\pm\phi(0)$. The different signs of b gives the left- and right- "kidneys" and "boojums", respectively.

$\kappa_s = 0$. By choosing $s = 0$ at a certain extreme point and letting $\phi(0) = \pi/2$, we have the apparent solution of

$$\kappa(s) = -\frac{d\phi}{ds} = -a\sqrt{1 + b \sin \phi(s)}\,, \tag{11}$$

where $a = \sqrt{\Lambda_0/\alpha_0}$, $b = (\kappa_0/a)^2 - 1$ and $\kappa_0 = \kappa(0)$. Obviously, if $b = 0$, then the above equation gives a circle with radius a. Moreover, Eq. (11) can describe both round shapes with convex or concave cusp. As illustrated, Fig. 2 shows numerical results of the equation with $a = 1$, $b = \pm 0.7$, and $\pm\phi(0)$. Here, we have chosen $s = 0$ at the cusp of the boundary and $\vec{m}(0)$ along y. We see that $\phi(0)$ with opposite signs can yield very distinct domain shapes (the upper and lower curves in the figure). These shapes were indeed observed at the Π_c plateau in the dipamitoylphosphatidylcholine (DPPC) monolayer experiments,[6] and the upper shapes in Fig. 2 were called boojum-like therein. Hence, the solution (11) is not only exact, but also satisfactorily predicts the interesting shapes observed in the experiment.

3. Exact Shape Equation

To our best knowledge, no exact shape equation derived by direct variation of the free energy equation (1) has been formally reported in the literature.

One of the possible reasons is that this free energy is in fact divergent. In mathematics, it can be readily seen from the presence of $1/|\mathbf{r}(l) - \mathbf{r}(s)|$ in the integrals. This nonphysical result arises from the use of a continuous description of dipoles distribution. Interestingly, this circumstance is not new in electromagnetics in that $F_\perp$ has the same formula as the Neumann's formula for the coefficient of self-inductance of a single coil except for the prefactors. To prevent the divergence of Eq. (1), in all existing literature, including the references, different kinds of cutoff in the distance were employed.[13–17,27] For instance, the McConnel group modified the free energy formulas by changing $|\vec{r}(l) - \vec{r}(s)|$ to $\sqrt{(\vec{r}(l) - \vec{r}(s))^2 + \Delta^2}$.[14,15] On the other hand, however, a divergent energy formula does not mean the absence of finite shape equation. In the following, we derive a general shape equation by a direct variation of $\delta F = 0$. We think that this derivation is meaningful. First of all, no exact shape equation of the solid domain has been obtained until now. In all previous works, the discussed equilibrium domain shapes were either predetermined according to observation, or were derived by approximated free energies, as we did previously. Second, an exact shape equation would be useful in practice, e.g., to investigate whether a well-defined shape equation exists, to evaluate the quality of various approximation schemes. Finally, we cannot exclude the possibility that the general shape equation can be solved in future.

To derive the shape equation for the equilibrium domain with $\vec{r}(s)$, we consider the slightly distorted domains

$$\vec{r}'(s) = \vec{r}(s) + \psi(s)\vec{m}(s)\,, \tag{12}$$

where $\vec{m} = (\sin\phi(s), -\cos\phi(s))$ is the unit outward normal vector of $\vec{r}(s)$, and $\psi(s)$ is a sufficiently small smooth variation function. With the help of the formulas shown in our previous work[22] (see the Appendix therein), the first variations of $F_\perp$ and $F_\parallel$ are

$$\delta^{(1)} F_\perp = -\mu_\perp^2 \oint\oint \frac{[\mathbf{r}(l) - \mathbf{r}(s)] \cdot \mathbf{m}(l)}{|\mathbf{r}(l) - \mathbf{r}(s)|^3} \psi(s) dl ds \tag{13}$$

and

$$\delta^{(1)} F_\parallel = \mu_\parallel^2 \oint\oint \frac{[\mathbf{r}(l) - \mathbf{r}(s)] \cdot \hat{x}_0 \mathbf{m}(l) \cdot \hat{x}_0}{|\mathbf{r}(l) - \mathbf{r}(s)|^3} \psi(s) dl ds\,, \tag{14}$$

respectively, where $\hat{x}_0$ is the unit vector of the x axis. The first variations of $\gamma \oint dl$ and $\Delta P \int dA$ are direct and they are $-\gamma \oint \kappa(s)\psi(s)ds$ and $\Delta P \oint \psi(s)ds$, correspondingly.[22] If a curve $\vec{r}(s)$ describes a boundary of an equilibrium domain, the first variation of its corresponding free energy must vanish, namely, $\delta^{(1)}F = 0$ for any infinitesimal function $\psi(s)$. Hence, combining these variation equations, we finally obtain the exact shape equation as

$$\Delta P - \gamma\kappa(s) - \mu_{\perp}^2 \oint \frac{[\mathbf{r}(l) - \mathbf{r}(s)] \cdot \mathbf{m}(l)}{|\mathbf{r}(l) - \mathbf{r}(s)|^3} dl + \mu_{\parallel}^2 \oint \frac{[\mathbf{r}(l) - \mathbf{r}(s)] \cdot \hat{x}_0 \mathbf{m}(l) \cdot \hat{x}_0}{|\mathbf{r}(l) - \mathbf{r}(s)|^3} dl = 0 \,. \tag{15}$$

This equation is a differential and integral equation, and seems to be impossible to solve analytically. Moreover, the presence of $1/\left|\mathbf{r}(l) - \mathbf{r}(s)\right|^3$ there implies that the exact shape equation might be ill-defined for any continuous boundary curves. As an illustration, we substitute a test circular solution of $\mathbf{r} = \rho_{\mathbf{0}}(\sin\phi(\mathbf{s}), -\cos\phi(\mathbf{s}))$ with radius ρ_0, Equation (15) gives the following relation:

$$\Delta P + \gamma\rho_o^{-1} - \mu_{\perp}^2 2^{-3/2}\rho_0^{-1}\alpha + \mu_{\parallel}^2 2^{-3/2}\rho_0^{-1}\beta \sin\phi(s) = 0 \,, \tag{16}$$

with two infinite integrals $\alpha = \int_0^{2\pi}(1/\sqrt{1-\cos\theta})d\theta$ and $\beta = \int_0^{2\pi}(\cos\theta/\sqrt{1-\cos\theta})d\theta$. Hence, Eq. (16) reveals that the circle is not a rigorous solution to Eq. (15). Even if the exact shape equation (15) has some flaws in definition, it is still exact if we introduce a certain cutoff in the integrations, which is always plausible in physics. Of course, the potential of the equation in application would largely depend on whether we can solve it numerically or analytically.

4. Conclusion

In this work, we report two new results in the study of the shapes of solid lipid monolayer domains at the air-water interface. We find an exact analytical solution to an approximated shape equation. This solution can well-describe the kidney- and boojum-like domains that abound in lipid monolayers. In addition, we first derive an exact domain shape equation by a direct variation of the domain energy without any approximation. Our results would be useful in the real experimental studies of the solid domain.

References

1. V. Tscharher and H. M. McConnel, *Biphys. J.* **36**, 409 (1981).
2. M. Losche, E. Sackmann and H. Mohwald, *Ber. Bunsen-Ges. Phys. Chem.* **87**, 848 (1983).
3. S. Henon and J. Meunier, *Rev. Sci. Instrum.* **62**, 936 (1991).

4. C. W. McConlogue and T. K. Vanderlick, *Langmuir* **13**, 7158 (1991).
5. C. W. McConlogue and T. K. Vanderlick, *Langmuir* **15**, 234 (1999).
6. K. Kjaer and J. A. Niesen, *Phys. Rev. Lett.* **58**, 2224 (1987).
7. K. J. Stine and D. T. Stratmann, *Langmuir* **8**, 2509 (1992).
8. D. K. Schwartz, M. W. Tsao, C. M. Knobler and J. C. Knobler, *J. Chem. Phys.* **101**, 8258 (1994).
9. M. Weis and H. M. McConnel, *Nature (London)* **310**, 47 (1984).
10. M. Weis and H. M. McConnel, *J. Phys. Chem.* **89**, 4453 (1985).
11. D. Andelman, F. Brochard, P. G. De Gennes and J. F. Joanny, *Acad. Sci. Ser. C* **301**, 675 (1985).
12. D. Andelman, F. Brochard and J. F. Joanny, *J. Chem. Phys.* **86**, 3673 (1987).
13. D. J. Keller, J. P. Korb and H. M. McConnel, *J. Phys. Chem.* **91**, 6417 (1987).
14. H. M. McConnell and V. T. Moy, *J. Phys. Chem.* **92**, 4520 (1988).
15. H. M. McConnel, *J. Phys. Chem.* **94**, 4728 (1990).
16. T. K. Vanderlick and H. Mohwald, *J. Phys. Chem.* **94**, 886 (1990).
17. M. A. Mayer and T. K. Vanderlick, *Langmuir* **8**, 13131 (1992).
18. J. M. Deutch and F. E. Low, *J. Phys. Chem.* **96**, 7097 (1990).
19. J. M. Miranda, *J. Phys. Chem. B* **103**, 1303 (1999).
20. J. M. Miranda and M. Widom, *Phys. Rev. E* **55**, 3758 (1993).
21. M. Iwamoto and Z.-C. Ou-Yang, *Phys. Rev. Lett.* **93**, 206101 (2004).
22. M. Iwamoto, F. Liu and Z.-C. Ou-Yang, *J. Chem. Phys.* **125**, 224701 (2006).
23. W. H. Heckel and H. Mohwald, *Ber. Bunsenges. Phys. Chem.* **90**, 1159 (1986).
24. C. W. McConlogue and T. K. Vanderlick, *Langmuir* **13**, 7158 (1997).
25. *Die mathmatischen Gesetze der inducirten eekrischen strome*, Berlin Acad. Wissen., Abh. 1845; Ostwald's Klassiker d. exakten Wissen, W. Engeman, Leipzig, 1892, No. 36, P. Moon and D. E. Spencer, Foundations of Electrodynamics (D. Van Nostrand Co., New Jersey, 1960).
26. W. Pauli, *Pauli Lecture on Physics, Vol.1: Electrodynamics*, ed. C. P. Enz; Trans. by S. Margulies and H. R. Lewis (MIT Press, Cambridge, MA, 1979), p. 93.
27. P. D. Duncan and P. J. Camp, *J. Chem. Phys.* **121**, 11322 (2004).

REVISITING THE HYDRODYNAMIC BOUNDARY CONDITION: NEW RESULTS ON AN OLD PROBLEM*

PING SHENG

Hong Kong University of Science and Technology
E-mail: sheng@ust.hk

The no-slip hydrodynamic boundary condition (NSBC) states that there can be no relative motion at the fluid-solid interface. In spite of the lack of first-principles support, either from kinetic theory or otherwise, NSBC has been regarded as a pillar of continuum hydrodynamics for more than a century. However, over the past fifty years, it has been realized that in immiscible two-phase flows, the moving contact line (MCL), defined as the intersection of the fluid-fluid interface with the solid wall, is incompatible with the NSBC. In 1988, molecular dynamic (MD) simulations for the first time explicitly demonstrated the breakdown of NSBC in the vicinity of the MCL. Since that time it was realized that without a new framework to treat the boundary condition(s), continuum hydrodynamics would be unable to accurately model nanoscale or microscale hydrodynamic behavior. This is sometimes described as an example of the "breakdown of continuum."

By applying Onsager's principle of minimum energy dissipation, we have recently derived the boundary conditions that for the first time enabled quantitative prediction of the MD simulation results from continuum hydrodyanmics, and in so doing revised the NSBC [1]. Our results indicate that the hydrodynamic boundary condition(s) should be consistent with the general statistical mechanic principle with underlie all linear response phenomena in dissipative systems. Implications of our results will be presented.

*Work done in collaboration with Tiezheng Qian and Xiaoping Wang.

RECENT PROGRESS IN ATOMIC IONISATION THEORY

I. BRAY*, D. V. FURSA, A. S. KADYROV and A. T. STELBOVICS

ARC Centre for Antimatter-Matter Studies, Curtin University of Technology, GPO Box U1987 Perth WA 6845, Australia
**E-mail: I.Bray@curtin.edu.au*
www.curtin.edu.au

We give a brief review of the considerable progress made recently in the field of electron-impact ionisation of atoms. Such processes require the solution of three (or more) particles interacting via the Coulomb potential. Though historically the formulation of such problems has been particularly difficult, the calculated fully differential cross sections are in remarkable agreement with experiment for both H and He from near threshold through to high energies.

Keyword: Electron-impact ionisation.

1. Introduction

The Coulomb three-body problem that is electron-impact ionisation of atomic hydrogen has defied adequate formulation, let alone solution, for a very long time. While Rudge[1] suggested a formulation for calculating the ionisation amplitude it required special choices of effective charges to ensure that the phase was not divergent, but even then it was not clear that the phase was uniquely defined. Little progress on the formalities of the problem has been reported for almost four decades, but the problem is now being revisited due to the rather surprising case of computational techniques yielding extraordinary agreement with experiment despite it not being clear why this is so.

First, we have the Exterior Complex Scaling (ECS) method of Rescigno et al.,[2,3] who obtained the total wavefunction by rotating the coordinates into the complex plane and then used Rudge's integral for the amplitudes, but with $Z = 1$ Coulomb waves thereby ensuring a divergent phase. Nevertheless, despite the divergent phase, absolute agreement with measurements of the ionisation fully differential cross sections, which depend only on the magnitude of the amplitude, was obtained. Another computational approach, the convergent close-coupling (CCC) method, expands the total wavefunction using square-integrable states and associates ionisation processes with exciting positive-energy pseudostates, which are obtained by diagonalising the target Hamiltonian. The other electron is treated as a plane wave, irrespective of the energies involved. This would seem inappropriate, particularly for the cases where the outgoing electrons have the same energy, and yet excellent agreement with experiment is obtained for these and all other kinematics.

In this paper we give an overview of the CCC method, which has been developed by the authors over the last fifteen years. Though the close-coupling method was originally designed with just discrete excitation processes in mind it soon became clear that it had something to say about the ionisation processes also.

The field of Coulomb few-body physics is fundamental and applies to high-energy physics just as it does in atomic physics. In the latter case we have the advantage of the potentials being pure Coulomb and have access to an immense database of detailed experiments making this an ideal testing ground of the underlying foundations generally.

2. Theory

Though the formulation of Coulomb few-body problems is particularly problematic when two or more particles are in the continuum, when such processes are energetically accessible they also affect elastic scattering and inelastic excitation. Hence we first consider discrete excitation and then turn to ionisation.

2.1. *Excitation*

The close-coupling method goes back a long way with one of the first applications to e-H scattering being presented by Massey and Mohr.[4] The idea is to expand the total wavefunction using a few discrete eigenstates of the target and then solve the close-coupling equations, resulting from a variational formulation, for the scattering amplitudes. The expectation is that such an approach will be accurate for elastic scattering and excitation of the lowest lying discrete states at sufficiently low energies where electron flux is restricted to just a few open channels. As the computational hardware developed, account of higher energies became possible by increasing the expansion to incorporate higher lying negative- and positive-energy pseudostates obtained from diagonalising the target Hamiltonian in some square-integrable basis.[5–8] Here the idea is that the higher energy states will absorb the electron flux available which will allow for still accurate results for say elastic scattering or $n = 2$ excitation at all incident energies. Unfortunately, at the intermediate energy of 54.4 eV, where the total ionisation cross section is at a maximum, there was no agreement between any theory and experiment[9,10] for angular correlation parameters of 2p excitation. The convergent close-coupling (CCC) method[11] was developed specifically to address this problem. The CCC method improves on the earlier close-coupling with pseudostates approaches by using a Laguerre basis for the generation of target states. This is a complete basis and hence allows for the testing of convergence systematically by simply increasing the basis size.[12] While the details of the CCC theory are given elsewhere[11] here we give just an overview of the approach. We will restrict this overview to just the H target, but the method has been applied to many others also.[13]

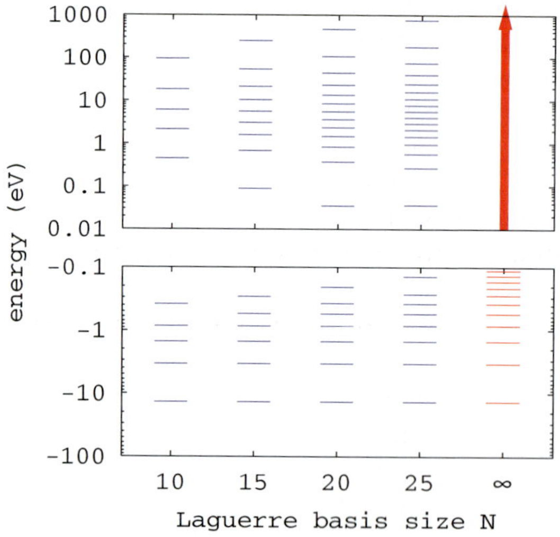

Fig. 1. Hydrogen $l = 0$ state energies for $\lambda = 1$ Laguerre bases.

Using a Laguerre basis $\xi_{nl}^{(\lambda)}(r)$, with λ being the exponential fall-off we write the N_l target states for each $l \le l_{\max}$ as

$$\phi_{nl}^{(\lambda)}(r) = \sum_{n'=1}^{N_l} C_{nl}^{n'} \xi_{n'l}^{(\lambda)}(r), \tag{1}$$

where the coefficients are obtained by diagonalising the target Hamiltonian

$$\langle \phi_f^{(\lambda)} | H_{\mathrm{T}} | \phi_i^{(\lambda)} \rangle = \varepsilon_f^{(\lambda)} \delta_{fi}. \tag{2}$$

The total number of states is $N = \sum_l N_l$. A typical example of how the state energies vary with N is given in Fig. 1.

Having defined the target state these are now used to expand the total wavefunction

$$|\Psi_i^{(+)}\rangle \approx \mathcal{A} \sum_{n=1}^{N} |\phi_n^{(\lambda)}\rangle \langle \phi_n^{(\lambda)} | \psi_i^{(+)} \rangle \equiv \mathcal{A} I_N |\psi_i^{(+)}\rangle, \tag{3}$$

where $\mathcal{A}$ is the antisymmetrisation operator. The usage of a complete basis ensures that as N is increased the operator $I_N \to I$ the identity operator. From $(H - E)|\Psi_i^{(+)}\rangle = 0$ we can define the close-coupling equations in momentum space for the transition matrix[11]

$$\begin{aligned} \langle \boldsymbol{k}_f \phi_f^{(\lambda)} | T | \phi_i^{(\lambda)} \boldsymbol{k}_i \rangle &= \langle \boldsymbol{k}_f \phi_f^{(\lambda)} | V | \phi_i^{(\lambda)} \boldsymbol{k}_i \rangle \\ &\quad + \sum_{n=1}^{N} \int d^3k \frac{\langle \boldsymbol{k}_f \phi_f^{(\lambda)} | V | \phi_n^{(\lambda)} \boldsymbol{k} \rangle \langle \boldsymbol{k} \phi_n^{(\lambda)} | T | \phi_i^{(\lambda)} \boldsymbol{k}_i \rangle}{E + i0 - \varepsilon_n - k^2/2}, \end{aligned} \tag{4}$$

where the electron of momentum $\boldsymbol{k}_i$ is incident on the target in state $\phi_i^{(\lambda)}$ leading to excitation of state $\phi_f^{(\lambda)}$ and the outgoing electron of momentum $\boldsymbol{k}_f$. The potential V contains the interaction potentials and the symmetrization operator. Eq. (4) is solved upon partial wave expansion and the T_{fi} are the scattering amplitudes used to directly calculate the experimentally measured observables.

Having developed the most sophisticated computer code to date we were ready to address the above-mentioned discrepancies between theory and experiment. Unfortunately, we found that our demonstrably convergent results were much the same as other theories and so disagreed with experiment. However, the strong foundations of the method were such that the experiments were repeated[14] and the new measurements were found to be in much better agreement with theory, see Fig. 2. Hence we can be reasonably confident that in solving Eq. (4) the T_{fi} amplitudes are reasonably accurately calculated for the discrete transitions, and so we turn to ionisation.

2.2. *Ionisation*

At energies above the ionisation threshold solution of Eq. (4) yields T_{fi} for both negative- and positive-energies ε_f. From Fig. 1 we see that with

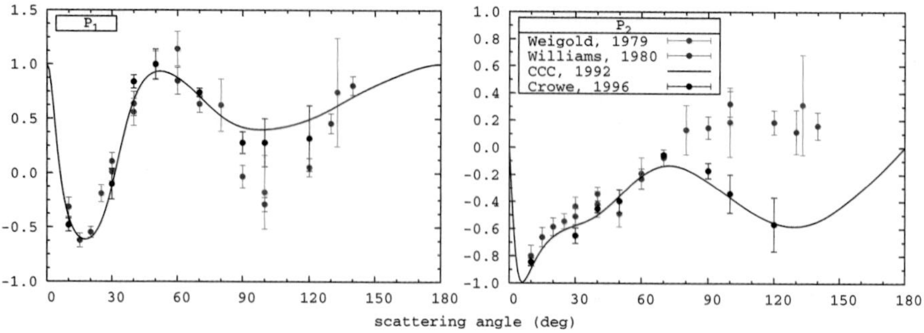

Fig. 2. The 54.4 eV e-H 2p excitation angular correlation parameters. The measurements are due to Weigold et al.,[9] Williams[10] and Yalim et al.[14] The theory is due to Bray and Stelbovics.[11] Note that the presented data and theory are just a small sample of the many theories and experiments that have considered this particular problem.

increasing basis size the negative energies (and states) converge to the corresponding true eigenstate energies (and states). However, the positive-energy states cannot converge to any true continuum eigenstate due to the square-integrable nature of the Laguerre basis. Nevertheless, it is clear from the optical theorem that if the total cross section can be obtained from the accurate elastic amplitude then the positive-energy states must correspond to the ionisation processes. The simplest thing we can do is to define the total ionisation cross section as

$$\sigma_i = \sum_{f:0<\varepsilon_f<E} |T_{fi}|^2. \tag{5}$$

In Fig. 3 we compare the CCC-calculated total ionisation cross section with experiment. The agreement with experiment is remarkable from near threshold through to high incident energies. This gave us the first hint that the CCC method contained some detailed ionisation information. The question was: how to extract it?

2.2.1. *Ionisation amplitudes*

The first step is to identify as to what might be the ionisation amplitudes arising in a theory that uses square-integrable states. Following the standard theory for discrete states we define the excitation amplitudes via[17]

$$\begin{aligned}\langle \Phi_f | \overleftarrow{H} - E | \Psi_i^{(+)} \rangle &\approx \langle \boldsymbol{k}_f \phi_f | I_N (\overleftarrow{H} - E) \mathcal{A} I_N | \psi_i^{(+)} \rangle \\ &= \langle \phi_f | \phi_f^{(\lambda)} \rangle \langle \boldsymbol{k}_f \phi_f^{(\lambda)} | T | \phi_i^{(\lambda)} \boldsymbol{k}_i \rangle,\end{aligned} \tag{6}$$

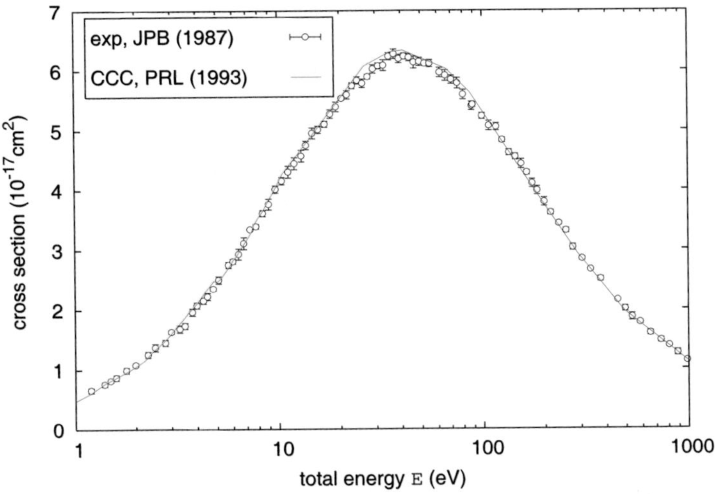

Fig. 3. The total electron-impact ionisation cross section of atomic hydrogen. The experiment is due to Shah et al.[15] and the CCC theory is due to Bray and Stelbovics.[16]

where for consistency with the close-coupling expansion the asymptotic state Φ_f is assumed to be a product of the plane wave of momentum $\boldsymbol{k}_f$ and the target eigenstate (discrete or continuous) ϕ_f. Note that $\langle\phi_f|I_N = \langle\phi_f|\phi_f^{(\lambda)}\rangle\langle\phi_f^{(\lambda)}|$ for $\varepsilon_f = \varepsilon_f^{(\lambda)}$. Then for states with $\varepsilon_f^{(\lambda)} < 0$ the overlap in Eq. (6) is simply unity, but for $\varepsilon_f^{(\lambda)} > 0$ the overlap essentially provides the continuum normalisation of the square-integrable state $\phi_f^{(\lambda)}$ i.e.,

$$\langle\phi_f| \approx \langle\phi_f|\phi_f^{(\lambda)}\rangle\langle\phi_f^{(\lambda)}|. \tag{7}$$

While the above definition of ionisation amplitudes via Eq. (6) arises from consistency arguments it clearly has several problems. For a start, just like it is not correct to take the asymptotic state of the total wavefunction to be a product of two Coulomb waves, as is done in the ECS method,[3] nor is it correct to take this state to be a product of a plane wave and a Coulomb wave. Furthermore, the non-symmetric treatment of the two outgoing electrons in the CCC method apparently leads to double counting of the ionisation processes, not withstanding the agreement with experiment in Fig. 3. From the denominator in Eq. (4) we see that the open channels will comprise states for which $E = \varepsilon_f^{(\lambda)} + k_f^2/2$. Given an ionisation process with outgoing electrons of energies $e_1 + e_2 = E$ we have two independent amplitudes corresponding to $e_1 = \varepsilon_f^{(\lambda)}, e_2 = k_f^2/2$ and $e_2 = \varepsilon_f^{(\lambda)}, e_1 = k_f^2/2$. Another way to see this is that in the close-coupling method the total ionisation cross section, see Eq. (5), sums over the excitation of all states with energies less than E, whereas a symmetric treatment of the electrons would yield a summation that ended at $E/2$.

The resolution of the double-counting problem turns out to be quite interesting.[18] Every solution of Eq. (4) yields step-function behaviour with $T_{fi} \approx 0$ whenever $\varepsilon_f^{(\lambda)} > k_f^2/2$. Not only does this effectively reduce the summation in Eq. (5) to $0 < \varepsilon_f^{(\lambda)} < E/2$, but is also not so surprising as such processes correspond to the unphysical case of the fast electron completely shielding the slow one. Furthermore, at the step $\varepsilon_f^{(\lambda)} = k_f^2/2$ the amplitude T_{fi} converges to half the step height with the conclusion that solving Eq. (4) is effectively like taking a Fourier expansion of a step function.[19]

2.2.2. *Differential ionisation cross sections*

With the above understanding application of the CCC method to ionisation became straightforward.[20] Counter-intuitively convergence at the lower energies is more rapid at the equal-energy sharing case rather than asymmetric energy sharing. At the higher energies convergence is more rapid generally because the size of the step is relatively much smaller. In the final analysis the CCC method is able to give accurate predictions of the ionisation cross sections irrespective of the kinematics involved or the geometrical orientation of the detectors, and thereby fulfilling the promise of being a complete scattering theory.[21] To demonstrate this graphically we step through the case of 25 eV electron-impact ionisation of atomic hydrogen. We have already seen in Fig. 3 that the total ionisation cross section is correct at this energy. This value is obtained from Eq. (5) which can be rewritten as an integral of the singly differential cross section (SDCS), and is presented in Fig. 4. We see that the 93-state CCC results yield what looks like a Fourier expansion of a step function, which can be estimated reasonably accurately and denoted as CCC(∞). The estimate is obtained with the knowledge of the integral, the convergence of the cross section at the step to a quarter (amplitude to a half) of the step size, and the assumption that the oscillations are about the smooth underlying cross sections. The accuracy of the estimation process can also be enhanced by comparing the results of other calculations with larger N that must yield the same integral and step height, but having different oscillations.

The experiments of Shyn[22] measured the doubly differential cross sections which are the angular distributions of the secondary electrons with energies 1, 2, 3, 4, and 5.7 eV. These are presented in Fig. 5. Upon integration over the solid angle these yield the SDCS given in Fig. 4.

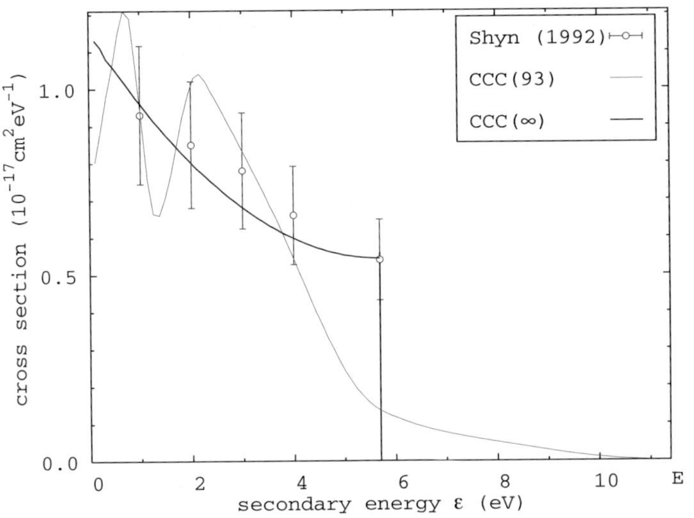

Fig. 4. Singly differential cross sections for 25 eV electron-impact ionisation of atomic hydrogen. The experiment is due to Shyn[22] and the theory is due to Bray et al.[13]

The CCC results given in Fig. 5 are the raw CCC results scaled to yield the estimated step-fuction SDCS of Fig. 4. If we were not to do this then calculations of different basis sizes would jump up and down according to the oscillation of Fig. 4. The ECS method[23] does not have these problems and yet the calculated DDCS are in good agreement with each other and the experiment.

Even more detail is provided by the measurements of the fully differential cross sections. These require coincidence measurements and specify the spacial distribution of one electron given the detection of another at a specified point in space. Accordingly the two outgoing electrons can be in or out of the plane of the incident electron. For the case of 25 eV incident energy fully differential measurements are only available for the case of equal-energy sharing in the coplanar geometry. They are presented in Fig. 6 as a study of the cross sections as the angle θ_{AB} between the two electrons is varied. Such a presentation is interesting because the cross sections rise rapidly with θ_{AB} indicating that equal-energy electrons prefer to exit in opposite directions. In the figure we see that there is good agreement between

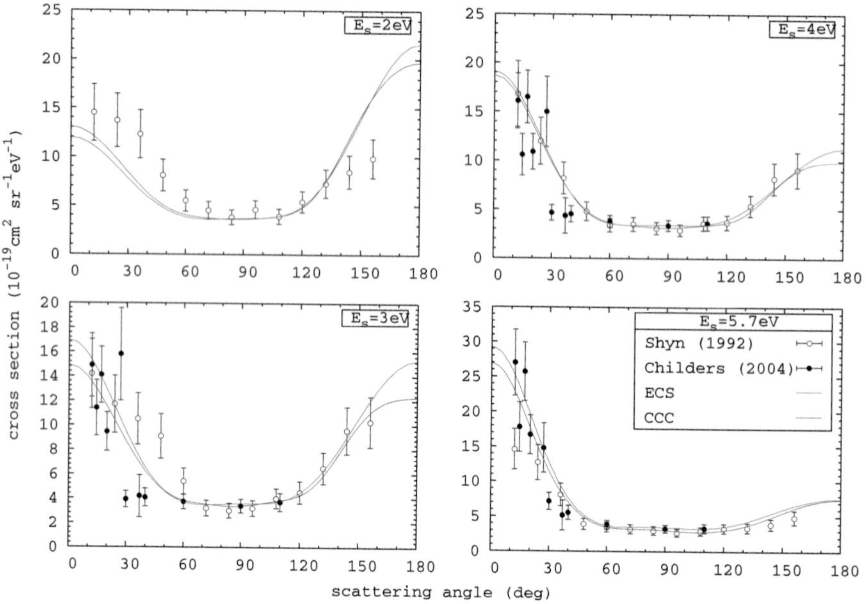

Fig. 5. Doubly differential cross sections for 25 eV electron-impact ionisation of atomic hydrogen for secondary electron energies of 1, 2, 3, 4, and 5.7 eV. The measurements are due to Shyn[22] and Childers et al.[24] The ECS theory is due Isaacs et al.[23] and the CCC theory due to Bray et al.[13]

the calculations and the experiment, and presently there is no other theory that can claim to yield the same quality agreement with such experiments. For targets such as helium there is a good deal more data and only the CCC theory is able to reproduce it fully as the ECS method has not yet been extended to helium. Incidentally, double photoionisation is a process that is very closely linked to electron-impact ionisation and there the CCC method has been equally successful.[13]

3. Formal Ionisation Theory Considerations

Given the success of the ECS and CCC theories for the description electron-impact ionisation of atoms we have devoted some time to try and understand why this is so. The Rudge formalism mentioned in the very beginning is based on so-called Peterkop's integral.[26] This integral was originally introduced as a trial function due to the absense of the post form of the ionisation amplitude. Rudge and Peterkop tried to relate this trial integral to the ionisation amplitude using effective charges z_1 and z_2. However, the condition on the effective charges turned out to be impossible to satisfy in

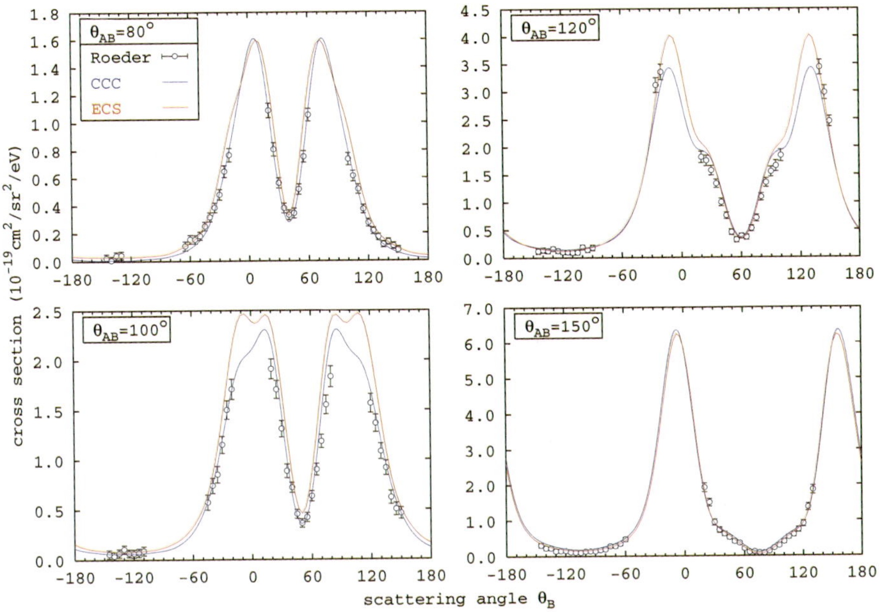

Fig. 6. Coplanar fully differential cross sections for 25 eV electron-impact ionisation of atomic hydrogen at equal-energy sharing. The internormalised relative measurements of Röder et al.[25] have been normalised using best visual fit to the theories. The ECS theory is due Baertschy et al.[3] and the CCC theory due to Bray et al.[13]

practice. Without satisfying this condition the representation for the ionisation amplitude turned out to have an ambiguos and divergent phase. The ionisation amplitudes emerging both in the ECS and CCC methods correspond to certain choice of these effective charges which does not satisfy the aforementioned condition. To be specific the ECS method sets $z_1 = z_2 = 1$ while the CCC method takes $z_1 = 0$ for the more energetic of the electrons and $z_2 = 1$ for the slower one. Such an arbitrary choice of the effective charges is formally justified.[27] It is shown that in fact there is no need for the effective charges to satisfy any condition and that they can be set equal to 1 or 0 if this is practically convenient. In this case, the only difference from the ionisation amplitude (in the ideal infinite-size calculation) would be a factor which asymptotically transforms into a phase.[27] Therefore, the ECS and CCC amplitudes are well justified approximations to the magnitude of the exact amplitude. The correct phase can be extracted in the S-wave model of e-H ionization,[28] but whether the convergent CCC phase is correct in general is still under investigation.

Since both ECS and CCC methods are based on the Rudge formalism, it is of extraordinary importance to answer the question: where does Peterkop's trial integral, the cornerstone of the formalism, come from and what is its relationship with formal theory of scattering? We derived a new post form of the ionisation amplitude[29]

$$T_{fi} = \langle \Phi_f^- | \overleftarrow{H} - E | \Psi_i^{(+)} \rangle, \tag{8}$$

which enabled us to answer the aforementioned question, where Φ_f^- is the exact three-body asymptotic state. This new form rigorously follows from the prior form of the ionisation amplitude. It is equivalent to the prior form known to be exact. Consequently, this form is not a testing function like Peterkop's integral but a proper post-form definition of the ionisation amplitude. Comparison between this form of the amplitude and Peterkop's integral with $z_1 = z_2 = 1$ shows that they are similar with the only difference being the electron-electron correlation term in Φ_f^-. This is not to say that in the resulting amplitude extracted using Peterkop's integral there is no electron-electron correlation. As stated above the only difference between the amplitude resulting from Peterkop's integral and the exact ionisation amplitude reduces to a phase factor. Therefore, Peterkop's integral can be interpreted as a possible practical approximation to the aforementioned post form of the ionisation amplitude which only affects the phase while leaving the magnitude of the amplitude untouched. And that is the reason for the extraordinary success of the ECS and CCC methods based on it. Therefore, the post form of the ionisation amplitude[29] establishes an intimate relationship between Peterkop's trial integral and rigorous scattering theory. In other words, it explains the meaning of Peterkop's integral as a specific approximation to the exact amplitude in the post form that is capable of leaving its magnitude exact. With this knowledge the ECS and CCC methods can be called complete and ab initio practical solutions of the electron-impact ionisation problem.

Acknowledgments

This work was supported by the Australian Research Council. We are grateful for access to the Australian Partnership for Advanced Computing and its Western Australian node iVEC.

References

1. M. R. H. Rudge, *Rev. Mod. Phys.* **40**, 564 (1968).
2. T. N. Rescigno, M. Baertschy, W. A. Isaacs and C. W. McCurdy, *Science* **286**, 2474 (1999).

3. M. Baertschy, T. N. Rescigno and C. W. McCurdy, *Phys. Rev. A* **64**, p. 022709 (2001).
4. H. S. W. Massey and C. B. O. Mohr, *Proc. Roy. Soc. A* **136**, 289 (1932).
5. J. Callaway, *Phys. Rep.* **45**, p. 89 (1978).
6. J. Callaway, *Phys. Rev. A* **32**, 775 (1985).
7. W. L. van Wyngaarden and H. R. J. Walters, *J. Phys. B* **19**, 929 (1986).
8. T. T. Scholz, H. R. J. Walters, P. G. Burke and M. P. Scott, *J. Phys. B* **24**, 2097 (1991).
9. E. Weigold, L. Frost and K. J. Nygaard, *Phys. Rev. A* **21**, 1950 (1980).
10. J. F. Williams, *J. Phys. B* **14**, 1197 (1981).
11. I. Bray and A. T. Stelbovics, *Phys. Rev. A* **46**, 6995 (1992).
12. I. Bray and A. T. Stelbovics, *Phys. Rev. Lett.* **69**, 53 (1992).
13. I. Bray, D. V. Fursa, A. S. Kheifets and A. T. Stelbovics, *J. Phys. B* **35**, R117 (2002).
14. H. Yalim, D. Cvejanovic and A. Crowe, *Phys. Rev. Lett.* **79**, 2951 (1997).
15. M. B. Shah, D. S. Elliot and H. B. Gilbody, *J. Phys. B* **20**, 3501 (1987).
16. I. Bray and A. T. Stelbovics, *Phys. Rev. Lett.* **70**, 746 (1993).
17. I. Bray and D. V. Fursa, *Phys. Rev. A* **54**, 2991 (1996).
18. I. Bray, *Phys. Rev. Lett.* **78**, 4721 (1997).
19. A. T. Stelbovics, *Phys. Rev. Lett.* **83**, 1570 (1999).
20. I. Bray, *Phys. Rev. Lett.* **89**, p. 273201 (2002).
21. I. Bray and D. V. Fursa, *Phys. Rev. Lett.* **76**, 2674 (1996).
22. T. W. Shyn, *Phys. Rev. A* **45**, 2951 (1992).
23. W. A. Isaacs, M. Baertschy, C. W. McCurdy and T. N. Rescigno, *Phys. Rev. A* **63**, p. 030704 (2001).
24. J. G. Childers, K. E. James, I. Bray, M. Baertschy and M. A. Khakoo, *Phys. Rev. A* **69**, p. 022709 (2004).
25. J. Röder, J. Rasch, K. Jung, C. T. Whelan, H. Ehrhardt, R. J. Allan and H. R. J. Walters, *Phys. Rev. A* **53**, 225 (1996).
26. R. K. Peterkop, *Opt. Spektrosk.* **13**, p. 153 (1962).
27. A. S. Kadyrov, A. M. Mukhamedzhanov, A. T. Stelbovics and I. Bray, *Phys. Rev. Lett.* **91**, p. 253202 (2003).
28. P. L. Bartlett, I. Bray, S. Jones, A. T. Stelbovics, A. S. Kadyrov, K. Bartschat, G. ver Steeg, M. P. Scott and P. G. Burke, *Phys. Rev. A* **68**, p. 020702(R) (2003).
29. A. S. Kadyrov, A. M. Mukhamedzhanov, A. T. Stelbovics and I. Bray, *Phys. Rev. A* **70**, p. 062703 (2004).

CHEN NING YANG AND THE FORMATION OF AAPPS (ASSOCIATION OF ASIA PACIFIC PHYSICAL SOCIETIES)

MICHIJI KONUMA

Keio University and Musashi Institute of Technology, Japan
mkonuma254@m4.dion.ne.jp

The AAPPS is an umbrella organization of physical societies in the Asia Pacific region. Chen Ning Yang was the most active and important key person for the formation of the AAPPS.

1. Encounters with Chen Ning Yang

The first time I saw Chen Ning Yang directly from a distance was at the International Conference on Theoretical Physics held in Tokyo and Kyoto, Japan in September 1953. I remember well his report, showing the latest information just arrived from the Brookhaven National Laboratory, where he stayed in summer of the year, on the first artificial production of V particles (strange particles) and on the first confirmation on the associate production of V particles (strange particles) by pi minus beam of 1.37 GeV colliding with proton. It was the latest experimental result from the Cosmotron, a proton synchrotron of 3.3 GeV.[1,2]

Since then, I have met Yang at rather many international conferences. The most intensive collaboration with Yang was to establish and to manage the Association of Asia Pacific Physical Societies (AAPPS) from 1988 to 1994. This will be described in details.

Before that, I add another comment. I was a visiting professor at the University of the Air, giving a nationwide TV lecture on Modern Physics from 1992 to 2001. One day I read Yang's article "Symmetry and Physics", published in AAPPS Bulletin, Vol. 1, No. 3, December 1991, enthusiastically.[3] I asked him to give a Special TV Lecture for the University. He accepted. Finally his lecture on "Symmetry and Physics" was broadcasted all over Japan repeatedly for many years from 1992. His lecture covered Beauty and Harmony → Symmetry → painting, decorative art, architec-

Fig. 1. C. N. Yang with his family in Japan in 1953. (Courtesy of Hideki Yukawa's family.)

ture, poetry, music and science including physics. This lecture was widely acknowledged.

It is announced in this Conference that a Dialogue between C. N. Yang and Fan Zeng (Chinese painter and calligrapher, Professor of Nankai University, Tianjin, China) will be held on 3 November 2007 at Nanyang Technical University. In the announcement a sentence is cited from Zhuang Zi (Chuang-tse) as "Nature shows its beauty without saying a word". Zhuang Zi was an ancient Chinese philosopher. This sentence is the first line of his following text:

莊子 (**Zhuang Zi**)
天地有大美而不言
四時有明法而不義
萬物有成理而不說
聖人者原天地之美
而達萬物之理

Hideki Yukawa, the first Japanese Nobel Laureate, loved Zhuang Zi from his childhood to his later years. I show you Yukawa's calligraphy in Fig. 2.

Fig. 2. Hideki Yukawa's calligraphy, reading from top to bottom and right to left. (Photo taken by the author.)

My translation of this text is as follows:

> There is the greatest beauty in Heaven and Earth, which say, however, nothing on the beauty;
> There are clear laws in four seasons, which discuss, however, nothing on the laws;
> There is the reason in all things, which explain, however, nothing on the reason;
> Outstanding persons reach, based on the beauty in Heaven and Earth, to the reason in all things.

Please pay attention to the last line. Chen Ning Yang is really one of these kinds of persons.

2. AAPPS

The Association of Asia Pacific Physical Societies, abbreviated as AAPPS, is an umbrella organization of physical societies in the Asia Pacific region which consists of 18 physical societies.[4] AAPPS is one of the co-organizers of this Conference in Honor of C. N. Yang's 85th Birthday. It was constituted in 1989 and the first General Meeting was held in 1990. Chen Ning Yang was the first President from 1990 to 1994.

According to its Constitution, the objectives of AAPPS are that the AAPPS aims to promote the advancement of knowledge in physics in the Asia Pacific region, including research, application and teaching, especially through international collaboration in these activities in this region. Further the AAPPS is defined to organize conferences and workshops, establish specialist divisions and subcommittees, and publish newsletters and journals.

The most important activities are the publication and conferences. AAPPS Newsletter was published in December 1989, April and December 1990. AAPPS Bulletin started in 1991. It continues until now. The series of the Asia Pacific Physics Conferences had been held in 1983 (Singapore), 1886 (2nd, Bangalore, India) and 1988 (3rd, Hong Kong). From the 4th Conference in 1990 (Seoul, Korea) this series of the Conferences has been organized under the auspices of the AAPPS.[5]

The series has continued: in 1992 (5th, Kuala Lumpur, Malaysia), 1994 (6th, 1994, Brisbane, Australia), 1997 (7th, Beijing, China), 2000 (8th, Taipei, Taiwan), 2004 (9th, Kuala Lumpur, Malaysia) and 2007 (10th, Pohang, Korea).

The Asia Pacific Center for Theoretical Physics (APCTP) was founded in Korea in 1996 to serve as a focal point for performing joint academic research and international cooperation in Asia Pacific region and beyond. AAPPS has supported and encouraged the creation and activities of APCTP. Chen Ning Yang was invited as the Founding President and Chairman of the Board of Trustees of the Center.[6]

AAPPS has kept close relations with International Union of Pure and Applied Physics (IUPAP), American Physical Society (APS), European Physical Society (EPS), etc. After its inauguration AAPPS was invited to the General Meetings of IUPAP as an observer together with APS and EPS. AAPPS organized the Second Conference of Research and Communications in Physics (World Congress of Physical Societies) with Physical Society of Japan (JPS) and Japan Society of Applied Physics (JSAP) at the United Nations University, Tokyo, Japan on 17–22 September 1995. Representatives from Physical Societies gathered from all over the world.[7]

The last comment in this section is present and future of AAPPS. Working Group on Women in Physics has been organized. Division of Plasma

Physics is in preparation and other Divisions may follow. Possibilities of Asia Pacific physics journals and specialized conferences and workshops are discussed.

3. The Road to the Formation of AAPPS

3.1. *An early attempt*

In 1980 an idea to establish the Asian-Pacific Physical Society was informally proposed by K. K. Phua and C. N. Yang. Phua sent a letter to Yang, C. A. Hurst (Australia), A. Salam (Pakistan) and Y. Yamaguchi (Japan) on 17 September 1980 mentioning that he discussed and agreed this matter with Yang, sounding out them on their ideas and asking support and contribution. On 20 October 1980 Yang wrote a letter to 16 physicists in Australia, China, Hong Kong, Japan, Korea, Malaysia, Singapore, Taiwan, Thailand, and Trieste. He asked their comments and opinions of their physical societies.[8,9]

Phua and Yang seem not to have known about the Asian Physical Society (APSO) which had been proposed in 1977 and established in 1979. The promoting group of APSO was physicists in South-East and South Asian countries. Its main aim and activities oriented in promotion of physics in developing countries and application to technology.

Phua visited Japan in March–June 1981 as a visiting professor at Research Institute for Fundamental Physics, Kyoto University and Yang came to Japan in June 1981 by invitation of the Japan Society of Promotion of Science. They discussed the matter with the Japanese group. Later Yang contacted Chinese Physical Society and discussed with the Chinese group on a possible Asia Pacific Physical Society. Yang wrote to Akito Arima, President of the JPS, on 1 September 1981 a letter mentioning that it was too early to establish the regional society and withdrew the previous proposal waiting for a favorable opportunity.

3.2. *Steps to the Asia Pacific Physics Conferences (APPC)*

K. K. Phua also told to the JPS to organize a Topical Conference on Physics in Singapore in 1982 or 1983 during his stay in Japan in 1981. The original idea of holding a conference was that it is important first to know from each other and it will pave a way for making the regional society. He wrote a letter to the JPS on 3 July 1981 asking the co-sponsor of a Topical Conference on Physics to be held in Singapore in either August 1982 or March 1983.

Finally APPC1 was co-organized by Singapore Institute of Physics, and JPS, Singapore National University, South East Asia Theoretical Physics Association and ICTP, Trieste. (Almost) all Asia and Pacific countries and regions were invited, including PR China and Taiwan; South and North Korea. Yang contacted North Korea through Chinese Academy of Science.

The Asia Pacific Physics Conference (APPC) was held on 12–18 June 1983 in Singapore. Plenary Speakers included C. N. Yang (Fields and Symmetries), N. P. Chang and Parallel Session Speakers included A. Tonomura and T. Eguchi attending at the present Session. C. N. Yang (Contribution of Matteo Ricci ⋯ to China ⋯) was also among the speakers at Forum on "Physics and Development".

Ad Hoc Meeting was held to review the Conference just being held. It was chaired by Yang and the attendance was 21 in total. They were A. Arima (Japan), S. P. Chia (Malaysia), S. Chandrasekhar (India), W. K. Chung (Canada?), C. A. Hurst (Australia), G. D. Jones (New Zealand), R. Kubo (Japan), Y. K. Lim (Malaysia), C. K. Majundar (India), Z. Maki (Japan), B. H. J. McKellar (Australia), M. Oda (Japan), K. K. Phua (Singapore), K. P. Wang (Taiwan), G. V. H. Wilson (Australia), T. Y. Wu (Taiwan), T. Yamazaki (Japan), C. N. Yang (USA and Hong Kong), K. Young (Hong Kong), G. Z. Zhou (China), and P. Y. Zhou (China).

At this Meeting the Second Conference was determined to be held in India. APPC2 was held on 13–17 January 1986 in Bangalore, India. Plenary Speakers included C. N. Yang (Some New Developments in High Energy Collisions).

Following to the experience in the First Conference, the Ad Hoc Meeting chaired by C. N. Yang was held. Participants were 12 in total. They were A. Arima (Japan), Y. W. Chan (Hong Kong), S. Chandrasekhar (India), Il-Tong Cheon (Korea), C. A. Hurst (Australia), Y. H. Ichikawa (Japan), Y. Kim (Korea), T. Kobayashi (Japan), R. Kubo (Japan), Mo Dang (China), C. N. Yang (USA and Hong Kong), and Z. G. Zhao (China). The Third Conference was decided to be held in Hong Kong.

3.3. *From dream to reality*

On 16 January 1987, C. N. Yang, Y. W. Chan and K. Young of Chinese University of Hong Kong sent a letter to Physical Societies in the Asia and Pacific Region. In this letter it was written that "it is felt that the time is ripe for discussions to explore the possibility of creating a more formal institutional framework for regional cooperation. A meeting will be called to explore this issue in depth at the APPC3 in Hong Kong".

Yoshio Yamaguchi, President of JPS, visited Hong Kong in January 1987 and told them his personal opinions.

As I was a member of the Executive Board of JPS, I was involved in discussions on this letter. In March 1987 I was elected as the next President of JPS (My term was from September 1987 to August 1988). My idea was that Japanese physicists should know physics and physicists in Asia and Pacific region as equally as they do those in US and Europe, and that Japanese scientific community should contribute to regional development of science more in return for great help received from foreign communities after the opening of the country in the 19th Century and after the 2nd World War in the 20th Century.[10]

On behalf of the JPS Y. Yamaguchi (President, JPS) sent a letter to C. N. Yang, Y. W. Chan and K. Young on 2 April 1987, explaining discussions at Executive Meetings of JPS:

(1) JPS is happy to see a movement to promote regional cooperation in physics.
(2) We need to take account of Asian Physical Society which was founded in 1979.
(3) In Japan several societies are relevant to physics.

On 1 September 1987 the President of the JPS changed from Yamaguchi to me. The Executive Meeting of JPS continued discussions on the matter. As we had received no information on further progress, I decided to contact C. N. Yang and Kenneth Young. On 3 March 1988 I made a telephone call to Kenneth Young of Hong Kong and we agreed that it was preferable to make an informal frank exchange of views with Chen Ning Yang in advance to APPC3. On 15 March I sent a letter to C. N. Yang of Stony Brook. He agreed to meet me and representatives of the JPS in Hong Kong.

Yang consulted with Zhou Guang-Zhao (China), President of the Chinese Academy of Sciences, Kun Huang, President of the Chinese Physical Society and K. K. Phua (Singapore). I also encouraged Zhou G.-Z. to come to Hong Kong.

In parallel, I learned experiences of other similar regional organizations: Federation of Asian Chemical Societies (FACS), the Asian Crystallographic Association (ASCA), the Federation of Engineering Institutions of Southeast Asia and the Pacific (FEISAP), Asian Physical Society (APSO), Asian Physics Education Network (ASPEN) and the American Physical Society (APS) and the European Physical Society (EPS) and others.

An informal meeting among China, Hong Kong, Japan and Singapore was realized on 19 June, 1988 in Hong Kong just before the APPC3. Participants were Y. W. Chan, C. N. Yang, K. Young of Hong Kong, Li S.-N. (Vice-President, Chinese Physical Society); Zhou G.-Z. (President, Chinese Academy of Science and Vice-President of Chinese Physical Society) M. Konuma (President, JPS); J. Tsujiuchi (President, JSAP); Y. Yamaguchi (Vice-President, IUPAP, Council member, JPS) and K. K. Phua (Nat. Univ. Singapore).

At this informal meeting we discussed very extensively:

- Priority among possible activities;
- Geographical areas to be covered;
- Members: no restriction such as "one society from a country or region"; individual members?
- Beijing and Taipei; Seoul and Pyongyang;
- General Meeting: no. of representatives, no. of votes;
- Finance: Annual fees (several categories or fixed amount?);
- Simulation towards the formation of AAPPS;
- Many other topics.

Except for a few points we arrived at common understanding.

In the APPC3 held on 20–24 June 1988 an Ad Hoc Meeting was called on 23 June 1988, following to the tradition of conferences in this series. Participants were 29. The meeting was chaired by C. N. Yang. The participants were S. H. Ahn (Korea), Y. W. Chan (Hong Kong), S. Chandrasekhar (India), N. H. Cheung (Hong Kong), S. P. Chia (Malaysia), Y. M. Cho (Korea), H. L. Huang (Taiwan), C. A. Hurst (Australia), S. Hyodo (Japan), M. A. Khan (Pakistan), T. Kobayashi (Japan), T. Koda (Japan), Y. S. Ko (Korea), M. Konuma (Japan), J. Lekner (New Zealand), S. N. Li (China), S. C. Lin (Malaysia), Y. K. Lin (Singapore), Y. C. Liu (Taiwan), P. K. Mackeon (Hong Kong), Y. Nishina (Japan), T. C. Ponce (Philippines), C. S. Shen (Taiwan), C. Singh (India, Asian Physical Society), J. Tsujiuchi (Japan), J. F. Williams (Australia), G. V. H. Wilson (Australia), C. N. Yang (Hong Kong and USA), K. Young (Hong Kong).

In this meeting the venue of the fourth Conference was decided as Seoul, Korea. Yang proposed AAPPS (exactly speaking the proposed name being Federation of Asia Pacific Physical Societies). After lengthy discussions the different role was confirmed from the existing Asian Physical Society. The Ad Hoc Group for the Formation of AAPPS was proposed. Members were recommended.

The final members of the Ad Hoc Group were C. N. Yang (Chair, USA and Hong Kong), K. Young (Secretary, Hong Kong), S. Chandrasekhar (India), Y. M. Cho (Korea), H. L. Huang (Taiwan), C. A. Hurst (Australia), M. Konuma (Japan), J. Lekner (New Zealand), S.-N. Li (China), G. Murtaza (Pakistan), K. K. Phua (Singapore), A. Salam (ICTP, Trieste), V. Singh (India), T. Sugano (Japan), B. C. Tan (Malaysia) and G. V. H. Wilson (Australia). It was understood that all members were by personal capacity.

The Ad Hoc Group was called at the lunch time next day. Seven members (Chandrasekhar, Huang, Hurst, Konuma, Li, Yang and Young) plus Ahn (the next President, Korean Physical Society) attended. Discussions of the Ad Hoc Meeting of the previous day were clarified and confirmed. The Ad Hoc Group meeting was decided to be held in Hong Kong in the coming December or January. Yang suggested that he and Young should summarize and send proposals to the members. Chandrasekhar was asked to help with this drafting. After the meeting was closed Yang asked me to collaborate and invited me to his office. We talked various matters, including how to step ahead and to prepare drafts of Constitution and By-Laws until the next Ad Hoc Group meeting.

Intensive works of this Ad Hoc Group were started.

On 26-30 August 1988, I visited Hong Kong. Yang, Young and I made *Draft 1.0 of Constitution* and sent it to members asking their responses. Kenneth Young was an extraordinarily able and efficient Secretary.

On November 1988 C. N. Yang and K. Young visited Japan and discussed with officers of two physical Societies. We three modified the materials. *Draft 2* was sent to members.

On 3-4 January 1989, Meeting of Ad Hoc Group was held in Hong Kong. *Draft 3.0 of the Constitution and Draft 1.0 of By-Laws Draft* was made. *Draft 1.0 of "Memorandum of Intent"*, including "It is our intention that the Association, and particularly its Council, shall be broadly representative of the physicists of the entire region and not to be dominated by those from any one country or region.", was written and sent to members.

It was confirmed that we are an NGO on science and we have no political prejudice. On the relation between Beijing and Taipei and the names of their physical societies, the Ad Hoc Group, Beijing Group and Taipei Group agreed to follow the procedure of ICSU (International Council of Scientific Unions; now International Council for Science) and IUPAP. The agreed name of member society from Taipei should be "The Physical Society located in Taipei, (Taiwan) China". The Korean Physics Association at Pyongyang, DPR of Korea (North Korea, NK) showed interests from the beginning and expressed future collaboration. But the NK group has not yet joined. India has many Physical Societies. Indian Physics Association was chosen to be invited to be a Founding Member.

Finally on 10 April 1989, on behalf of the Ad Hoc Group for Formation of AAPPS, Yang sent the invitation letter to prospective physical societies in the region to become a Founding Member of the AAPPS, asking to return response by 15 October 1989.

Yang wrote a message that "The formation of AAPPS is a consequence of the rising importance of the Asia Pacific region in research, education and application in the science of physics. I am confident that the combined efforts of its members, the Association will fulfil its mission of promoting the advancement of knowledge in physics in the region".

Seventeen Physical Societies agreed to become the Founding Member. AAPPS was formally constituted on 15 October 1989.

4. Inauguration of the AAPPS and After

The First General Meeting of AAPPS was held on 10 August 1990 at Yonsei University, Seoul, Korea. The meeting was chaired by Yang as the Chairman of the Ad Hoc Group for the Formation of AAPPS. The General Meeting elected 12 Council members. The next day the First Council Meeting was held and Yang was elected as President unanimously.

He worked for AAPPS as President for 4 years and as the ex-officio member of the Council until the end of 1997. The Council acknowledged his long and enthusiastic efforts and presented the title of the Honorary President of AAPPS. Further the "AAPPS Chen Ning Yang Award" was created in 1997.

I remember well his Opening Address as the President of AAPPS at the Fourth Asia Pacific Physics Conference on 13 August 1990. He explained the "Congress of Art and Science" held in 1904 at St. Louis, USA and continued as follows:[11]

> "When the St. Louis Congress met, the USA was not yet a world power; not a world power in commerce, not a world power in military might, certainly not a world power in knowledge and research ....
>
> The American Physical Society was formed in 1899. It rapidly became the most important professional physics society in the world until recent years, when there appeared a strong challenge from the European Physical Society ....

> The formation of the AAPPS is a historic event of great long range significance. I say this for four reasons:
>
> (a) The rapidly accumulating knowledge resulting from education and research in all branches of physics;
> (b) The increasing importance of physics for the industrial, economic and environmental developments of all countries;
> (c) The rapid increase in the weight of the Asia Pacific region in world economy and world politics; and
> (d) The shifting realignment of world economy and regional politics. The formation of the AAPPS is the right event at the right time ...."

This last sentence shows Chen Ning Yang's real intentions which are at the base of his eager and tenacious efforts for many years.

Further developments by the younger generation for the promotion of regional and international collaboration are highly expected.

References

1. C. N. Yang, Proceedings of the International Conference of Theoretical Physics, Kyoto and Tokyo, September 1953 (The Science Council of Japan, 1954), p. 137.
2. See, also, W. B. Fowler, R. P. Shutt, A. M. Thorndike and W. L. Whittemore, Phys. Rev. **91** (1953) p. 1287 and **93** (1954) p. 861.
3. C. N. Yang, AAPPS Bulletin, **1** (1991) (3), p. 3.
4. AAPPS Newsletter, **1** (1989) (1), **1** (1990) (2 and 3).
5. T. Kobayashi, The 20th Anniversary of Asia Pacific Physics Conference, BUTSURI (Bulletin of the Physical Society of Japan), **59** (2004) p. 48 (in Japanese).
6. APCTP's 10th Anniversary, AAPPS Bulletin, **16** (2006) (5), p. 2.
7. Proceedings of the Second International Conference on Research and Communications in Physics (RACIP2), Tokyo, September 18–22, 1995 (Physical Society of Japan, 1996).
8. M. Konuma's Collected Documents on AAPPS.
9. Y. Yamaguchi, On the Asian Physical Society, BUTSURI (Bulletin of the Physical Society of Japan), **37** (182) p. 326 (in Japanese).
10. T. Kobayashi and M. Konuma, Asia-Pacific Physics News, **5** (1990) (1), p. 30.
11. C. N. Yang, AAPPS Newsletter, **1** (1990) (3), p. 5.

BANQUET ADDRESS: A COMMON TARGET

MICHIJI KONUMA

First I wish to congratulate Professor Chen Ning Yang in Japanese.[a]

> *Yang Chen Ning Sensei, hachiju-go-sai no otanjobi to parity hi-hozon no hakken goju-nen omedeto gozaimasu.*

Professor Chen Ning Yang, please accept my sincere congratulations on your happy eighty-fifth birthday and on the fiftieth anniversary of your discovery on parity violation.

I saw Professor Yang first in Tokyo and Kyoto, Japan, in 1953. It was his first visit to Japan at the occasion of the International Conference on Theoretical Physics. I mentioned this and other encounters with him at my talk in today's afternoon session. Our close collaboration continued in 1980's and 1990's for the formation of the AAPPS, Association of Asia-Pacific Physical Societies.

One day in the late 1980's, he told me of his willingness to climb Mt. Fuji, the highest mountain in Japan. I said I had climbed Mt. Fuji four times, and if he wished we should go together. Since then, it became our common target. Time after time he confirmed this unfinished promise. It is not an easy task because it can be done only on fine days in July and August. Even yesterday and today he reminded me so.

I admire that he is healthy, active, enthusiastic and curious. He is young as ever. I wish to keep my health to accomplish our common target. We shall reach at the top of the mountain some future time.

Professor Yang, this is now a public commitment. Please keep your good health for a long time.

---

[a] 小沼通二
楊振寧先生、85 歳のお誕生日とパリティ非保存の発見からの 50 年おめでとうございます。

THE MODELING AND FUNCTIONAL CONNECTIVITY OF THE BRAIN

SEUNGHWAN KIM

Department of Physics,
POSTECH, Korea
E-mail: swan@postech.ac.kr

The brain is considered to be the most complex system, a fertile ground for understanding the complexity of its functions through dynamical modeling. In this talk, we present some biophysical models that help to reveal the complexity of visual functions of the brain through functional self-organization processes. We also present some recent results on how the functional connectivity arises and changes in the brain, reflecting the underlying dynamics of nervous systems. The implications of our work to the brain function are discussed.

SYMMETRY EFFECTS IN COMPUTATION

ANDREW CHI-CHIH YAO

Center for Advanced Study,
Tsinghua University, China
E-mail: andrewcyao@tsinghua.edu.cn

The concept of symmetry has played a key role in the development of modern physics. For example, using symmetry, C.N. Yang and other physicists have greatly advanced our understanding of the fundamental laws of physics. Meanwhile, computer scientists have been pondering why some computational problems seem intractable, while others are easy. Just as in physics, the laws of computation sometimes can only be inferred indirectly by considerations of general principles such as symmetry. The symmetry properties of a function can indeed have a profound effect on how fast the function can be computed. In this talk, we present several elegant and surprising discoveries along this line, made by computer scientists using symmetry as their primary tool.

APPLICATIONS OF NANOTECHNOLOGY TO THE DEVELOPMENT OF ENERGY-RELATED TECHNOLOGIES

MAW-KUEN WU

Institute of Physics, Academia Sinica
E-mail: mkwu@phys.sinica.edu.tw

In response to current energy and environmental issues the development of renewable energy sources becomes an urgent need. On the other hand, fabrication of better performance devices with much higher energy efficiency, such as LED lightings, has also become the focus of current research. Nanotechnology has been demonstrated to be a valuable and important path for the development of technologies in both the renewable energy sources and energy-saving devices. In this presentation, I shall summarize the energy related research activities in Taiwan, including our own effort on the development of high performance (high ZT value) thermoelectric materials, that supported by the national initiatives on nanotechnology.

CHEN-NING YANG AND MY STUDENT DAYS AT STONY BROOK

ALEX CHAO

Stanford Linear Accelerator Center,
2575 Sand Hill Road, Menlo Park, CA 94025, USA

The time was 35 years ago. Chen-Ning Yang, my Ph.D. advisor at Stony Brook, thoughtfully led me through a career decision path that I followed, and have enjoyed since then. This is my reminiscence of that time so critical to my career.

I went to Stony Brook in 1971 – my first trip out of Taiwan. My reason of choosing Stony Brook for my graduate study was simple, Stony Brook had Chen-Ning Yang. Yang's name inspired a generation of young students in China and Taiwan, and I was one among them. I was not dreaming necessarily to become his student, but I went to Stony Brook because he was there. "If Yang is there, it must be a good place to go." As I found out later, Yang and Stony Brook had even more to offer.

Those were cold-war days. US-China relation was thawing, however, and Yang had just returned from his historical first visit to China. The world was changing rapidly. Soon after I arrived Stony Brook, Yang was to give a university-wide colloquium about the New China. All students, me included, flocked to hear him. Stony Brook was at the forefront of this changing tide, and Yang was leading the way. Those were exciting times. And what a shock to a young student who just arrived from Taiwan!

I did not meet him in person, however, until after I passed my qualifying exam a year later. I assembled my courage, went to his office, and introduced myself to him. I am sure I was a bit awkward, but his patient and encouraging manner soon led us to a nice conversation. Soon after that, he accepted me as his Ph.D. student. Today, 35 years later, I consider myself still one of his students, learning not only his physics, but also his inexhaustible knowledge on society, people, politics, or literature and arts.

Yang is kind to his students. He cares for them. Being his student, I had a number of occasions when he made suggestions or arrangements that pleasantly surprised me. In 1973, for example, he arranged for me to visit the then-sparkling-new Fermilab for a summer, a visit that became my honeymoon because I got married just then.

Another more significant surprise came in 1973 when he urged me to take a graduate course in accelerator theory with Professor Ernest Courant. Little did I know at the time that this was to become the beginning of my career choice. Not fully realizing his deeper reasons, I followed the advice happily, and as it turned out, totally enjoyed the course. Before then, I had not thought that accelerator theory was so rich and so much fun. It was fascinating.

The topic I worked on with Yang was high energy theory. We wrote three papers together. My last year at Stony Brook, 1973, was a good year working with Yang. I learned tremendously from him. (I also learned from Professors Chris Quigg and Fred Goldhaber on two other topics on high energy theory.) Now looking back, I feel I was too busy trying to learn the technical aspects of the physics that related to my thesis problems. I was too young and narrow-minded to not have taken advantage of the opportunity to learn his other thoughts, physics or otherwise. On the physics topics, we actually worked very hard, together, both he and I. For a long period of time, I lingered in his office for hours almost daily. What an opportunity was I given! I did not appreciate it then, but thinking back, by occupying way too much of his time, I think I must have delayed the progress of modern physics by three months. He was of course lightening fast, but he was patient with me. On one occasion, I found empirically by using a computer a neat formula for the solution to a rather complex recurrence equation. When I proudly presented him the finding, he was delighted, paused for a minute, and came up with an elegant proof of it using group theory, which to this day, still amazes me. Sometimes we also chatted about non-physics. Once he mentioned fondly of a cross-country trip with friends, including T. D. Lee, and he seemed very happy recalling those memories. All these were tremendously fun.

In early 1974, I was going to graduate and it was time to look for a job. One day, he told me, across his big desk in his office, that I could find a job in high energy theory and that I would do well, but he said high energy physics was not a rewarding field, having too many people competing on too few meaningful topics. He advised me to go into accelerator theory, that field being less populated and much more rewarding. I had enjoyed both high energy theory and accelerator theory, but it was then when I finally realized he was advising me to change career! I debated, and hesitated, I recall, for a few weeks. While I was still hesitating, he questioned me again about my decision. I hardly finished saying "accelerator theory is great . . . , but how about all that I have learned in high energy theory, which would all be wasted . . ." than he became impatient with me, for the one and only time that I could recall. He said in a raised voice, "I strongly dispute this", and continued to give his reasons again. This time his advice worked. It was after this meeting when I finally made up my mind to go into accelerator theory. The time was spring 1974. At the time, in the back of my mind, I thought I could return to high energy theory if it didn't work out. Fortunately, it did work out, and I have not thought of returning to high energy physics ever since.

On several occasions over the later years, Yang talked about my finally switching to accelerator theory. He told me once that he considered that to be one of his important accomplishments. Now that is quite a statement, considering what his other important accomplishments are! I think, however, he is first happy for me finding a rewarding field for my career, and secondly, he is happy that it was his vision that caused the difference, not only in my case, but also in several other cases that I will mention next.

Let me now talk about accelerator physics at Stony Brook. In those days, accelerator physics was not as broadly recognized as an independent research field in physics, and there was basically no universities offering courses on it, not to mention having any accelerator physicist on the faculty. It was Yang's vision that Stony Brook was one exception (another exception was Cornell). Yang had invited and established an accelerator physics faculty in the Institute of Theoretical Physics. The faculty was Ernest Courant, inventor of the strong focusing principle that laid the foundation of all modern accelerators. I mentioned earlier that Yang advised me to take a course taught by Courant and that I totally enjoyed it.

I recall there were three students taking Courant's course that semester. To the students, Courant was received most fittingly as professorial. Sometimes I wondered whether it mattered to him how many students were in the class. He would lecture in his seemingly disorganized manner using his classical paper as text. I soon learned that in his class, the way to learn was to ask questions. Without questions, he would continue on with his lecture seemingly not paying undue attention to whether the students were following, but as soon as a question was raised, his eyes lit up and he uttered words of wisdom. In a few words, he would either answer your question exactly head-on — and for me at least, those were the moments I treasured most in his class — or in case of a bad question, he mumbled a bit and continued on with his lecture. He also amused the audience with his low-keyed sense of humor. Most importantly, I learned from Courant that accelerator physics was really exciting, in spite of its engineering sounding name, and that was a great revelation for me.

I was not the only Stony Brook student who was inspired by Yang and Courant to take on accelerator physics as career. After me, Ron Ruth, now a distinguished faculty at Stanford, was also inspired by Yang and became Courant's student. Ruth began his work with Courant on polarized beams in synchrotrons, one of Courant's favorite topics. Ruth's thesis was on collective beam instability, where he collaborated with Juinn-Ming Wang, another one-time Stony Brook faculty. Courant has maintained his keen interest in polarized beams up to today. Retired, he still travels from New York City, where he resides with his wife Sara - a most intelligent lady, to Brookhaven on a weekly basis to work on polarized beam physics.

Other than Ron Ruth and Juinn-Ming Wang, two more distinguished accelerator physicists inspired by Yang and switched field to accelerator physics were Bill Weng and Shyh-Yuan Lee. Weng is now a senior member at Brookhaven, and Lee is a distinguished professor at Indiana University. Following them, Courant's Ph.D. students from Stony Brook included Jie Wei, who is now a senior accelerator physicist at Brookhaven, and Steve Tepikian, who worked on polarized beams under the supervision mainly of Shyh-Yuan Lee. Yet another prominent Stony Brook accelerator physicist is Lihua Yu, who was a student of Janos Kirz's. Also inspired by Yang, he is now a world's leader in the physics of free electron lasers.

I should mention that Yang was himself interested in accelerator theory. I recall in one seminar by Courant, he showed quite detailed knowledge on the intricate nonlinear dynamics under discussion, and as Courant emphasized some intricate observations, he appeared very delighted. Yang also showed great interest in free electron laser. In fact, recently he has been pushing for a free electron laser project in China.

I was most fortunate and privileged that at a critical juncture of my career, I was given the opportunity to receive an advice so thoughtful and far reaching. I, and several others like me from Stony Brook, owe Professor Yang our deep gratitude.

Contributed Talks

QUASICRYSTALS AND PARTIAL DIFFERENTIAL EQUATIONS*

TIAN YOU FAN

Department of Physics, School of Science, Beijing Institute of Technology, P. O. Box 327, Beijing, 100081, P R China

This paper gives an introduction to some partial differential equations appeared in the study of quasicrystals. In addition the analytic solutions and weak solutions of the equations under complicated boundary conditions and initial conditions are also discussed.

*This work is supported by the National Natural Science Foundation of China through grants 10372016, 10672022.

PREPOTENTIAL APPROACH TO EXACT AND QUASI-EXACT SOLVABILITIES OF HERMITIAN AND NON-HERMITIAN HAMILTONIANS

CHOON-LIN HO

Department of Physics, Tamkang University,
Tamsui 251, Taiwan, Republic of China

In this talk I present a simple and unified approach to both exact and quasi-exact solvabilities of the one-dimensional Schrödinger equation. It is based on the prepotential together with Bethe ansatz equations. This approach gives the potential as well as the eigenfunctions and eigenvalues simultaneously. In this approach the system is completely defined by the choice of the change of variables, and the so-called zero-th order prepotential. We illustrate the approach by several examples of Hermitian and non-Hermitian Hamiltonians with real energies. The method can be easily extended to the constructions of exactly and quasi-exactly solvable Dirac, Pauli, and Fokker-Planck equations, and to quasinormal modes.

PROF. C. N. YANG AND QUANTUM ENTANGLEMENT IN PARTICLE PHYSICS

YU SHI

Department of Physics, Fudan University, Shanghai 200433, China
E-mail: yushi@fudan.edu.cn

Some of Prof. C. N. Yang's works are related to quantum entangled states in particle physics. Quantum entanglement leads to quantum information, which has been discussed in almost all areas of physics reigned by quantum mechanics, except particle physics. Here we demonstrate that the concept of quantum information can be extended to the regime of particle physics. Specifically, we will describe a scheme of quantum teleportation using neutral kaons. We also explain an idea on how to define quantum entanglement in the setting of relativistic quantum field theory.

IS NON-ABELIAN GAUGE THEORY RELEVANT TO THE TECHNOLOGY OF SPINTRONICS*

S. G. TAN

Data Storage Institute, DSI Building, 5 Engineering Dr 1, (Off Kent Ridge Crescent, NUS), Singapore 117608

M. B. A. JALIL

Information Storage Materials Laboratory, ECE Department, National University of Singapore, 4 Engineering Drive 3, Singapore 117576

X.-J. LIU

Department of Physics, Texas A&M University, College Station, Texas 77843-4242, USA

T. FUJITA

Data Storage Institute, DSI Building, 5 Engineering Dr 1, (Off Kent Ridge Crescent, NUS), Singapore 117608

Non-Abelian gauge field approach is applied to condensed matter system with time-reversal invariance. In this article, we show its relevance to the technologically-important area of spintronics. We consider specifically the spin Hall modulation and magnetic moment dynamic and discuss the important roles they play in the physics of magnetic recording and magnetic memory.

*Work partially supported by NUS ARF Grant No. R-263-00-329-112.

A TWO-PARAMETRIC GRADED R-MATRIX SATISFYING THE YANG-BAXTER EQUATION

VO THANH CUONG

Faculty of Physics, Vinh University, Vinh, Nghe An, Vietnam

NGUYEN ANH KY* and NGUYEN THI HONG VAN*

Institute of Physics and Electronics,
10 Dao Tan, Ba Dinh, Hanoi, Vietnam
and
Laboratory of High Energy Physics and Cosmology, Department of Physics,
Hanoi National University, 334 Nguyen Trai, Hanoi, Vietnam

A two-parametric $gl(2/1)$-graded R-matrix satisfying the Yang-Baxter and Hecker equations is suggested. The associated quantum supergroup is a two-parametric quantum deformation of $U[gl(2/1)]$ which by construction is a quasi-triangular Hopf superalgebra.

*Also The Abdus Salam ICTP, Trieste, Italy and CERN, Geneva, Switzerland.

A NEW TWO-PARAMETRIC DEFORMATION OF $U[osp(1/2)]$

NGUYEN ANH KY

Institute of Physics and Electronics,
10 Dao Tan, Ba Dinh, Hanoi, Vietnam
&
Laboratory of High Energy Physics and Cosmology, Department of Physics,
Hanoi National University, 334 Nguyen Trai, Hanoi, Vietnam

and

LE BA NAM

Institute of Engineering Physics, Hanoi University of Technology,
1 Dai Co Viet, Hai Ba Trung, Hanoi, Vietnam

A new kind of two-parametric deformations is suggested. Applying it to the superalgebra $osp(1/2)$ we obtain a new two-parametric deformation of $U[osp(1/2)]$ and then its grade-star representations. The latter are either finite-dimensional (for integral highest weights) or infinite-dimensional (otherwise).

THE KONTSEVICH INTEGRAL AND COVERING SPACES

ANDREW KRICKER

Division of Mathematical Sciences, Nanyang Technological University,
E-mail: AJKricker@ntu.edu.sg

A new field was created in 1984 when, famously, Vaughan Jones discovered a new polynomial invariant of knots. In the years that immediately followed there appeared a confusing proliferation of similar constructions. The Yang-Baxter equation proved to be the essential tool that led to the systematization of these ideas within the algebraic theory of quantum groups. In this talk I want to give a taste of what has happened in the field in the twenty years since these first developments by describing a key recent result: the rationality of the Kontsevich integral.

BYERS AND YANG'S THEOREM ON FLUX QUANTIZATION

KESHAV N. SHRIVASTAVA

Department of Physics, University of Malaya,
Kuala Lumpur, 50603, Malaysia
Also at: The School of Physics, University of Hyderabad,
Hyderabad 500046, India
E-mail: keshav@um.edu.my

In 1961 Byers and Yang have discovered a theorem of periodicity of energy levels in units of a flux quantum. In superconductors, the quantum unit is hc/2e. In the semiconductors, the pairing does not occur so that the flux quantum is hc/e. We show that Byers and Yang's flux quantization is applicable to the fractional charge, experimentally observed in the quantum Hall effect. It particularly explains the observation of 1/3 charge. The flux quantization also occurs for spin other than 1/2. We find that for spin 3/2 unusual series of large charges are predicted. The charge becomes very large such as 10e, 14e,..., etc. For example, for S = 3/2, (1/2)g = (1/2)+s = 2 so that g = 4, and the factor (5/2)g becomes 10 which corresponds to the effective charge of 10e. The flux quantization gives rise to plateaus and energies become independent of magnetic field. Apparently, the problem becomes independent of charge and the plateaus are then quantized in units of h^2/m, the square of the Planck's constant divided by the mass of the electron.

VOLTAGE-CONTROLLED BERRY PHASES IN TWO COUPLED QUANTUM DOTS

KA-DI ZHU

Department of Physics, Shanghai Jiao Tong University
E-mail: zhukadi@sjtu.edu.cn

The Berry phases controlled by voltage in two coupled quantum dots are investigated theoretically. The coupled quantum dots are embedded in a high-Q single mode cavity and coupled to the common phonon fields. It is found that the Berry phase is different from zero even in the vacuum state of the cavity. The Berry phase can change obviously from 0 to 2π via simply tuning the gate voltage. We also show how the Berry phase depends on the environmental temperatures.

A COMMENT ON THE WAVE FUNCTION OF NEUTRINO AND P-NONCONSERVATION

VO VAN THUAN

Vietnam-Auger Cosmic Ray Laboratory (VATLY)
Institute for Nuclear Science and Technology (INST)
Hoang Quoc Viet Street, Nghiado, Hanoi, Vietnam
E-mail: vvthuan@vaec.gov.vn

Based on the space-time symmetry, we proposed a generalized superluminal Lorentz transformation (GSLT) of the lepton sector which leads to physics of Majorana particles. In the frame of this model, despite of proving non-zero mass in recent oscillation experiments, neutrino may conserve strictly its helicity. The space-time structure of neutrino shows that its wave function may be a dual appearance of the geodesic evolution of a material point twisting in the 3D-space. In a simplified GSLT model the geodesic cycling defines the structure of neutrino wave function. The statistical feature of the quantum wave function as well as the P-violation are commented from a point of view of the space-time symmetry and asymmetry.

THEORETICAL MODELING OF B-Z DEOXYRIBONUCLEIC (DNA) TRANSITION

WILBER LIM

National Institute of Education, Nanyang Technological University,
1 Nanyang Walk, Singapore 637616
E-mail: wilber.lim@aya.yale.edu

Experimental efforts to elucidate the molecular mechanism of B-Z DNA transition have often led to controversial and inconclusive results. Some experiments suggest that the nucleobases flip out of the stack before rotating about their respective glycosyl bonds while others suggest that interbase hydrogen bonds remain intact during the transition. Recent computer simulations even suggest that the DNA oligomer stretches and unwinds during the transition. To account for the myriad of mechanisms proposed, a nonlinear DNA Hamiltonian based on three conformational variables, the glycosidic bond angle, base extrusion angle and the longitudinal displacement of the nucleotide, was recently developed. In this article, we will examine some of the theoretical results derived from this Hamiltonian and explore the possibility of including helicity in the model.

APPLICATION OF GAUGE THEORY TO ACOUSTIC FIELDS — REVOLUTIONIZING AND REWRITING THE WHOLE FIELD OF ACOUSTICS

W. S. GAN

Acoustical Technologies Singapore Pte Ltd, 209-210 Innovation Centre, NTU, 16 Nanyang Drive, Singapore 637722

This paper is to be dedicated to Prof C N Yang's 85th birthday celebration because the idea here was inspired by Prof Yang's public lecture in Singapore in 2006. There are many similarities between electromagnetic waves and acoustic waves. Maxwell's equations for em waves is the oldest gauge theory. We discover symmetries in the pair of wave equations in the acoustic stress field and the velocity field. We also derive a new equation in terms of the stress field for sound propagation in solids. This is different from the Christoffel's equation which is in term of the velocity field. We feel that stress field can better characterize the elastic properties of the sound waves. We also derive the acoustic gauge field condition and gauge invariance and symmetries for the acoustic fields. We also apply symmetries to study negative refraction.

BONDING ELECTRONICS AND ENERGETICS: AN APPROACH CROSSING THE BARRIERS OF CLASSICAL AND QUANTUM APPROXIMATIONS

CHANG QING SUN

School of Electrical and Electronic Engineering, Nanyang Technological University, Singapore
E-mail: ecqsun@ntu.edu.sg

The physical properties of a macroscopic system can be well described using classical approaches in terms of the Gibbs free energy or the continuum medium mechanics, for instances, that relate the detectable quantities directly to the external stimulus such as the temperature, pressure, chemical composition, electric and magnetic field, etc, without needing consideration of atomistic origins of the property change. At the atomic scale, quantum effect becomes dominant and the physical properties of a small object can be reliably optimized in computations by solving the Schüdinger equations or Newtonian motion of equations with a sum of averaged interatomic potentials as key factors. However, for a small system at the nanometer regime, both the classical and the quantum approaches encountered some difficulties.

Recent exercises [1, 2] showed that the difficulties could be solved by considering the interatomic bond formation, dissociation, relaxation, and vibration and the associated energetic response of atoms and electrons and their consequences on the detectable quantities. It is possible to establish the functional dependence of a detectable quantity $Q = Q(z_i, d_i, E_i, m)$ on the bonding identities such as bond order, z_i, bond length, d_i, bond energy, E_i, and bond nature indicator, m, and the response of the bonding identities to the external stimulus such as the coordination environment, temperature, pressure, etc. Progress made insofar is quite encouraging, which may evidence the essentiality of the approach that could describe the true situation and predict the performance of nanostructures.

TOPOLOGICAL QUANTUM PHASE TRANSITIONS OF THE KITAEV MODEL

GUANG-MING ZHANG

Physics Department, Tsinghua University
E-mail: gmzhang@tsinghua.edu.cn

Applying the Jordan-Wigner transformation to spin-1/2 operators on special quasi-one dimensional paths, we show that the two-dimensional Kitaev model can be exactly mapped to a free Majorana fermion model without any redundant degrees of freedom. Via duality transformation, it can be further shown that the quantum phase transitions of the Kitaev model are described by non-local topological order parameters, which become Landau type local order parameters in the dual space. A closed relationship between conventional and topological quantum phase transitions is revealed, and the validity of conventional Landau phase transition theory with spontaneous symmetry breaking and local order parameters is also extended.

WORLDLINE INSTANTONS AND PAIR PRODUCTION

QING-HAI WANG

Physics, University of Connecticut
E-mail: qhwang@phys.uconn.edu

The imaginary part of the QED effective action can be approximated by the contribution of a worldline instanton, a solution to the classical Euclidean worldline equations of motion. In this talk, I will briefly review this formalism and compute also the prefactor arising from quantum fluctuations about this classical path. I will show the excellent agreement between our semiclassical approximation, conventional WKB, and numerical results using numerical worldline loops.

I will also show the extension of the worldline instanton technique to multidimensional spatially inhomogeneous electric background fields, for which, WKB failed to apply.

EXACT GROUND STATES AND CORRELATION FUNCTIONS OF INTERACTING SPINLESS FERMIONS ON TWO-LEGGED LADDER

S. A. CHEONG

School of Physical and Mathematical Sciences,
Nanyang Technological University,
21 Nanyang Link, Singapore 637371

In this talk, I will describe how the ground state of a two-legged ladder of spinless fermions with infinite nearest-neighbor repulsion can be mapped onto a Fermi sea in three limiting cases: (i) very strong correlated hopping; (ii) very strong hopping along the legs; and (iii) very strong hopping along the rungs. I will then how ground-state correlations in the two-legged ladder can be computed in terms of Fermi sea correlation functions via the use of an intervening particle expansion, before discussing the asymptotic spatial dependence of Fermi-liquid-type, charge-density-wave-type, and superconducting-type correlations in each of the three limiting cases.

NONLINEAR SUPERSYMMETRIC GENERAL RELATIVITY AND UNITY OF NATURE

KAZUNARI SHIMA* and MOTOMU TSUDA+

Laboratory of Physics, Saitama Institute of Technology,
Fukaya, Saitama 369-0293, Japan
**E-mail: shima@sit.ac.jp*
+E-mail: tsuda@sit.ac.jp

The basic idea and some physical implications of nonlinear supersymmetric general relativity(NLSUSY GR) are discussed, which give new insights into the origin of mass and the mysterious relations between the cosmology and the low energy particle physics, e.g., the spontaneous SUSY breaking scale, the cosmological constant, the (dark) energy density of the universe and the neutrino mass.

NOVEL NANO-DEVICE ANDSYMMETRYROLE IN CHEMICAL EVENT OF C60 AND CARBON NANOTUBE COMPOSITE

HAIBIN SU

Division of Materials Science, Nanyang Technological University
E-mail: HBSU@NTU.EDU.SG

The discovery of carbon nanotubes and C-60 fullerenes has created an enormous impact on next generation nano devices. Wepropose a new generation of fullerene nano-oscillators: a single-walled carbon nanotube with one buckyball inside with an operating frequency in the tens-of-gigahertz range. A quantitative characterization of energy dissipation channels in the peapod pair has been performed via molecular dynamics simulation. Edge effects are found to be the dominant cause of dynamic friction in the carbon-peapod oscillators. A comparative study on the energy dissipation also reveals the significant impact of temperature and impulse velocity on the frictional force.

Then we apply the recently developed Reactive Force Field (ReaxFF) to study the dynamics of tubular fullerene formation process starting from C_{60}-buckyball/nanotube peapod structures. We find that the space confinement provided by the single wall nanotube encapsulating the buckyballs is of critical importance to this coalescence reaction. More importantly, we propose one effective approach to break the symmetry fortuning theenergy barrier of forming a 4-membered ring between adjacent buckyballs. This research can help the community to gain better understanding of the complicateddynamic processes in fullerene systems.

SINGULAR GAUGE TRANSFORMATION AND WU-YANG SINGULAR-FREE MONOPOLE

J.-Q. LIANG

Institute of Theoretical Physics, Shanxi University, Taiyuan, Shanxi 030006, China

The topological quantum effect of a magnetic string in the field-free manifold of nontrivial topology is investigated in terms of the Dirac nonintegrable phase, which is a special kind of function on a principal fibre bundle over the manifold. For the quantized string the nonintegrable phase is shown to become a gauge transformation, however, with a singularity, which we call the singular gauge transformation. The Wu-Yang singular-free monopole is obtained naturally with the singular gauge transformation.

LABORATORY PLASMA ASTROPHYSICS RESEARCH WITH INTENSE LASERS

HIDEAKI TAKABE,* TSUNEHIKO KATO, YASUHIRO KURAMITSU and YUICHI SAKAWA

Institute of Laser Energetics, Osaka University, Japan
**E-mail: takabe@ile.osaka-u.ac.jp*

Large scale laser facilities mainly constructed for fusion research can be used to produce high-energy-density plasmas like the interior of stars and planets. They can be also used to reproduce the extreme phenomena of explosion and high Mach number flow in mimic scale in laboratory. With advanced diagnostic technique, we can study the physics of plasma phenomena expected to control a variety of phenomena in Universe. The subjects studied so far are reviewed, for example, in [1], [2].

The project to promote the laboratory astrophysics with Gekko XII laser facility has been initiated from April 1st this year as a project of our institute. It consists of four sub-projects. They are

1. Physics of collisionless shock and particle acceleration,
2. Physics of Non LTE (local thermodynamic equilibrium) photo-ionized plasma,
3. Physics of planets and meteor impact,
4. Development of superconducting Terahertz device.

I will briefly explain what the laser astrophysics means and introduce what are the targets of our project. Regarding the first sub-project, we have carried out hydrodynamic and PIC simulation to design the experiments with intense laser. We clarified the physical mechanism of generation of the magnetic field in non-magnetized plasma and the collsionless shock formation caused by the ion orbit modifications by the magnetic fields generated as the result of plasma instability.

NEW RF HELICON-PLASMA DEVICES FOR VARIOUS APPLICATIONS

TAKAO TANIKAWA,*[,1] S. SHINOHARA,[2] T. MOTOMURA,[2]
K. TANAKA,[2] K. TOKI[3] and I. FUNAKI[4]
Tokai University[1], *Kyushu University*[2],
Tokyo University of Agriculture and Technology[3], *ISAS/JAXA*[4]
**E-mail: tnth@keyaki.cc.u-tokai.ac.jp*

Helicon-plasma discharge is a convenient, rather easy way to produce weakly to strongly magnetized plasmas with good efficiency; helicon-plasmas have been utilized in various plasma applications. We have recently developed several different types of helicon-plasma devices for basic plasma physics experiments (mainly to simulate phenomena in space plasmas) and research on developing advanced electric propulsion systems.

In a large device (2 m^3) at ISAS/JAXA, a very high production efficiency of 10^{15} ion-electron pairs per 1 W of rf input power has been achieved using a flat spiral antenna for Ar plasma. At Tokai, a new type of flat segmented circular antenna has been tested for plasma production as well as excitation of various plasma waves for basic physics experiments. At TUAT, a very small (2.5 cm in diameter) high-density helicon plasma device has been constructed to develop a small plasma thruster without electrodes that come to direct contact with high-density plasma. Recent new results will be presented and some of application studies using these devices will be briefly discussed.

THz RADIATION GENERATION VIA LASER PLASMA INTERACTION EXPERIMENTS

NOBORU YUGAMI* and TAKESHI HIGASHIGUCHI

Utsunomiya University, Japan
**E-mail: yugami@cc.utsunomiya-u.ac.jp*

Recently radiation generation from the interaction between laser and plasma is studied. Terahertz radiation from photo-conductive antenna which is based on semiconductor technology is widely used, The power is in the order of nano-watt level so that it is hard to use for application. On the other hand, terahertz radiation from laser plasma interaction is much higher than that of semiconductor technology. In our experiments, we have studied by use DARC (dc to ac radiation converter) mechanism by using YAG laser with nano-second pulse duration. DARC is novel radiation source using the interaction between laser-created ionization front and static electric field. The frequency of radiation is determined by both plasma density of ionization front and the geometry of DARC structure. We observed radiation pulse of frequency of 1.2 THz and pulse duration of 2 ps with ZnSe crystal as media detected by EO (electro-optics) sampling technique.

NON-MAXWELLIAN VELOCITY DISTRIBUTION: A CHARACTERISTIC OF SPACE PLASMA

L.-N. HAU

Institute of Space Science, National Central University, Taiwan R.O.C.
E-mail: linhau@jupiter.ss.ncu.edu.tw

Although most of the matter in the Universe is in the state of ionized gas, the solar system remains to be the only natural laboratory for spacecraft to conduct the in-situ measurement of cosmic plasma. As revealed by many satellite observations, the velocity distribution of charged particles in the interplanetary space and Earth's space plasma environment usually exhibits non-Gaussian form. In this presentation the origin and consequence of non-Maxwellian distribution are addressed and the inter-comparison between theory and observation is illustrated.

DYNAMO MECHANISM BY TRANSPORT FLOW

CHANG-MO RYU

POSTECH, Pohang, 790-784, Korea
E-mail: ryu201@postech.ac.kr

This paper discusses a novel magnetic field generation mechanism by a transport flow. The distinguishing feature of this mechanism compared with the conventional one by the convective flow lies in the intrinsic characteristics of the random motion, responsible for such transport flow, to break the line-tying effect. It is shown that anomalous plasma diffusion can induce a substantial amount of magnetic flux under certain conditions. It is also discussed that when the Bohm diffusion, typically observed in laboratory plasmas, is incorporated, the magnetic induction equation can be transformed to a type of heat equation which allows exact analytical solutions, admitting magnetic field amplification solutions.

DEGRADABLE CHANNELS IN QUANTUM INFORMATION THEORY

RAJDEEP S. RAWAT

National Institute of Education, Nanyang Technological University,
1 Nanyang Walk, Singapore 637616
E-mail: Rajdeep.Rawat@nie.edu.sg

The dense plasma focus (DPF) device is a coaxial plasma gun that heats and compresses the plasmas to high temperatures (1–2 keV) and densities (10^{25-26} m^{-3}). Under such extreme conditions, the plasma radiates copious X-rays and particle beams such as relativistic electrons and ion beams and also neutrons if the device is operated with deuterium gas. At Plasma Radiation Sources Laboratory (PRSL), NIE, Singapore our group has six plasma focus devices and our research efforts encompass a very wide range of topics covering various fundamental aspects of plasmas to the application of this device to lithography, soft and hard x-ray imaging, material modification and thin film deposition. In the proposed talk the details of our groups' efforts in developing dense plasma focus device for these various applications will be discussed. The proposed talk will contain details of our research work on

1. Soft X-ray lithography and micromachining using neon filled plasma focus device
2. Soft and hard X-ray imaging using DPF device
3. Synthesis of nanophase magnetic materials using focus device
4. Processing of thin films using energetic ions of plasma focus
5. Portable miniature plasma focus device development for on-site neutron diagnostics for explosive detection.

PLASMA NANOSCIENCE: FROM ASTRONUCLEOSYNTHESIS TO ORIGIN OF LIFE AND INDUSTRIAL NANOMANUFACTURING

KOSTYA OSTRIKOV*,[1,3] and SHUYAN XU[2,3]

[1] *Plasma Nanoscience, School of Physics, The University of Sydney, Australia*
[2] *Plasma Sources and Applications Center, NIE, Nanyang Technological University, Singapore*
[3] *Institute of Advanced Studies, Nanyang Technological University, Singapore*
* *E-mail: K.Ostrikov@physics.usyd.edu.au*

Plasma nanoscience is an emerging research area at the cutting edge of the physics of plasmas and gas discharges, nanoscience and nanotechnology, materials science and engineering, structural chemistry, and life sciences. The existing approaches to fabricating exotic nanostructures and functional nanofilms are mostly process-specific and suffer from cost-inefficient "trial and error" practices. One of the reasons is that the ability to control the generation, transport, deposition, and structural incorporation of the building units of such films and structures, still remains elusive. On the other hand, the pioneering concept of deterministic plasma nanoscience is treated with extreme caution due to inherent chaotic nature of the plasma at the microscopic level. This contribution shows how to challenge one of the previously intractable problems of bridging nine orders of magnitude between the sizes of plasma nanofabrication facilities (~ 0.5 m) and self-organization of building units on solid surfaces (~ 0.2 nm). One of the possibilities is to manipulate a variety of building blocks in the plasma sheath that separates the plasma and solid surfaces and control self-organization of nanostructure building blocks on plasma-exposed surfaces and their insertion into the nanoassemblies. The desired nanoassemblies can be engineered by using hybrid multi-scale numerical simulations and sophisticated experimentations. Recent experimental and computational results obtained within the International Research Network for Deterministic Plasma-Aided Nanofabrication suggest the possibility of deterministic synthesis of a large variety of nanostructures and their functional arrays, and are overviewed in this talk. An issue of creation of self-assembled nanodevices on plasma-exposed surfaces is discussed as well. Finally, this talk reveals how the Nature's mastery works in the assembly of nanometre-sized particles in the Universe via the astronucleosynthesis and ion-induced nucleation pathway, exotic nanoassemblies in laboratory plasmas and in possible creation of building blocks of life in primordial Earth.

PLUME DYNAMICS IN TEA CO_2 LASER ABLATION OF POLYMERS AND GRAPHITE

T. Y. TOU

Faculty of Engineering, Multimedia University, Cyberjaya, Selangor, Malaysia

O. H. CHIN

Department of Physics, University of Malaya, Kuala Lumpur, Malaysia

Streak photography of TEA-CO_2 laser ablation of graphite and polymers in helium and argon background gases showed two luminous fronts lifting off the ablation targets at delayed intervals. The first luminous front was driven by the plume from the immediate ablation of top surface layer of polymer, while the second front by the delayed ejection of laser-decomposed, heavy fragment of polymeric materials which lasted for more than a few hundred μs. The dynamics of these two luminous fronts were complicated by the onset of the laser-supported absorption wave at higher background gas pressure of > 50 mbar Ar, as seen in nanosecond dye-laser shadowgraphy. Optical emission spectroscopy of spatial-temporal distribution of the first plume fitted with the shifted Maxwellian-Boltzmann distribution and implied plume splitting or the presence of two-velocity components of plume specie.

PLASMA HOLE — A SINGULAR VORTEX IN A MAGNETIZED PLASMA

M. Y. TANAKA

Kyushu University, Japan
E-mail: mytanaka@aees.kyushu-u.ac.jp

A vortex with a density cavity in its core has been observed in a magnetized cylindrical plasma. It is called "plasma hole" from the visual impression when viewed along the axis of the vortex. The flow velocity measurements revealed that the plasma hole accompanies with supersonic azimuthal flow and radial flow toward the center, on a plane perpendicular to the magnetic field. The vorticity distribution evaluated from the flow velocity field is localized near the vortex center axis. This vorticity localization is identified as a Burgers vortex, which is the first observation of Burgers vortex in a plasma. The plasma hole is divided into two regions; in the peripheral regions the Lorentz force is balanced with the electric force (ExB drift), and in the core regions the Lorentz force is balanced with the centrifugal force. Rotation driven by centrifugal force is called fast rotation, and is realized only in non-neutral plasmas so far. It is found that charge neutrality condition in the core region breaks down by three order of magnitude compared with the case without plasma hole (10^{-6}). The effective viscosity in the core region exhibits an anomaly as well. The detailed experimental results on the plasma hole and the implication from the viewpoint of basic plasma physics will be presented.

QUANTUM PLASMAS — SPACE CHARGE LIMITED ELECTRON FLOWS

L. K. ANG

Nanyang Technological University
E-mail: ELKAng@ntu.edu.sg

In this talk, we will discuss the quantum effects of high current space charge limited (SCL) electron flows from field emission in a nangap at both steady state and ultrashort pulse conditions at 1D and 2D models. It is found the traditional models and scaling of field emission law and SCL electron flow (Child-Langmuir law) are no longer when quantum effects become important, such as limiting current density, quantum transit time, electron shot noise, quantum capacitance, etc. The extension of the models to solids with low mobility (like organic materials) will be discussed.

GENERATION AND APPLICATION OF HIGH DENSITY LOW-FREQUENCY INDUCTIVELY COUPLED PLASMAS

SHUYAN XU*[,1,2] and KOSTYA OSTRIKOV[2,3]

[1]*Plasma Sources and Applications Center, NIE, Nanyang Technological University, Singapore*
[2]*Institute of Advanced Studies, Nanyang Technological University, Singapore*
[3]*Plasma Nanoscience, School of Physics, The University of Sydney, Australia*
**E-mail: Shuyan.xu@nie.edu.sg*

This talk focus on the generation and application of steady-state, high density plasmas in two forms of low-frequency inductively coupled plasma (LF ICP) sources, namely a conventional ICP with external flat coil configuration and an internal planar unidirectional RF current driven through a specially designed internal antenna configuration. The underlying physics, results of the experimental investigation and applications of both types of LF ICP sources are discussed. The main properties of the LF ICP processed materials can be efficiently controlled by the plasma parameters and discharge operating regimes. The examples include fabrication of crystalline thin-film photovoltaic solar cells, assembly of ordered nanostructures and growth of uniform quantum dots and superlattices.

LIST OF PARTICIPANTS

AH QUNE, LLOYD FOONG NIEN	g0202706@nus.edu.sg
AHUJA, TANVI	Z070235@ntu.edu.sg
AN, JUNHONG	phyaj@nus.edu.sg
ANG, LAY KEE	ELKAng@ntu.edu.sg
AW, CHEE YANG	aw0003ng@ntu.edu.sg
AZAR, EYVAZOV	AZAR0001@ntu.edu.sg
BAAQUIE, BELAL E.	phybeb@nus.edu.sg
BALACHANDRAN, VINITHA	g0601145@nus.edu.sg
BASU, BANASRI	—
BATCHELOR, MURRAY	Murray.Batchelor@anu.edu.au
BRAY, IGOR	I.Bray@curtin.edu.au
CAI, YICHANG	CAIY0012@ntu.edu.sg
CHAKRABARTI, SHRABANA	—
CHAN, PHIL	phycahp@nus.edu.sg
CHAN, WEI PING	CHAN0378@ntu.edu.sg
CHANG, CHING RAY	crchang@phys.ntu.edu.tw
CHANG, LAY NAM	laynam@vt.edu
CHANG, NGEE PONG	npccc@sci.ccny.cuny.edu
CHAU, LING-LIE	chau@physics.ucdavis.edu
CHEN, HESHENG	chenhs@post.ihep.ac.cn
CHEN, ISABEL	phycmqi@nus.edu.sg
CHEN, YU	g0600513@nus.edu.sg
CHIANG, TSAI CHIEN	—
CHING, CHEE LEONG	phyccl@nus.edu.sg
CHIU, SHEAU MING	CHIU0014@ntu.edu.sg
CHO, YONGMIN	ymcho@phya.snu.ac.kr
CHONG, YOU QUAN	CHON0128@ntu.edu.sg
CHOO, WEIQIANG DANIEL	CHOO0043@ntu.edu.sg
CHRISTOPHER, RAYMOND	RAYM0008@ntu.edu.sg
CHU, PAUL CHING WU	ophkust@ust.hk
CHU, SENG	—

CIRAC, IGNACIO	Ignacio.Cirac@mpq.mpg.de
COHEN-TANNOUDJI, CLAUDE	cct@lkb.ens.fr
DEWANTO, ANDREAS	phyda@nus.edu.sg
DING, KEQUAN	kqding@yahoo.com
DING, LIANG	—
EGUCHI, TOHRU	eguchi@hep-th.phys.s.u-tokyo.ac.jp
ENG, SUKY	ENGS0005@ntu.edu.sg
ENGLERT, B. G.	phyebg@nus.edu.sg
ER, CHERN HAN	ERCH0001@ntu.edu.sg
FADDEEV, LUDWIG	faddeev@euclid.pdmi.ras.ru
FAN, JUNJIE BERTRAND	bertrand_fan@hotmail.com
FAN, TIAN YOU	litongtzb@263.net
FAN, ZENG	—
FANG, BESS	u0401938@nus.edu.sg
FANG, XIAOQIN	XQFANG@NTU.EDU.SG
FENG, XUN-LI	phyfxl@nus.edu.sg
FENG, Y. P.	phyhead@nus.edu.sg
FISHER, MICHAEL E.	claremon@ipst.umd.edu
FOONG, SOOK CHING	FOON0013@ntu.edu.sg
FU, RONG-QIANG	—
FU, SHENG	MSFu@ntu.edu.sg
FUJIKAWA, KAZUO	fujikawa@phys.cst.nihon-u.ac.jp
FUNG, MING KONG	mkfung@phy03.phy.ntnu.edu.tw
GALAPON, E. A.	Eric.galapon@up.edu.ph
GAN, WOON SIONG	wsgan@acousticaltechnologies.com
GAO, HONGMEI	—
GE, MOLIN	geml@nankai.edu.cn
GOH, BOON CHONG	GOHB0014@ntu.edu.sg
GOO, YIK WEN	GOOY0002@ntu.edu.sg
GOOLAUP, SARJOOSING	—
GU, MINGXIA	GUMI0002@ntu.edu.sg
HAM CHAU, NGUYEN	hamchau2007@gmail.com
HAO, BAILIN	hao@itp.ac.cn
HAU, L.-N.	linhau@jupiter.ss.ncu.edu.tw
HO, CHOON-LIN	hcl@mail.tku.edu.tw
HONG, LIYING	HONG0081@ntu.edu.sg
HU, BAMBI	bhu@hkbu.edu.hk
HUANG, KERSON	kersonhuang@aol.com
HWANG, PAUCHY W.-Y.	wyhwang@phys.ntu.edu.tw

JIANG, YUN	jy_1344@yahoo.com.cn
JUNIWATI, FITRI	Z060172@ntu.edu.sg
KANG, KEAN LEE	KANG0025@ntu.edu.sg
KAUFFMAN, LOUIS	kauffman@uic.edu
KE, CHANG	KECH0001@ntu.edu.sg
KIM, SEUNGHWAN (KIM SWAN)	swan@postech.ac.kr
KIRILLY, DANIEL	—
KOH, WEI PING VERONICA	VERO0001@ntu.edu.sg
KOHN, WALTER	kohn@physics.ucsb.edu
KONUMA, MICHIJI	mkonuma254@m4.dion.ne.jp
KOREPIN, VLADIMIR	korepin@insti.physics.sunysb.edu
KRICKER, A.	AJKricker@ntu.edu.sg
KWONG, CHANG CHI	g0700681@nus.edu.sg
KYAW, ZIN HTET	KYAW0057@ntu.edu.sg
LAI, CHOY HENG	phylaich@nus.edu.sg
LAN, BOON LEONG	lan.boon.leong@eng.monash.edu.my
LAU, LAWRENCE	lawrencelau@cuhk.edu.hk
LAU, SHU PING DANIEL	ESPLau@ntu.edu.sg
LEE, HWEE KUAN	leehk@bii.a-star.edu.sg
LEE, YOU JING EUNICE	EUNI0003@ntu.edu.sg
LEW, KING LOONG	LE0023NG@ntu.edu.sg
LI, BAOWEN	phylibw@nus.edu.sg
LI, CUIPING	cuipingli@buaa.edu.cn
LI, JUN	—
LI, JUN	LIJU0004@ntu.edu.sg
LI, ZHIPENG	g0403192@nus.edu.sg
LIANG, JIUQING	jqliang@sxu.edu.cn; jqliang516@yahoo.com
LIM, CHOON SIANG WILBER	wilber.lim@aya.yale.edu
LIM, HOCK	tslhead@nus.edu.sg
LIM, JUE RU AUDREY	LI0003EY@ntu.edu.sg
LIN, HAI QING	hqlin@phy.cuhk.edu.hk
LIU, SHAOTAO	Z060056@ntu.edu.sg
LIU, SHI KUN	—
LIU, YANG	—
LO, SHUI YIN	syinlo@ateg.com
MATSUYAMA, KIMI	kimi.matsuyama.up@hitachi.com
MCKELLAR, BRUCE H. J.	mckellar@physics.unimelb.edu.au
MINIATURA, CHRISTIAN	phymc@nus.edu.sg

NEGI, O. P. S.	ops_negi@yahoo.co.in
NG, CHEE MENG	phyncm@nus.edu.sg
NG, JUN SIANG	NGJU0004@ntu.edu.sg
NG, LAY YONG	NG0086NG@ntu.edu.sg
NG, TIEN TJUEN	g0700689@nus.edu.sg
NG, WEI KHIM	g0500437@nus.edu.sg
NGUYEN, THI KIM NGAN	V060010@ntu.edu.sg
NI, XIAOXI	g0700690@nus.edu.sg
NIEH, HUA TONG	nieh@tsinghua.edu.cn
NIEMI, ANTTI	Niemi@teorfys.uu.se
NINOMIYA, MASAO	ninomiya@yukawa.kyoto-u.ac.jp
OH, CHOO HIAP	phyoch@nus.edu.sg
ONG, EDDIE	ias@ntu.edu.sg
ONG, YENCHIN	—
OSTRIKOV, K.	K.Ostrikov@physics.usyd.edu.au
OUYANG, ZHONGCAN	oy@itp.ac.cn
PANG, SI QI	PA0001QI@ntu.edu.sg
PERL, MARTIN	martin@slac.stanford.edu
PETRASHYK, ANDRIY	PETR0005@ntu.edu.sg
PRAYUDI, UTOMO	PRAY0001@ntu.edu.sg
QI, DONGCHEN	g0304910@nus.edu.sg
QIU, Z. Q.	qiu@socrates.berkeley.edu
RAKHIMOV, ABDULLA	rakhimovabd@yandex.ru
RAMANATHAN, RAVISHANKAR	phyrr@nus.edu.sg
RAMATHAN, RAVISHANKAR	g0605463@nus.edu.sg
RAWAT, R. S.	Rajdeep.Rawat@nie.edu.sg
RIEPER, ELISABETH	elisabeth.rieper@gmail.com
RUSKAI, MARY BETH	Marybeth.Ruskai@tufts.edu
RYU, C. M.	ryu201@postech.ac.kr
SATHE, DILEEP V.	DileepSathe@vsnl.net
SANDUK, M. I.	m_sanduk@physics.org
SAW, VEE LIEM	P070025@ntu.edu.sg
SEOW, KAI LING CHERYL	SEOW0028@ntu.edu.sg
SHENG, PING	sheng@ust.hk
SHERRINGTON, DAVID	d.sherrington1@physics.ox.ac.uk
SHI, YU	yushi@fudan.edu.cn
SHIMA, KAZUNARI	shima@sit.ac.jp
SHRIVASTAVA, KESHAV N.	keshav_shri@yahoo.com
SONG, JIANJUN	edward.song@family.ust.hk

SU, HAIBIN	HBSU@NTU.EDU.SG
SUN, YING	—
TAKABE, HIDEAKI	takabe@ile.osaka-u.ac.jp
TAKHTAJAN, LEON	leontak@math.sunysb.edu
TAN, MEE CHENG	Z070292@ntu.edu.sg
TAN, SENG GHEE	Tan_Seng_Ghee@dsi.a-star.edu.sg
TAN, VEE RAO	TANV0004@ntu.edu.sg
TAN, YI CHIEN MABEL	TANY0184@ntu.edu.sg
TAN, YING SHAN VIVIEN	TA0008EN@ntu.edu.sg
TAN, ZHENG WEN	TANZ0038@ntu.edu.sg
TANAKA, M. Y.	mytanaka@aees.kyushu-u.ac.jp
TANIKAWA, TAKAO	tnth@keyaki.cc.u-tokai.ac.jp
TEE, KOK HENG	TE0006NG@ntu.edu.sg
TENG, SUNG SHIUAN	TENG0047@ntu.edu.sg
TEO, EDWARD	phyteoe@nus.edu.sg
THOULESS, DAVID	thouless@u.washington.edu
THUAN, VO VAN	vvthuan@vaec.gov.vn
TIAN, GANG	tian@math.mit.edu
TING, SAMUEL C.C.	ting@lns.mit.edu
TOH, BOON TECK	TOHB0005@ntu.edu.sg
TONG, MUN LING	Z070150@ntu.edu.sg
TONOMURA, AKIRA	tonomura@harl.hitachi.co.jp
TOU, TECK YONG	tytou@mmu.edu.my
TSONG, TIEN TZOU	tsongtt@phys.sinica.edu.tw
VAN HIEU, NGUYEN	nvhieu@ims.ncst.ac.vn
VAN HOVE, MICHEL A.	vanhove@cityu.edu.hk
VERMA, CHANDRA SHEKHAR	chandra@bii.a-star.edu.sg
VINH, LE THE	l.t.vinh@vnn.vn
WADATI, MIKI	wadati@rs.kagu.tus.ac.jp
WAHYU SAPUTRA	WAHY0003@ntu.edu.sg
WANG, DAWEI	WANG0365@ntu.edu.sg
WANG, DUANWEI	WANG0099@ntu.edu.sg
WANG, ENGE	egwang@aphy.iphy.ac.cn
WANG, QING-HAI	qhwang@phys.uconn.edu
WANG, SHENGTAO	wang0425@ntu.edu.sg
WEE, ANDREW	scidean@nus.edu.sg
WEE, PEIXIN	WEEP0005@ntu.edu.sg
WONG, AI XIN LEI LEI	WONG0476@ntu.edu.sg
WOO, WENG FOONG	WOOW0002@ntu.edu.sg

WU, ALFRED C. T.	actwu@umich.edu
WU, CHUNFENG	phywc@nus.edu.sg
WU, FA YUEH	fywu@neu.edu
WU, MAW KUEN	mkwu@phys.sinica.edu.tw
WU, TAI TSUN	ttwu@wisconsin.cern.ch
WU, WEIMIN	weimin@fnal.gov
WU, WEISHAN	—
WU, XIANG	g0600525@nus.edu.sg
XU, GANGQIN	g0600526@nus.edu.sg
XU, SHUYAN	Shuyan.xu@nie.edu.sg
XUE, KANG	—
YAN, JIE	phyyj@nus.edu.sg
YANG, HUIYING	hyyang@pmail.ntu.edu.sg
YANG, LIMIN	YANG0158@ntu.edu.sg
YANG, SHIN-NAN	snyang@phys.ntu.edu.tw
YANG, TZYH HAUR	u0601744@nus.edu.sg
YANG, XU	YANG0167@ntu.edu.sg
YAO, ANDREW CHI-CHIH	andrewcyao@tsinghua.edu.cn
YAO, DONG LAI	G0600528@@nus.edu.sg
YAO, EDWARD YORK-PENG	yyao@umich.edu
YEN, EDWARD	edyen@phys.nthu.edu.tw
YEO, WEE SING	—
YEO, YE	phyyy@nus.edu.sg
YONG, ZHIHUA	g0501208@nus.edu.sg
YOUNG, BINGLIN	young@iastate.edu
YOUNG, KENNETH	kyoung@cuhk.edu.hk
YU, LU	lyu@aphy.iphy.ac.cn
YU, SHI	yushi@fudan.edu.cn
YU, XIXIA	phyys@nus.edu.sg
YUGAMI, NOBORU	yugami@cc.utsunomiya-u.ac.jp
ZEE, ANTHONY	zeeitp@gmail.com
ZHANG, GUANG-MING	gmzhang@tsinghua.edu.cn
ZHANG, SHOUCHENG	sczhang@stanford.edu
ZHANG, TIANHUI	g0301202@nus.edu.sg
ZHAO, XIAODAN	g0600530@nus.edu.sg
ZHAO, ZHENTANG	zhaozt@ssrc.ac.cn
ZHOU, ZHEN	g0600535@nus.edu.sg
ZHU, BANG FEN	bfz@mail.tsinghua.edu.cn
ZHU, KA-DI	zhukadi@sjtu.edu.cn